Manufacturing Emulsion Polymers for Adhesives, Paints, and Coatings

Manufacturing Emulsion Polymers for Adhesives, Paints, and Coatings

Basil E. Shomali
Amman, Jordan

WILEY

Published by John Wiley & Sons, Inc., Hoboken, New Jersey.

Library of Congress Cataloging-in-Publication Data:

Names: Shomali, Basil E. author
Title: Manufacturing emulsion polymers for adhesives, paints, and coatings / Basil E. Shomali.
Description: Hoboken, New Jersey : Wiley, [2026] | Includes bibliographical references and index.
Identifiers: LCCN 2026003494 (print) | LCCN 2026003495 (ebook) | ISBN 9781394420001 hardback | ISBN 9781394420025 adobe pdf | ISBN 9781394420018 epub
Subjects: LCSH: Emulsion polymerization | Polymer engineering | Adhesives | Paint | Coatings
Classification: LCC TP156.P6 S56 2026 (print) | LCC TP156.P6 (ebook)
LC record available at https://lccn.loc.gov/2026003494
LC ebook record available at https://lccn.loc.gov/2026003495

Cover Design: Wiley
Cover Image: Courtesy of Basil Elias Shomali

Set in 9.5/12.5pt STIXTwoText by Straive, Chennai, India

To my beloved family—for your patience, love, and understanding as I spent countless hours working on this book instead of by your side.

And to all the trainees and professionals—who allowed me to pass on my knowledge and, in doing so, helped it grow into something greater.

Contents

About the Author

Basil Elias Shomali is a formulator and chemical consultant specializing in the design and manufacture of emulsion polymers—covering PVAc homopolymers, VAE, VAc–acrylate and VeoVa-acrylate copolymers, and waterborne alkyd emulsions—as well as waterborne adhesives and coatings. His work integrates polymer science with industrial practice, supporting manufacturers in Jordan, the Middle East, and international markets with technology packages, process optimization, and customized formulations. He has guided plants through scale-up and modernization programs and delivered training that strengthens technical teams and ensures long-term operational success.

A graduate of the American University of Beirut with a degree in Chemistry, Shomali began his career in leadership roles, including Deputy General Manager at The Union Vegetable Oils and Chemical Industries and Director of the King Hussein Industrial Complex, where he introduced modern production techniques, enhanced manufacturing standards, and directed research and development activities.

As an independent practitioner, he has authored and presented more than 200 formulations across binder families and application spaces. In response to frequent questions about testing, he emphasizes that his approach is that of a professional formulator: by understanding the roles of all ingredients and their effective dosage ranges, he deduces robust, scalable starting points and then validates them through mechanism-based adjustment and targeted performance testing against international benchmarks. This method allows rapid translation from laboratory concepts to industrially viable products while meeting regulatory and sustainability expectations.

In addition to this book, Shomali is authoring volumes on waterborne alkyd resins, powder and liquid detergents, and pool chemicals and maintenance, aimed at providing practical, formulation-driven guidance for industrial practitioners and R&D professionals.

Preface

This book, *Manufacturing Emulsion Polymers for Adhesives, Paints, and Coatings*, has been written as a field-oriented guide for chemists, formulators, engineers, and industry professionals involved in the development and production of emulsion polymers and adhesives.

Drawing on decades of hands-on industrial experience, this work bridges the gap between academic principles and the realities of factory operations. Throughout my career, I have worked closely with manufacturers and research teams, solving formulation challenges, optimizing processes, and implementing scalable solutions under real-world conditions. These experiences have shaped the practical insights presented here. The perspective is that of a practicing formulator: over the course of my consultancy and plant work, I have authored and presented more than 200 formulations across PVAc homopolymers, vinyl acetate–ethylene (VAE), and VAc–acrylate and VeoVa–acrylate copolymers. The formulations included are not lists to be copied uncritically; they are starting points derived from first principles—understanding the role and interaction of each ingredient and establishing effective dosage ranges—then refined by mechanism-based adjustment and verified against appropriate benchmarks.

The book combines fundamental concepts, representative formulations, and proven industrial practices into a single reference. It addresses not only how to design and manufacture emulsion polymers and PVAc-based adhesives but also how to adapt technologies for sustainability and regulatory compliance—two imperatives driving today's adhesives and coatings industry. Each chapter is structured to deliver clear technical explanations, practical examples, and troubleshooting guidance, making this book suitable for both R&D environments and production settings. Whether you are scaling up from the lab to full-scale manufacturing, improving adhesive performance, or addressing quality issues, the aim is to provide robust decision criteria and actionable solutions.

Beyond industrial practice, the knowledge presented in this book reflects years of study and reference to scientific literature, textbooks, patents, and technical brochures. These resources have been a continuous part of my professional development, shaping my understanding of emulsion polymerization, adhesive formulation, and process design. By combining published research with factory-floor experience, I have sought to integrate academic rigor with practical application. This dual approach—drawing equally on the accumulated expertise of the field and on hands-on formulation work—has been essential in authoring a book that is both technically reliable and industrially relevant.

Finally, this volume marks the first in a continuing series on modern water-based technologies. Future volumes will expand into waterborne alkyd resins, hybrid coating systems, detergents, and specialized functional coatings, reflecting the industry's shift toward sustainable and innovative solutions.

Basil Elias Shomali
Amman, Jordan
10 November 2025

Disclosure of Artificial Intelligence (AI) Use

Book Title: *Manufacturing Emulsion Polymers for Adhesives, Paints, and Coatings*

Author: Basil Elias Shomali
Date: 10 November 2025

This declaration is provided in accordance with Wiley's guidelines for the disclosure and declaration of artificial intelligence (AI) use in authored books. The author confirms that the following information accurately describes the nature and scope of AI-assisted tools used during manuscript preparation and revision. In this work, artificial intelligence was used solely as a tool to support language refinement, content organization, and figure concept development under the author's supervision. It was not used as a research or analytical technology.

1.1 Identification of AI Tool and Purpose of Use

OpenAI's ChatGPT was employed under the author's direct supervision to support language refinement, document structuring, and consistency of terminology. Occasional assistance included drafting outlines, reshaping section flow, converting lists to continuous prose to comply with Wiley's house style, and generating preliminary figure captions where appropriate. The tool was not used to generate unverified scientific claims, original research data, or analysis, and it did not replace the author's judgment at any stage.

1.1.1 Alignment with Wiley's Disclosure Criteria

This declaration follows Wiley's four main disclosure categories for artificial intelligence use. The author confirms that no verbatim AI-generated output was included in the manuscript. AI tools were used to substantially edit and refine existing content under the author's supervision. AI tools assisted in generating supplementary materials, including preliminary figure concepts and the formatting of tables, but not in producing final data or artwork. No AI system performed analysis, modeling, or data interpretation that could influence the author's scientific conclusions. Accordingly, this work involves AI solely as an editorial and structural tool, not as a research or analytical technology.

1.2 Scope and Limits of AI Assistance

AI assistance was confined to drafting, visualization guidance, and editorial support. All scientific and technical content—including formulations, process conditions, testing methods, and regulatory interpretations—was authored, validated, and approved by the author. For illustrative figures, the AI tool was used only to generate preliminary concept images under the author's explicit direction. Each image was subsequently revised through multiple iterations guided by the author's technical feedback until the final version accurately reflected the intended chemical or process representation. The AI system did not independently design or interpret scientific data, nor did it create any figure, equation, or graphic without the author's approval. All resulting visual content therefore remains the intellectual creation of the author. Where external facts, standards, or regulations are cited, they are referenced to primary sources within the book's reference lists.

AI also assisted in consolidating and summarizing background literature while standardizing the structure and accuracy of reference lists. All references were verified, corrected, and formatted manually by the author to ensure compliance with Wiley's numeric style and citation integrity.

1.2.1 Additional Clarification on Technical Assistance

At various stages of manuscript development, the author used AI tools to obtain preliminary technical explanations or contextual information on specific chemical, physical, or process-related topics. These interactions were employed solely to clarify concepts or terminology. All such information was subsequently reviewed, cross-checked, and validated against authoritative primary references and the author's own technical expertise before being integrated into the text. No AI-generated material was accepted without independent verification, and all final technical interpretations remain the intellectual responsibility of the author.

1.2.2 Case Studies and Troubleshooting Sections

The industrial case studies and troubleshooting FAQs included in this book are based on the author's professional experience and verified consulting practice.

AI tools were used only to improve narrative clarity, consistency, and presentation structure, for example, organizing content into sections such as Background, Root Cause, Solution Implemented, and Lessons Learned. No AI system was used to create or interpret any technical scenario, process data, or problem analysis. All findings and conclusions are derived from the author's direct expertise and remain his intellectual work.

1.2.3 Tables and Data Presentation

The book contains approximately 240 tables that summarize formulations, testing parameters, and comparative performance data. All tabulated content was authored and verified by the author based on practical formulations, industrial references, and validated literature sources. AI tools were used solely to assist in linguistic phrasing, typographical consistency, and conversion of tabular layouts into boxed Wiley style, ensuring compliance with the publisher's formatting requirements. No table contains AI-generated numerical data, and all technical values and relationships were reviewed and confirmed by the author.

1.3 Verification, Accuracy, and Author Accountability

The author personally reviewed, edited, and verified all AI-assisted text and visuals for accuracy, originality, and appropriateness prior to inclusion. All references were checked for existence and correctness, and no AI-generated citations were used without verification. Full responsibility for the content and its compliance with ethical and publishing standards lies solely with the author; no AI system is credited as an author or co-author.

1.4 Confidentiality, Privacy, and Intellectual Property

No confidential, proprietary, or third-party unpublished materials were uploaded to the AI system. The author ensured that any content derived from external sources is properly cited and used in accordance with permissions and fair-use provisions. The AI tool was not used to evaluate peer-review materials or handle any content that would breach confidentiality obligations.

1.5 Images, Figures, Tables, and Media

The AI tool was not used to generate images, figures, or tables as final artwork. Any figures and tables in the book were created and curated by the author and are sourced, designed, or redrawn as appropriate with attribution where required.

The author supervised all related tasks, and all scientific and technical statements were authored and verified by the author prior to finalization.

1.6 AI-use Tracking Summary by Chapter

The following chapter-by-chapter summary details the specific scope and extent of AI involvement across the manuscript. All assistance was provided under the author's direct supervision and within the ethical guidelines established by Wiley.

Chapter 1—Introduction (7 May–10 May 2025)

1) AI assistance limited to editorial and structural tasks.
2) Supported text refinement, technical-note drafting, and visualization guidance under author supervision.
3) All technical statements, data interpretations, and formulations authored, verified, and approved by the author.

Chapter 2—Key Definitions and Terminology (10 May–13 May 2025)

1) AI assistance limited to editorial and structural tasks.
2) Supported text refinement, technical-note drafting, and visualization guidance under author supervision.
3) All technical statements, data interpretations, and formulations authored, verified, and approved by the author.

Chapter 3—Main Ingredients of Emulsion Polymers (13 May–20 May 2025)

1) AI assistance limited to editorial and structural tasks.
2) Supported text refinement, technical-note drafting, and visualization guidance under author supervision.
3) All technical statements, data interpretations, and formulations authored, verified, and approved by the author.
4) AI assistance in tabular content was restricted to phrasing, typographical consistency, and boxed-format alignment in Wiley style; all technical values and data were authored and verified by the author.

Chapter 4—Fourth Ingredient: Surfactants (21 May–26 May 2025)

1) AI assistance limited to editorial and structural tasks.
2) Supported text refinement, technical-note drafting, and visualization guidance under author supervision.

3) All technical statements, data interpretations, and formulations authored, verified, and approved by the author.
4) AI assistance in tabular content was restricted to phrasing, typographical consistency, and boxed-format alignment in Wiley style; all technical values and data were authored and verified by the author.

Chapter 5 – Surfactants in Emulsion Polymerization and Performance Impact (27 May–1 June 2025)

1) AI assistance limited to editorial and structural tasks.
2) Supported text refinement, technical-note drafting, and visualization guidance under author supervision.
3) All technical statements, data interpretations, and formulations authored, verified, and approved by the author.
4) Where applicable, AI support was limited to drafting preliminary figure captions and improving clarity of schematic descriptions under the author's direct supervision.
5) AI assistance in tabular content was restricted to phrasing, typographical consistency, and boxed-format alignment in Wiley style; all technical values and data were authored and verified by the author.

Chapter 6 – Initiators and Auxiliary Ingredients (2 June–5 June 2025)

1) AI assistance limited to editorial and structural tasks.
2) Supported text refinement, technical-note drafting, and visualization guidance under author supervision.
3) All technical statements, data interpretations, and formulations authored, verified, and approved by the author.
4) AI assistance in tabular content was restricted to phrasing, typographical consistency, and boxed-format alignment in Wiley style; all technical values and data were authored and verified by the author.

Chapter 7 – Process, Mechanisms, and Variables in Vinyl Acetate-Based Emulsion Polymerization (6 June–15 June 2025)

1) AI assistance limited to editorial and structural tasks.
2) Supported text refinement, technical-note drafting, and visualization guidance under author supervision.
3) All technical statements, data interpretations, and formulations authored, verified, and approved by the author.

4) Where applicable, AI support was limited to drafting preliminary figure captions and improving clarity of schematic descriptions under the author's direct supervision.
5) AI assistance in tabular content was restricted to phrasing, typographical consistency, and boxed-format alignment in Wiley style; all technical values and data were authored and verified by the author.

Chapter 8—PVAc-based Adhesives (16 June–20 June 2025)

1) AI assistance limited to editorial and structural tasks.
2) Supported text refinement, technical-note drafting, and visualization guidance under author supervision.
3) All technical statements, data interpretations, and formulations authored, verified, and approved by the author.
4) Where applicable, AI support was limited to drafting preliminary figure captions and improving clarity of schematic descriptions under the author's direct supervision.
5) AI assistance in tabular content was restricted to phrasing, typographical consistency, and boxed-format alignment in Wiley style; all technical values and data were authored and verified by the author.
6) This chapter forms a strategic cornerstone of the book, defining the technological framework for waterborne adhesive systems. All concepts, structural–property correlations, and performance strategies were developed by the author through verified industrial practice and research experience. AI assistance was confined to linguistic clarity and document organization, without influence on conceptual design or technical conclusions.

Chapter 9—VAM-based Copolymers with Acrylates and VeoVa 10 (21 June–25 June 2025)

1) AI assistance limited to editorial and structural tasks.
2) Supported text refinement, technical-note drafting, and visualization guidance under author supervision.
3) All technical statements, data interpretations, and formulations authored, verified, and approved by the author.
4) Where applicable, AI support was limited to drafting preliminary figure captions and improving clarity of schematic descriptions under the author's direct supervision.
5) AI assistance in tabular content was restricted to phrasing, typographical consistency, and boxed-format alignment in Wiley style; all technical values and data were authored and verified by the author.

6) This chapter extends the author's original framework into paints and coatings applications, demonstrating strategic formulation routes that connect polymer architecture with film performance. All concepts, structure–property analyses, and application strategies were developed by the author based on verified industrial experience and research insight. AI assistance was restricted to improving clarity, phrasing, and structural flow, without influencing technical reasoning or formulation design.

While the AI tool assisted in refining the language and structure of conceptual discussions, the core ideas, strategic frameworks, and proposed pathways for R&D in adhesives, paints, and coatings were entirely conceived, developed, and substantiated by the author based on his own research, industrial experience, and professional insight.

1.6.1 Continuation of AI-use Tracking Summary by Chapter

Chapter 10—Advanced PVAc–PVOH Adhesives, Tackification, Reactive PVOH, and Crosslinking (26 June–2 July 2025)

1) AI assistance limited to editorial and structural tasks.
2) Supported text refinement, technical-note drafting, and visualization guidance under author supervision.
3) All technical statements, data interpretations, and formulations authored, verified, and approved by the author.
4) AI assistance in tabular content was restricted to phrasing, typographical consistency, and boxed-format alignment in Wiley style; all technical values and data were authored and verified by the author.

Chapter 11—VAE Copolymers in Adhesives and Coatings (3 July–8 July 2025)

1) AI assistance limited to editorial and structural tasks.
2) Supported text refinement, technical-note drafting, and visualization guidance under author supervision.
3) All technical statements, data interpretations, and formulations authored, verified, and approved by the author.
4) AI assistance in tabular content was restricted to phrasing, typographical consistency, and boxed-format alignment in Wiley style; all technical values and data were authored and verified by the author.

Chapter 12—Special Binders and Cold-climate Paints (9 July–12 July 2025)

1) AI assistance limited to editorial and structural tasks.
2) Supported text refinement, technical-note drafting, and visualization guidance under author supervision.

3) All technical statements, data interpretations, and formulations authored, verified, and approved by the author.
4) AI assistance in tabular content was restricted to phrasing, typographical consistency, and boxed-format alignment in Wiley style; all technical values and data were authored and verified by the author.

Chapter 13—Testing Methods and Durability (13 July–20 July 2025)

1) AI assistance limited to editorial and structural tasks.
2) Supported text refinement, technical-note drafting, and visualization guidance under author supervision.
3) All technical statements, data interpretations, and formulations authored, verified, and approved by the author.
4) AI assistance in tabular content was restricted to phrasing, typographical consistency, and boxed-format alignment in Wiley style; all technical values and data were authored and verified by the author.

Chapter 14—Design, Safety, and Instrumentation of a 5-Ton Plant (21 July–5 August 2025)

1) AI assistance limited to editorial and structural tasks.
2) Supported text refinement, technical-note drafting, and visualization guidance under author supervision.
3) All technical statements, data interpretations, and formulations authored, verified, and approved by the author.
4) Where applicable, AI support was limited to drafting preliminary figure captions and improving clarity of schematic descriptions under the author's direct supervision.
5) AI assistance in tabular content was restricted to phrasing, typographical consistency, and boxed-format alignment in Wiley style; all technical values and data were authored and verified by the author.

Chapter 15—Frequently Asked Questions (6 August–20 August 2025)

1) AI assistance limited to editorial and structural tasks.
2) Supported text refinement, technical-note drafting, and visualization guidance under author supervision.
3) All technical statements, data interpretations, and formulations authored, verified, and approved by the author.
4) AI assistance in tabular content was restricted to phrasing, typographical consistency, and boxed-format alignment in Wiley style; all technical values and data were authored and verified by the author.

5) All troubleshooting questions and answers were based on the author's professional expertise and verified industrial practice; AI assistance was confined to improving clarity, phrasing, and logical organization, without altering technical meaning or recommendations.

Chapter 16—Industrial Case Studies (21 August–10 September 2025)

1) AI assistance limited to editorial and structural tasks.
2) Supported text refinement, technical-note drafting, and visualization guidance under author supervision.
3) All technical statements, data interpretations, and formulations authored, verified, and approved by the author.
4) Where applicable, AI support was limited to drafting preliminary figure captions and improving clarity of schematic descriptions under the author's direct supervision.
5) AI assistance in tabular content was restricted to phrasing, typographical consistency, and boxed-format alignment in Wiley style; all technical values and data were authored and verified by the author.
6) All industrial case studies and troubleshooting scenarios were drawn from verified professional experience and validated practice data; AI tools were used only to refine narrative structure and language, with no role in data generation, analysis, or interpretation.

Chapter 17—Regulatory Aspects and Sustainability (11 September–25 September 2025)

1) AI assistance limited to editorial and structural tasks.
2) Supported text refinement, technical-note drafting, and visualization guidance under author supervision.
3) All technical statements, data interpretations, and formulations authored, verified, and approved by the author.

Chapter 18—Toward the Future of Waterborne Alkyd Emulsions (26 September–7 October 2025)

1) AI assistance limited to editorial and structural tasks.
2) Supported text refinement, technical-note drafting, and visualization guidance under author supervision.
3) All technical statements, data interpretations, and formulations authored, verified, and approved by the author.

4) AI assistance in tabular content was restricted to phrasing, typographical consistency, and boxed-format alignment in Wiley style; all technical values and data were authored and verified by the author.

1.6.2 General Author's Note on Conceptual Development

While the AI tool assisted in refining the language and structure of conceptual discussions throughout this work, the core ideas, strategic frameworks, and proposed pathways for research and development in adhesives, paints, and coatings were entirely conceived, developed, and substantiated by the author. All intellectual concepts originate from the author's professional experience, industrial research, and consulting practice. The AI system functioned solely as an editorial and linguistic aid and had no role in the creation of scientific, technical, or conceptual content.

1.7 AI Use Category Table

In accordance with Wiley's 2025 disclosure framework, the author confirms that AI use in this work falls under the following categories. AI uses requiring disclosure included drafting and editing manuscript sections, formatting technical tables and figures, synthesizing background literature, and harmonizing terminology across chapters. AI uses not requiring disclosure were limited to basic grammar and punctuation correction, style alignment to Wiley format, and minor image or table resizing. These activities align with Wiley's AI Use Category Table, in which substantive drafting, analysis, or visual generation are declared, while editorial aids and immaterial linguistic adjustments are exempt. All substantive scientific content, analyses, and conclusions remain entirely the author's own and were independently verified prior to submission.

1.8 AI Acknowledgment (for Front Matter)

Portions of this text were prepared with assistance from ChatGPT (OpenAI, 2025). The tool supported language refinement, organization, and clarity under the author's supervision. All AI-assisted content was reviewed, edited, and approved by the author. The analyses, interpretations, and conclusions presented reflect the author's professional expertise, who retains full responsibility for the accuracy and integrity of this work.

For transparency to readers, Wiley recommends a concise acknowledgment within the book's Preface or Acknowledgments section. The above statement will be inserted there during final proofreading.

1.9 Final Author Affirmation

The author affirms that this disclosure is complete and accurate to the best of his knowledge and that the manuscript complies with Wiley's current guidance on responsible AI use, including disclosure, verification of outputs, and author accountability.

Basil Elias Shomali
Amman, Jordan
10 November 2025

Acknowledgments

I wish to express my deepest gratitude to everyone who supported this work in different ways.

To my family for their patience and understanding during the many long hours of writing and research; to the professionals, trainees, and industry colleagues who shared their experiences and challenges, inspiring the practical approach reflected in these pages; and to those who reviewed and guided the development of this manuscript, ensuring its clarity and relevance.

Finally, I acknowledge the publishing team for their guidance in preparing this book for a global audience.

List of Abbreviations and Acronyms

1K	One-component (single-pack)
2-EHA	2-ethylhexyl acrylate
2K	Two-component (two-pack)
4–20 mA	Standard analog current loop for instrumentation
AA	Acrylic acid
AAEM	Acetoacetoxyethyl methacrylate
AE	Acrylic emulsion
AES	Alkyl ether sulfate
AESn	Alkyl ether sulfate, sodium salt
AFM	Atomic force microscopy
AI	Analog input
AIBN	Azobisisobutyronitrile (free-radical initiator)
$AlCl_3$	Aluminum chloride
AMPS	2-acrylamido-2-methyl-1-propanesulfonic acid
AN	Acrylonitrile
AO	Analog output
APE	Alkylphenol ethoxylate
APG	Alkyl polyglucoside
APS	Ammonium persulfate
AS	Alkyl sulfate (e.g., sodium lauryl sulfate)
ASE	Alkali-swellable emulsion (associative thickener)
ASTM	American Society for Testing and Materials
AT	Analyzer transmitter
ATEX (from the French ATmosphéres EXplosibles)	European Union directives for equipment and workplace safety in potentially explosive atmospheres
ATRP	Atom transfer radical polymerization
BA	Butyl acrylate

BAT	Best available techniques
BET	Brunauer–Emmett–Teller (specific surface area)
BIT	Benzisothiazolinone
BOPP	Biaxially oriented polypropylene
CARB	California Air Resources Board
CDI	Carbodiimide
CFD	Computational fluid dynamics
CFR	Code of Federal Regulations (US)
China MEE	Ministry of Ecology and Environment (China)
CHW	Chilled water
CIP	Clean-in-place
CLP	Classification, Labelling and Packaging (EU)
CMC	Critical micelle concentration
Cobb60	Cobb 60 water absorptiveness test
CO_2e	Carbon dioxide equivalent
CT	Cooling tower
CTA	Chain transfer agent
CW	Condenser water
D10	Particle diameter at 10th percentile
D3	EN 204 wood-adhesive water-resistance class D3
D4	EN 204 wood-adhesive water-resistance class D4
D50	Median particle diameter
D90	Particle diameter at 90th percentile
DBP	Dibutyl phthalate
DCS	Distributed control system
DEGDB	Diethylene glycol dibenzoate
DI	Deionized water
DINP	Diisononyl phthalate
DLS	Dynamic light scattering
DMA	Dynamic mechanical analysis
DOTP	Dioctyl terephthalate
DPnB	Dipropylene glycol *n*-butyl ether
DSC	Differential scanning calorimetry
E	Ethylene (comonomer)
EA	Ethyl acrylate
EIS	Electrochemical impedance spectroscopy
EP	Emulsion polymerization
EPS	Expanded polystyrene
ESD	Emergency shutdown (system or valve)

EtherNet/IP	Industrial Ethernet protocol developed and maintained by ODVA, an international industrial automation standards organization
FAE	Fatty alcohol ethoxylate
FT	Flow transmitter
FTIR	Fourier transform infrared spectroscopy
GHG	Greenhouse gas
GHS	Globally Harmonized System of Classification and Labelling of Chemicals
GMA	Glycidyl methacrylate
GPC	Gel permeation chromatography
GU	Gloss unit
GWR	Guided wave radar (level measurement)
HALS	Hindered amine light stabilizer
HART	Highway Addressable Remote Transducer
HASE	Hydrophobically modified alkali-swellable emulsion (associative thickener)
HEC	Hydroxyethyl cellulose
HEMA	2-hydroxyethyl methacrylate
HEUR	Hydrophobically modified ethoxylated urethane (associative thickener)
HLB	Hydrophile–lipophile balance
HMI	Human–machine interface
HPLC	High-performance liquid chromatography
HX	Heat exchanger
I/O	Inputs and outputs
IA	Itaconic acid
ICI	Cone-and-plate high-shear viscosity measurement (named after Imperial Chemical Industries)
IECEx	International Electrotechnical Commission certification system for equipment used in explosive atmospheres
IFT	Interfacial tension
IP (rating)	Ingress protection rating
IPA	Isopropyl alcohol
IR	Infrared spectroscopy
ISO	International Organization for Standardization
KPS	Potassium persulfate
KU	Krebs units (Stormer viscosity)
LCA	Life cycle assessment
LC-MS	Liquid chromatography–mass spectrometry
LD	Laser diffraction (particle sizing)

LEL	Lower explosive limit
LT	Level transmitter
MAA	Methacrylic acid
MCC	Motor control center
MDA	Methylenedianiline (4,4′-diaminodiphenylmethane)
MEK	Methyl ethyl ketone (2-butanone)
MF	Melamine–formaldehyde (resin)
MFFT	Minimum film-forming temperature
MIT	Methylisothiazolinone
MMA	Methyl methacrylate
MOAH	Mineral oil aromatic hydrocarbons
Modbus	Industrial communication protocol used over serial and Ethernet networks
MOSH	Mineral oil saturated hydrocarbons
MOV	Motor-operated valve
MSDS	Material safety data sheet (legacy; now "SDS")
MVTR	Moisture vapor transmission rate
NCO	Isocyanate functional group (—N=C=O)
NIAS	Non-intentionally added substances
NIOSH	National Institute for Occupational Safety and Health
nm	Nanometer
NMR	Nuclear magnetic resonance spectroscopy
NPE	Nonylphenol ethoxylate
NV	Non-volatile content (solids by weight)
NVP	*N*-vinyl-2-pyrrolidone
O/W	Oil-in-water (emulsion)
OECD	Organisation for Economic Co-operation and Development
OEL	Occupational exposure limit
OPC UA	Open Platform Communications Unified Architecture
OTR	Oxygen transmission rate
P&ID	Piping and instrumentation diagram
PAA	Poly(acrylic acid)
PDI	Polydispersity index (particle size; by DLS)
PEG	Poly(ethylene glycol)
PET	Polyethylene terephthalate
PLC	Programmable logic controller
PMN	Premanufacture notice (TSCA)
PP	Polypropylene
PPE	Personal protective equipment
PPG	Poly(propylene glycol)

PROFIBUS	Process Field Bus; an industrial fieldbus communication protocol used in automation systems
PROFINET	Industrial Ethernet protocol (Siemens)
PRV	Pressure relief valve
PSA	Pressure-sensitive adhesive
PSV	Pressure safety valve
PT	Pressure transmitter
PTFE	Polytetrafluoroethylene
PU	Polyurethane
PVAc	Poly(vinyl acetate)
PVC	Poly(vinyl chloride)
PVOH	Poly(vinyl alcohol)
QA	Quality assurance
QC	Quality control
QR	Quick response (matrix barcode)
QUV	Accelerated weathering tester using fluorescent ultraviolet exposure and condensation
R&D	Research and development
RAFT	Reversible addition–fragmentation chain transfer polymerization
REACH	Registration, Evaluation, Authorisation and Restriction of Chemicals (EU)
RH	Relative humidity
RT	Room temperature
RTD	Resistance temperature detector
SCBA	Self-contained breathing apparatus
SDBS	Sodium dodecylbenzene sulfonate
SDS	Safety data sheet
SDS	Sodium dodecyl sulfate
SEC	Size-exclusion chromatography
SEM	Scanning electron microscopy
SF	Starved-feed (semi-batch feeding strategy)
SFS	Sodium formaldehyde sulfoxylate
SIL	Safety integrity level
SIS	Safety instrumented system
SLS	Sodium lauryl sulfate (sodium dodecyl sulfate)
SNUR	Significant New Use Rule under TSCA (Toxic Substances Control Act, US)
SVHC	Substance of very high concern
SVOC	Semi-volatile organic compound
TAPPI	Technical Association of the Pulp and Paper Industry

TBHP	*Tert*-butyl hydroperoxide
TC	Thermocouple
Td	Thermal decomposition temperature
TDA	Toluenediamine (2,4-/2,6- isomers)
TDR	Time-domain reflectometry (level measurement)
Tg	Glass transition temperature
TGA	Thermogravimetric analysis
TMA	Thermomechanical analysis
TMPTA	Trimethylolpropane triacrylate
t-PVOH	Tackified polyvinyl alcohol
TR	Ton of refrigeration
TSCA	Toxic Substances Control Act (U.S.)
TT	Temperature transmitter
UV	Ultraviolet
UVA-340 lamp	Ultraviolet A fluorescent lamp for weathering tests; here "A" denotes the longer-wavelength, lower-energy ultraviolet region relative to UVB
UV-Vis	Ultraviolet–visible spectroscopy
VA	Vinyl acetate
VAc	Vinyl acetate (monomer)
VAE	Vinyl acetate–ethylene (copolymer/binder)
VAM	Vinyl acetate monomer
VC	Vinyl chloride
VeoVa	Vinyl Versatate; trademark for vinyl esters of Versatic acids
VFD	Variable frequency drive
VSD	Variable speed drive
W/O	Water-in-oil (emulsion)
WVTR	Water vapor transmission rate
XPS	Extruded polystyrene
XRD	X-ray diffraction
ZnO	Zinc oxide
Zr	Zirconium (element)

1

Introduction

Adhesives, coatings, and emulsion polymers form the backbone of countless industrial and consumer applications, shaping how we build, protect, and decorate materials. From the earliest use of natural binders like tree resins and starch pastes to the development of sophisticated synthetic polymers, the field has undergone a remarkable transformation. This evolution reflects not only advances in chemistry but also the persistent demand for materials with improved performance, durability, and environmental responsibility.

1.1 Historical Background

The origins of adhesive and coating technology can be traced to ancient civilizations. Early Egyptians used protein-based animal glue for woodworking, while the Greeks and Romans developed rudimentary paints from natural oils, waxes, and mineral pigments obtained from naturally occurring ores such as iron oxides, carbonates, and silicates. By the 19th century, industrialization and the petrochemical revolution accelerated innovation, giving rise to synthetic resins such as phenol–formaldehyde and alkyds. The mid-20th century marked the emergence of emulsion polymerization, enabling the large-scale production of polymers like polyvinyl acetate (PVAc) and acrylics in waterborne form. This shift revolutionized manufacturing by reducing reliance on solvents, lowering volatile organic compounds (VOC) emissions, and opening pathways to safer, more sustainable technologies.

1.2 Evolution of Emulsion Polymers and Adhesives

Following the historical foundations, the mid-20th century marked a pivotal era for emulsion polymerization technology. Advances in surfactant chemistry, reactor design, and polymerization control strategies transformed the industry from small-scale batch operations to highly sophisticated, continuous, and semicontinuous production systems.

In recent decades, environmental regulations and market demands have accelerated innovation in waterborne adhesives, coatings, and polymer dispersions, replacing solvent-borne systems and enabling the development of low-VOC, high-performance products.

The past decade has also seen significant integration of green chemistry principles, biobased raw materials, and advanced functional monomers, ensuring sustainability without sacrificing performance.

1.3 Scope and Objectives of the Book

This book is designed as a comprehensive, hands-on reference for professionals, researchers, and technical staff engaged in the formulation, production, and application of emulsion polymers and waterborne adhesives based on polyvinyl acetate (PVAc) and its copolymers. It bridges the gap between academic theory and industrial practice, ensuring that the content is not only scientifically sound but also directly applicable in manufacturing environments.

The objectives of this book are fourfold:

1) To provide a deep understanding of the chemistry, process technology, and formulation principles underlying PVAc-based emulsion polymers and adhesives.
2) To offer practical guidance on raw material selection, process optimization, and quality control methods, enabling readers to produce consistent, high-quality products.
3) To address real-world challenges in manufacturing through more than 80 documented industrial case studies, illustrating both common and advanced troubleshooting scenarios.
4) To explore market trends, sustainability considerations, and global regulatory requirements, ensuring the reader's knowledge remains relevant in a rapidly evolving industry.

In addition to covering fundamental principles, this book includes detailed formulations, process flow diagrams, and explanations of best practices in scaling laboratory concepts to commercial production. Special attention is given to environmental compliance, market competitiveness, and the development of innovative, sustainable products.

In preparing this work, I have drawn not only on decades of factory and consultancy experience but also on an extensive body of scientific literature, textbooks, patents, and technical brochures. These references provide the framework to connect industrial practice with established knowledge, ensuring that the formulations and processes presented here rest on both practical experience and the accumulated expertise of the field.

Author's note: This book is written from the standpoint of a practicing formulator. Over the course of my consultancy and plant work, I have authored and presented more than 200 formulations across PVAc homopolymers, vinyl acetate–ethylene (VAE), and VAc–acrylate and VeoVa–acrylate copolymers. The formulations presented here are derived from first principles—understanding the function and interactions of each ingredient, establishing effective dosage ranges, and translating those relationships into robust starting points—then refined by mechanism-based adjustment and verified against appropriate performance benchmarks. This practical orientation is intended to help R&D teams move efficiently from concept to plant-ready recipes while meeting regulatory and sustainability expectations.

1.4 Book Coverage and Practical Focus

This book covers the full breadth of emulsion polymers, adhesives, and coatings, with an emphasis on practical, industrially relevant methods. It is intended as a working reference for researchers, formulation chemists, production managers, and technical sales professionals.

1) To describe the chemistry, key raw materials, and practical methods of emulsion polymerization, with emphasis on vinyl acetate homopolymers, copolymers, and relevant hybrid systems.
2) To provide detailed formulation guidance for water-based adhesives and coatings, including solutions for hard-to-bond substrates and approaches that leverage crosslinkers where appropriate.
3) To consolidate 71 industrial case studies that demonstrate real-world problem solving, process optimization, and product development across adhesives, polymers, and coatings (Chapter 16).
4) To integrate a focused treatment of design and safety for a 5-ton emulsion-polymerization plant, translating practical engineering considerations for chemists (Chapter 14).
5) To document testing methodologies together with durability and aging concepts, linking standardized procedures to formulation choices and long-term performance (Chapter 13).
6) To deepen understanding through professional technical diagrams and mechanistic explanations tied to variables that govern stability and process optimization (Chapters 5–7).

7) To provide decision frameworks on component selection—monomers, surfactants, thickeners, initiators, and auxiliary ingredients—and on the use of reactive PVOH, AA/MAA, and other crosslinkable chemistries for advanced PVAc-based adhesives (Chapters 3, 4, 6, 8, 10, and 11).
8) To present regulatory aspects, market trends, and sustainability strategies relevant to emulsion polymers, adhesives, and coatings, enabling alignment with industry best practice (Chapter 17).
9) To orient readers toward forthcoming developments by outlining the transition to water-based alkyd technologies and future directions for the second volume (Chapter 18).

1.5 Structure of the Book

The sequence proceeds chapter by chapter, from terminology and raw materials to polymerization mechanisms, formulation practice, design and safety, frequently asked questions, and industrial case studies, and concludes with a forward-looking chapter on water-based alkyd resins. Figures and diagrams are integrated within individual chapters and cross-referenced where appropriate.

Chapter 1: Introduction and context for emulsion polymers, adhesives, and coatings.

Chapter 2: Key definitions and terminology used throughout the book.

Chapter 3: Main ingredients—monomers and their roles in emulsion polymerization and performance.

Chapter 4: Main ingredients—thickening agents and surfactants, properties and selection.

Chapter 5: Surfactants in emulsion polymerization and their impact on stability and performance.

Chapter 6: Initiators and auxiliary ingredients for formulation and process control.

Chapter 7: Emulsion polymerization of vinyl acetate and its copolymers—mechanism, variables, stability, and process optimization.

Chapter 8: PVAc-based adhesives—copolymerization with acrylic and methacrylic acids (AA/MAA) and the role of crosslinkers in enabling new applications.

Chapter 9: VAc-based copolymers with acrylates and VeoVa in coatings.

Chapter 10: Advanced PVAc-PVOH adhesives—tackification, reactive PVOH, and expanded applications with AA/MAA and crosslinkers.

Chapter 11: VAE copolymers in adhesives and coatings.

Chapter 12: Special binders and adhesives for cold climates.

Chapter 13: Testing methods, durability, and aging in emulsion polymers, adhesives, and coatings.

Chapter 14: Design and safety considerations in a 5-ton emulsion-polymerization reactor.

Chapter 15: Frequently asked questions—emulsion polymers, adhesives, and coatings.
Chapter 16: Industrial case studies—71 examples across EP, adhesives, and coatings.
Chapter 17: Regulatory aspects, market trends, and sustainability in emulsion polymers, adhesives, and coatings.
Chapter 18: Toward the future—water-based alkyd resins and the second volume.

1.6 How to Use This Book Effectively

This book functions as both a technical reference and a practical guide for professionals, researchers, and students working with emulsion polymers, adhesives, and coatings. Readers may enter at any chapter according to their immediate need—whether to review fundamentals, resolve a manufacturing issue, or explore advanced applications—while using cross-references to connect concepts across the text.

New readers should begin with Chapter 2 (Key Definitions and Terminology)

1) and then build core knowledge in Chapters 3 and 4 (main ingredients: monomers, thickeners, and surfactants), Chapter 5 (surfactants in emulsion polymerization and performance impact), and Chapter 6 (initiators and auxiliary ingredients). Chapter 7 then consolidates the fundamentals through emulsion polymerization of vinyl acetate and its copolymers, highlighting mechanism, variables, stability, and process optimization.
2) For adhesive applications, consult Chapter 8 (PVAc-based adhesives with AA/MAA and crosslinkers for new applications), Chapter 10 (advanced PVAc–PVOH adhesives, tackification, reactive PVOH, and expanded uses with AA/MAA and crosslinkers), and Chapter 11 (VAE copolymers in adhesives and coatings). These chapters translate composition choices into performance outcomes and provide practical decision pathways.
3) For coatings, use Chapter 9 (VAc-based copolymers with acrylates and VeoVa in coatings) together with Chapter 11 (VAE copolymers in adhesives and coatings) to connect binder architecture to film properties, durability, and end-use requirements.
4) For climate-sensitive designs and field reliability, refer to Chapter 12 (special binders and adhesives for cold climates). Formulation tactics and processing guidance are aligned with low-temperature application constraints.
5) For measurement, qualification, and life-cycle performance, use Chapter 13 (testing methods, durability, and aging). Standardized protocols are linked to degradation mechanisms and to formulation levers that improve long-term stability.

6) For plant implementation and safe operation, see Chapter 14 (design and safety considerations in a five-ton emulsion-polymerization reactor). This chapter translates design choices—reactor configuration, agitation, utilities, and interlocks—into actionable guidance for chemists working with engineers.
7) For rapid troubleshooting and knowledge checks, Chapter 15 (FAQ) provides concise, problem-oriented answers spanning emulsion polymerization, adhesives, and coatings.
8) For benchmarking and technology transfer, Chapter 16 presents 71 industrial case studies that connect theory to practice across production and application settings.
9) For strategy and compliance, Chapter 17 consolidates regulatory aspects, market trends, and sustainability in emulsion polymers, adhesives, and coatings, with practical implications for formulation, labeling, and commercialization.
10) For forward-looking directions, Chapter 18 outlines the transition toward water-based alkyd resins and frames the scope of the second volume, including hybrid and functional coating technologies.

Figures and diagrams appear within their respective chapters and are cross-referenced in the text. Readers are encouraged to use Chapter 2 as a standing glossary while navigating later sections, and to consult formulation tables, procedures, and decision frameworks where indicated.

1.7 Target Audience

This book serves a broad readership across emulsion polymers, adhesives, and coatings. It is written to bridge theory and industrial practice so that concepts translate directly into manufacturing and R&D workflows.

1) R&D scientists and formulation chemists developing new adhesive, coating, and polymer systems.
2) Process engineers and production managers responsible for scaling laboratory formulations to plant operations.
3) Technical managers and quality-assurance specialists focused on performance optimization and regulatory compliance.
4) Consultants and technical advisors supporting problem-solving in adhesives, coatings, and emulsion polymerization.
5) Industrial chemists in training programs or academic settings who require a concise, practice-oriented reference.
6) Equipment manufacturers and suppliers seeking a clearer view of formulation demands and process constraints.

7) Entrepreneurs and investors evaluating opportunities in adhesives, coatings, and polymer manufacturing.
8) Application and technical-service engineers supporting customers across packaging, wood, paper, construction, and industrial coatings.
9) *University audience*: Senior undergraduate and graduate students in chemical engineering, chemistry, materials and polymer science, coatings technology, and packaging science. For these readers, the text can serve as a primary reference or course-adjacent resource, supporting laboratories, capstone projects, and research; its integration of mechanisms, formulation decisions, and 71 industrial case studies enables direct linkage between classroom learning and plant practice.

By addressing entry-level professionals, a dedicated university cohort, and experienced specialists, the book functions as a comprehensive reference that supports problem-solving, innovation, and decision-making in real-world contexts.

1.8 Scope and Limitations

The scope is intentionally broad, covering raw materials, polymerization processes, formulation strategies, performance enhancement, troubleshooting, and emerging trends in regulatory compliance and sustainability—all within water-based systems.

Several boundaries are defined to keep the focus practical and actionable:

1) The emphasis is on water-based technologies, notably PVAc, PVAc–acrylic copolymers, styrene–acrylics, and specialty hybrids.
2) Industrial production and commercially viable technologies take precedence; laboratory procedures are included when they function as prototypes for industrial translation.
3) Formulations are provided as guidelines and starting points for development rather than prescriptive solutions; users should adapt to local raw-material availability, regulatory requirements, and target performance, and conduct appropriate validation prior to commercialization.
4) Testing methodologies are described by purpose and principles; readers should consult relevant ASTM, ISO, or national standards for exact procedures.
5) The content reflects professional experience, market observations, and verified literature at the time of writing; rapid developments in raw materials and regulations may require periodic updates.

These limits are intended to provide depth and clarity in the chosen areas while preserving the book's central aim: practical, directly applicable knowledge for formulation, manufacturing, and application.

1.9 Conclusion

The book is structured to link scientific foundations with industrial applications across emulsion polymers, adhesives, and coatings. A detailed technical core is complemented by extensive practical examples, including 71 industrial case studies in Chapter 16 and additional examples embedded elsewhere, so that readers can apply the material to production challenges.

Environmental considerations, regulatory frameworks, market trends, and sustainability are consolidated in Chapter 17 to align technical innovation with compliance and long-term market viability. Durability, aging, troubleshooting, and manufacturing guidance are integrated to distinguish this work from purely academic treatments.

Whether the reader is an R&D chemist, production manager, technical consultant, supplier, university student, or investor, the following chapters are designed to be both self-contained and interconnected—suitable for targeted consultation or comprehensive study.

2

Key Definitions and Terminology

This chapter provides a consolidated reference of essential terms used throughout the book, ensuring consistency in understanding and application. Emulsion polymerization and adhesive technology encompass multiple disciplines, including chemistry, materials science, and process engineering. In such an environment, terminology can vary between academic literature, industrial practice, and regulatory documentation.

The primary objective of this chapter is to establish a clear and practical terminology framework. Readers—whether students, laboratory chemists, production engineers, or technical sales professionals—will find this section useful for clarifying the meaning of common terms and abbreviations encountered in subsequent chapters. By doing so, the chapter enhances comprehension and facilitates accurate communication across different functional roles.

Where applicable, definitions are based on recognized standards and authoritative sources. However, in many cases, they have been adapted to reflect industrial usage and practical interpretation. This approach bridges the gap between formal textbook definitions and the language commonly used in manufacturing, quality control, and technical marketing.

Readers are encouraged to revisit this chapter whenever they encounter unfamiliar terms or need to confirm their meaning in context. Doing so will help reinforce conceptual understanding and facilitate precise communication within multidisciplinary teams.

2.1 Main Definition—Emulsion Polymerization

Emulsion polymerization is a free-radical process in which water-insoluble monomers are dispersed in water with the aid of surfactants (and, where applicable, protective colloids) and polymerized using water-soluble initiators.

After nucleation, propagation occurs primarily inside stabilized micelles or within growing polymer particles suspended in the aqueous phase.

A key advantage of the method is the ability to obtain high molecular weight at high reaction rates under controlled thermal conditions. The continuous water phase provides effective heat removal for the highly exothermic reaction, which reduces the risk of thermal runaway and enhances process safety.

The product is a polymer latex: a colloidal dispersion of polymer particles in water used widely in adhesives, paints and coatings, textiles, and paper finishing. In practice, the technology is valued for:

1) Environmental benefits, including low or zero volatile-organic-compound emissions.
2) Formulation flexibility, with broad compatibility across comonomers and additives.
3) Process efficiency, enabling high solids without excessive viscosity increase.
4) Performance advantages, notably film formation, durability, and adhesion to diverse substrates.

Emulsion polymerization underpins the technologies discussed in subsequent chapters and is central to the study of waterborne adhesives and coatings.

2.2 Key Definitions

2.2.1 Emulsion Polymerization-Related Definitions

1) Emulsions and types.
   a) *Macroemulsions*: Opaque dispersions with particle sizes typically greater than 300 nm.
   b) *Microemulsions*: Optically clear or translucent systems with droplets below about 100 nm; thermodynamically stable in the presence of sufficient surfactant and co-surfactant.
   c) *Nano-/mini-emulsions*: Blue-white, kinetically stable dispersions with particle sizes between roughly 100 and 300 nm; commonly used as feeds for mini-emulsion polymerization.
2) Monomers and polymers.

   Monomers are small molecules that combine to form macromolecules. Examples in this book include vinyl acetate (VAM), butyl acrylate (BA), and VeoVa 10.

   Polymers are long-chain molecules made from repeating monomer units, such as polyvinyl acetate (PVAc) or VAc–BA copolymers.

3) Homopolymers and copolymers.
   A homopolymer consists of a single monomer type (e.g., PVAc from VAM).
   A copolymer contains two or more different repeat units (e.g., VAc–BA from VAM and BA).
4) Polymerization types.
   a) *Addition (chain-growth) polymerization*: Monomers with carbon–carbon multiple bonds add to growing radicals; vinyl acetate forms PVAc in this way.
   b) *Condensation (step-growth) polymerization*: Bifunctional or tri-functional monomers react with the release of a small by-product (e.g., water or methanol); examples include nylons and alkyd resins.
5) Degree and rate of polymerization.
   Degree of polymerization: The number of monomer units in a polymer chain.
   Rate of polymerization: The speed at which monomers are converted to polymer; depends on monomer and initiator concentrations, temperature, and radical population.
6) Hydrophobic compounds.
   Substances that avoid the aqueous phase (e.g., oils and VAM). Vinyl acetate is slightly soluble in water (about 2–3% at room temperature) but behaves as a hydrophobic monomer in emulsion polymerization.
7) Hydrophilic compounds.
   Substances that prefer the aqueous phase and are readily solvated by water.
8) Amphiphilic compounds.
   Molecules bearing both hydrophilic and hydrophobic segments (e.g., sodium dodecyl sulfate).
9) Emulsifiers and solubilizing agents.
   Emulsifiers are amphiphilic compounds that stabilize two immiscible phases as a dispersion.
   Solubilizing agents increase the apparent solubility of otherwise insoluble materials in a single phase (e.g., polysorbate 20 for essential oils).
   Emulsification involves two phases; solubilization yields one phase.
10) Surfactants.
   Surface-active agents that reduce surface and interfacial tension, orient at interfaces, and are critical for droplet formation, particle nucleation, and latex stability in emulsion polymerization.
11) Polarity in chemistry.
   Polar molecules exhibit uneven electron distribution and partial charges (e.g., water); nonpolar molecules have more uniform electron distribution (e.g., hydrocarbons). Large polarity differences limit mutual solubility and motivate the use of surfactants in emulsion systems.

12) *Surface tension of water*: The tendency of a liquid surface to minimize area due to cohesive forces; in water, hydrogen bonding creates a strong surface layer because surface molecules have fewer neighbors than those in the bulk.
13) *Surface tension vs. interfacial tension (IFT)*: Surface tension refers to a liquid–gas interface (e.g., water–air); interfacial tension refers to a liquid–liquid interface between immiscible liquids (e.g., water–oil).
14) *Glass transition temperature (T_g)*: The temperature range where an amorphous polymer changes from a rigid, glassy state to a softer, rubbery state.
15) *Minimum film-formation temperature (MFFT)*: The lowest temperature at which a wet dispersion forms a continuous, defect-free film on drying.
16) *Newtonian vs. thixotropic viscosity*: Newtonian fluids (e.g., water, mineral oil) show viscosity independent of shear rate; thixotropic fluids show a reversible decrease in apparent viscosity under shear and recover when shear is removed.
17) *Emulsion polymerization (process)*: Free-radical polymerization carried out in water to produce a stable, milky colloidal dispersion (latex).
18) *Stability of emulsion polymers (latex stability)*: The ability of a latex to resist undesirable change over time (e.g., flocculation, coagulation, sedimentation, phase separation, viscosity drift).
19) *Electrostatic stabilization*: Stabilization arising from electrical double-layer repulsion between like-charged particle surfaces; when repulsion exceeds van der Waals attraction, particles remain dispersed.
20) *Steric stabilization*: Stabilization provided by adsorbed macromolecules (e.g., nonionic surfactants or protective colloids) that create an entropic barrier preventing close particle approach.
21) *Latex coagulation*: Irreversible agglomeration of particles into macroscopic clumps, often triggered by increased ionic strength, large pH shifts, freezing, excessive shear, contamination, or depletion of stabilizer.
22) *Ethoxylation*: Step-growth addition of ethylene oxide (an epoxide) to nucleophilic groups (e.g., alcohols, phenols, amines), producing ethoxylates that typically increase water solubility or surface activity.
23) *Particle size and distribution*: Latex particles can be monodisperse (narrow size spread) or polydisperse (broad size spread).

    Particle-size spread: Dynamic light scattering (DLS) reports a polydispersity index (PDI) for particle size. Values around 0.1–0.2 indicate a narrow distribution; larger values indicate a broader distribution.

    Molecular-weight spread (a different measure): Size-exclusion chromatography (SEC), also called gel-permeation chromatography (GPC), reports the dispersity (Đ), written as $Đ = M_w/M_n$, where M_w is the *weight-average molecular weight* and M_n is the *number-average molecular weight*. Dispersity is at least 1; values closer to 1 indicate a narrower molecular-weight distribution.

24) *Emulsion types*: (i) oil-in-water (O/W): oil droplets dispersed in a continuous water phase; (ii) water-in-oil (W/O): water droplets dispersed in a continuous oil phase.
25) *Crosslinking*: Formation of covalent (or strong physical) links between polymer chains to create a three-dimensional network; vulcanization of rubber is a classic example.
26) *Homogeneous vs. heterogeneous systems*: Homogeneous systems exhibit uniform composition and properties (e.g., salt dissolved in water); heterogeneous systems contain more than one phase (e.g., polymer particles dispersed in water).
27) *Reduction of crystallinity (in EP films)*: Deliberate disruption of regular chain packing in dried films—using branched surfactants, bulky comonomers (e.g., VeoVa, isobornyl acrylate), or plasticizers—to hinder crystallite formation and modify mechanical/optical properties.
28) *Critical micelle concentration (CMC)*: The minimum surfactant concentration above which micelles form in solution.
29) *Nucleation (in EP)*: Initial formation of polymer particles from micelles, monomer-swollen micelles, or oligomeric radicals in the aqueous phase.
30) *Seeded polymerization*: Use of preformed seed particles to control final particle size, distribution, and morphology.
31) *Semibatch (semicontinuous) polymerization*: Controlled feeding of monomer and/or initiator into a reactor over time to manage heat release, particle growth, and composition.
32) *Batch polymerization*: Charging all reactants (or an initial charge) at the start, with no feed during the reaction.
33) *Conversion rate*: Fraction (or percentage) of monomer converted to polymer as a function of time.
34) *Gel content*: Insoluble fraction of a polymer network determined by solvent extraction; reflects extent of crosslinking.
35) *Serum*: The continuous aqueous phase of a latex, excluding polymer particles.
36) *Electrolyte tolerance*: Ability of a latex to remain stable in the presence of added salts; often assessed via a critical electrolyte concentration.
37) Particle morphology: Internal structure of latex particles (e.g., homogeneous, core–shell, gradient, raspberry), which governs film formation and properties.
38) *Protective colloid*: Water-soluble polymer (e.g., polyvinyl alcohol) that adsorbs on particle surfaces to provide steric stabilization and often beneficial rheology.
39) *Mechanical stability*: Resistance of a latex to coagulation or excessive viscosity change under shear, pumping, and handling.
40) *Freeze–thaw stability*: Resistance to coagulation and property loss after repeated freezing and thawing cycles.

41) *Foam stability*: Persistence of foam once formed, governed by surface elasticity, drainage, and stabilizer chemistry; excessive stability can hinder processing, while controlled defoaming may be required.
42) *Surfactant migration*: Movement of surfactant to the surface or interfaces during drying/film formation, which can influence surface tack, water sensitivity, and dirt pickup.
43) *Particle agglomeration*: Clustering of dispersed particles into larger assemblies; typically stronger (often irreversible) than loose flocculation and detrimental to latex stability and film quality.
44) *Residual monomer*: Unreacted monomer remaining in a latex after polymerization and any stripping/purification steps; reported as ppm or weight percent.
45) *Film coalescence*: Fusion of individual polymer particles as water/solvent evaporates to form a continuous film; effective coalescence generally requires temperatures at or above the MFFT.
46) *Chain transfer agent*: Compound that reacts with growing radicals to limit average molecular weight and control branching (e.g., mercaptans, halogenated hydrocarbons).
47) *Solids content*: Mass fraction of nonvolatile material in a latex, typically determined by oven solids; includes polymer and nonvolatile additives.
48) *Zeta potential*: Electrical potential at the hydrodynamic slipping plane around a particle; widely used as an indicator of colloidal stability (higher magnitude often correlates with better dispersion).
49) *Plasticizer compatibility*: Ability of a plasticizer to blend uniformly with a polymer without phase separation, exudation, or significant haze over the intended temperature range.
50) *One-component (1K) and two-component (2K) systems*: Terminology describing packaging and cure. (i) 1K: supplied as a single, ready-to-use product that cures upon exposure to triggers such as moisture, heat, or light, without on-site mixing. (ii) 2K: supplied as two separate parts (e.g., resin and hardener) that must be mixed immediately before application; cure starts after mixing, with a defined pot life. The “K” originates from German technical usage, where it denotes Komponente (component); the 1K/2K notation is now standard in adhesives and coatings.

2.2.2 Adhesive–Related Terminology

1) *Chemical adhesive*: A formulated material capable of bonding two substrates; bonding may involve wetting, interfacial forces, and/or chemical reaction.
2) *Chemical adhesion*: The interfacial forces that hold an adhesive and a substrate together.

3) *Adherent*: A substrate that is bonded to another using an adhesive.
4) *Adhesive bond strength*: The force required to separate the adhesive from the substrate under specified geometry and rate. Cohesive strength is the internal strength of the adhesive itself.
5) *Setting time*: The duration from joining substrates to the point at which the adhesive reaches handling strength or is considered set.
6) *Setting rate*: The rate at which bond strength increases over time; often slows as cure approaches completion.
7) *Tack*: The ability of an adhesive to form an initial bond immediately upon contact under light pressure.
8) *Creep*: Time-dependent deformation of an adhesive joint under sustained load (sometimes called cold flow at ambient temperature).
9) *Curing process*: Transformation of an adhesive to its final state by chemical reaction (e.g., crosslinking) or physical change (e.g., evaporation, cooling).
10) *Curing time*: The total time required to achieve the specified degree of cure; insufficient cure can lead to bond failure.
11) *Thixotropy*: A time-dependent shear-thinning behavior in which apparent viscosity decreases under shear and recovers when shear is removed.
12) *Open time*: The usable time window after application during which substrates can be joined before skinning or curing, which prevents proper wetting.
13) *Sagging*: Downward flow of an applied adhesive film before setting, commonly on vertical surfaces.
14) *Blocking*: Unwanted adhesion between contacting surfaces due to pressure, heat, or incomplete cure.
15) *Wet bonding*: Bonding with a water-based adhesive while at least one porous substrate remains wet with adhesive.
16) *Peel strength*: The force required to peel an adhesive from a substrate at a specified angle and rate.
17) *Shear strength*: The ability of an adhesive joint to resist forces parallel to the bond line.
18) *Substrate preparation*: Cleaning, abrasion, priming, or other treatments used to improve adhesion.
19) *Priming*: Application of a primer or chemical treatment to enhance adhesion to the substrate.
20) *Bondline thickness*: The thickness of the adhesive layer between substrates.
21) *Viscoelasticity*: Time-dependent mechanical behavior that exhibits both elastic (recoverable) and viscous (time-dependent) responses.
22) *Plasticizer migration*: Movement of plasticizer from the adhesive into the substrate or surrounding environment.
23) *Service temperature range*: The temperature range over which the adhesive maintains specified performance.

24) *Heat resistance*: The ability of a bonded joint to retain strength at elevated temperatures.
25) *Moisture resistance*: The ability of a bonded joint to maintain performance under humid or wet conditions.
26) *Aging*: Gradual change of adhesive or bond properties over time due to physical or chemical processes.
27) *Failure mode*: The manner in which a bonded joint fails—adhesive failure (at interface), cohesive failure (within adhesive), substrate failure, or mixed.
28) *Repositionability*: The ability to adjust bonded parts before cure completion; limited by open time and tack.
29) *Hot-melt adhesive*: A thermoplastic adhesive applied in the molten state and solidifying on cooling.
30) *Pressure-sensitive adhesive (PSA)*: An adhesive that forms a bond under light pressure at room temperature without external activation.

3

Main Ingredients of Emulsion Polymers

3.1 Monomers

In emulsion polymerization, monomers are the fundamental building blocks from which polymer chains are formed. The selection of monomers determines the chemical, physical, and application-related properties of the final product. While all monomers used in this process share a common feature—the presence of carbon–carbon double bonds enabling free-radical polymerization—they differ significantly in structure, reactivity, and the specific performance attributes they impart to the polymer. Following is a list of common monomers used in emulsion polymerization:

1) *Vinyl acetate (VAc)*: Produces polyvinyl acetate with good adhesion, film-forming ability, and compatibility with other monomers.
2) *Butyl acrylate (BA)*: Provides softness, flexibility, and excellent low-temperature film formation.
3) *2-Ethylhexyl acrylate (2-EHA)*: Imparts extreme softness, elasticity, and good water resistance.
4) *Methyl methacrylate (MMA)*: Delivers hardness, transparency, and weather resistance.
5) *Acrylic acid (AA)*: Adds hydrophilicity, improves adhesion to polar substrates, and enables crosslinking.
6) *Methacrylic acid (MAA)*: Similar to AA but produces harder films and offers enhanced durability.
7) *Vinyl versatate (VeoVa 10)*: Increases hydrophobicity, improves alkali resistance, and enhances durability in exterior applications.
8) *Styrene*: Contributes to hardness, gloss, and chemical resistance but may reduce flexibility.
9) *Butadiene*: Imparts flexibility and resilience, commonly used in elastomeric applications.

10) *Ethylene*: Improves water resistance and flexibility in vinyl acetate–ethylene copolymers.
11) *Vinyl chloride (VC)*: Offers excellent chemical resistance and barrier properties.
12) *Acrylamide*: Enhances hydrophilicity and is used in specialty functional polymers.
13) *Acrylonitrile (AN)*: Provides chemical resistance, toughness, and weatherability.

3.1.1 Key Notes

1) All common monomers used in emulsion polymerization are unsaturated compounds suitable for free-radical polymerization.
2) Each monomer has a distinct chemical structure that contributes specific properties to the resulting polymer.
3) Carbon–carbon double bonds are the primary reactive sites enabling polymeri zation.
4) Structural differences among monomers influence both polymerization behavior and final product performance (Table 3.1).

Table 3.1 Molecular and structural formulas of common monomers used in emulsion polymerization.

Monomer name	Molecular formula (MF)	Structural formula (SF)
Vinyl acetate (VAM)	$C_4H_6O_2$	$CH_2{=}CHOCOCH_3$
Butyl acrylate (BA)	$C_7H_{12}O_2$	$CH_2{=}CHCOO(CH_2)_3CH_3$
2-Ethylhexyl acrylate (2-EHA)	$C_{11}H_{20}O_2$	$CH_2{=}CHCOOCH_2CH(C_2H_5)(CH_2)_3CH_3$
Methyl methacrylate (MMA)	$C_5H_8O_2$	$CH_2{=}C(CH_3)COOCH_3$
Acrylic acid (AA)	$C_3H_4O_2$	$CH_2{=}CHCOOH$
Methacrylic acid (MAA)	$C_4H_6O_2$	$CH_2{=}C(CH_3)COOH$
Styrene	C_8H_8	$CH_2{=}CHC_6H_5$
Butadiene	C_4H_6	$CH_2{=}CHCH{=}CH_2$
Ethylene	C_2H_4	$CH_2{=}CH_2$
Vinyl chloride (VCM)	C_2H_3Cl	$CH_2{=}CHCl$
Acrylamide	C_3H_5NO	$CH_2{=}CHCONH_2$
Acrylonitrile	C_3H_3N	$CH_2{=}CHCN$
VeoVa 10	Proprietary	Not disclosed—Vinyl ester of Versatic Acid (C10 grade)

Note: Acrylamide is included here solely for completeness and industry context. It is not used or recommended in any formulations in this book due to safety and regulatory concerns.

Table 3.1 compares the molecular formulas (MF) and simplified structural formulas (SF) of various monomers commonly employed in emulsion polymerization. These monomers, all containing unsaturated carbon–carbon double bonds, are capable of undergoing free-radical polymerization. Their unique molecular structures impart distinct physical and chemical characteristics to the resulting polymers. Note: VeoVa 10, a vinyl ester of a proprietary Versatic™ acid mixture, does not have a fully disclosed molecular or structural formula in the public domain; only an approximate empirical composition is available.

3.2 Solubility Classification of Monomers

Monomers can be classified according to their solubility in water at room temperature. Solubility influences monomer distribution in the aqueous phase during emulsion polymerization, affecting reaction kinetics, particle nucleation, and final polymer properties (Table 3.2).

Table 3.2 Solubility classification of selected monomers at room temperature.

Group	Description	Monomer	Approx. solubility (% w/w in water)
A	Completely water-soluble	Acrylic acid	100
A		Methacrylic acid	100
B	Good solubility in water	Acrylonitrile	~8
B		Methyl acrylate	~5
C	Slightly soluble	Vinyl acetate	~2.5
C		Methyl methacrylate	~1.5
C		Ethyl acrylate	~1.5
D	Poorly soluble	Butyl acrylate	~0.16
D		2-Ethylhexyl acrylate	~0.10
E	Practically insoluble	Styrene	~0.03
E		Butadiene	~0.09
E		Ethylene	~0.06
E		Vinyl versatate (VeoVa 10)	<0.001

Monomers are grouped according to their approximate solubility in water at room temperature. Solubility influences monomer distribution between the aqueous phase and micelles/droplets during emulsion polymerization, thereby affecting nucleation and polymerization kinetics.
Note: Generally, the lower the water solubility, the more hydrophobic the monomer, resulting in a greater tendency to polymerize within micelles or monomer droplets rather than in the aqueous phase.

3.3 Classification by Polymer Hardness

Monomers may also be classified according to the hardness of the homopolymers they produce.

Soft monomers: Yield rubber-like, flexible polymers. This includes butyl acrylate, butadiene, ethylene.
Hard monomers: Yield rigid, glassy polymers. This includes methyl methacrylate, vinyl acetate, styrene, and vinyl chloride.

3.4 Vinyl Acetate Monomer (VAM)

3.4.1 General Characteristics

1) Highly flammable; vapors may cause flash fire.
2) Can polymerize spontaneously if exposed to heat, radiation, or strong acids/bases.
3) May cause skin, eye, and respiratory irritation.
4) Harmful if inhaled.
5) Potential carcinogen based on animal studies.

3.4.2 Polymerization of VAM

Vinyl acetate monomer (VAM) exhibits distinctive polymerization characteristics in emulsion systems due to its moderate water solubility, high reactivity, and specific radical chemistry. These factors strongly influence particle nucleation, growth, molecular architecture, and final latex performance.

Table 3.3 shows a summary of VAM polymerization characteristics and industrial implications.

3.4.3 Homopolymerization of Vinyl Acetate Monomer

The homopolymerization of VAM proceeds via free-radical initiation. The double bond of the vinyl acetate molecule opens to form repeating units, resulting in polyvinyl acetate (PVAc) as per the following reaction scheme:

$$n\mathrm{CH_2{=}CH{-}OCOCH_3} \rightarrow -[\mathrm{CH_2{-}CH(OCOCH_3)}]_n- \qquad (3.1)$$

Note: Initiators (e.g., persulfates or redox systems) generate free radicals that trigger chain growth, while surfactants or protective colloids (such as polyvinyl alcohol) stabilize the growing polymer particles in emulsion polymerization.

Table 3.3 Summary of VAM polymerization characteristics and industrial implications.

Characteristic	Effect on polymerization	Industrial implication
Use of protective colloids	Steric stabilization; modifies particle size distribution; enhances water resistance and film strength	Preferred for PVAc homopolymers; improves adhesion and reduces the need for high surfactant
Moderate water solubility (~2.5%)	Increases aqueous-phase polymerization and radical desorption rates	Higher polymerization rates; requires control to avoid excessive water-phase homopolymer
High swelling ratio (up to 7:1)	Significant monomer uptake by particles, accelerating growth	Enables high conversion rates but may cause particle size broadening if uncontrolled
High chain transfer Constant	Limits molecular weight; moderates gel content	Facilitates processing; reduces excessive molecular weight, but may reduce tensile strength
Branching tendency	Hydrogen abstraction from polymer backbone increases chain branching	Improves cohesion, but excessive branching reduces clarity and flexibility
Reactive polymer radicals	High reactivity toward other vinyl monomers	Useful in copolymerization; requires balance to avoid gelation
High MFFT (~30 °C)	Limits film formation at low temperatures	Necessitates plasticizers, coalescents, or soft comonomers for cold-weather applications
pH sensitivity	Rate and stability depend on pH; acidic to neutral preferred	Avoids hydrolysis and instability during polymerization
Hydrolysis susceptibility	PVAc hydrolyzes to PVOH and acetic acid under alkaline conditions	Affects emulsion stability and performance; requires pH control during storage and use

Key physical and chemical attributes of VAM and their practical consequences in emulsion polymerization.

3.4.4 VAM Physiological Properties and Exposure Limits

VAM is moderately toxic, causing irritation to the eyes, skin, and respiratory system, and may have chronic health effects. Proper ventilation and Personal

Table 3.4a Exposure limits of VAM.

Agency	Exposure limit	Notes
OSHA	10 ppm (8-h TWA)	Time-weighted average
NIOSH	4 ppm (ceiling)	Do not exceed at any time

Protective Equipment (PPE) are required to minimize exposure. Such exposure limits are shown in Table 3.4a.

3.4.5 Commercial Specifications and Manufacturing of Vinyl Acetate Monomer

Vinyl acetate monomer (VAM) is a vital industrial intermediate primarily used in the production of polyvinyl acetate (PVAc) and its copolymers with acrylates, ethylene, and other monomers.

The performance and consistency of these polymers depend heavily on the purity and stability of the VAM used.

Commercial production requires the monomer to meet stringent specifications for parameters such as purity, acetaldehyde content, acidity, and moisture, as even trace impurities can affect polymerization rate, molecular weight distribution, and product quality.

This section provides an overview of typical commercial specifications, standard analytical methods for their determination, and the main industrial manufacturing routes for VAM, along with their economic implications.

3.4.6 Commercial Specifications of VAM

Typical specification ranges for commercially available VAM are summarized in Table 3.4b.

These specifications are generally verified using ASTM standard test methods, such as ASTM D-2190 (distillation range), ASTM D-2086 (purity), ASTM D-2191

Table 3.4b Typical commercial specifications of vinyl acetate monomer (VAM).

Property	Specification
Purity (wt%)	99.8–99.99
Acetaldehyde (max wt%)	0.010–0.013
Acidity (as acetic acid, wt%)	0.005–0.007
Water content (wt%)	≤0.04
Distillation range (°C)	72.5–73.0
Specific gravity @ 20 °C	0.9335–0.9340
Color (APHA)	0–5
Suspended matter	None
Activity time	130–150 min (polymerization onset via dilatometer)

(specific gravity), and ASTM D-2193 (acidity). Purity may also be confirmed via gas chromatography (GC), liquid chromatography (LC), and ultraviolet/infrared (UV/IR) spectroscopy, or titration. Table 3.4b shows the typical commercial specifications of VAM.

3.4.7 Methods for Determining VAM Specifications

The following summarizes typical laboratory methods for determining VAM specifications:

1) *Specific gravity*: Determined using a pycnometer accurate to four decimal places at $20 \pm 0.05\,°C$.
2) *Acetaldehyde content*: React VAM with sodium bisulfite in an ice bath, titrate with 0.1 N iodine using a starch indicator.

$$\text{wt\% aldehyde} = (\text{blank} - \text{sample titration}) \times N(I_2) \times 0.02202 \times 100/(\text{specific gravity} \times \text{mL sample}).$$

3) *Acidity (as acetic acid)*: Add bromothymol blue to methanol and ice, titrate after adding the VAM sample.

$$\text{wt\% acetic acid} = \text{mL NaOH} \times N(\text{NaOH}) \times 0.060 \times 100/(\text{mL sample} \times \text{specific gravity}).$$

4) *Water content*: Determined via Karl Fischer titration.

$$\text{wt\% water} = (\text{blank} - \text{sample titration}) \times F \times 100/(\text{mL sample} \times \text{specific gravity}),\ \text{where } F \text{ is reagent strength.}$$

5) *Activity time*: Measured with a dilatometer at $50\,°C$ using 0.5% benzoyl peroxide, reflecting polymerization onset under standard test conditions.

3.4.8 Synthesis and Manufacture of VAM

Vinyl acetate monomer (VAM) is usually produced by two principal methods: the historical acetylene route and the modern ethylene route, as outlined below.

3.4.8.1 Historical Acetylene Route

In the historical process, acetylene reacted with acetic acid over a zinc acetate catalyst at approximately $200\,°C$. The reaction was exothermic and required cooling. Unreacted components were recovered by distillation.

3.4.8.2 Modern Ethylene Route

In the modern process, ethylene is bubbled through acetic acid at about 120 °C in the presence of palladous chloride and cuprous chloride catalysts.

Key reactions:

$$C_2H_4 + CH_3COOH + Pd^{2+} \rightarrow CH_3COOCH = CH_2 + 2HCl \quad (1)$$

$$C_2H_4 + H_2O + Pd^{2+} \rightarrow CH_3CHO + 2HCl \quad (2)$$

The Pd^{2+} and Cu^{+} catalytic cycle enables continuous reaction. Distillation removes by-products such as acetaldehyde and esters.

3.4.8.3 Economic Advantages of Ethylene Route

Transitioning from the acetylene to the ethylene route has reduced raw material costs by approximately 20%, making the ethylene route more economical for large-scale production.

3.5 Acrylate Monomers

Acrylate monomers are esters, salts, or conjugate bases of acrylic acid and its derivatives. They typically contain a vinyl group ($-CH{=}CH_2$) directly attached to the ester's carbonyl carbon, giving them high reactivity in radical polymerization.

Common examples include: Butyl acrylate (BA), 2-hydroxyethyl acrylate (2-HEA), ethyl acrylate (EA), methyl methacrylate (MMA), trimethylolpropane triacrylate (TMPTA), sodium acrylate, and sodium methacrylate.

Acrylates are typically produced via acid-catalyzed esterification of acrylic or methacrylic acid with the desired alcohol, followed by stabilization with polymerization inhibitors.

3.5.1 Molecular Formula (MF) and Structural Formula (SF) of Selected Acrylates

Table 3.5 summarizes the molecular formulas (MF) and structural formulas (SF) of selected acrylate monomers, providing a concise reference for their chemical composition and structural representation.

3.5.1.1 Butyl Acrylate (BA)

Butyl acrylate (BA) is one of the most widely used soft acrylate monomers in emulsion polymerization. Its long butyl side chain imparts flexibility and softness to the resulting polymer, significantly lowering the glass transition temperature (T_g). This makes BA a vital component in pressure-sensitive adhesives, flexible coatings, and sealants where elasticity and tack are required.

Table 3.5 Common acrylates with their molecular and structural formulas.

Acrylate	Molecular formula (MF)	Structural formula (SF)
Butyl acrylate (BA)	$C_7H_{12}O_2$	CH_2=CH-COO-$(CH_2)_3$-CH_3
2-Hydroxyethyl acrylate (2-HEA)	$C_5H_8O_3$	CH_2=CH-COO-CH_2-CH_2-OH
Ethyl acrylate (EA)	$C_5H_8O_2$	CH_2=CH-COO-CH_2-CH_3
Methyl methacrylate (MMA)	$C_5H_8O_2$	CH_2=C(CH_3)-COO-CH_3
Trimethylolpropane triacrylate (TMPTA)	$C_{15}H_{20}O_6$	$C(CH_2OCOCH{=}CH_2)_3CH_2OH$
Sodium acrylate	$[C_3H_3O_2]^- \cdot Na^+$	CH_2=CH-$COO^- \cdots Na^+$
Sodium methacrylate	$[C_4H_5O_2]^- \cdot Na^+$	CH_2=C(CH_3)-COO^- Na^+

In emulsion polymerization, BA is frequently copolymerized with harder monomers such as methyl methacrylate (MMA) or vinyl acetate (VAM) to balance hardness and flexibility, resulting in polymers with a broad application spectrum. Due to its versatility, BA is often selected as the main representative when discussing acrylates in emulsion polymerization. Table 3.6 summarizes the key molecular properties and technical data of butyl acrylate, highlighting its role as a flexible, low-T_g acrylate monomer widely employed in emulsion polymerization.

Table 3.6 Properties and technical data for butyl acrylate.

Parameter	Details
Molecular formula	$C_7H_{12}O_2$ [$C_4H_9O_2C$-CH=CH_2]
Physical state	Clear, colorless liquid with strong fruity odor
Monomer type	Soft monomer (butyl ester of acrylic acid)
Production	Acid-catalyzed esterification of acrylic acid with butanol
Inhibitors	MEHQ, phenothiazine, or hydroquinone ethyl ether
Density	Less than water; may form surface slick
Water solubility	~2 g/L; water in BA ~0.7 g/100 g BA
Solubility in solvents	Miscible with alcohol, ether, and most organic solvents
Purity (commercial)	≥99.5% w/w; typical impurities: ≤0.05% water, ≤0.01% acid (as acrylic acid)
Hydrolysis	Very slow at neutral or low pH; rapid at pH ≥11

3.5.1.2 Functional Role of Butyl Acrylate in Polymers

1) Acts as a soft monomer to adjust hardness, tack, flexibility, and durability.
2) Balances properties in copolymers for adhesives, sealants, and coatings.

3.5.1.3 Safety and Handling of Butyl Acrylate

1) Flammable and combustible.
2) Irritant to skin, mucous membranes, and respiratory tract.

3.5.1.4 Inhibition and Storage of Butyl Acrylate

1) Standard inhibitor: 15 ppm MEHQ.
2) Store at ≤25 °C, away from light, in oxygen-rich conditions to maintain stability.

3.5.1.5 Application of Butyl Acrylate

Butyl acrylate is commonly copolymerized with acrylic acid and its salts, methacrylates, acrylonitrile, maleic esters, vinyl acetate, vinyl chloride, styrene, butadiene, and unsaturated polyesters. These copolymers are widely applied in coatings, adhesives, inks, sealants, elastomers, textiles, and plastics.

3.6 Vinyl Versatate (VeoVa 10)–Vinyl Neodecanoate

VeoVa 10 is the trade name for vinyl neodecanoate, a vinyl ester of a branched neodecanoic acid. It is used primarily as a hydrophobic comonomer in vinyl acetate-based emulsions to improve water resistance, alkali resistance, and durability.

Table 3.7 shows the key properties of VeoVa 10.

Table 3.7 Key properties of VeoVa 10.

Property	Details
Chemical type	Vinyl ester of branched neodecanoic acid
Homopolymer T_g	~−3 °C
Water solubility	~0.0001%
Reactivity	Similar to VAM; copolymerizes with acrylates and ethylene
Structure	Vinyl group for polymerization; bulky alkyl group for hydrophobicity, UV stability

3.6.1 Performance Benefits of VeoVa 10 in Coatings

1) High alkali and hydrolysis resistance
2) Low water absorption and high water repellency
3) Excellent outdoor durability and UV resistance
4) Strong adhesion to nonpolar substrates

3.6.2 Typical Applications of VeoVa 10

1) Interior/exterior decorative paints
2) Wood coatings
3) Industrial finishes
4) Construction adhesives
5) Redispersible powders
6) Elastomeric roof coatings

3.6.3 Processing and Stability of VeoVa 10

1) Compatible with vinyl/acrylic systems
2) Provides shear stability during paint processing
3) Long shelf-life due to hydrolytic stability
4) Improves viscosity and pH control during formulation
5) Contributes to uniform film formation

3.6.4 Copolymerization Advantages of VeoVa 10

1) Enhances pigment wetting and scrub resistance
2) Enables low-VOC decorative and industrial coatings
3) Copolymerizes readily with vinyl acetate, forming stable emulsions
4) Reduces reactor fouling, supports batch reproducibility

3.6.5 Processing and Paint Properties of VeoVa 10

1) Compatible with various raw materials
2) High shear stability during paint manufacturing
3) Long shelf life and stable viscosity/pH
4) *Excellent rheology*: Good film build, hiding power, and application feel

3.6.6 Enhanced Alkali Resistance in Coatings of VeoVa 10

1) The bulky nonpolar structure of VeoVa 10 shields VA units from saponification
2) VA/VeoVa 10 copolymers resist alkaline degradation (e.g., from cement/masonry)

3) Higher VeoVa content improves alkali resistance and scrub durability
4) Optimal performance plateau around 15–25% VeoVa content

3.6.7 Particle Size and Film Properties of VeoVa 10

VeoVa 10 promotes the formation of latexes with small particle sizes, improving dispersion stability and pigment binding. In interior paints, this translates into higher wet scrub resistance. In exterior masonry systems, VeoVa 10 enhances alkali and efflorescence resistance, reduces fade and dirt pick-up, improves chalking resistance and color retention, and contributes to better UV durability.

3.6.8 Environmental and Safety Aspects of VeoVa 10

VeoVa 10 is generally not classified as hazardous under OSHA's Hazard Communication Standard; however, prudent handling is required. It is combustible, so storage and use should avoid sparks, open flames, and hot surfaces, with appropriate grounding/bonding and ventilation. The monomer can irritate skin and eyes and is harmful if swallowed; gloves, goggles, and good hygiene practices are recommended. From an environmental standpoint, VeoVa 10 may cause long-term aquatic toxicity, so releases to drains and surface waters should be prevented, and spills contained with absorbents suitable for organic liquids. Always consult the supplier's current Safety Data Sheet (SDS) for jurisdiction-specific classifications and detailed first-aid, firefighting, and disposal guidance.

3.7 Styrene Monomer

Styrene is one of the most important monomers in the polymer industry and plays a central role in both bulk plastics and emulsion polymers. It is an aromatic vinyl compound characterized by its reactive double bond and hydrophobic benzene ring, which together confer a unique balance of rigidity, processability, and compatibility with a wide range of comonomers. In emulsion polymerization, styrene is rarely used alone; rather, it is primarily valued as a comonomer for tailoring the performance of latexes. Its rigid aromatic backbone imparts hardness, gloss, and chemical resistance, making styrene-based copolymers particularly attractive for coatings, adhesives, sealants, and inks.

The versatility of styrene arises from its ability to copolymerize with acrylates, methacrylates, butadiene, acrylonitrile, maleic esters, and vinyl acetate. This allows manufacturers to design polymers with customized balances of hardness and flexibility, transparency and toughness, or gloss and weatherability, depending on the target application. Styrene also offers economic advantages, as it is widely available and relatively low in cost compared to many specialty monomers.

For these reasons, styrene remains a cornerstone monomer in modern emulsion polymerization, and it is essential to understand its chemical nature,

Table 3.8 Physical and chemical properties of styrene chemical.

Parameter	Details
Chemical formula	$C_6H_5CH{=}CH_2$ (aromatic monomer with vinyl side chain)
Appearance and odor	Colorless oily liquid; may appear yellowish upon aging; sweet, distinctive odor
Volatility	Evaporates readily
Production source	Almost entirely derived from ethylbenzene (produced specifically for styrene)
Primary uses	Monomer for polystyrene, ABS, SBR, SIS, SEBS, styrene-acrylic emulsions

polymerization behavior, and contribution to copolymer properties before considering its practical applications.

3.7.1 Physical and Chemical Properties and Uses

Table 3.8 shows the physical and chemical properties of styrene.

3.7.2 Polymerization Characteristics of Styrene

Styrene polymerizes readily via free-radical mechanisms initiated thermally or by radical initiators such as organic peroxides and persulfates (e.g., benzoyl peroxide,

Table 3.9 Key styrene-based copolymer systems and their applications.

Copolymer system	Key properties	Main applications
Polystyrene (PS)	Transparency, rigidity, brittleness, electrical insulation	Packaging, disposable containers, insulation foams, electronics
Styrene–acrylates	Gloss, hardness, chemical resistance, weatherability	Architectural coatings, industrial paints, adhesives, inks
Styrene–butadiene (SBR)	Elasticity, toughness, abrasion resistance	Tires, paper coatings, carpet backing, adhesives
Styrene–acrylonitrile (SAN)	Transparency, rigidity, heat resistance	Plastics, appliances, automotive components
Acrylonitrile–butadiene–Styrene (ABS)	Impact resistance, toughness, processability	Automotive parts, consumer goods, electronics housings
Styrene–vinyl acetate	Flexibility, adhesion, balance of hardness and toughness	Adhesives, coatings, construction materials
Styrene–maleic esters	Film formation, hardness, solubility control	Paper coatings, specialty adhesives, binders

potassium persulfate). Industrially important styrene-based polymers include polystyrene; acrylonitrile–butadiene–styrene (ABS); styrene–butadiene rubber (SBR); styrene–isoprene–styrene (SIS); and styrene–ethylene/butylene–styrene (SEBS). Table 3.9 summarizes key styrene-based copolymer systems and representative applications.

3.7.3 Safety and Reactivity of Styrene

Styrene can undergo auto-polymerization at elevated temperatures, and the reaction rate rises sharply as the temperature increases. Uncontrolled polymerization generates heat and pressure, which introduces thermal runaway and explosion risk; as an order of magnitude, rates on the order of ~2% per hour have been reported at 100 °C. From a health perspective, styrene exhibits acute and chronic toxicity. It has been discussed in the literature as toxic, mutagenic, and potentially carcinogenic, with reported effects on the gastrointestinal, renal, and respiratory systems. Appropriate engineering controls, personal protective equipment, inhibitor management, and strict adherence to the current Safety Data Sheet are essential for safe handling and storage.

3.7.4 Emulsion Polymerization Applications of Styrene

Styrene–acrylic emulsions are typically produced by copolymerizing styrene with acrylic comonomers such as acrylic acid (AA) or methacrylic acid (MAA), soft acrylates like ethyl acrylate and butyl acrylate, and harder monomers such as methyl methacrylate (MMA) or 2-ethylhexyl acrylate (2-EHA). These binders offer a favorable balance of performance and cost: they can reduce raw-material cost relative to pure acrylics, improve water resistance through the hydrophobic contribution of styrene, and—when formulated for small particle size—enhance primer and binder performance. They also show good weatherability and stain resistance, high pigment-binding capacity, and can be crosslinked to increase film durability. Adhesion to substrates such as aluminum, wood, and galvanized steel is typically strong when surface preparation and formulation are optimized.

Limitations include a tendency toward UV-induced yellowing—especially at higher styrene content—and the possibility that residual radical activity can accelerate discoloration if stabilization is inadequate. Representative applications include ceramic-tile adhesives, fillers and putties, elastomeric roof coatings, and architectural coatings.

3.7.5 Formulation Considerations of Styrene

Vinyl carboxylic acids such as acrylic acid (AA) and methacrylic acid (MAA) are sometimes incorporated into styrene–acrylic emulsion formulations at low levels (typically 1–5% by weight of total monomer) to:

1) *Enhance emulsion stability*: The ionizable -COOH groups tend to localize at the particle surface, improving colloidal stability, dispersibility, and pigment wetting.
2) *Improve adhesion*: Carboxyl functionality promotes hydrogen bonding or covalent bonding with polar substrates such as wood, metal, or mineral surfaces.
3) *Enable crosslinking with special crosslinkers*: The -COOH groups serve as reactive sites for one-component (1K) crosslinkers such as carbodiimides, aziridines, or polyaziridines. The purpose is to increase water, solvent, and scrub resistance by forming covalent bonds during film formation or curing, as shown in the table below:
   Safety principle (book-wide): As a standing principle of this book, any crosslinkers—internal or external—that are not considered safe for routine industrial use are excluded. This includes acrylamide, *N*-methylolacrylamide (NMA), aziridines, and any formaldehyde-containing or formaldehyde-releasing systems.

Table 3.10 lists representative crosslinkers for styrene–acrylic emulsions in which -COOH functionality is introduced exclusively via acrylic acid (AA) or methacrylic acid (MAA). The listed crosslinkers are formulated to remain storage-stable and subsequently react with the -COOH groups during film formation or curing. The "addition stage" indicates whether the crosslinker is incorporated during polymerization (internal) or added later during formulation (post-added). Only the main types are shown here; more advanced systems are discussed in later chapters.

Safety principle (book-wide): Unsafe crosslinkers are excluded—whether internal or external—including acrylamide, *N*-methylolacrylamide (NMA), aziridines, and any formaldehyde-containing or formaldehyde-releasing systems.

Condition for external crosslinkers: The polymer must contain -COOH functionality introduced via acrylic acid (AA) and/or methacrylic acid (MAA) for effective ionic or covalent crosslinking.

Table 3.10 Crosslinkers for –COOH-functional styrene–acrylic emulsions.

Crosslinker type	Addition stage	Reaction with –COOH groups	Cure conditions	Key performance benefits
Glycidyl methacrylate (GMA) comonomer	Internal (copolymerized)	Epoxy ring opens; esterification with –COOH creates covalent crosslinks within the film	Room temperature to mild heat (ambient to 80 °C)	Water/chemical resistance, hardness build without brittle failure, improved block/scrub
Zirconium complexes (e.g., zirconium ammonium carbonate, zirconium acetate)	Post-added	Zr^{4+} coordinates with $-COO^-$ (ionic/coordination crosslinking)	Room temperature; best at mildly acidic pH (~4.5–5.5)	Humidity and heat resistance, faster set-through-dry, improved wet adhesion
Aluminum chloride ($AlCl_3$)	Post-added	Al^{3+} forms ionic bridges with $-COO^-$ groups	Room temperature; acidic pH (~3.5–4.5); avoid overdose	Higher early strength, moisture resistance; economical (watch corrosion)
Zinc oxide/zinc salts	Post-added	Zn^{2+} coordination with $-COO^-$ groups (ionic crosslinks)	Room temperature; neutral to mildly alkaline pH (~7–9)	Wet adhesion, anti-blocking, stain resistance; good clarity when dispersed well
Carbodiimides	Post-added	Addition to –COOH to form *N*-acylurea linkages (covalent)	Room temperature	Hydrolytic stability, humidity resistance, long-term durability
Blocked polyisocyanates	Post-added	Upon unblocking, reacts with –COOH to form amide linkages (CO_2 release); it can also react with –OH where present	Mild heating (60–90 °C) to bake (≥120 °C)	High abrasion resistance, tunable flexibility–hardness balance, solvent/chemical resistance

3.8 Workplace Safety and Handling of Various Monomers

Safe handling of reactive monomers such as VAM, acrylate esters, and styrene requires strict adherence to personal protective measures, storage and handling protocols, firefighting readiness, and first-aid procedures.

3.8.1 Personal Protective Equipment (PPE) and Training

Provide comprehensive, competency-based training on the selection, fit, use, limitations, maintenance, and storage of PPE. Personnel working with monomers and auxiliary chemicals should wear chemical-resistant gloves, splash goggles or a face shield, flame-resistant coveralls, and respiratory protection selected according to the hazard assessment and the current Safety Data Sheet. Contaminated garments must be removed immediately; leather items such as shoes and belts cannot be reliably decontaminated and should be discarded. Maintain unobstructed eyewash stations and emergency showers close to work areas, verify access routes remain clear, and function-test them on a routine schedule.

3.8.2 Storage and Handling of Flammable Monomers

Store flammable monomers away from heat, sparks, open flames, and direct sunlight, and control temperature as required by inhibitor specifications. Segregate from oxidizing agents and strong acids or bases. Use compatible equipment—typically carbon steel or stainless steel—and avoid copper, brass, and other copper-containing alloys. Ground and bond all containers and transfer lines to prevent static discharge. Use explosion-proof electrical equipment and nonsparking tools. Prefer top unloading of tank cars or tank trucks using dedicated pumps or nitrogen-blanketed systems; never unload by air pressure. Empty drums should be thoroughly cleaned and dried, sealed, and stored in a cool, well-ventilated area away from ignition sources.

3.8.3 Firefighting Procedures

For incipient, small-scale fires involving flammable monomers, portable carbon dioxide or dry chemical extinguishers are appropriate. For larger fires or those involving storage vessels, use dry chemical or alcohol-resistant foam applied through fixed systems or unmanned monitors when conditions permit. Apply water spray as a cooling medium to protect exposures and to cool containers, but avoid high-pressure jets or solid streams that can spread burning liquid.

Responders must wear full structural firefighting protective clothing together with a self-contained breathing apparatus (SCBA). Control runoff to prevent entry into drains and waterways, and contain contaminated water for proper disposal. When tank fires occur, establish wide isolation and exclusion zones per the incident commander's direction and local emergency guidelines; in some circumstances, large-radius isolation (on the order of hundreds of meters) may be warranted.

After knockdown, continue cooling of exposed equipment until well after extinguishment, monitor for re-ignition, and ventilate to disperse vapors. All actions should follow the facility's emergency response plan and the current Safety Data Sheets for the materials involved, and be coordinated with local fire authorities.

3.8.4 First-Aid Measures

First-aid actions should follow the facility's emergency response plan and the current Safety Data Sheets for the materials involved. For inhalation exposure, move the person to fresh air and keep them at rest; provide oxygen if trained personnel deem it necessary, and administer CPR if breathing has stopped. For skin contact, remove contaminated clothing promptly and wash the affected skin with soap and water for at least 15 min. For eye contact, rinse cautiously with water for at least 15 min, lifting upper and lower eyelids; remove contact lenses if present and easy to do, and seek urgent medical care. If ingestion occurs, do not induce vomiting; rinse the mouth and give small amounts of water or milk if the person is conscious; obtain immediate medical attention. For thermal or chemical burns, cool the area with water; do not attempt to remove clothing that is adhered to the skin; keep the victim warm and monitor until professional help arrives.

3.9 Second Ingredient—Water

Water is the principal ingredient in emulsion polymerization, functioning as the continuous phase and dispersion medium for monomer droplets, micelles, and polymer particles. Its presence enables low-viscosity reaction systems, allowing effective heat and mass transfer while providing enhanced safety and controllability during polymerization.

3.9.1 Functional Role in Emulsion Polymerization

Water facilitates the transfer of monomer molecules from dispersed droplets to the loci of polymerization (micelles and growing polymer particles). It also serves as the solvent for water-soluble formulation components such as surfactants, initiators, buffers, protective colloids, and other functional additives.

3.9.2 Preferred Water Quality

Deionized (DI) water is preferred for most emulsion polymerization systems. Although softened water can be used, it often contains residual electrolytes (e.g., Na^+, Ca^{2+}, Mg^{2+}) that can destabilize the latex by reducing electrostatic repulsion between particles. This destabilization increases particle aggregation, leading to coalescence and a larger average particle size.

3.9.3 Water Quality Parameters

Beyond deionization, several water attributes influence emulsion polymerization outcomes. The working pH should be neutral to slightly alkaline (about 6.5–8.5) to minimize premature hydrolysis of sensitive monomers or surfactants. Dissolved oxygen can inhibit free-radical initiation and propagation; oxygen-sensitive systems often benefit from deaeration or nitrogen purging prior to charge and during make-up. Microbial contamination can degrade surfactants and other organics, destabilizing the latex; routine monitoring and appropriate biocide programs are therefore recommended. Trace organic contaminants—such as oils, lubricants, or residual process organics—may disrupt micelle formation and alter polymerization kinetics, so equipment cleanliness and upstream filtration are critical. Finally, conductivity reflects ionic content; for high-purity DI water used in sensitive latexes, maintaining conductivity below roughly 50 μS/cm helps limit parasitic electrolyte effects on stability, particle size control, and conversion behavior.

3.9.4 Effect of Electrolytes on Polymer Properties

Some electrolytes are unavoidable due to the presence of ionic initiators, anionic surfactants, and buffer systems. These ionic species influence the electrical double layer surrounding polymer particles, impacting particle stability and size distribution. Excessive electrolyte concentration can lead to coagulation or secondary nucleation.

3.9.5 Sequestering Agents and Buffers

Sequestering agents and buffers help stabilize the aqueous phase and, by extension, the developing latex. Sodium hexametaphosphate chelates multivalent cations such as Ca^{2+} and Mg^{2+}, mitigating ion-induced destabilization. Because it alters the electrostatic balance of the dispersion, however, its use can increase polymer particle size in some systems. Sodium bicarbonate functions primarily as a buffering agent to maintain pH stability; in certain PVAc-based emulsions, use in the range of approximately 0.15–0.29% of total formulation weight has been reported to yield the smallest particle size.

3.9.6 Temperature Considerations

The temperature of the process water at the start of polymerization influences reaction control and early kinetics. Cold water can help suppress premature polymerization but may slow initial reaction rates. Conversely, warm water (about 30–40 °C) can reduce viscosity and improve monomer dispersion—useful in cold-climate operation—yet must be balanced against the risk of premature initiator decomposition and inhibitor depletion. Select the starting temperature with reference to initiator half-life, inhibitor specifications, and the reactor's heat-removal capability.

3.9.7 Industrial Practice

In large-scale operations, water is typically filtered, deionized, and stored in stainless steel or lined tanks to prevent metal contamination. A recirculation loop with UV sterilization and inline filtration (e.g., 5 μm) is often used to maintain purity during continuous plant operation.

3.10 Third Ingredient—Thickening Agent

Thickening agents are critical in maintaining colloidal stability during emulsion polymerization. They enhance viscosity in the aqueous phase, reduce particle settling, and help control rheology during both polymerization and end-use application. The most common are polyvinyl alcohol (PVOH) and hydroxyethyl cellulose (HEC)—both functioning as stabilizing colloids and nonionic surfactants.

3.10.1 Polyvinyl Alcohol (PVOH)

Polyvinyl alcohol (PVOH) is widely used as the principal protective colloid in VAM polymerization. Its effects on latex formation and downstream properties are governed primarily by molecular weight and degree of hydrolysis.

3.10.1.1 Effect of Molecular Weight

Higher-molecular-weight grades increase viscosity at a given solids level, accentuate shear-thinning behavior, and typically raise tensile strength, wet tack, and thixotropy of the resulting latex and films. Lower-molecular-weight grades reduce solution and latex viscosity, lessen chain entanglement effects, and may improve processability—particularly in high-solids or high-throughput operations.

3.10.1.2 Effect of Degree of Hydrolysis

Lower degrees of hydrolysis generally produce smaller particle sizes and improved colloidal stability but increase water sensitivity; they also tend to enhance freeze–thaw resistance. Higher degrees of hydrolysis yield larger particle sizes and lower viscosity with more nearly Newtonian flow behavior, while improving the water resistance of the final film.

3.10.2 Best Practices in Semicontinuous PVAc Polymerization Using PVOH

In semicontinuous PVAc polymerization, PVOH serves as the main stabilizer, with emulsifiers playing a supporting role. Grades near ~88% hydrolysis often provide a practical balance between conversion efficiency and colloidal stability. An operating window of approximately 82.6–91.5% hydrolysis at about 4% based on total monomer is commonly effective for minimizing precipitates.

Fully hydrolyzed PVOH (95–98%) can improve the water resistance of the final film, but may compromise the storage stability of the latex. Because VAM-derived radicals may graft onto PVOH during polymerization, the stabilizer–surfactant balance should be tuned to maintain particle stability and avoid excessive gelation.

Higher adsorption of PVOH at the particle surface generally drives smaller particle sizes and higher latex viscosity. Compared with protective-colloid-dominated systems, ionic surfactants such as sodium dodecyl sulfate (SDS) can produce finer particles, though the influence on overall polymerization rate is typically limited. PVOH can also depress the glass-transition temperature (T_g) through hydration-shell formation. High PVOH-to-surfactant ratios tend to increase particle size and viscosity; blended systems are often used to optimize stability and rheology.

3.10.3 Hydroxyethyl Cellulose (HEC)

Hydroxyethyl cellulose (HEC) is a nonionic, water-soluble cellulose ether used as part of the stabilizer system in emulsion polymerization. While applicable to polyvinyl acetate (PVAc) homopolymer manufacture, its stabilizing and rheology-modifying functions are most pronounced in copolymer emulsions—particularly those incorporating softer comonomers such as butyl acrylate, ethyl acrylate, and VeoVa.

3.10.3.1 Chemical Structure and Functional Groups

HEC is derived from cellulose by controlled reaction with ethylene oxide, introducing hydroxyethyl side chains along the anhydroglucose backbone. The degree of substitution (DS) and molar substitution (MS) define the number and average length of these hydroxyethyl groups. These modifications increase water solubility relative to unmodified cellulose, reduce interchain hydrogen bonding so the polymer behaves more flexibly in solution, and introduce acetal hydrogens that can be susceptible to abstraction by free radicals under polymerization conditions.

3.10.3.2 Role as a Protective Colloid

HEC stabilizes latex particles through complementary chemical and physical mechanisms. Chemically, initiator-derived radicals can abstract relatively labile acetal hydrogens from the HEC backbone to generate reactive macroradicals; grafting of growing polymer chains onto HEC then anchors the polysaccharide to the particle surface, providing robust steric stabilization. Physically, where grafting is less favored, HEC can adsorb via hydrogen bonding and van der Waals interactions. This adsorption is reversible and typically affords less durable stability than covalent grafting.

3.10.3.3 Advantages of HEC in Emulsion Polymerization

The nonionic character of HEC makes it broadly compatible with both anionic and cationic additives. It provides shear stability that helps protect particles under high-shear processing, and exhibits good tolerance to electrolytes and ionic initiators. As a rheology modifier, HEC increases viscosity and thereby moderates particle–particle collisions. During film formation, it enhances water retention in the wet film, supporting complete coalescence where required.

3.10.3.4 Impact of Molecular Weight and Substitution Level

Higher-molecular-weight grades generally deliver greater viscosity and stronger stabilization at lower use levels, whereas lower-molecular-weight grades are easier to handle but may require higher concentrations to achieve the same effect. Moderate molar substitution (approximately MS 1.3–2.5) often provides a favorable balance of water solubility and surface interaction for latex stabilization.

3.10.3.5 Processing Considerations

For consistent performance, HEC should be fully dissolved in deionized water prior to reactor addition. It is typically charged to the initial water phase ahead of surfactants and initiators. Common use levels are on the order of 0.3–1.0% of total formulation weight. HEC is stable over a wide pH range, though prolonged exposure to extreme pH can degrade viscosity; pH control of make-up solutions is therefore advisable.

3.10.3.6 Limitations and Precautions

HEC can promote foaming and may require compatible antifoams. Aqueous HEC solutions are susceptible to microbial growth and should be preserved appropriately. Over-concentration can raise viscosity excessively and increase gelation risk; dosing should be validated against target rheology and mixing capability.

3.10.3.7 Applications in PVAc and Copolymer Systems

In PVAc wood adhesives, HEC can improve wet tack and storage stability. In vinyl acetate–acrylic copolymers, it enhances stability in the presence of low-T_g comonomers. For architectural coatings, HEC contributes to sag resistance and pigment suspension, and in paper and textile binders, it aids binder distribution and can improve mechanical properties.

4

Fourth Ingredient—Surfactants

4.1 Introduction

Surfactants—surface-active agents—are essential to emulsion polymerization because they lower surface and interfacial tensions, allowing otherwise immiscible phases (water and hydrophobic monomers) to form and maintain finely dispersed systems. By controlling how new interfaces are created and stabilized, surfactants enable micelle formation, particle nucleation, and the long-term colloidal stability of the resulting latex.

Two terms are used throughout this chapter. Surface tension refers to the energy at the water–air interface, while interfacial tension refers to the energy at the boundary between two immiscible liquids—here, the aqueous phase and the hydrophobic monomer phase. Reducing these tensions promotes emulsification, narrows particle-size distributions, and improves process control during polymerization and film formation.

Surfactants are amphiphiles: a polar, hydrophilic head interacts with water, and a nonpolar, hydrophobic tail associates with monomers and other organic phases. At low concentrations, they adsorb at interfaces; above the critical micelle concentration, they self-assemble into micelles that solubilize monomer and provide loci for radical entry and particle nucleation. The balance of head-group chemistry and tail length thus governs wetting, emulsification efficiency, micelle structure, and downstream latex properties.

A familiar example is sodium dodecyl sulfate (SDS), $CH_3(CH_2)_{11}OSO_3^- \ Na^+$. The sulfate group serves as the anionic, hydrophilic head, while the dodecyl chain is the hydrophobic tail—an archetype that illustrates how molecular architecture dictates interfacial behavior in emulsion polymerization.

4.2 Types of Surfactants Used in Emulsion Polymerization

Surfactants are indispensable in emulsion polymerization because they dictate not only the formation and stabilization of latex particles but also the ultimate performance of the polymer in its end-use applications. Their amphiphilic structure enables the reduction of interfacial tension, facilitating the dispersion of monomers in water and providing colloidal stability throughout the reaction. The type and concentration of surfactant selected directly influence particle size, particle size distribution, viscosity, polymer molecular weight, and even film properties such as water resistance and adhesion.

In practice, formulators rarely rely on a single surfactant type. Instead, combinations of anionic, nonionic, or polymerizable surfactants are often employed to achieve a balance between process stability and application performance. For example, an anionic surfactant may provide excellent nucleation efficiency, while a nonionic surfactant improves freeze–thaw stability. This synergistic use reflects the dual challenge in emulsion polymerization: ensuring smooth reaction kinetics in the reactor while simultaneously tailoring the polymer for downstream requirements in adhesives, coatings, or paints.

The following sections classify surfactants according to their charge characteristics, beginning with anionic surfactants, which remain the most widely used in industrial practice.

4.2.1 Anionic Surfactants Used in Emulsion Polymerization

Anionic surfactants comprise a negatively charged hydrophilic head group—typically sulfate, sulfonate, phosphate, or carboxylate—attached to a nonpolar hydrophobic tail such as a straight or ethoxylated alkyl chain. In emulsion polymerization, they lower interfacial tension, promote micellization and nucleation, and stabilize polymer particles over a wide range of shear and solids.

Alkyl sulfates (AS) and alkyl ether sulfates (AES) are workhorse emulsifiers. The AS types are obtained by sulfation of fatty alcohols followed by neutralization, as shown in (4.1). The AES types are produced by sulfating ethoxylated fatty alcohols, followed by neutralization, as shown in (4.2). The ethoxylate length (n) tunes HLB, wetting, and foam; blends with nonionic co-surfactants or protective colloids are common to moderate foaming and improve electrolyte tolerance.

$$\mathrm{ROH + SO_3 \rightarrow ROSO_3H \rightarrow neutralization \rightarrow ROSO_3^{-}\,M^{+}} \tag{4.1}$$

$$\mathrm{R(CH_2CH_2O)_{\mathit{n}}OH + SO_3 \rightarrow R(CH_2CH_2O)_{\mathit{n}}OSO_3H \rightarrow neutralization \rightarrow R(CH_2CH_2O)_{\mathit{n}}OSO_3^{-}\,M^{+}} \tag{4.2}$$

Linear alkylbenzene sulfonates (LAS) are prepared by sulfonating linear alkylbenzene. A common example is sodium dodecylbenzene sulfonate (SDBS), with structure C_6H_4–$C_{12}H_{25}$–SO_3^- Na^+ (empirical formula $C_{18}H_{29}NaO_3S$). Because the sulfonic group is bonded directly to the aromatic ring, LAS exhibits high thermal and hydrolytic stability relative to sulfate esters. In polymerization, they deliver strong wetting and fine particles; dosage control is important to manage foaming and avoid rate suppression at high levels.

Diphenyl oxide disulfonates (e.g., sodium dodecyl diphenyl oxide disulfonate, $C_{24}H_{33}O_7S_2Na_2$) carry two sulfonate groups, providing high charge density per molecule. They offer rapid micelle formation, excellent colloidal stability, and good electrolyte tolerance where salts, initiator residues, or metal ions are present.

Phosphate ester surfactants are obtained by phosphorylating alcohols or ethoxylated alcohols with polyphosphoric acid (monoesters) or phosphorus pentoxide, P_2O_5 (mixed mono/diesters), followed by neutralization to the sodium, potassium, or triethanolamine salts. The mono/diester ratio governs HLB and emulsifying character. Acidic forms can be neutralized in situ, and phosphate esters often aid pigment wetting and adhesion to metal substrates.

Sulfosuccinate esters are produced by reacting maleic anhydride with fatty alcohols, followed by sulfonation. Sodium dioctyl sulfosuccinate (DOSS, $C_{20}H_{37}NaO_7S$) is notable for exceptionally low equilibrium and dynamic surface tensions, promoting rapid wetting and small-particle nucleation. Hydrolytic lability under strong acid or base conditions often limits use to co-surfactant levels.

Carboxylate soaps have the general formula R–COO^- M^+ (R = long-chain alkyl; M^+ = Na^+, K^+, NH_4^+). Examples include sodium laurate ($C_{11}H_{23}COO^-$ Na^+), sodium palmitate ($C_{15}H_{31}COO^-$ Na^+), and sodium stearate ($C_{17}H_{35}COO^-$ Na^+). Performance is strongly pH-dependent; soaps are effective under alkaline conditions but are susceptible to precipitation with Ca^{2+}/Mg^{2+} and can impair water resistance. In humidity-resistant or crosslinkable binders, soaps are typically minimized in favor of more robust anionics or nonionics.

Within the ether-sulfate and ether-sulfonate families, ethoxylate length provides a convenient handle to tune HLB, wetting speed, foam, and electrolyte tolerance. Short EO chains favor rapid nucleation and small particles; moderate EO improves pigment compatibility and process latitude; long EO reduces foam but may raise the CMC. These materials are widely used as primary emulsifiers or as co-surfactants paired with protective colloids to balance kinetics and final film properties.

Alkyl ether sulfate ethoxylates (AES) and alkyl ether sulfonate ethoxylates (AESn) are among the most widely used anionic surfactants in emulsion polymerization due to their excellent emulsifying efficiency, high foaming capacity, and compatibility with a wide range of monomers. Their amphiphilic structure allows them to lower interfacial tension effectively, facilitating the formation of fine and stable latex particles.

4.2.1.1 Alkyl Ether Sulfate Ethoxylates (AES) and Alkyl Ether Sulfonate Ethoxylates (AEnS)

Alkyl ether sulfate ethoxylates (AES) and alkyl ether sulfonate ethoxylates (AEnS) are among the most widely used anionic surfactants in emulsion polymerization. Their amphiphilic architecture lowers interfacial tension efficiently, promotes micellization, and facilitates the formation of fine, stable latex particles. In practice, these surfactants provide high emulsifying efficiency with narrow particle-size distributions, improve stability against coagulation during polymerization, and show excellent compatibility with other anionic or nonionic surfactants. AEnS is often preferred when higher salt or acid resistance is required.

Author's Note on Nomenclature In industrial practice, alkyl ether sulfates are commonly abbreviated as AES, referring to sulfated ethoxylated fatty alcohols. Sulfonate-based surfactants, although chemically distinct, may also be derived from ethoxylated precursors in certain formulations.

For clarity in this section, the abbreviation AEnS is used to denote ethoxylated sulfonate-type surfactants, where "n" qualitatively indicates the presence of ethylene oxide units. This notation is adopted solely for internal consistency in this discussion and does not represent a standardized industry abbreviation.

4.2.1.2 Alkyl Ether Sulfate Ethoxylates (AES)

AES surfactants are typically produced by ethoxylating a fatty alcohol, followed by sulfation and neutralization. They contain a sulfate ($-OSO_3^-$) group attached to an ethoxylated alkyl chain. A general formula is $R\text{-}(OCH_2CH_2)_n\text{-}OSO_3^-\ M^+$, where M^+ is Na^+, K^+, or NH_4^+.

Step 1: Ethoxylation

$$R - OH + n(EO) \rightarrow R - (OCH_2CH_2)_n - OH \tag{4.3}$$

Step 2: Sulfation

$$R - (OCH_2CH_2)_n - OH + X \rightarrow R - (OCH_2CH_2)_n - OSO_3H \tag{4.4}$$

where $X = SO_3$ or $ClSO_3H$

Step 3: Neutralization

$$R - (OCH_2CH_2)_n - OSO_3H + MOH \rightarrow R - (OCH_2CH_2)_n - OSO_3^-\ M^+ \tag{4.5}$$

where $M^+ = Na^+, K^+,$ or NH_4^+

4.2.1.3 Alkyl Ether Sulfonate Ethoxylates (AEnS)

AEnS differ from AES by bearing a sulfonate ($-SO_3^-$) group instead of a sulfate group. This substitution generally enhances electrolyte tolerance and stability under extreme pH, which is advantageous in emulsion polymerization systems operating at higher ionic strength or with variable pH.

Example synthesis

$$R-(OCH_2CH_2)_n-OH+Y \rightarrow R-(OCH_2CH_2)_n-SO_3H \tag{4.6}$$

where Y = sulfonating agent

$$R-(OCH_2CH_2)_n-SO_3H+MOH \rightarrow R-(OCH_2CH_2)_n-SO_3^-\,M^+ \tag{4.7}$$

General formula: $R\text{-}(OCH_2CH_2)_n\text{-}SO_3^-\,M^+$

4.2.1.4 Role of Anionic Surfactants in Emulsion Polymerization

In polymerization practice, alkyl ether sulfate ethoxylates (AES) and alkyl ether sulfonate ethoxylates (AEnS) are used as primary emulsifiers or as co-surfactants alongside protective colloids. By tuning the ethoxylate length n, formulators balance wetting speed, foam, electrolyte tolerance, and pigment compatibility. These choices translate into high emulsifying efficiency with correspondingly narrow particle-size distributions, improved stability against coagulation during feed and conversion, and broad compatibility with other anionic or nonionic surfactants within typical process windows. Where higher salt content or lower pH is expected, AEnS is often preferred because the sulfonate headgroup offers greater electrolyte and acid resistance than the sulfate analog. Taken together, surfactant identity and EO chain length govern nucleation kinetics, initial particle size, and the stability window during feed, leading to reproducible latex quality.

4.2.2 Nonionic Surfactants Used in Emulsion Polymerization

Nonionic surfactants lack charged head groups; water-solubility is instead provided by polar moieties such as poly(ethylene oxide) (EO) chains. They are widely used in emulsion polymerization for their high tolerance to electrolytes, compatibility with anionic surfactants, and the ability to stabilize emulsions over a broad temperature range.

4.2.2.1 Fatty Alcohol Ethoxylates (FAE)

Fatty alcohol ethoxylates can be represented by the general formula

$$RO(CH_2CH_2O)_nH$$

where R is a long-chain fatty alcohol (C_{12}–C_{18}), and n is the number of EO units. FAEs are produced by reacting fatty alcohols with ethylene oxide. In emulsion polymerization, they are valued for low foaming, excellent freeze–thaw stability, and improved control of particle size. Shorter EO chains ($n \approx 3$–6) yield more hydrophobic surfactants, whereas longer EO chains increase HLB, enhance water solubility, and electrolyte tolerance.

4.2.2.2 Alkylphenol Ethoxylates (APE)—e.g., Nonylphenol Ethoxylates (NPE)

Nonylphenol ethoxylates can be represented by the general formula

$$C_9H_{19} - C_6H_4 - O(CH_2CH_2O)_nH$$

Nonylphenol ethoxylates are produced by the ethoxylation of nonylphenol. They offer strong emulsifying power for hydrophobic monomers such as styrene and butyl acrylate. NPEs provide excellent particle size control and compatibility with anionics. However, due to environmental restrictions in many regions, they are being phased out in favor of linear alcohol ethoxylates.

4.2.2.3 EO–PO Block Copolymers (Pluronic-Type Surfactants)

EO-PO block copolymers can be represented by the general formula

$$HO - (CH_2CH_2O)_x - (CH_2CH(CH_3)O)_y - (CH_2CH_2O)_x - H$$

These nonionic surfactants are synthesized by sequential polymerization of propylene oxide (PO) and ethylene oxide (EO) onto a propylene glycol core. The hydrophobic PO block is flanked by hydrophilic EO blocks. In emulsion polymerization, they improve freeze–thaw stability, control particle size, and can act as co-surfactants to anionics for enhanced stability in high-electrolyte systems.

4.2.2.4 Ethoxylated Fatty Acid Esters

Ethoxylated fatty acid esters can be represented by the general formula $RCOO(CH_2CH_2O)_nH$. Produced by ethoxylation of fatty acid esters, these surfactants provide good emulsification of soft monomers like butyl acrylate and 2-ethylhexyl acrylate. They are often used in combination with anionic surfactants to balance HLB and enhance monomer dispersion.

Performance in Emulsion Polymerization: Nonionic surfactants generally yield latexes with improved electrolyte stability, better adhesion to hydrophobic substrates, and enhanced freeze–thaw resistance compared to systems stabilized solely by anionics. The choice of EO chain length and hydrophobic group strongly influences particle size, film formation, and water resistance.

4.2.2.5 Narrow Range Ethoxylate

Narrow-range ethoxylates are nonionic surfactants manufactured with catalysts and processes that yield a tightly peaked distribution of ethylene-oxide homologs rather than the broad distribution typical of conventional alcohol ethoxylates. The narrower homolog profile provides a more precise hydrophile–lipophile balance at a given average ethoxylation level and reduces the fraction of very low- or very high-EO species that can destabilize latexes or remain mobile in dried films.

In VAc–acrylic systems, narrow-range ethoxylates function as nonreactive co-surfactants in the monomer preemulsion. Their peaked ethoxylate distribution provides a tighter, more predictable hydrophile–lipophile balance and higher interfacial efficiency, allowing equivalent colloidal stability at lower dosage with stronger adsorption to particle interfaces. Consequently, the amount of unbound surfactant remaining in the serum and in the dried film is reduced relative to conventional broad-range nonionics, yielding cleaner film formation with lower water whitening, improved early scrub resistance, and fewer surfactant-related defects. Minimizing very low-EO species decreases the tendency toward premature flocculation under redox or high-solid conditions, while limiting very high-EO species reduces postfilm water sensitivity.

Narrow-range Ethoxylate Selection Guidance For VAc-rich binders, a practical HLB window is approximately ten to thirteen, using C12–C13 or C12–C15 alcohol backbones with six to nine ethylene-oxide units on average. The narrow-range nonionic is paired with an anionic primary emulsifier in the seed to establish particle number, after which the nonionic dominates interfacial stabilization during growth. Typical active dosage is approximately 0.2–0.6% of the total formulation in the preemulsion. Post-additions at very low levels can be used for substrate and pigment wetting, where low foam and clean recoat are required. When crosslinking is planned by carbodiimide, zirconium, or epoxy chemistries, nonamine narrow-range grades are preferred to avoid side reactions and to maintain the target pH window of approximately 4.5–5.5 in VAc systems.

Use of Narrow-range Ethoxylates Narrow-range ethoxylates are selected when a formulator needs tighter process robustness and superior film appearance without increasing the total surfactant level. Typical triggers include persistent water whitening on drawdowns, inadequate early scrub, surfactant migration causing blocking or dirt pickup, reactor foaming under high shear, or sensitivity to electrolyte spikes in the feed. They are also valuable in cold-to-moderate temperature polymerizations where the nonionic cloud point could otherwise be crossed; the narrow range provides greater buffer between operating temperature and cloud point, so long as temperature is controlled below the measured cloud point of the chosen grade.

Practical Cautions Narrow-range ethoxylates complement but do not replace the primary anionic surfactant used to define particle number in the seed stage. Excess nonionic can increase interfacial mobility and water uptake; therefore, the objective is to use the lowest level that delivers stability and target appearance. Polymerization temperature should remain below the cloud point of the selected grade throughout the feed to avoid transient destabilization. In D3 and D4 adhesive formulations, free surfactant should be kept modest to protect wet strength; therefore, narrow-range grades are favored because they allow lower total nonionic at equivalent stability.

Outcomes and Benefits When properly selected and dosed, narrow-range ethoxylates provide reproducible HLB, lower foam, enhanced shear and electrolyte tolerance, reduced surfactant migration, and cleaner film formation. In practice, the result is improved early scrub and stain resistance in coatings and higher wet-adhesion retention in crosslinkable adhesives, with fewer defects related to surfactant bleed or water whitening (Tables 4.1 and 4.2).

4.2.2.6 High-HLB Nonionic Surfactants for Seed Nucleation and Cold/Redox Starts

Scope and Rationale This section addresses highly hydrophilic nonionic surfactants with HLB values in the approximate range of fifteen to eighteen, used strategically for seed nucleation and for polymerizations initiated at low temperature or under redox conditions. These grades are not narrow-range ethoxylates and

Table 4.1 Trade names of common anionic surfactants used in emulsion polymerization.

Chemical type	Example trade name	Supplier	Notes/application
Sodium lauryl sulfate (SLS)	Texapon® N70	BASF	High-foaming, good particle size control in vinyl acetate systems
Ammonium lauryl sulfate	Empicol® ESB/3 AM	Innospec	Ammonium form, stable in acidic systems
Fatty alcohol ether sulfates (AES)	Texapon® ASV	BASF	Ethoxylated for improved electrolyte tolerance
Sodium dioctyl sulfosuccinate	Aerosol® OT	Cytec/Solvay	Powerful wetting agent, used at low levels to control particle size
Linear alkylbenzene sulfonates (LAS)	Nacconol® 90G	Stepan	Strong detergency, mainly in blends; not for low-foam systems
Phosphate esters	Rhodafac® RS-610	Solvay	Excellent for adhesion promotion and pigment compatibility

Representative trade names of anionic surfactants commonly used in emulsion polymerization. These commercial products illustrate the diversity of suppliers and chemistries available, supporting latex stabilization and performance tuning.

Table 4.2 Trade names of common nonionic surfactants used in emulsion polymerization.

Chemical type	Example trade name	Supplier	Notes/application
Alcohol ethoxylates (C12–C15)	Lutensol® AO series	BASF	Broad HLB coverage; common replacement for NPEs. Select grade to meet HLB target and cloud point
Alkyl polyglucosides (APG)	Glucopon® series	BASF	Renewable, high biodegradability; often blended with anionic for stability
Ethoxylated fatty acid esters	Tween® 20/ Tween® 80	Croda	Co-emulsifier or post-add wetting aid; supports film formation and can reduce foam in some systems
Fatty alcohol ethoxylates (branched)	Tomadol® series	Sasol	Improved cold-water solubility; useful with hydrophobic monomers; pick grade for target HLB
Ethoxylated alkyl phenols (restricted)	Triton™ X-100 (legacy)	Dow	Legacy example; restricted due to NPE phase-out
Narrow-range alcohol ethoxylates (peaked distribution)	Lutensol® XP series	BASF	Narrow/peaked homolog distribution; low odor and low residual alcohol. Choose grade to hit HLB ~10–13 for VAc preemulsions; check cloud point > process T
Narrow-range alcohol ethoxylates (single-branch substrates)	Lutensol® XA series	BASF	Narrow distribution for predictable HLB; often enables lower total dosage at equivalent stability; select grade by HLB.
Narrow-range alcohol ethoxylates	NOVEL® series (e.g., NOVEL® G12 family)	Sasol	Proprietary catalyst yields peaked distribution; select grade to meet HLB target; useful where low foam and migration control are desired
Narrow-range natural alcohol ethoxylates	Berol® 360/Berol® 366	Nouryon	Vegetable-based narrow-range AEs; low-foam wetting and stable preemulsions; select grade to meet HLB and cloud-point targets

Representative trade names of nonionic surfactants used in emulsion polymerization. Selection should be based on HLB and cloud-point targets rather than nominal EO alone. For VAc-rich preemulsions, a practical starting window is HLB ~10–13 (often achieved with ~6–9 EO on C12–C13 alcohols). Families listed below span multiple EO cuts; choose the grade that meets the required HLB and process temperature.

Note: Supplier families span several ethoxylation cuts. For VAc-rich preemulsions, start with a grade targeting HLB ~10–13 and verify cloud point above the polymerization temperature. Outside this HLB window, grades may still be appropriate for seed nucleation or specific wetting roles; confirm by screening and QC.

therefore are not covered in Section 4.2.2.5. Their strong hydrophilicity promotes efficient micelle formation and robust particle number control when electrolyte levels are elevated, temperatures are modest, or shear is high. The tradeoffs are a greater tendency to foam and an increased risk of postfilm water sensitivity if overdosed.

Formulation Role High-HLB nonionics are best applied at low levels during the seed stage to define particle number and to stabilize the early nuclei. After nucleation, the preemulsion feed should transition to a nonionic suited to growth—typically a narrow-range alcohol ethoxylate selected to deliver an HLB near ten to thirteen, as discussed in Section 4.2.2.5—paired with the primary anionic emulsifier. This staged approach balances early-stage robustness with low migration and clean film formation during growth and drying.

Practical Guidance A typical starting dosage for the high-HLB seed aid is approximately 0.05–0.15% active on total formulation, adjusted to the minimum level that secures a stable seed. The growth-phase nonionic is commonly dosed at approximately 0.2–0.6% active, with the exact grade chosen to meet hydrophile–lipophile balance and cloud-point targets. The polymerization temperature should remain at least 5–10 °C below the measured cloud point of the chosen nonionics. Where reduction of free, mobile surfactant is critical, a polymerizable surfactant or a reactive protective colloid can be incorporated in a small proportion to anchor interfacial species without increasing the total nonionic burden.

Application Note Plant experience shows that certain high-HLB alcohol ethoxylates can deliver excellent seeding behavior and cold-start reliability in VAc-rich systems when used sparingly and paired with a lower-HLB narrow-range nonionic for the growth feed. A representative example is Sasol's Nonidac family, which has been used successfully in seed nucleation at low dosage; selection and performance should be verified by screening under plant conditions.

Selection should be based on HLB and cloud-point targets rather than nominal ethoxylation alone. For VAc-rich preemulsions, the growth-phase nonionic typically targets HLB near 10–13; high-HLB nonionics in the range of 15–18 may be advantageous in the seed or in cold/redox starts at low dosage. Validate foam formation, water whitening, and surfactant migration in the intended process window.

4.2.3 Alternatives to Nonylphenol Ethoxylates (NPEs)

Nonylphenol ethoxylates (NPEs) have been widely used as nonionic surfactants in emulsion polymerization due to their strong emulsifying power, adjustable HLB values, and low cost. However, their degradation product, nonylphenol, is persistent, bioaccumulative, and an endocrine disruptor. Many regions, including the EU, the United States, and Canada, have banned or restricted their use, driving the demand for biodegradable, environmentally friendly alternatives.

4.2.3.1 Desired Properties in NPE Alternatives

When replacing NPEs, the alternative surfactant should meet the following requirements to ensure stable polymerization and high-quality latex products:

1) *HLB value (hydrophilic–lipophilic balance)*: The HLB determines micelle size and stability, which directly influences latex particle size distribution. For vinyl acetate homopolymerization, an HLB of approximately 13–15 ensures optimal dispersion and stable nucleation. For acrylate-rich systems, a slightly higher HLB (around 14–16) may be needed due to the lower polarity of the monomers. If the HLB is too low, monomers may not emulsify well, leading to large, unstable particles. If the HLB is too high, overstabilization can cause excessive water sensitivity in the final film.
2) *Cloud point*: The cloud point is the temperature at which a nonionic surfactant solution becomes cloudy; above this temperature, surfactant solubility drops. A cloud point at least 10–20 °C higher than the polymerization temperature is preferred to maintain micelle integrity during the entire process. If the cloud point is too close to the polymerization temperature, phase separation may occur, leading to instability and a broad particle size distribution.
3) *Surface tension reduction*: Surfactants must lower the interfacial tension between water and monomer to promote micelle formation and monomer droplet stabilization. Latex systems typically require a surface tension in the range of 28–35 mN/m for efficient nucleation. Insufficient reduction of interfacial tension prevents proper micelle formation, leading to poor particle nucleation and increased coagulum formation.
4) *Electrolyte tolerance*: High-solids latexes often accumulate salts from initiators, buffer systems, and pH control agents during polymerization. The surfactant system must maintain particle stabilization under these ionic conditions; otherwise, electrostatic repulsion is reduced, leading to micelle collapse and coagulum formation. Fatty alcohol ether sulfates and certain alkyl polyglucosides generally exhibit good electrolyte tolerance, whereas some alcohol ethoxylates may require blending with anionic surfactants to enhance stability.
5) *Compatibility*: The surfactant must be compatible with radical initiators, monomers, and other surfactants in the formulation. It should not reduce initiator activity or destabilize the system. The surfactant must be stable within the pH range of the process (often 4.0–6.5 for vinyl acetate polymerization) and blend well with co-surfactants to prevent flocculation or excessive foaming.

Table 4.3 Practical alternatives to NPEs.

Alternative	Chemical class	Typical HLB range	Cloud point (°C)	Advantages	Limitations/notes
Alcohol ethoxylates (AEs)	Primary linear C12–C15 alcohol +7–9 EO	12–15	60–70	Biodegradable, adjustable HLB, widely available	Can foam more than NPEs; may require antifoam
Alkyl polygluco-sides (APGs)	C8–C16 alkyl chain + glucose	10–14	>80	Renewable, biodegradable, low toxicity, stable in alkaline conditions	Higher cost; limited solubility in highly hydrophobic monomer feeds
Branched alcohol ethoxylates	C13–C15 branched alcohol +7–10 EO	11–14	55–65	Good cold-water solubility, stable emulsions	Slightly less biodegradable than linear AEs
Ethoxylated fatty acid esters	Sorbitan ester +10–20 EO	12–16	50–65	Low irritation, good co-emulsifier for blends	Not as strong as primary emulsifier; often used with fatty alcohol ether sulfates
AES (fatty alcohol ether sulfates)	C12–C14 AE sulfate salt	12–15	N/A (ionic)	Excellent emulsification, high foaming, good synergy with nonionics	Can cause excess foaming; may need silicone defoamer

Table 4.3 presents practical alternatives to nonylphenol ethoxylates (NPEs), highlighting commonly used surfactant classes, their typical HLB ranges, cloud points, advantages, and limitations.

4.2.3.2 Selection Guidelines

1) Match HLB. Start with a replacement surfactant whose HLB is within about ±1 unit of the NPE being replaced, then fine-tune based on foam, particle size, and stability.
2) Normalize to active content. Compare performance on an active-matter basis. Adjust dose for differences in purity, solvent/water content, and counter-ions so that the comparison reflects equal actives.
3) Track kinetics and particle size. Monitor conversion versus time and measure particle-size distribution (e.g., D50 and span) during and after feed; note any coagulum or viscosity drift.
4) Evaluate latex and film properties. Assess water resistance, viscosity/rheology, film formation (including MFFT/open time), and any impacts on adhesion, block, scrub, or stain resistance.

4.2.3.3 Practical Application Note

Replacing a nonylphenol ethoxylate with a fatty alcohol ethoxylate in vinyl acetate–acrylate emulsion polymerization, at a dosage of 2.5% on total monomer (normalized to active content), yielded a similar median particle size (~200 nm) and a slightly higher monomer conversion (98% vs. 96%). The system showed improved biodegradability. Foam increased by ~15%, which was effectively controlled by adding 0.1% of a suitable defoamer.

4.2.3.4 Blending Strategies for Vinyl Acetate/Vinyl Versatate (VAc/VeoVa) Copolymerization

Vinyl acetate/vinyl versatate (VAc/VeoVa) copolymer latexes place higher demands on surfactant selection than VAc homopolymerization because VeoVa is highly hydrophobic. Replacing nonylphenol ethoxylates (NPEs) in these systems often requires blending nonionic and anionic surfactants to achieve stable nucleation, controlled particle size, and good water resistance without excessive foaming.

Blending Principles Specific to VAc/VeoVa

1) Pair a high-HLB nonionic surfactant (e.g., alcohol ethoxylate such as Berol® 25) with an anionic surfactant (e.g., phosphate ester or alkyl ether sulfate such as Daclor®).
2) Increase nonionic fraction as VeoVa content rises; reduce overall surfactant as larger particle sizes are targeted.
3) Maintain cloud point at least 10–20 °C above the polymerization temperature (typically 60–80 °C in semibatch).
4) Assess electrolyte tolerance when persulfate initiators or neutralizing agents are used; phosphate esters or AES blends often improve robustness.

Table 4.4 presents recommended surfactant blends for VAc/VeoVa copolymerization, showing how nonionic and anionic ratios are adjusted depending on VeoVa content.

4.2.3.5 Common Issues When Replacing NPEs in VAc/VeoVa and How to Fix Them

This section outlines the main formulation challenges encountered when replacing nonylphenol ethoxylates (NPEs) in VAc/VeoVa copolymer systems, along with practical strategies for mitigation. Table 4.5 represents these issues and their recommended solutions.

4.2.3.6 Case Study—Replacement of NPE in a VAc/VeoVa (20 wt% VeoVa) Semibatch Process

A vinyl acetate–VeoVa emulsion (20 wt% VeoVa) was polymerized in semibatch mode at 70–72 °C using a persulfate initiator. The incumbent surfactant was a

Table 4.4 Recommended surfactant blends by VeoVa content.

VeoVa in feed (wt%)	Target particle size (nm)	Recommended blend (wt% of total surfactant)	Typical total surfactant (% on monomer)	Rationale
5–10	200–300	AE-9 EO 60% + AES 40%	1.8–2.2	Balanced HLB; AES improves electrolyte tolerance and film water resistance
10–20	180–250	AE-9–10 EO 70% + AES 25% + Fatty acid ester 5%	2.0–2.6	Higher nonionic for hydrophobicity; small ester fraction lowers foam and aids film formation
20–30	150–220	AE-10–12 EO 65% + APG 20% + AES 15%	2.3–3.0	More hydrophilic nonionics to disperse VeoVa; APG boosts stability and biodegradability
>30	120–180	AE-12 EO 60% + APG 20% + AES 20%	2.8–3.5	High EO nonionics for hydrophobic VeoVa; AES ensures ionic stabilization at small sizes

Table 4.5 Common issues and solutions when replacing NPEs in VAc/VeoVa copolymer systems.

Observed issue	Likely cause	Corrective action
Broad or bimodal particle size	HLB mismatch; cloud point too low	Increase EO number of AE or add APG; ensure cloud point ≥10–20 °C above polymerization temperature
Coagulum formation	Insufficient ionic stabilization; poor electrolyte tolerance	Increase AES fraction by 5–10%; check initiator salt load; add phosphate ester (0.2–0.4% on monomer)
Excessive foaming	High AES or high-shear feed strategy	Introduce fatty acid ester (3–5% of surfactant blend) and silicone defoamer (0.05–0.10% on total batch)
Low conversion at high VeoVa	Insufficient micellar solubilization	Raise nonionic level by 0.2–0.4% on monomer; consider AE with higher EO (10–12 EO)
Poor water resistance of film	Overstabilization by high-HLB nonionic	Reduce total nonionic by 0.2–0.3%; increase AES by 5%

nonylphenol ethoxylate with 10 mol of ethylene oxide, dosed at 2.6% on total monomer. The target median particle size was 180–200 nm.

As a replacement, a blend was employed consisting of an alcohol ethoxylate with 10 mol of ethylene oxide, an alkyl ether sulfate, and a fatty acid ester at a 70/25/5 ratio, with the total surfactant level maintained at 2.6% on monomer. A silicone-based defoamer was included at 0.08% in the batch.

Initial trials produced an average particle size of approximately 210 nm with increased foam. Optimization was achieved by reducing the alcohol ethoxylate fraction by 0.2% points and increasing the alkyl ether sulfate fraction by 0.2% points (at constant total surfactant). The adjusted blend yielded an average particle size of about 190 nm, a monomer conversion of 98.5%, Brookfield viscosity within specification, and improved resistance to water whitening in the resulting films.

4.2.3.7 Additional Considerations in Selecting NPE Alternatives for VAc/VeoVa Copolymerization

The surfactant blend strongly influences the water-whitening resistance of the dried film. Higher VeoVa content increases hydrophobicity and generally improves water resistance. Raising the anionic fraction (e.g., a larger alkyl ether sulfate share) can further reduce water whitening, but may modestly hinder film coalescence—especially in low-MFFT systems. High-HLB nonionics facilitate dispersion and broaden process latitude, yet can yield films that soften or whiten more readily in water. For adhesives and coatings requiring high water resistance, a practical approach is to bias slightly toward anionic surfactant and/or include a hydrophobic co-surfactant (such as a fatty acid ester), while keeping the nonionic level at the minimum needed for colloidal stability.

Storage stability can also be affected by the choice of NPE alternative. High-EO alcohol ethoxylates and alkyl polyglucosides, in particular, may promote a gradual viscosity rise in stored latex due to hydrogen-bonding interactions and structural rearrangement in the aqueous phase. During manufacture, this tendency can be mitigated by incorporating protective colloids—such as low-hydrolysis PVOH at about 0.5–1.0% on total polymer—or by adding small amounts of HEC (approximately 0.1–0.3%) postpolymerization. Alternatively, blending a shorter-EO nonionic with the primary surfactant can limit viscosity drift without sacrificing stability (Table 4.6).

4.2.3.8 Additional Practical Notes

Compatibility with other additives—Some NPE alternatives can interact with coalescing agents, plasticizers, or thickeners, altering latex rheology and film

Table 4.6 Additional considerations in selecting NPE alternatives for VAc/VeoVa copolymerization.

Surfactant type	Relative cost	Environmental profile	Performance in VAc/VeoVa	Notes
Alcohol ethoxylates (AEs)	Low to medium	Readily biodegradable	Good nucleation and dispersion; HLB adjustable via EO number	May foam more than NPEs; watch cloud point
Alkyl polyglucosides (APGs)	High	Renewable, excellent biodegradability	High stability, good electrolyte tolerance	Higher cost; may need blending for hydrophobic monomers
Fatty alcohol ether sulfates (AES)	Low	Readily biodegradable	Excellent particle stability and water resistance	Can cause high foaming; may require defoamer
Ethoxylated fatty acid esters	Medium	Biodegradable	Improves film formation and reduces foam	Weaker emulsifier; best as co-surfactant

formation. For example, high-EO alcohol ethoxylates can slightly increase the demand for coalescents in low-MFFT formulations, while fatty alcohol ether sulfates may destabilize certain associative thickeners. It is advisable to check compatibility during laboratory trials by evaluating viscosity, particle size, and film performance after all additives are introduced.

Environmental marketing advantage—Many customers are actively seeking NPE-free products for compliance with eco-certifications such as EU Ecolabel, Blue Angel, or Green Seal. Using alternatives like alkyl polyglucosides or certain alcohol ethoxylates not only improves the environmental profile but can also be highlighted in technical datasheets and marketing materials to enhance product positioning in environmentally sensitive markets.

4.2.3.9 Dosage Guidance for Surfactant Use in Vinyl Acetate (VAc)/VeoVa 10 Copolymers

The table below provides practical starting ranges (as % of total monomer) for VAc/VeoVa copolymer systems at different VeoVa contents. These ranges reflect industry practice and vendor consultations, as well as published trends on non-ionic/anionic balances in VAc/VeoVa emulsions. They are guidelines only and must be verified by laboratory and pilot trials for specific raw materials, reactor configurations, and end-use requirements (Table 4.7).

Table 4.7 Surfactant dosage guidelines by VeoVa content in copolymerization.

VeoVa content (wt%)	Total surfactant (% on monomer)	Nonionic/ anionic ratio	Special notes
5–10	1.8–2.2	≈ 60/40	—
10–20	2.0–2.6	≈ 70/30	Optional 3–5% fatty acid ester
20–30	2.3–3.0	≈ 80/20	Consider alkyl polyglucoside (10–20% of blend)
>30	2.8–3.5	Nonionic rich with anionic	Prefer high-EO nonionics + AES for stability

Note: These quantitative ranges are author derived but align with published qualitative trends on surfactant ratio adjustment when hydrophobic comonomer content increases.

4.3 Polymerizable Surfactants in Emulsion Polymerization

4.3.1 What Are Polymerizable Surfactants?

Polymerizable surfactants are surface-active molecules that carry a polymerizable functional group—typically vinyl, allyl, methacrylate, or acrylate—so they can be covalently incorporated into the growing polymer during emulsion polymerization. As with conventional surfactants, they reduce surface and interfacial tension to enable emulsification and particle stabilization; however, their reactive moieties allow them to become chemically bonded to the polymer backbone.

Functionally, they operate first as classical surfactants prior to and during particle nucleation, stabilizing monomer droplets and nascent polymer particles. As polymerization proceeds, the reactive group copolymerizes so that the surfactant remains attached to the particle surface or within the particle, providing long-term interfacial activity with reduced desorption or migration after film formation.

4.3.2 Common Reactive Functional Groups of Polymerizable Surfactants

Polymerizable surfactants represent a unique class of surface-active molecules that not only stabilize emulsion polymerization systems but also become chemically incorporated into the growing polymer network. Unlike conventional surfactants, which may later migrate or leach out of the final film, polymerizable surfactants carry reactive groups capable of covalently bonding to the polymer backbone.

This dual role—surface activity and chemical participation—significantly enhances latex stability, water resistance, and long-term performance of the emulsion polymer.

At the heart of this functionality lies their reactive moieties. These groups determine how the surfactant interacts with vinyl acetate, acrylates, or other monomers, and they influence the strength of anchoring within the polymer matrix. Understanding these reactive groups is therefore essential for formulators aiming to design more durable adhesives, paints, and coatings.

The most widely used reactive groups include acrylate and methacrylate esters, allyl groups, vinyl groups, and epoxy functionalities, each providing a distinct balance of reactivity, stability, and compatibility with specific polymerization systems.

The following table summarizes the most common reactive groups of polymerizable surfactants, which are also called surfmers, highlighting their chemical nature, reactivity, and typical application benefits (Table 4.8).

Table 4.8 Common reactive functional groups of polymerizable surfactants (surfmers).

Type	Reactive group	Representative surfmers	Remarks/benefits
Anionic surfmers	Vinyl (–$CH{=}CH_2$)	Sodium vinyl sulfonate; sodium dodecyl vinyl sulfonate	Copolymerizes with VAc and acrylates; enhances latex stability
Anionic surfmers	Allyl (–CH_2–$CH{=}CH_2$)	Sodium allyl sulfonate; sodium dodecyl allyl sulfosuccinate	Mild reactivity with gradual anchoring; improves flexibility
Anionic surfmers	Acrylate/ methacrylate	Maximul® 6000 (Croda); Sipomer COPS-series	Benchmark industrial surfmers; strong anchoring; reduces surfactant migration; improves water resistance
Anionic surfmers	Phosphate acrylates	Sipomer PAM series	Improved adhesion, wet scrub resistance, and corrosion protection on metal and mineral substrates
Nonionic surfmers	Acrylate/ methacrylate	PEG-monoacrylate; MPEG-methacrylate; polypropylene glycol methacrylate	High reactivity and efficient incorporation; minimizes free surfactant migration; useful in low-foam systems
Styrenic surfmers	Styrenic anchor group	Vinylbenzyl PEG ether sulfate; styrenic surfmers (vendor-specific)	Styrene-like polymerizability with amphiphilic stabilization improves gloss, hardness, and chemical resistance

Table 4.9 Features and advantages of using surfmers instead of conventional surfactants.

Feature	Benefit
Permanent anchoring	Eliminates desorption and migration
Improved colloidal stability	Even under stress (shear, freeze–thaw, electrolyte presence)
Reduced surfactant leaching	Better water resistance in coatings and adhesives
Enhanced film properties	Higher gloss, lower tack, better adhesion
Environmentally friendlier	Less free surfactant in effluents

4.3.3 Advantages of Polymerizable Surfactants

Polymerizable surfactants offer distinct advantages over conventional surfactants in emulsion polymerization. Because they become covalently anchored into the polymer backbone during polymerization, they remain permanently bound within the latex particles, reducing the risk of migration, foaming, or film defects. This structural integration improves water resistance, mechanical durability, and long-term stability of the final product. Moreover, the use of surfmers contributes to enhanced environmental performance by minimizing free surfactant content, a critical factor in modern low-VOC and high-performance coating and adhesive formulations. The following table shows such advantages (Table 4.9).

4.3.4 Application Considerations

While polymerizable surfactants (surfmers) offer clear advantages in terms of film stability and long-term performance, their successful use in emulsion polymerization requires a careful formulation strategy. Unlike conventional surfactants that remain free in the aqueous phase, surfmers must not only stabilize the system during polymerization but also copolymerize effectively with the growing latex particles. This dual role means that their dosage, timing of addition, and compatibility with the monomer system must be precisely controlled. Failure to optimize these parameters may compromise nucleation, reduce copolymer reactivity, or lead to phase separation.

Key Considerations

1) *Dosage:* Must be carefully optimized; excessive levels can lower reactivity or cause particle inhomogeneities, while insufficient levels may not provide adequate stabilization.

2) *Timing of incorporation:* While the bulk of surfmers is best introduced in the later stages of polymerization to ensure efficient anchoring and avoid interference with micelle nucleation, a small controlled portion may be included in the seeding stage. This early addition can help stabilize the initial micelles and suppress secondary nucleation, but the dosage should be kept low (typically ≤10–15% of the total surfmer) to prevent disruption of particle formation.
3) *Copolymer compatibility:* The reactive functional group of the surfmer should be chemically compatible with the monomer system (e.g., acrylate surfmers for acrylic-based systems, vinyl sulfonate surfmers for VAc-based systems).

Table 4.10 presents selected commercial surfmers, their trade names, chemical names, and typical copolymer systems where they are most effectively applied.

4.3.5 Limitations for the Use of Polymerizable Surfactants

Polymerizable surfactants ("surfmers") offer clear advantages, but several practical limitations should be addressed during selection and formulation.

1) *Cost*: Surfmers are typically more expensive than conventional surfactants due to specialized synthesis. This can affect formulation economics, especially in commodity products where cost competitiveness is critical.
2) *Shelf-life and storage stability*: Because surfmers contain polymerizable groups (e.g., acrylate or vinyl), they are prone to premature reaction if not properly stabilized. Appropriate inhibitors, careful packaging, and controlled storage conditions are required for consistent performance.
3) *Formulation complexity*: Unlike conventional surfactants that remain free in the aqueous phase, surfmers must be balanced with the monomer system. An appropriate surfmer-to-monomer ratio is essential; otherwise, particle-size distribution, latex stability, or copolymer properties may be adversely affected.

4.3.6 Explanation and Practical Considerations Regarding Formulation Complexity

Polymerizable surfactants, also called reactive surfactants or surfmers, are unlike conventional stabilizers. They carry two functional components within a single molecule:

1) A surface-active group that ensures emulsification and particle stabilization.
2) A polymerizable functional group (often vinyl or acrylate) that becomes covalently incorporated into the growing polymer chain.

This dual role is the reason surfmers are valued: they minimize or even eliminate free surfactant residues in the final latex. However, it is also the source of their

Table 4.10 Selected commercial surfmers and their suitable copolymer systems.

Surfmers (type)	Trade name	Supplier	Chemical nature/main functional group	Best copolymer compatibility
Anionic (vinyl sulfonate)	– (generic)	Various suppliers	Sodium vinyl sulfonate	VAc, VAc/Veova
Anionic (allyl sulfonate)	– (generic)	Various suppliers	Sodium allyl sulfonate; sodium dodecyl allyl sulfosuccinate	VAc, VAc/Veova
Phosphate-functional surfmers	Maxemul™ 6106	Croda	Anionic phosphate ester acrylate (copolymerizable)	VAc/Veova, acrylics, styrene-acrylics
Phosphate-functional surfmers	Maxemul 6112	Croda	Anionic phosphate ester acrylate (copolymerizable)	VAc/Veova, acrylics, styrene-acrylics
Phosphate-functional surfmers	Maxemul 6112/20	Croda	Anionic phosphate ester acrylate (modified for enhanced film properties)	VAc/Veova, acrylics, styrene-acrylics
Phosphate-functional surfmers	Sipomer® PAM-200	Arkema/Sartomer	Acrylated phosphate ester (copolymerizable)	VAc/Veova, acrylics, styrene-acrylics
Phosphate-functional surfmers	Sipomer® PAM-300	Arkema/Sartomer	Acrylated phosphate ester (higher functionality)	VAc/Veova, acrylics, styrene-acrylics
PEG-acrylate surfmers	Sipomer® COPS-1	Arkema/Sartomer	Poly(ethylene glycol) monoacrylate	Acrylic copolymers, VAc/acrylics
PEG-acrylate surfmers	Sipomer® COPS-2	Arkema/Sartomer	Poly(ethylene glycol) dimethacrylate derivative	Acrylic copolymers, VAc/acrylics
PEG-acrylate surfmers	Sipomer® COPS-3	Arkema/Sartomer	Allyl-functional acrylic surfmer	VAc/acrylics, styrene-acrylics
Phosphate-functional surfmer	TREM® LF-40	Dow	Phosphate-functional acrylate surfmer	VAc/Veova, acrylics, styrene-acrylics
Nonionic PEG acrylates	– (generic)	BASF, Evonik, others	Polyethylene glycol monoacrylate	Acrylics, VAc/acrylics
Nonionic PPG methacrylates	– (generic)	BASF, Evonik, others	Polypropylene glycol methacrylate	Acrylics, styrene-acrylics
Styrenic surfmer	Surfmers-St (literature)	Academic sources	Vinylbenzyl PEG ether sulfate	Styrene-acrylic copolymers, specialty coatings

formulation complexity. Striking the right balance between surface activity and polymerizability is rarely straightforward, and fine-tuning is essential.

4.3.6.1 Why Fine-tuning Is Required

Accurate dosing of polymerizable surfactants is essential because both under- and over-addition, as well as monomer–surfmer reactivity, directly affect nucleation, particle stability, and the properties of the resulting latex and film.

1) *Insufficient surfmer*: If the surfmer level is too low, early stabilization of micelles and primary particles is compromised, leading to poor colloidal stability, coagulum formation, and a broadened particle-size distribution.
2) *Excess surfmer*: At high levels, surfmers can oversaturate the particle interface or undergo partial homopolymerization. Consequences include reduced surface-activity efficiency, formation of insoluble or gel-like species, and altered film properties such as excessive stiffness or brittleness.
3) *Monomer–surfmer interaction*: The extent of incorporation depends on the reactivity ratio of the surfmer with the main monomers (e.g., VAc, MMA, BA, styrene). Low-reactivity surfmers may remain partially unreacted, accumulating at the interface rather than becoming fully anchored in the polymer backbone, which can diminish long-term stability benefits.

4.3.7 Case Study–Effect of Surfmers on VAc/VeoVa Copolymer Stability

In a formulation trial, a vinyl acetate–VeoVa 10 copolymer (85/15 by weight) was prepared using Maximul® 6112 as the primary anionic surfmer. Three surfmer levels were evaluated while keeping initiator, temperature, and other conditions constant.

Case A: Low surfmer level (0.2 wt% on total monomer). The latex became unstable during the early stage of polymerization. Significant coagulum formed, particle size was uneven with a population of larger particles, and the resulting film showed low gloss and water whitening on immersion.

Case B: Excess surfmer level (1.0 wt% on total monomer). The system remained stable during polymerization, but a portion of the surfmer underwent self-reaction, leading to insoluble gel particles in the latex. Films prepared from this latex were overly stiff and brittle.

Case C: Optimized surfmer level (0.45 wt% on total monomer; approximately 10% charged in the seed with the remainder fed later). The latex was stable throughout polymerization, with a uniform, fine particle-size distribution. Films exhibited high gloss, good resistance to water whitening, and retained flexibility.

Lesson learned: Fine-tuning both the surfmer dosage and the timing of its addition is essential. Too little surfmer leads to instability; too much promotes gel formation and brittleness; the appropriate balance yields a stable latex and high-performance film properties.

Practical Considerations

1) As a starting range, use approximately 0.5–3.0 wt% surfmer based on total monomer and adjust on an active-matter basis.
2) Maintain a co-surfactant or small amount of conventional surfactant (e.g., SDS or a nonionic) during seed formation to support nucleation.
3) Optimize initiator type and timing; early crosslinking at the interface can interfere with micelle formation and particle growth.
4) Monitor the final latex for particle-size distribution, film clarity, surface energy, and adhesion under the intended cure and test conditions.

4.3.8 Performance Requirements for Polymerizable Surfactants

To function effectively, an optimal polymerizable surfactant should meet the following criteria:

1) *Low or no homopolymerization tendency*: It should not self-polymerize, which could lead to instability or unwanted gel formation.
2) *Comparable behavior to conventional surfactants*: It must effectively reduce surface/interfacial tension and stabilize monomers and particles throughout the reaction.
3) *Controlled reactivity timing*: Ideally, the surfactant should remain unreacted in the early stages of polymerization, enabling sufficient particle growth. It should be incorporated primarily during the later stages, where it anchors to particle surfaces, enhancing latex stability and final film properties.

4.3.9 Advantages Offered by Polymerizable Surfactants

The use of polymerizable surfactants provides several distinct benefits over conventional surfactants, particularly in emulsion polymerization systems designed for high-performance coatings and adhesives.

4.3.9.1 Enhanced Latex Stability

Conventional surfactants may desorb from the surface of latex particles under conditions such as high shear, electrolyte addition, or freeze–thaw cycling. This desorption weakens particle stabilization and may result in coagulation. Polymerizable surfactants, by contrast, are covalently anchored into the polymer backbone, ensuring long-term colloidal stability even under demanding conditions.

4.3.9.2 Improved Film Performance

During film formation, conventional surfactants often migrate within the drying matrix. Such migration can:

1) Increase water sensitivity by forming hydrophilic aggregates within the film.
2) Reduce gloss and gloss retention by accumulating at the air–film interface.
3) Diminish adhesion strength by concentrating at the substrate–film interface.

Reactive surfactants overcome these drawbacks by being chemically immobilized within the polymer matrix, resulting in superior water resistance, durability, and surface esthetics.

4.3.10 Comparison Between Conventional and Polymerizable Surfactants

A practical way to highlight the unique benefits of polymerizable surfactants is to compare them directly with conventional surfactants. Conventional surfactants remain physically adsorbed on the particle surface, whereas polymerizable surfactants (surfmers) become covalently bound to the polymer backbone. This fundamental difference leads to distinct behaviors in latex stability, film properties, and long-term performance (Table 4.11).

4.3.11 Real World Applications of Polymerizable Surfactants

Polymerizable surfactants (also called reactive surfactants or surfmers) have moved beyond laboratory interest and are increasingly applied in commercial emulsion polymerization systems. Their covalent incorporation into the polymer backbone offers advantages in product performance, regulatory compliance, and long-term stability. Below are selected real-world application areas:

4.3.11.1 Architectural Coatings

Conventional surfactants tend to migrate to the film surface, causing water sensitivity, efflorescence, and reduced gloss. Efflorescence refers to the whitish, crystalline salt deposit that appears on coating or masonry surfaces when water migration carries dissolved salts to the surface, where they crystallize after evaporation. Polymerizable surfactants remain locked within the polymer matrix, resulting in coatings with enhanced water and scrub resistance, higher gloss and gloss retention, and reduced surfactant leaching under humid conditions. For example, acrylic- or vinyl acetate-based architectural paints formulated with surfmers achieve longer durability in exterior applications.

Table 4.11 Comparison between conventional surfactants and polymerizable surfactants (surfmers).

Feature	Conventional surfactants	Polymerizable surfactants (surfmers)
Mode of action	Adsorbed on particle surface (physical stabilization)	Covalently bonded into the polymer backbone (chemical stabilization)
Latex stability	May desorb under shear, dilution, or freeze–thaw cycling → coagulation risk	Strongly anchored; improved resistance to shear, dilution, and freeze–thaw
Film formation	Surfactants migrate to air–film or substrate–film interface, affecting gloss and adhesion	Immobilized; better gloss, adhesion, and reduced water sensitivity
Water resistance	Limited, due to surfactant migration and hydrophilicity	Improved, as reactive groups remain trapped in polymer
Cost	Lower, widely available	Higher, requires careful optimization
Formulation complexity	Easier to balance surfactant levels	Requires fine-tuning monomer/surfmer ratios
Long-term performance	Gloss loss, adhesion reduction, surfactant leaching possible	Stable properties with less migration and leaching

4.3.11.2 Adhesives

In adhesives, the migration of conventional surfactants at the adhesive–substrate interface can lower adhesion strength. Polymerizable surfactants improve wet adhesion and reduce sensitivity to moisture. Pressure-sensitive adhesives (PSAs) and laminating adhesives containing reactive surfactants demonstrate stronger bonding to difficult substrates such as polyethylene and polypropylene.

4.3.11.3 Textile and Nonwoven Binders

When conventional surfactants are used in textile binders, residual surfactant can affect fabric handle, cause yellowing, and reduce wash resistance. Polymerizable surfactants provide a softer hand feel, less yellowing, and better resistance to laundering. Medical nonwovens such as surgical gowns and masks employ surfmers to meet stringent wash and sterilization requirements.

4.3.11.4 Paper and Packaging Coatings

In paper and packaging coatings, conventional surfactants may migrate and cause printability issues, foaming, and reduced barrier properties. Polymerizable surfactants provide stronger binding, smoother surfaces, and improved ink holdout. Water-based barrier coatings containing reactive surfactants also show enhanced grease and moisture resistance in food packaging applications.

4.3.11.5 Regulatory and Environmental Advantages

Formulations containing polymerizable surfactants can be designed with lower volatile organic compound (VOC) levels since they reduce the need for external emulsifiers. Their immobilization within the film also minimizes free surfactant migration into the environment, supporting compliance with ecolabel standards.

4.3.11.6 Summary

The use of polymerizable surfactants in coatings, adhesives, textiles, and packaging systems has demonstrated measurable performance gains, particularly in water resistance, adhesion, durability, and regulatory compliance. These real-world applications illustrate how surfmers transform laboratory concepts into industrially significant technologies.

4.3.12 Formulation Guidelines for Polymerizable Surfactants

The successful incorporation of polymerizable surfactants (surfmers) into emulsion polymerization systems requires careful consideration of both chemical and process variables. While their covalent bonding to the polymer backbone offers unique advantages, inappropriate selection or dosing may compromise stability, conversion, or film properties. The following guidelines summarize practical strategies for formulators.

4.3.12.1 Selection of Surfmers

The choice of a polymerizable surfactant depends strongly on the nature of the monomer system. Hydrophilic monomers such as vinyl acetate typically require moderately hydrophobic surfmers to balance particle nucleation and stabilization. Conversely, more hydrophobic systems, for example, pure acrylates or VeoVa copolymers, benefit from surfmers with higher water solubility to ensure adequate micelle formation. The functional group of the surfmer (sulfate, sulfonate, phosphate, or acrylate) should also be matched with the initiator system to avoid undesired side reactions.

4.3.12.2 Dosage Considerations

In most practical formulations, surfmer concentrations between 1 and 5 wt% based on monomer provide an effective balance of stabilization and reactivity. Too low a level may lead to insufficient particle stabilization, whereas excessive amounts can cause increased viscosity or gel formation during polymerization.

Pilot trials are therefore recommended to determine the optimum concentration for a specific recipe.

4.3.12.3 Interaction with Conventional Surfactants

Although surfmers can replace conventional surfactants entirely, many industrial formulations employ a blend. A small proportion of traditional anionic or nonionic surfactants is often included to improve initial emulsification of monomers and to enhance short-term colloidal stability. The surfmer then becomes incorporated into the polymer, ensuring long-term stability and film performance. This hybrid approach is particularly useful in large-scale reactors where shear conditions vary.

4.3.12.4 Influence on Film Properties

The immobilization of surfmers within the polymer matrix significantly reduces surfactant migration. As a result, coatings exhibit superior water resistance, gloss retention, and adhesion. However, formulators should be cautious: overly high incorporation of ionic surfmers may increase the hydrophilicity of the final film, which can negatively impact barrier properties. Balancing surfmer content with coalescents, crosslinkers, or hydrophobic monomers is often required.

4.3.12.5 Process Parameters

Temperature, pH, and initiator feed strategy all influence the reactivity of surfmers. Elevated polymerization temperatures may accelerate their copolymerization but can also lead to side reactions. Maintaining a controlled pH, typically between 4 and 6 for vinyl acetate systems, minimizes hydrolysis of functional groups. Continuous or semicontinuous feeding of surfmers alongside monomers generally ensures a more uniform distribution within the polymer particles.

4.3.12.6 Summary

Formulating with polymerizable surfactants is not a "one-size-fits-all" approach. Success depends on aligning surfmer chemistry with the monomer system, controlling dosage, and adjusting process conditions to achieve the desired balance of stability and performance. These guidelines provide a framework for industrial practice while leaving room for optimization in each application.

4.3.13 Formulation Examples with Polymerizable Surfactants—(Semibatch, Preemulsified)

This revision corrects the surfmer charging strategy and clarifies the semibatch, preemulsified emulsion polymerization (EP) procedures. In all three formulations below, the seed stage uses a conventional surfactant only; the polymerizable surfactant (surfmer) is added exclusively via the monomer preemulsion during the feed. Each subsection includes: (i) an introductory paragraph, (ii) a boxed formulation table with a caption, (iii) a clear, stepwise semibatch procedure indicating when and how the surfmer is added, and (iv) a formulation note on recommended surfmer class and justification.

4.3.13.1 Vinyl Acetate–Butyl Acrylate Latex (Architectural Coating Binder)

A semicontinuous VAc–BA process offers a versatile binder platform for architectural coatings. During nucleation, stabilization primarily depends on the conventional surfactant present in the reactor charge. A small aliquot of monomer preemulsion is introduced to initiate particle formation; this may contain a minimal amount of polymerizable surfactant, but the bulk of the surfmer is fed gradually with the main monomer stream. This approach minimizes premature copolymerization of the surfmer while ensuring its effective incorporation into the growing polymer particles during the primary feed phase (Table 4.12).

Procedure: Semibatch, preemulsified (seed stabilized primarily by conventional surfactant with a small amount of surfmer)

1) *Preemulsion preparation*: In a separate tank, combine vinyl acetate monomer (VAM) and butyl acrylate (BA) with the surfmer and the preemulsion water. Shear-mix for 5–10 minutes to form a uniform preemulsion. Do not add initiator to the preemulsion.

Table 4.12 Representative formulation for a VAc–BA binder using a polymerizable surfactant (semibatch, preemulsified).

Ingredient	**Parts by weight**
Deionized water (reactor/seed charge)	80
Conventional anionic surfactant (seed; e.g., AES or SDS, 30% active)	1.0
Nonionic surfactant (emulsification aid, e.g., fatty alcohol EO)	0.5
Sodium bicarbonate (buffer)	0.3
Deionized water (preemulsion tank)	40
Vinyl acetate monomer (VAM)	70
Butyl acrylate (BA)	25
Polymerizable surfactant (surfmer)	2.0
Ammonium persulfate (initiator, in separate aqueous solution)	0.5
Optional coalescent (post-add)	2.0
Defoamer/biocide (as needed)	q.s.

2) *Seed charge and nucleation*: Charge reactor water, buffer, and a conventional anionic surfactant to the reactor. Heat to 70–75 °C under nitrogen. Add a small aliquot (about 3–5% of total monomer) from the preemulsion and initiate with persulfate solution to nucleate particles; maintain 70–80 °C.
3) *Semicontinuous feeds*: Begin simultaneous feeds of (i) the remaining monomer preemulsion (containing the surfmer) over 3–4 h and (ii) initiator solution at a rate that maintains low residual monomer (starved feed). Maintain 70–80 °C and moderate agitation.
4) *Hold and chase*: After feeds, hold for 30–45 min. If needed, add a short chaser to reduce residual monomer.
5) *Finish*: Cool to 30–35 °C, adjust pH to 4.5–5.5, and filter through a 100–200 μm mesh. Post-add coalescent, defoamer, and biocide as required.

Formulation note (recommended surfmer class): For VAc–BA architectural binders, sulfate/sulfonate-type surfmers (e.g., allyl ether sulfates or AMPS-type-2-acrylamido-2-methylpropane sulfonic acid. Provide strong anchoring and improved water resistance while minimizing surfactant migration.

Expected properties: Solids 50–55%, particle size ~0.15–0.20 μm, high gloss, and good water/scrub resistance.

Formulation note (recommended surfmer class): Phosphate- or sulfonate-functional surfmers (e.g., vinylphosphonate or vinyl sulfonate types) reduce interfacial migration while preserving tack, improving peel on PE/PP, and moisture resistance.

Expected properties: Solids ~50%, balanced tack/peel, enhanced adhesion to low-energy films, lower surfactant migration.

4.3.13.2 Representative Formulation for an Acrylic PSA, Semibatch, Preemulsified, Containing Polymerizable Surfactant

Acrylic PSA–Semibatch Process (Seed Without Surfmer) This procedure ensures that the surfmer (polymerizable surfactant) is excluded from the seeding stage and introduced only during the main feed. The materials are listed as parts by weight (w/w) (Table 4.13).

Procedure (Semibatch, Preemulsified)

1) *Seed and nucleation*: Charge reactor with seed water, buffer, and conventional surfactant only. Heat to 70–75 °C under nitrogen. Add a small seed monomer shot (BA, 2-EHA, AA) and initiate with persulfate solution. This forms clean seed particles without surfmer.
2) *Main preemulsion preparation*: In a separate tank, prepare the bulk preemulsion with BA, 2-EHA, AA, surfmer, conventional surfactant, and water. Shear-mix to form a stable emulsion.

Table 4.13 Representative formulation for an acrylic PSA, semibatch, preemulsified, containing Polymerizable surfactant.

Component	Parts by weight (w/w)
Seed charge (reactor)	
Deionized water	90
Conventional anionic surfactant (30% active)	1.2
Sodium bicarbonate (buffer)	0.2
Nonionic surfactant (optional)	0.8
Seed monomer shot (no surfmer)	
Butyl acrylate	2–3
2-Ethylhexyl acrylate	1–2
Acrylic acid	0.1–0.2
Main preemulsion (with surfmer)	
Deionized water	50
Butyl acrylate	52–54
2-Ethylhexyl acrylate	28–29
Acrylic acid	2.8–3.0
Polymerizable surfactant (surfmer)	1.5
Conventional surfactant (balance)	0.5–0.8
Initiator feed	
Ammonium persulfate in water	0.6

3) *Feeds*: Cofeed the main preemulsion (with surfmer) and initiator solution over 3–4 h at 70–80 °C under starved-feed conditions. Maintain steady agitation to avoid coagulation.
4) *Hold/chase*: Hold for 30–45 min and apply a brief initiator chase if residual monomer remains.
5) *Finish*: Cool, neutralize to pH 6.5–7.5, filter, and add post-additives as required.

°C, filter, and apply post-add wax or barrier aids if specified.

Formulation note (recommended surfmer class): For paper/packaging barrier applications, nonionic polymerizable surfmers (e.g., PEG-methacrylate types) or mildly ionic phosphate-functional surfmers reduce free-surfactant migration and preserve printability while maintaining barrier properties.

Expected properties: Solids 55–60%, low migration, good grease, and moisture resistance with an improved surface for printing.

General formulation note: When selecting a surfmer for a given formulation, the choice depends on monomer polarity, desired film properties, and regulatory requirements. Sulfonate and sulfate surfmers are preferred for VAc–BA architectural binders, phosphate surfmers for acrylic PSAs, and nonionic PEG-methacrylate surfmers for food-contact barrier coatings. Optional alternatives such as sodium vinyl sulfonate or styrene sulfonate may also be employed where cost or availability dictates.

Although the preceding formulations specify the use of polymerizable surfactants in general terms, the practical question remains: which type of surfmer is most suitable for a given system?

4.3.13.3 Paper and Packaging Barrier Coating Latex (VAc–E)

Barrier coatings must resist grease and moisture while maintaining printability. A VAc–ethylene latex stabilized with a reactive surfactant minimizes free-surfactant migration. The seed is generated with a conventional surfactant only; the surfmer is introduced via the monomer preemulsion during the semibatch feed. Redox initiation provides temperature control (Table 4.14).

When selecting a surfmer for a given formulation, the choice depends on monomer polarity, desired film properties, and regulatory requirements. Sulfonate and sulfate surfmers are preferred for VAc–BA architectural binders, phosphate surfmers for acrylic PSAs, and nonionic PEG-methacrylate surfmers for food-contact barrier coatings. Optional alternatives such as sodium vinyl sulfonate or styrene sulfonate may also be employed where cost or availability dictates.

Although the preceding formulations specify the use of polymerizable surfactants in general terms, the practical question remains: which type of surfmer is most suitable for a given system?

Table 4.15 summarizes representative classes of polymerizable surfactants, together with their generic chemical names, indicative trade names, and typical application areas. This overview provides formulators with guidance for surfmer selection based on monomer composition and end-use requirements.

4.3.14 Optimization and Practical Considerations in Using Surfmers

Polymerizable surfactants (surfmers) are powerful tools for stabilizing latex particles and improving the performance of the final film. Their effectiveness depends strongly on how they are used in the emulsion polymerization process. This section outlines the critical strategies and practical considerations required for successful application.

Table 4.14 Representative formulation for a paper/packaging barrier coating latex (VAc–E; semibatch, preemulsified; target solids 54–56%, pH 4.8–5.2).

Component	Parts (pbw)	Feed step	Note
Deionized water	60–68	IC	Aqueous phase for initial charge
PVOH (10% soln; 88–92% hydrolysis)	30–35 (soln)	IC	Protective colloid; 3.0–3.5 pbw as solids
Sodium bicarbonate (buffer)	0.15–0.25	IC	pH control limits hydrolysis
Defoamer (mineral-oil/siloxane)	0.15–0.25	IC	Foam control
Conventional anionic surfactant (seed only)	0.35–0.50	IC	Seed generation; no surfmer in seed
Vinyl acetate (VAc)	100	PE (fed)	Base monomer
Reactive surfactant (surfmer, 100%)	0.6–1.0	PE (fed)	Meth(acrylate/allyl)-functional surfactant
Nonionic wetting aid (EO 6–10)*	0–0.3	PE (fed)	Optional; keep minimal for barrier
Deionized water (for preemulsion)	25–35	PE (fed)	Preemulsion carrier
Ethylene (gas)	—	IF (press)	Adjust pressure to achieve 12–15 wt% in copolymer; typical 6–10 bar at 50–60 °C
Potassium persulfate (KPS, 5–10% soln)	0.35–0.45	IF (fed)	Oxidant in redox pair
Sodium metabisulfite (SMBS, 5–10% soln)	0.25–0.35	IF (fed)	Reductant; temperature control
Ferrous sulfate ($FeSO_4 \cdot 7H_2O$, 0.1% soln)	0.003–0.008	IF (shot)	Catalyst for redox (ppm Fe^{2+})
t-Butyl hydroperoxide (TBHP, 5% soln)	0.15–0.25	PF (shot)	Postredox for residual monomer
Sodium formaldehyde sulfoxylate (SFS, 5% soln)	0.10–0.20	PF (shot)	Pair with TBHP
Sodium hydroxide (10% soln)	q.s.	PF (adj)	Adjust pH to 4.8–5.2 (avoid ammonium)
Biocide (in-can)	0.10–0.20	PF (shot)	Preservation
Defoamer (finish)	0.05–0.15	PF (shot)	Final foam control
Deionized water	q.s. to 100	PF	Adjust to target solids

Abbreviations: IC, initial charge; PE, monomer preemulsion feed; IF, initiator/ethylene feed; PF, postfinish.
Process notes (concise): Conduct a semibatch at 50–60 °C under ethylene pressure; start with IC and seed, ramp pressure, and begin concurrent PE and redox feeds (starved-feed). Maintain gentle to moderate agitation to avoid air entrainment. Finish with brief TBHP/SFS redox to reduce residual monomer, cool, degas, and adjust pH and solids. Target viscosity 1,500–2,500 mPa·s (Brookfield, 20 rpm).
*Keep nonionic wetting aid as low as possible to preserve barrier performance; the reactive surfactant provides anchoring and reduces migratory free surfactant.

Table 4.15 Examples of polymerizable surfactants (surfmers) for emulsion polymerization.

Surfmers (class)	Chemical name (generic)	Trade name examples*	Typical applications
Sulfonate-functional	2-Acrylamido-2-methylpropane sulfonic acid (AMPS)	Lubrizol AQUAMER® AMPS, Toagosei Sumikaacryl AMPS	VAc–BA binders, water-resistant coatings
Sulfate-functional	Allyl polyoxyethylene sulfate	Clariant HOSTAPAL®, BASF Lutensol® A-series (allyl-modified)	Architectural coatings
Phosphate-functional	Vinylphosphonic acid derivatives	Solvay Sipomer® VSP, Rhodia Sipomer® PAM	Acrylic PSAs, adhesion promoters
Nonionic surfmer	Polyethylene glycol methacrylate (PEG-MA)	Sartomer SR-PEGMA series	Barrier coatings, food-contact
Alternatives (previously listed)	Sodium vinyl sulfonate, sodium styrene sulfonate	Sipomer® VSS, VSS-Na	General latex stabilization

4.3.14.1 Dosage Optimization

The level of surfmer must be controlled carefully. At insufficient levels, the particle surface may not be adequately covered, which increases the risk of coagulation and colloidal instability. Excessive levels can promote particle aggregation, reduce conversion efficiency, and alter film morphology. In most systems, effective dosages fall within 1–5 wt% relative to total monomer, but optimization trials are recommended.

4.3.14.2 Interactions with Conventional Surfactants

Although surfmers provide covalent anchoring to the particle surface, conventional surfactants are typically required during nucleation and preemulsification to ensure reliable particle initiation and early stability. A practical approach is to rely on a small amount of conventional anionic or nonionic surfactant in the seed, then introduce the surfmer progressively with the semibatch monomer preemulsion. This strategy minimizes free surfactant migration while preserving robust processability.

4.3.14.3 Influence on Final Properties

Appropriate selection and dosing of surfmers can enhance water and solvent resistance because the emulsifier is immobilized within the polymer matrix. Coatings frequently show improved scrub resistance and gloss retention, and adhesives benefit from stronger interfacial adhesion to difficult substrates due to reduced surfactant migration. In porous or mineral-containing systems, careful use of surfmers can also help mitigate efflorescence by limiting mobile surfactant species at the film surface.

4.3.14.4 Limitations and Challenges

Surfmers can be more expensive than conventional surfactants and are not universally compatible with all monomers. Some systems require higher polymerization temperatures or specific redox conditions to achieve full incorporation. If the reactive group does not completely react, residual free surfmer may remain and compromise performance. These constraints should be evaluated against the targeted property gains.

Table 4.16 summarizes the potential consequences of not following the practical guidance for each parameter.

Table 4.16 Key considerations in the use of surfmers.

Parameter	Practical guidance	Potential consequences if ignored
Dosage	Maintain within 1–5 wt% of total monomer; confirm by pilot optimization.	Instability, coagulation, or poor film properties
Addition stage	Introduce the semibatch monomer preemulsion; avoid full seed charging.	Inhomogeneous incorporation or loss of reactivity
Blend with surfactants	Use a small amount of conventional surfactant in the seed for reliable nucleation.	Incomplete stabilization and broad particle size distribution
Monomer compatibility	Match surfmer functionality with monomer polarity and reactivity.	Residual-free surfmer and inconsistent performance
Cost and availability	Assess commercial supply and cost-benefit versus property gains.	Higher cost without commensurate performance benefits

4.3.15 Case Studies

The following case studies translate the optimization principles into practical outcomes. Each case compares a baseline latex prepared with conventional surfactants only to an otherwise identical system in which a polymerizable surfactant is introduced via the semibatch monomer preemulsion. Test conditions are indicated beneath each table to facilitate reproduction and benchmarking.

4.3.15.1 Architectural Coating Binder (VAc–BA)

Objective: Improve water whitening and scrub resistance without sacrificing gloss.

Approach: A VAc–BA binder was prepared in two variants. The baseline used a conventional anionic surfactant in both seed and feed. The reactive system used a conventional surfactant in the seed only, while a sulfate-functional surfmer was added exclusively via the monomer preemulsion during the semibatch feed. All other recipe and process parameters were held constant (Table 4.17).

Key Learnings

Introducing a sulfate-functional surfmer only via the preemulsion feed, while keeping the seed purely conventional, reduced water whitening by approximately 60% and increased scrub resistance by about 45% without a gloss penalty. The modest decrease in median particle size indicates improved particle stabilization rather than over-crosslinking.

Table 4.17 Performance comparison for architectural binder with and without a polymerizable surfactant.

Parameter	Unit	Baseline	With surfmer	Change
Solids content	%	52.0	52.1	≈ equal
Particle size (D50)	μm	0.195	0.180	−8%
Water whitening (24 h immersion)	% ΔL*	8.2	3.1	−62%
Scrub resistance	Cycles to failure	1,100	1,600	+45%
Gloss at 60°	GU	82	85	+3

Note: The asterisk in L* refers to the **CIELAB color space**, where L* denotes lightness. Accordingly, **% ΔL*** represents the percentage change in lightness and is correctly used here to quantify water whitening after 24 h immersion.

Method Notes

Drawdowns were made at 45% PVC on primed panels and conditioned for seven days at 23 °C and 50% RH. Scrub resistance followed a modified ASTM D2486 method. Water whitening was assessed after 24 h immersion and 24 h recovery.

4.3.15.2 Acrylic Pressure-sensitive Adhesive (PSA)

Objective: Increase adhesion to low-energy polyolefin films while maintaining tack.

Approach: An acrylic PSA was produced as a baseline with conventional surfactants only and as a reactive system using a phosphate-functional surfmer added solely through the preemulsion feed. The seed relied on a conventional anionic surfactant. All other variables, including monomer ratios and conversion, were held constant (Table 4.18).

Key Learnings

Feeding a phosphate-functional surfmer through the preemulsion improved peel on PE and PP by forty to 57% and moisture-aged peel by 68% while maintaining tack. The results indicate that covalent anchoring—not additional free emulsifier—drives durable adhesion to low-energy films.

Method Notes

Adhesive was coated at 25 g/m^2 (dry) onto the PET carrier and laminated after a 10 min open time. Peel adhesion was measured according to a modified Pressure Sensitive Tape Council (PSTC) Test Method PSTC-101 at a crosshead speed of 300 mm/min. Static shear adhesion was evaluated following PSTC-107. Moisture aging was conducted in a controlled-humidity chamber for 96 h at 40 °C and 95% RH.

Table 4.18 Adhesive performance on polyolefin substrates with and without a polymerizable surfactant.

Parameter	Unit	Substrate	Baseline	With surfmer	Change
180° peel (24 h dwell)	N/25 mm	PE film	5.8	8.1	+40%
180° peel (24 h dwell)	N/25 mm	PP film	4.2	6.6	+57%
Loop tack	N/25 mm	Stainless	9.5	9.3	≈ equal
Static shear (1 kg, 25 mm × 25 mm)	min	SS	980	1,340	+37%
Moisture-aged peel (96 h, 40 °C, 95% RH)	N/25 mm	PE film	4.0	6.7	+68%

4.3.15.3 Paper and Packaging Barrier Coating (VAc–E)

Objective: Reduce liquid water and oil penetration while preserving printability.

Approach: A VAc–E latex for paper coating was prepared in a baseline version using conventional surfactants only and in a reactive version in which a nonionic PEG-methacrylate surfmer was metered solely in the monomer preemulsion. The seed used a conventional surfactant and PVOH. Coating color and coat weight were kept constant (Table 4.19).

Key Learnings

Using a nonionic PEG-methacrylate surfmer in the preemulsion lowered Cobb60 by 40% and MVTR by 21% and raised the Kit rating by two steps with unchanged print density, indicating tighter barrier formation without surface contamination.

Method Notes

Coatings were applied at 6 g/m^2 dry onto 80 g/m^2 base paper and conditioned for 48 h at 23 °C and 50% relative humidity to ensure moisture equilibrium before testing.

Cobb60, reported in grams per square meter, quantifies water absorptiveness according to standard practice (e.g., TAPPI T 441 or ISO 535). A known volume of water contacts the paper surface for 60 s, the specimen is blotted and reweighed, and the mass gain per unit area is reported. Lower Cobb60 values indicate better resistance to liquid-water uptake.

The Kit test, performed in accordance with TAPPI T 559, assesses resistance to oil and grease. Standardized Kit solutions of increasing aggressiveness are applied briefly to the coated surface and the highest number that does not produce visible staining within the specified time is recorded. Ratings run from one to twelve, and higher numbers indicate stronger grease resistance.

Table 4.19 Barrier performance of paper coating with and without a polymerizable surfactant.

Parameter	Unit	Baseline	With surfmer	Change
Cobb60 (water)	g/m^2	30	18	−40%
Kit rating (TAPPI T559)	1–12 scale	7	9	+2
MVTR (38 °C, 90% RH)	g/m^2·day	210	165	−21%
Print density	OD units	1.32	1.35	+0.03

Moisture-vapor transmission rate was measured by a modified ASTM E96 upright-cup method. The specimen seals a test cup so that vapor must pass through the coating to reach the desiccant. The assembly is held in a controlled environment—typically 38 °C and 90% relative humidity—and the mass change over time is used to calculate the transmission rate in grams per square meter per day. Lower values indicate a tighter moisture barrier.

Print density was determined after applying a standard offset ink under controlled inking and nip settings and allowing the film to dry under the same conditioning used before testing. A spectrodensitometer measures optical density; higher values generally reflect better ink holdout on the coated surface.

These methods make the performance differences attributable to surfmer use observable and reproducible. The results illustrate the practical benefits of reactive-surfactant incorporation when the charging strategy, feed profile, and surfmers selected are aligned with the formulation's performance objectives.

Test Methods Notes

Cobb60 is a standardized water-absorptiveness test for paper and paperboard in which the coated surface is contacted with water for 60 seconds, blotted, and reweighed. Results are reported as mass gain per unit area (g/m^2). Common normative references include ISO 535 and TAPPI T 441. Lower Cobb60 values indicate greater resistance to liquid-water uptake.

TAPPI is the Technical Association of the Pulp and Paper Industry. Its "T" methods (for example, TAPPI T 559 for the Kit test) are consensus standard procedures widely used in the pulp and paper sector. The Kit test rates resistance to oil and grease on a 1–12 scale; higher ratings indicate stronger grease resistance.

MVTR (moisture-vapor transmission rate) quantifies the rate at which water vapor passes through a substrate, typically reported in $g/m^2 \cdot day$. It is commonly measured by ASTM E96 (upright-cup or inverted-cup variants) under specified temperature and relative-humidity conditions. Lower MVTR values denote a tighter moisture barrier.

4.3.16 Future Trends and Research Directions

In this section, we look ahead at where polymerizable surfactants are going. The aim is not to cover everything, but to give a clear idea of why people are still improving these materials. The basic idea stays the same: a surfactant that reacts with the polymer so it does not wash out or migrate to the surface.

Future work mainly tries to make these systems cleaner, easier to use, and better for the environment. That can mean using less solvent, making latexes with higher solids, or choosing raw materials that are safer and more sustainable. It can also mean making the surfactant do more than one job, such as helping adhesion or improving barrier properties, while still keeping the film stable.

As you read the next pages, keep one question in mind: what problem is each trend trying to solve for the formulator or the end user? If the reason is clear, the trend will be easy to remember. Details and advanced topics will appear in later chapters; here we focus on the main ideas and why they matter in practice.

5

Surfactants in Emulsion Polymerization and Performance Impact

5.1 Introduction to Surfactants in Latex Systems

Surfactants occupy a dual role in emulsion polymerization. In Chapter 4, they were presented primarily as one of the four fundamental ingredients—classified into anionic, nonionic, and reactive types, with emphasis on their chemistry, structures, and general functions in stabilizing latex dispersions. This foundational discussion provided the "what" of surfactants. The present chapter moves toward the "how" and "why," with a focus on their direct influence on polymerization behavior and end-use performance.

The central issue addressed here is that surfactants do not merely stabilize particles during synthesis; they actively determine the course of the polymerization and the properties of the final film. Their effects may be beneficial or detrimental, depending on type, dosage, and interaction with other ingredients. Anionic surfactants, for example, offer strong electrostatic stabilization and efficient nucleation, but excessive levels may compromise water resistance and promote surfactant migration to the film surface. Nonionic surfactants, in contrast, provide steric stabilization and compatibility across wide pH and electrolyte ranges, yet they may plasticize the film or interfere with coalescence. These contrasting behaviors reveal why surfactant choice in emulsion polymerization is never a neutral decision but rather a key process variable with far-reaching consequences.

Accordingly, this chapter examines surfactants not as generic stabilizers but as functional agents whose chemistry shapes nucleation, particle size distribution, polymerization kinetics, and ultimately film performance. It explores how dissolution, micelle formation, and surface activity link to latex stability, how anionic and nonionic surfactants compare in efficiency and limitations, and how their attributes influence film formation, adhesion, barrier properties, and durability. In this way, the discussion establishes a direct line between molecular structure and application performance, complementing the compositional overview of Chapter 4 with a performance-driven perspective.

5.2 Dissolution Behavior and Surface Tension Reduction

When a surfactant dissolves in water, its hydrophobic group disrupts the water's hydrogen bonding network. Surfactant molecules migrate to the water–air interface, aligning with hydrophobic tails away from water and hydrophilic heads into the aqueous phase. This reduces surface tension and stabilizes the interface.

5.3 Micelle Formation and Critical Micelle Concentration (CMC)

As surfactant concentration increases, surface adsorption reaches saturation. Excess surfactant molecules then self-assemble into micelles—spherical structures where hydrophobic tails are inward and hydrophilic heads face outward. The concentration at which this begins is known as the critical micelle concentration (CMC).

5.4 Surfactant Functionality in Emulsion Polymerization

Surfactants in emulsion polymerization serve multiple roles: micelle formation, monomer Solubilization, latex particle stabilization, and reduction of interfacial tension. Anionic surfactants are typically used at 0.5–3% based on monomer weight, and combinations with nonionic surfactants (0.5–2%) enhance thermal and mechanical stability.

5.5 Impact of Surfactant Type on Polymerization Performance

Surfactants affect polymerization kinetics, particle size, viscosity, and stability. Their influence extends to final latex and film properties—such as adhesion, wetting, foam control, and water resistance—making surfactant selection critical for performance optimization.

5.6 Micelle Formation Efficiency and Alkyl Chain Length

For anionic surfactants with the formula $C_nH_{2n-1}O_2Na$, micelle formation is weak when $n \leq 8$, optimal for $n = 12$–18, and declines for $n > 18$ due to excessive hydrophobicity. Each additional CH_2 group roughly halves the CMC, increasing micelle-forming efficiency.

5.7 Ethoxylation Effects on Surfactant Performance

In nonionic surfactants of the type $R\text{–}(CH_2CH_2O)_n\text{–}OH$, increasing the ethylene oxide (EO) chain length raises CMC and particle size, which can reduce micelle efficiency. An optimal range is typically $n = 7$–9, which balances water solubility, micelle formation, and particle size control in emulsion polymerization systems.

5.8 Molecular Structure, Solubility, and Surfactant Efficiency

Structural variations such as branching and unsaturation increase surfactant solubility. Bulky hydrophobes reduce micelle efficiency, while optimal alkyl chains for emulsification typically fall within C12–C18, combined with EO chains tailored to balance solubility and performance.

5.9 Surfactant Role in Latex Stability and Storage

Surfactants improve latex stability during production and application. They enhance freeze–thaw resistance, prevent coagulation, and support long-term mechanical stability. Polymeric surfactants further improve particle adhesion and film durability.

5.10 Hydrophilic-Lipophilic Balance (HLB): Concept and Applications

The hydrophilic–lipophilic balance (HLB) is a numerical expression that reflects the relative strength of the hydrophilic and lipophilic portions of a surfactant molecule. It serves as a useful guide for selecting nonionic surfactants based on their solubility behavior and suitability for different emulsion systems. The HLB concept is not applicable to ionic surfactants.

The HLB value is defined by the relationship:

$$HLB = 20 \times \frac{MW_{hydrophilic}}{MW_{total}} \tag{5.1}$$

Worked example (Span 80, sorbitan monooleate): molecular formula $C_{24}H_{44}O_6$. The total molecular weight is 428 g/mol, and the hydrophilic contribution from the sorbitan ring and hydroxyl groups is approximately 147 g/mol. Substituting into the equation gives:

$$HLB = 20 \times \frac{147}{428} = 6.9 \tag{5.2}$$

Interpreting and applying HLB values provides a practical framework for surfactant selection. Surfactants with HLB values between 1 and 6 are suited to water-in-oil systems, while those in the range of 8–18 are more effective in oil-in-water emulsions, detergents, and solubilizers. Values in the intermediate region are frequently associated with wetting and dispersing functions.

For optimal performance, the HLB of the nonionic surfactant system should match the required HLB of the formulation. Required HLB values for different oils and additives are available in the literature and are often used to guide the blending of surfactants in order to achieve stability in emulsification, solubilization, and wetting applications.

In the case of nonionic surfactants, the effective HLB is influenced by temperature. As temperature increases, the hydration of ethylene oxide head groups diminishes, reducing hydrophilicity and effectively lowering the HLB. At a characteristic temperature known as the cloud point, the surfactant becomes insoluble and phase separation occurs, an important consideration in the design of thermally stable formulations.

5.11 Selection of Nonionic Surfactant for Emulsion Polymerization

The choice of a nonionic surfactant in emulsion polymerization is critical because it directly influences latex particle size, polymerization kinetics, and overall stability of the dispersion. The surfactant's surface activity, compatibility with the monomer mixture, and ability to stabilize hydrophobic domains determine the efficiency of nucleation, particle stabilization, and solubilization of monomers.

Key parameters that require careful evaluation include the hydrophilic–lipophilic balance (HLB), the critical micelle concentration (CMC), the strength of the hydrophilic group, and the chemical nature of the hydrophobic chain. Together, these factors govern the surfactant's emulsifying power, stabilization capacity, and ability to act as a transport medium for poorly soluble monomers.

Structural variations within the hydrophobic group also influence surfactant behavior. Linear hydrophobic chains pack more efficiently at the interface, lowering the CMC and producing stronger surface activity. In contrast, branching or unsaturation disrupts packing, increases water solubility, and generally raises the CMC. Such branching can be advantageous in systems requiring higher solubility and reduced tendency to crystallize, but less favorable when extremely low CMC values are required.

A systematic selection process is recommended. The starting point for nonionic surfactants is an HLB value in the range of 13–18, which is generally appropriate for oil-in-water polymerization systems. Candidate surfactants should then be compared on the basis of their CMC and surface tension reduction capacity. The degree of ethoxylation should also be considered: higher ethoxylation increases water solubility but reduces surface activity. The optimum choice is therefore the surfactant that combines an appropriate HLB, a low CMC, and strong surface tension reduction.

5.12 Comparison of Nonionic Surfactants and Their Impact in Emulsion Polymerization

Nonionic surfactants exert a profound effect on polymerization outcomes, extending beyond emulsification. By adsorbing at particle surfaces, they control nucleation pathways, particle size distribution, and molecular weight development. In the final latex films, residual nonionic surfactants can influence gloss, water resistance, and adhesion. Selection is therefore guided not only by colloidal stability but also by anticipated effects on film performance.

To highlight these differences, it is useful to compare typical nonionic surfactants in terms of their structure, HLB values, CMC values, and performance characteristics. This comparison, which is manifested in Table 5.1, enables formulators to link molecular features of surfactants with their practical performance in polymerization systems.

5.13 Selection of Anionic Surfactants for Emulsion Polymerization

Anionic surfactants are among the most widely used stabilizers in emulsion polymerization because of their strong ability to reduce surface tension, form micelles

Table 5.1 Comparison of nonionic surfactants and their impact on emulsion polymerization.

Surfactant type	Example compound	HLB range	Typical CMC (mM)	Performance notes
Sorbitan ester ethoxylates	Polyoxyethylene (20) sorbitan monolaurate (Tween 20)	16–17	0.06–0.07	Produces fine particle size, strong O/W stabilization, good compatibility with VAc
Sorbitan ester ethoxylates	Polyoxyethylene (20) sorbitan monooleate (Tween 80)	15	0.012–0.015	Efficient emulsifier for hydrophobic monomers, stable O/W dispersions
Fatty alcohol ethoxylates	C_{12}–C_{14} fatty alcohol + 7EO	12–14	0.08–0.12	Balanced HLB, widely used for acrylates and vinyl esters
Fatty alcohol ethoxylates	C_{16}–C_{18} fatty alcohol + 10EO	14–15	0.03–0.05	Strong stabilizer, low foaming, suited for higher molecular weight polymers
Alkylphenol ethoxylates*	Nonylphenol + 9EO (NP-9)	13	0.2–0.3	Strong emulsifier, but restricted in many regions due to regulatory concerns
Polyoxy-ethylene ethers (Brij)	Brij 35 ($C_{12}H_{25}O(CH_2CH_2O)_{23}H$)	16.9	0.09–0.11	Excellent O/W emulsifier, produces narrow particle size distribution

*Note: Nonylphenol ethoxylates are provided for historical and comparative purposes only; their use is limited by regulatory restrictions.

at relatively low concentrations, and impart electrostatic stabilization to latex particles. Their negatively charged head groups ensure that dispersed particles repel one another, preventing coagulation and providing stable dispersions under a broad range of conditions.

The efficiency of anionic surfactants depends on their critical micelle concentration (CMC), the length and structure of the hydrophobic chain, the degree of branching, and the counterion associated with the head group. Longer alkyl chains typically lower the CMC, whereas branching tends to raise it and enhance solubility. The choice of counterion (sodium, ammonium, potassium) influences solubility, electrolyte tolerance, and foaming behavior.

In emulsion polymerization, anionic surfactants are selected primarily on the basis of their ability to lower interfacial tension and generate a sufficient number of micelles for nucleation. Surfactants with lower CMC values are more efficient, as they produce stable micellar systems at lower concentrations. Charge density is equally important: sulfonates, sulfates, and phosphates differ in their stabilization power, electrolyte sensitivity, and compatibility with monomer systems.

For practical use, formulators consider not only the polymerization stage but also the downstream film properties. Residual anionic surfactants can affect water sensitivity, adhesion, and foaming in coatings and adhesives. Thus, the choice involves balancing strong colloidal stabilization during polymerization with minimal negative impact on final film performance.

Table 5.2 presents guidelines for the selection of anionic surfactants for emulsion polymerization.

Table 5.2 Guidelines for the selection of anionic surfactants for emulsion polymerization.

Surfactant type	Example compound	Typical CMC (mM)	Key characteristics
Alkyl sulfates	Sodium lauryl sulfate (SLS)	8.0–9.0	Strong emulsifier, high foaming, sensitive to electrolytes, widely used in VAc and acrylate polymerizations
Alkyl ether sulfates	Sodium laureth sulfate (2EO)	6.5–7.5	Improved solubility compared with SLS, slightly lower foaming, more tolerant to hardness ions
Alkyl aryl sulfonates	Sodium dodecylbenzene sulfonate (SDBS)	1.6–1.8	Excellent detergency, effective at low concentrations, strong electrostatic stabilization
Sulfosuccinate esters	Dioctyl sodium sulfosuccinate (DOSS)	0.7–0.8	Powerful wetting and dispersing agent, highly efficient emulsifier at low levels
Phosphate esters	Mono/dialkyl phosphate esters	2.5–3.5	Provide stabilization and hydrophobic–hydrophilic balance, improve adhesion to polar substrates
Diphenyl oxide disulfonates	Dowfax 2A1	0.5–0.6	Extremely low CMC, very stable in high-temperature polymerizations, excellent electrolyte tolerance

5.14 Combined Use of Anionic and Nonionic Surfactants in Emulsion Polymerization

In industrial emulsion polymerization, the combined use of anionic and nonionic surfactants is often preferred over single-surfactant systems. Each class contributes complementary stabilization mechanisms: anionic surfactants provide strong electrostatic repulsion between latex particles, while nonionic surfactants impart steric stabilization by forming hydrated ethylene oxide layers at the particle surface. This dual mechanism enhances colloidal stability, particularly under conditions of high shear, elevated temperature, or variable ionic strength.

When used individually, anionic surfactants are typically employed at levels of 1–3% relative to monomer weight, while nonionic surfactants are used in the range of 1–2%. In combination, the total surfactant dosage can often be reduced by 20–30% because of the synergistic effect. Common practice is to use anionic to nonionic ratios between 70:30 and 50:50, although the optimum depends on the monomer system and desired latex properties.

The benefits of using mixed systems are numerous. They broaden the window of electrolyte tolerance, reduce the risk of coagulum formation, and improve particle size control. Mixed systems also enhance freeze–thaw stability, reduce foaming compared to purely anionic systems, and improve film properties by reducing surfactant migration. In adhesive formulations, the combination improves tack and adhesion on polar substrates, while in coatings it enhances gloss retention and water resistance.

The combined use of anionic and nonionic surfactants is summarized in Table 5.3, which lists representative examples, their typical ratios, usage levels, and the positive effects observed in emulsion polymerization. This tabulated overview highlights the synergistic benefits of mixed systems compared with single-surfactant formulations.

5.15 Quick Summary Table: Surfactant Attributes and Relevance

For clarity, the key attributes of surfactants and their practical implications in emulsion polymerization are summarized in Table 5.4. This overview highlights the parameters most frequently considered when selecting a surfactant system.

5.16 Biodegradability and Regulatory Considerations

Environmental safety and regulatory compliance are increasingly important criteria in the selection and application of surfactants for emulsion polymerization.

Table 5.3 Representative combinations of anionic and nonionic surfactants in emulsion polymerization.

Combination (anionic + nonionic)	Typical ratio (wt%)	Total usage (% of monomer)	Positive impact on EP
Sodium lauryl sulfate + Tween 20	70:30	1.5–2.0	Fine particle size distribution, improved O/W stabilization, reduced coagulum
SDBS + Brij 35*	60:40	1.0–1.5	Narrow particle size, excellent colloidal stability under electrolytes
DOSS + C_{12}–C_{14} alcohol ethoxylate (7EO)	50:50	1.0–1.5	Strong wetting and dispersion, enhanced adhesion, improved stability at low dosage
SLS + Tween 80	70:30	1.5–2.5	Stable dispersions of hydrophobic monomers, reduced surfactant migration in films
Dowfax 2A1 + C_{16}–C_{18} fatty alcohol + 10EO	50:50	0.8–1.2	Extremely low CMC, stable at high temperature, improved freeze–thaw resistance

*Brij 35 (Polyoxyethylene (23) lauryl ether) is a nonionic surfactant belonging to the polyoxyethylene alkyl ether class. It has the general structure R–$(CH_2CH_2O)_{23}$–OH, where R is a lauryl (dodecyl, C_{12}) group. With an average of 23 ethylene oxide (EO) units attached to a dodecyl alcohol hydrophobe, Brij 35 exhibits a high HLB value (~16.9), making it strongly hydrophilic. It is widely used in emulsion polymerization as a stabilizer, providing steric stabilization and solubilization. However, its high EO number increases the CMC and may result in larger particle sizes if used at high levels.

Table 5.4 Surfactant attributes and their relevance in emulsion polymerization.

Property	Relevance
HLB	Predicts emulsification type (oil-in-water or water-in-oil)
CMC	Indicator of efficiency and micelle formation
Surface tension (static/dynamic)	Governs film leveling, wetting, and adhesion
Charge type (anionic/nonionic)	Determines stabilization mechanism (electrostatic vs. steric)
Hydrophobic tail (C_{12}–C_{18} ideal)	Controls micelle efficiency and solubility
Molecular weight/structure	Influences migration, foaming tendency, and film properties
Reactivity (polymerizable or not)	Determines anchoring to the polymer and long-term performance

Table 5.5 Key regulatory and environmental considerations for surfactants.

Regulatory/certification framework	Relevance to surfactant selection
OECD 301 biodegradability test	Defines criteria for ready biodegradability
REACH (EU regulation)	Requires registration, evaluation, and authorization; restricts hazardous surfactants such as NPEs
GHS (Globally Harmonized System)	Provides hazard classification and labeling of surfactants
EU Ecolabel	Mandates use of readily biodegradable surfactants in certified formulations
Green Seal (US)	Encourages use of safe, environmentally preferable surfactants

Beyond technical performance, formulators must consider biodegradability, toxicity, and alignment with regional and international regulations. Surfactants that persist in the environment or are classified as hazardous face restrictions under several regulatory frameworks.

Of particular concern are nonylphenol ethoxylates (NPEs), which are restricted in the European Union due to their poor biodegradability and endocrine-disrupting potential. Modern practice favors alternatives that meet OECD 301 standards for ready biodegradability and that are fully compliant with REACH and GHS classifications. For eco-certified products, such as those bearing the EU Ecolabel or Green Seal, the use of readily biodegradable surfactants is mandatory (Table 5.5).

Incorporating these environmental and regulatory considerations ensures that surfactant systems not only perform technically but also align with modern sustainability and compliance requirements.

5.17 Film Formation Mechanism in Emulsion Polymers

Film formation is a fundamental process in emulsion-polymer systems and determines the ultimate performance of the resulting coating or adhesive. It is the transformation of dispersed polymer particles into a continuous film during drying. The process integrates physical, chemical, and thermodynamic factors and provides the basis for understanding how emulsions achieve their final functional properties.

5.17.1 Stages of Film Formation

Film formation from an emulsion proceeds through six stages.

1) Wet applied film (dispersion stage). Discrete latex particles are uniformly dispersed in water with surfactants and additives.
2) Evaporation and concentration. Water (and any volatile components) evaporates; the particle-volume fraction rises and viscosity increases.
3) Close packing. Particles come into contact and arrange into a densely packed array; capillary forces develop at the air–film interface.
4) Particle deformation and coalescence. At temperatures at or above the minimum film-formation temperature (MFFT)—or in the presence of coalescent—particles soften, deform, and fuse, eliminating most interstitial water and voids.
5) Polymer-chain interdiffusion. Chains diffuse across former particle boundaries; interfaces fade and mechanical integrity develops.
6) Vitrification and final film. Residual water and coalescent leave; the glass-transition temperature (T_g) rises to service conditions and a continuous, durable film is established.
7) Readers familiar with the traditional three-stage description may map stages 2–4 to evaporation, particle packing, and coalescence. Stages 1, 5, and 6 make explicit the initial dispersion, interdiffusion, and final vitrification that complete the process (Figure 5.1).

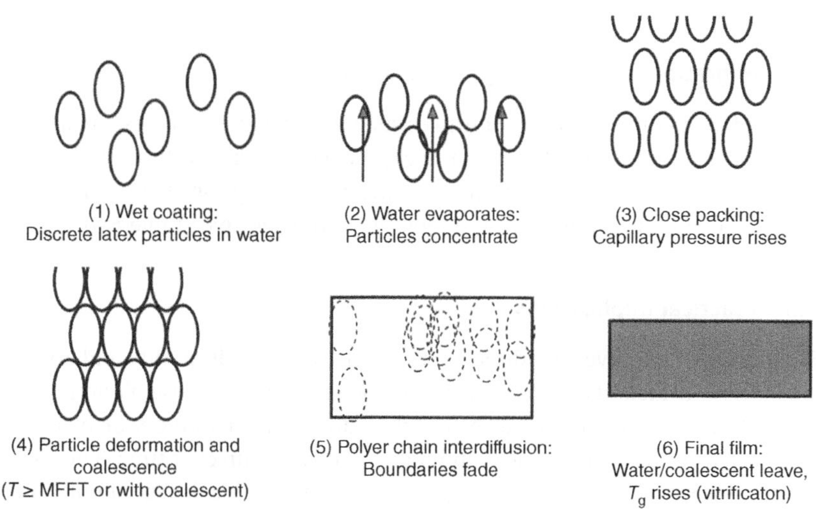

Figure 5.1 Mechanism of film formation in waterborne emulsion polymers.

Table 5.6 Behavior of additives during film formation.

Additive type	Primary role	Volatility	Migration tendency	Blooming risk	Gloss impact	Water resistance impact
Surfactants	Stabilize emulsion	Low to medium	Medium to high	High	Loss via haze	Negative if concentrated at surface
Coalescing agents	Aid particle fusion	Medium to high	High	Low	Slight loss	Negative if retained in the film
Plasticizers	Improve flexibility	Low	Medium	Medium	Possible dulling	Negative if migration occurs

Step 1: Wet coating: Discrete latex particles dispersed in water.
Step 2: Evaporation and concentration: Water leaves; particle volume fraction increases.
Step 3: Close packing: Capillary forces rise as particles touch.
Step 4: Particle deformation and coalescence: At T ≥ *MFFT (or with coalescent), particles deform and fuse.*
Step 5: Interdiffusion: Polymer chains interpenetrate; particle boundaries fade.
Step 6: Final film: Volatiles leave; T_g *rises (vitrification) and properties develop.*

5.17.2 Migration and Blooming of Additives

During and after film formation, additives such as surfactants, coalescing agents, and plasticizers may migrate within the polymer matrix or toward the film surface. Such migration can influence optical clarity, gloss, tack, and water resistance, depending on the type and mobility of the additive (Table 5.6).

5.17.3 Practical Implications

The migration of additives has direct consequences for the durability and appearance of films. Surfactants that crystallize on the surface may cause blooming and haze. Coalescing agents retained in the film can result in softness or blocking, while plasticizers may reduce gloss and increase dirt pickup over time. The use of polymerizable surfactants or careful optimization of additive dosage significantly reduces these risks and enhances long-term film quality.

5.17.4 Field Diagnosis of Migration Effects

Practical diagnosis of additive migration can often be performed in the field. A common test involves wiping the surface of a dried film with a clean, dry cloth. If the gloss is restored after wiping, the loss is likely caused by blooming from migrated additives rather than intrinsic deficiencies in the film-forming process.

5.17.5 Conclusion

An understanding of film formation and additive migration is essential for designing high-performance emulsion polymer systems. Effective formulations must balance the functional benefits of additives with the need to maintain long-term film integrity, thereby ensuring consistent performance under real-world service conditions.

5.18 Surfactants and Film Formation in Latex Systems

Surfactants play a defining role in latex film formation by stabilizing colloidal particles during polymerization and influencing additive distribution during drying. Their mobility, interaction with other additives, and potential for migration must all be considered when designing latex systems for adhesives and coatings.

5.18.1 Role of Surfactants in Film Formation

5.18.1.1 Positive Contributions of Surfactants

Surfactants play several beneficial roles during latex film formation. They promote uniform particle distribution by ensuring a homogeneous dispersion of polymer particles in the aqueous medium. This uniformity allows for even packing during drying, which minimizes the risk of defects such as craters, fisheyes, or voids.

By lowering interfacial tension, surfactants facilitate smoother deformation of polymer particles during coalescence, enabling the formation of dense and crack-free films. This contribution is particularly important when working near the minimum film formation temperature (MFFT).

Another important function is the improvement of pigment and filler dispersion. By modifying the surface compatibility between pigments and the latex, surfactants enhance pigment distribution, which results in smoother films with improved opacity, color uniformity, and gloss.

Throughout the drying process, surfactants also stabilize the latex against premature flocculation or phase separation, ensuring consistent film formation under variable humidity or temperature. Furthermore, surfactants improve substrate wetting, particularly on low-energy or slightly contaminated surfaces, thereby enhancing leveling and adhesion during film formation.

5.18.1.2 Potential Drawbacks of Surfactants

Despite their positive roles, surfactants may also introduce challenges. Nonpolymerizable surfactants can migrate toward the film surface, resulting in issues such as surface tackiness, gloss loss, or water whitening. In more severe cases, migrated surfactants may crystallize at the surface, causing visible blooming. This phenomenon is particularly associated with low molecular weight and highly water-soluble surfactants.

Polymerizable surfactants provide a solution to these drawbacks. Because they are covalently bound into the polymer backbone during polymerization, they do not migrate to the film surface. As a result, they enhance water resistance, gloss retention, and clarity in the final film.

5.18.1.3 Impact on Film Properties

The migration and distribution of surfactants directly affect key film properties. Gloss is reduced when surfactant residues scatter light at the surface; however, gloss can often be temporarily restored by wiping, which removes the migrated layer. Water resistance is diminished when surfactants at the surface absorb moisture, causing whitening or reduced barrier function. Adhesion and durability may also be compromised if surfactants accumulate at the film–substrate interface. The use of polymerizable surfactants or carefully optimized blends helps overcome these limitations and ensures improved film durability.

5.18.1.4 Conclusion

A comprehensive understanding of surfactant contributions and potential drawbacks during film formation is essential for designing robust latex systems. Optimal surfactant selection, combined with the appropriate use of coalescing agents, controlled drying conditions, and tailored polymer architectures, ensures the formation of smooth, defect-free films with the desired optical, mechanical, and barrier properties (Table 5.7).

5.18.2 Case Samples of Film Formation Challenges

Film formation in latex systems is influenced by the balance of surfactants, additives, and drying conditions. While surfactants enable stable dispersions and smooth coalescence, their migration or incompatibility can create challenges in practical applications. A few representative cases are outlined in Table 5.8, highlighting common issues, underlying causes, and approaches to mitigation.

Table 5.7 Surfactant effects on film properties in latex systems.

Aspect of film formation	Positive contributions	Potential drawbacks	Solutions and improvements
Particle distribution	Promotes uniform dispersion and even particle packing	—	Proper surfactant selection with balanced HLB
Coalescence and film continuity	Lowers interfacial tension, enables crack-free films near MFFT	Excess residual surfactant may soften films	Optimize dosage; combine with coalescing agents
Pigment and filler dispersion	Enhances pigment compatibility, opacity, color uniformity, gloss	—	Surface-active blends tailored to pigment type
Latex stability during drying	Prevents flocculation and phase separation	—	Maintain minimum stabilizer concentration
Substrate wetting	Improves adhesion and leveling on low-energy substrates	—	Tailored surfactant systems for substrate compatibility
Surface migration	—	Tackiness, gloss loss, water whitening, blooming	Use polymerizable surfactants; reduce water-soluble low-MW surfactants
Long-term performance	—	Reduced adhesion, lower water resistance, dirt pickup	Use optimized blends; incorporate reactive surfactants into polymer

5.18.3 Practical Guidelines for Minimizing Film Defects

Film formation in latex systems is not a purely theoretical process; it plays out under practical conditions where substrates vary, humidity fluctuates, and drying environments are rarely ideal. Surfactants, while indispensable, can either safeguard stability or introduce new problems depending on how they are chosen and applied. The following guidelines illustrate common real-world challenges, linking observed defects with their root causes and practical strategies for mitigation (Table 5.9).

These practical lessons reinforce the importance of selecting surfactants not only for their role in polymerization but also for their long-term impact on

Table 5.8 Representative case examples of film formation challenges in latex systems.

Case description	Observed film defect	Likely cause	Mitigation strategy
Vinyl acetate–acrylic copolymer latex applied on low-energy plastic substrate	Poor wetting, craters, fisheyes	Insufficient surfactant wetting power; low substrate compatibility	Incorporate mixed anionic–nonionic system with higher wetting capacity
Styrene–butadiene latex coating under high humidity	Whitening during drying	Migration of water-soluble surfactants; moisture retention	Use polymerizable surfactants; reduce dosage of highly hydrophilic types
Pigmented acrylic latex paint	Uneven gloss, haze at surface	Surfactant blooming and crystallization	Replace with higher MW nonionics; employ reactive surfactants
Adhesive latex film on porous paper	Blocking, tacky surface	Residual coalescing agent and surfactant migration	Optimize coalescent dosage; improve drying profile
Outdoor emulsion coating (freeze–thaw exposure)	Cracking, loss of adhesion	Inadequate stabilization during drying–rewetting cycles	Mixed surfactant system with freeze–thaw stabilizer; higher Tg binder

film performance. By anticipating environmental stresses, substrate interactions, and additive migration, formulators can minimize defects and ensure reliable performance in real-world applications.

5.19 Summary and Closing of Chapter 5

This chapter has examined the role of surfactants in emulsion polymerization with a focus on their impact on latex stability, polymerization kinetics, and final film properties. Beginning with fundamental concepts such as hydrophilic–lipophilic balance (HLB), critical micelle concentration (CMC), and surface tension, the discussion advanced to the selection criteria for both nonionic and anionic surfactants, as well as their synergistic use in combination systems.

Attention was also given to surfactant migration and its implications for film performance, including gloss retention, water resistance, and adhesion. Real-world examples illustrated how improper surfactant choice can result in

Table 5.9 Practical guidelines for minimizing film defects in latex systems.

Case description	Observed film defect	Likely cause	Mitigation strategy
Vinyl acetate–acrylic copolymer latex applied on low-energy plastic substrate	Poor wetting, craters, fisheyes	Insufficient surfactant wetting power; low substrate compatibility	Incorporate mixed anionic–nonionic system with higher wetting capacity
Styrene–butadiene latex coating exposed to high humidity during drying	Whitening and loss of clarity	Migration of water-soluble surfactants combined with moisture retention	Employ polymerizable surfactants; lower dosage of highly hydrophilic species
Pigmented acrylic latex paint	Uneven gloss and hazy surface	Surfactant blooming and crystallization at the air–film interface	Replace with higher MW nonionics; introduce reactive surfactants for anchoring
Adhesive latex film applied to porous paper	Tacky surface and blocking	Residual coalescing agent and surfactant migration	Optimize coalescent dosage; adjust drying conditions for full release
Outdoor emulsion coating subjected to freeze–thaw cycles	Cracking and adhesion loss	Inadequate stabilization under drying–rewetting stress	Use mixed surfactant systems with freeze–thaw stabilizers; adjust binder Tg

defects such as blooming, whitening, or blocking, while practical guidelines emphasized approaches to minimize these risks through optimized dosage, surfactant structure, and the use of polymerizable types.

The chapter further integrated practical case studies and mitigation strategies, reinforcing the direct connection between laboratory formulation decisions and performance outcomes in the field. By balancing surfactant functionality during polymerization with long-term film integrity, formulators can design latex systems that meet demanding requirements in coatings, adhesives, and other applications.

Overall, Chapter 5 highlighted the dual nature of surfactants as both enablers and potential sources of challenge. Understanding their multifaceted role is therefore essential for ensuring consistent quality, reliability, and durability in emulsion polymer systems.

6

Fifth Ingredient—Initiators, and Auxiliary Ingredients

6.1 Initiators in Emulsion Polymerization

An initiator in emulsion polymerization is a chemical compound that decomposes to generate free radicals, which are essential to initiate the polymerization of monomers in an emulsion system. Most initiators used in emulsion polymerization are water-soluble and play a critical role in starting the chain reaction that leads to the formation of polymer particles.

Initiators may be categorized according to their mechanism of radical generation. Table 6.1 summarizes the principal types used in emulsion polymerization.

6.2 Types of Radicals Formed

Radical formation differs with the initiator type.

Persulfates

$$S_2O_8{}^{2-} \rightarrow 2SO_4{}^{\cdot -} \tag{6.1}$$

Hydrogen peroxide

$$H_2O_2 \rightarrow 2HO^{\cdot} \tag{6.2}$$

Organic peroxides

$$ROOR' \rightarrow RO^{\cdot} + R'O^{\cdot} \tag{6.3}$$

Azo initiators

$$RN = N - R' \rightarrow R^{\cdot} + R'^{\cdot} + N_2 \tag{6.4}$$

Persulfate–bisulfite

$$S_2O_8{}^{2-} + HSO_3{}^{-} \rightarrow SO_4{}^{\cdot -} + SO_3{}^{\cdot -} + HSO_4{}^{-} \tag{6.5}$$

Table 6.1 Common initiators in emulsion polymerization.

Type	Examples	Characteristics and application notes
Thermal decomposition	Potassium persulfate (KPS), sodium persulfate (NaPS), ammonium persulfate (APS), hydrogen peroxide (H_2O_2), *tert*-butyl hydroperoxide (tBHP)[b], 2,2′-azobisisobutyronitrile (AIBN), benzoyl peroxide	Decomposes upon heating; widely used in high-temperature polymerizations
Redox systems	Persulfate–metabisulfite, H_2O_2–ascorbic acid, tBHP–ascorbic acid, tBHP–sodium formaldehyde sulfoxylate (SFS), tBHP–Bruggolite® FF6 M	Operate effectively at lower temperatures; reduce residual monomer and VOCs
Surface-active initiators (inisurfs)[a]	Bis[2-(4′-sulfophenyl)alkyl]-2,2′-azodiisobutyrate salts; 2,2′-azobis(N-2′-methylpropanoyl-2-amino-alkyl-1-sulfonate)	Combine initiating and emulsifying roles; allow surfactant-free high-solids polymerizations

a) Surface-active initiators ("inisurfs") have been investigated as dual-function molecules combining the roles of initiator and surfactant. While they can reduce free surfactant migration and introduce functionality at the particle surface, their use in large-scale emulsion polymerization remains rare. Cost, limited commercial availability, and regulatory uncertainty have restricted inisurfs to research and niche specialty applications rather than mainstream industrial practice.

b) Although AIBN (2,2′-azobisisobutyronitrile) is a widely cited radical initiator in academic literature, its use in large-scale emulsion polymerization is very limited. The primary reasons are its oil solubility (making it inefficient for micellar nucleation), higher cost compared to persulfates, and hazardous handling properties as a shock-sensitive solid. In practice, water-soluble persulfates and redox systems are preferred for emulsion polymerization due to their low cost, safety, and better control of particle formation. AIBN remains more relevant for solution and suspension polymerizations, not mainstream emulsion systems.

6.3 Initiator Selection Criteria

The choice of initiator depends on several factors, including temperature, monomer solubility, desired polymer molecular weight, and environmental considerations. Thermal initiators are most suitable for high-temperature processes targeting higher molecular weight polymers. Redox initiators are preferred when low-temperature polymerization or efficient monomer cleanup is required. Oil-soluble initiators are best suited for monomers with low water solubility, such as styrene. Additional considerations include phase solubility, pH stability, safety, and VOC control.

6.4 Initiator Efficiency and Dosage

Initiator concentration exerts a decisive effect on both the molecular weight of the polymer and the kinetics of particle formation. At higher levels of initiator, a greater number of free radicals are generated, producing more growing chains; this shortens the average chain length and yields polymers of lower molecular weight. Conversely, lower concentrations reduce the radical flux, resulting in fewer chains and higher molecular weights.

In industrial emulsion polymerization, initiators are commonly employed at concentrations in the range of 0.1–0.5 wt% of total monomer, although the optimal value depends on the monomer system, surfactant package, and process temperature.

During the nucleation stage, however, somewhat higher initiator levels are advantageous. Typical initiator concentrations of 0.3–0.7 wt% relative to monomer are employed in this stage to provide a strong radical flux, thereby increasing the probability of radical entry into micelles and improving particle formation efficiency. Higher initiator dosages at nucleation reduce the induction period, promote the formation of a larger number of seed particles, and often lead to smaller and more uniform particle sizes. Nevertheless, excessive initiator (>0.7 wt%) may reduce polymer molecular weight too sharply and increase the fraction of water-soluble oligomers.

An increased number of seed particles generated at high initiator flux also requires a corresponding increase in stabilizer coverage. Each additional particle presents a new surface to be coated with surfactant. For this reason, the total surfactant-to-monomer ratio is generally maintained between 2 and 4 wt%, with the lower end of the range (~2 wt%) adequate for systems with moderate particle numbers and the upper end (~3–4 wt%) preferred when higher initiator dosages are used to deliberately generate many seeds. Maintaining the correct balance of anionic surfactants, which drive micelle nucleation, and nonionic surfactants, which provide steric stabilization, ensures that the increased particle population remains colloidally stable.

Thus, while a higher initiator concentration during nucleation improves particle formation efficiency, it must always be matched with sufficient surfactant to preserve stability and prevent coagulum formation.

6.5 Initiator Storage and Stability

Proper storage is essential for safety and reliability. Persulfates are moisture- and heat-sensitive, while hydrogen peroxide requires vented containers to prevent pressure buildup. Organic peroxides and azo compounds such as AIBN must be stored under refrigeration in stable conditions to avoid premature decomposition.

6.6 Initiator Phase Location

The phase in which an initiator decomposes determines nucleation behavior. Water-soluble initiators act in the aqueous phase, favoring micellar nucleation. Oil-soluble initiators act in the monomer phase, promoting droplet nucleation.

6.7 Regulatory Considerations

Increasing emphasis is placed on safe and sustainable initiator systems. Formaldehyde-releasing agents such as sodium formaldehyde sulfoxylate (SFS) are being phased out due to toxicity and regulatory pressure. Safer alternatives such as Bruggolite® FF6 M are formaldehyde-free and widely accepted.

Residual ions, particularly sulfate species, must also be closely monitored. Conventional initiators such as ammonium persulfate (APS) and potassium persulfate (KPS) decompose into sulfate radicals during polymerization, and the resulting end products are sulfate ions (SO_4^{2-}) and bisulfate ions (HSO_4^{-}). These ionic residues remain in the aqueous phase of the latex. At low concentrations, they are generally harmless, but at higher levels, they can present several challenges:

Film stability and water sensitivity: Sulfate ions increase the ionic strength of the aqueous phase, which can destabilize particle–particle interactions during film formation. This often manifests as poorer water resistance, blistering, or increased tack in adhesive films.

Corrosion issues: For coatings applied to metallic substrates, residual sulfates can accelerate corrosion, particularly under humid conditions.

Environmental discharge limits: Wastewater streams containing elevated sulfate levels may exceed regulatory discharge thresholds, especially in regions where limits are set at ~250–500 mg/L.

Interaction with additives: Sulfates can interfere with certain thickeners, dispersants, or crosslinkers (e.g., zirconium complexes), reducing their effectiveness.

For these reasons, regulatory guidance increasingly emphasizes monitoring and reporting sulfate levels in finished dispersions, and in some cases requires demonstration of compliance with limits on extractable ionic species. Industrial practice often includes:

Optimizing initiator dosage to minimize unnecessary sulfate residue.

Postpolymerization purification (e.g., ultrafiltration or ion-exchange) in high-purity latexes.

Transition to redox systems where the reducing component limits sulfate accumulation, or to alternative initiators such as azo initiators or organic peroxides in specialized systems.

Thus, while persulfates remain the most widely used initiators in emulsion polymerization, awareness of residual sulfate species and their downstream effects is essential both for regulatory compliance and end-use performance of coatings and adhesives.

6.8 VAM Homopolymerization Strategy

In vinyl acetate homopolymerization, micellar nucleation is generally less dominant than in systems such as styrene or acrylates. Instead, polymerization occurs predominantly in the aqueous phase, where water-soluble initiators directly generate radicals that attack monomer molecules. Stabilization of the system is typically achieved by protective colloids such as polyvinyl alcohol (PVOH) or hydroxyethyl cellulose (HEC), which adsorb onto the growing particles and prevent coagulation.

The typical initiator dosage ranges from 0.1 to 0.4 wt% of total monomer. Careful control of radical generation is essential to minimize coagulum formation and undesired by-products, thereby ensuring stable latexes suitable for adhesives and coatings.

6.9 Initiator Selection Impact

The choice of initiator system in emulsion polymerization profoundly affects both the kinetics of the reaction and the characteristics of the resulting polymer. The initiator is not simply a trigger for the process but a determinant of polymer molecular weight, particle morphology, and final performance properties.

6.9.1 Radical Flux

The rate of free radical generation, or radical flux, is controlled by the initiator system. High radical flux, such as from fast-decomposing or redox initiators, promotes the formation of a large number of particles, leading to smaller average particle sizes. Lower radical flux, typically from slower thermal initiators at lower dosage, produces fewer particles and results in higher molecular weights. The balance between initiation and propagation governs both the polymerization rate and particle morphology.

6.9.2 Polymer Molecular Weight

The frequency of chain initiation is a primary determinant of polymer molecular weight. Rapid initiation generates a large number of growing chains simultaneously, each consuming monomer at a faster rate and terminating relatively early. This results in many short chains and a lower average molecular weight. By contrast, slower initiation yields fewer active chains, allowing each to propagate longer before termination, thereby producing higher molecular weights.

The molecular weight distribution (MWD) is as important as the average value. Narrow MWDs, typically obtained under controlled radical flux, favor more predictable rheology and uniform film properties. Broader distributions can impart toughness through the presence of long chains while still retaining processability from shorter fractions, but they also risk variability in performance.

In emulsion polymerization, initiator type, concentration, and redox balance strongly affect molecular weight. Water-soluble persulfates generally produce higher radical fluxes and lower molecular weights, while staged or low-temperature redox systems can extend chain length. Chain transfer agents, oxygen contamination, and surfactant composition also influence propagation and termination events, further modifying MWD.

Film formation is especially sensitive to molecular weight. For latex particles to form a strong and continuous film, the polymer chains must not only coalesce but also overlap with one another across particle boundaries. As the chains grow longer during polymerization, they begin to interpenetrate, creating a network that gives the film its strength and durability. If the chains are too short, they cannot overlap sufficiently, so when the particles fuse, the film appears continuous but has weak junctions. Such films are mechanically fragile and more permeable to water. As the chains become longer, the degree of overlap increases. This overlap produces stronger cohesion, better resistance to cracking, and improved water resistance. Very high molecular weight can further enhance strength, but at the expense of higher viscosity and sometimes slower film formation.

Thus, molecular weight control in emulsion polymers is a balancing act: too low and the film is weak and water-sensitive; too high and the latex may be unworkable or slow to coalesce. Careful optimization of initiator dosage, process temperature, and chain transfer strategy enables formulators to tailor polymer molecular weight to the performance requirements of paints, coatings, and adhesives.

6.9.3 Particle Formation and Size Distribution

The characteristics of the initiator system play a decisive role in determining how particles are nucleated and how uniform they become in size. In emulsion polymerization, particle formation proceeds through micellar nucleation, homogeneous nucleation, or a combination of both, depending on surfactant concentration, initiator solubility, and reaction conditions.

Initiator kinetics are especially critical: Redox initiators provide a sharp onset of radical generation, even at low temperatures. This rapid and well-defined flux of radicals favors simultaneous nucleation events across many micelles. The result is a large number of particles formed in a narrow time window, leading to a narrow particle size distribution (PSD).

Persulfates at moderate temperatures generate radicals more gradually. If the radical flux is not well controlled, nucleation may continue over an extended period. This causes new particles to form later in the process, while earlier-formed particles are already growing, resulting in a broader PSD.

Poorly controlled initiation—for example, due to oxygen inhibition, uneven dosing, or thermal spikes—can produce what is known as a bimodal distribution. A bimodal PSD means that two distinct populations of particles exist, one small and one significantly larger, rather than a single continuous distribution. Such latexes may show instability, increased viscosity, or poor film uniformity.

The particle size distribution directly influences application properties: Narrow PSDs lead to latexes with predictable rheology, stable storage behavior, and consistent film formation. Broad PSDs can sometimes be tolerated in adhesives where tack depends on multiple relaxation times, but they often impair gloss, clarity, and mechanical strength in coatings. Bimodal or multimodal PSDs may destabilize the dispersion and increase the risk of coagulation, sedimentation, or phase separation, ultimately compromising end-use performance.

Thus, careful matching of the initiator system (type, dosage, and feeding strategy) with the surfactant level and process conditions is essential to achieve controlled nucleation and a particle size distribution tailored to the requirements of paints, coatings, and adhesives.

Table 6.2 Formulator's tips for initiator selection.

Criterion	Considerations
Monomer system	VAc vs. acrylates vs. styrene–acrylic blends require tailored initiators
Reaction temperature window	Thermal initiators for heated systems; redox initiators for low-temperature
Desired final performance	Higher MW and film strength require slower initiation; redox enhances clean-up
Regulatory constraints	Ensure compliance with VOC and residual monomer limits; food-contact approvals

6.9.4 Residual Monomer Content

Inefficient initiation can leave high levels of residual monomer, which is problematic for regulatory compliance and product safety. Redox systems are particularly advantageous in this regard, as they allow efficient polymerization at lower temperatures, helping achieve stringent targets such as <500 ppm free monomer in finished latexes for adhesives and paints (Table 6.2).

6.10 Initiator Selection Guide Based on Application

Initiator systems must be selected with careful consideration of the intended end-use application, the operating reaction conditions, and relevant regulatory requirements. While the chemistry of initiators determines radical flux and polymerization kinetics, their practical impact is best judged by the performance of the final latex product in adhesives, coatings, and specialty applications (Table 6.3). Table 6.3 summarizes recommended initiator systems for common emulsion polymerization applications, together with typical reaction temperatures and key considerations.

Table 6.3 Recommended initiator systems for common emulsion polymerization applications.

Application type	Recommended initiator system	Typical reaction temperature (°C)	Key considerations
General-purpose adhesives (PVAc-based)	Ammonium persulfate (APS)	75–85	Simple, cost-effective; slower onset time
Heat-sensitive substrates (e.g., BOPP)	APS + sodium bisulfite (redox)	40–60	Low-temperature initiation; fast onset
High-solids paints or coatings	Potassium persulfate (KPS)	75–85	Reliable at high solids; requires careful inhibitor removal
Low-VOC adhesives (food packaging)	$H_2O_2 + Fe^{2+}$ (Fenton system)	25–35	Rapid initiation; excellent for residual monomer control
Construction primers (VAc/acrylic copolymers)	APS or redox system	60–80	Balances film strength with low odor requirements
Sprayable or cold-processed systems	Sodium persulfate + ascorbic acid	Room temperature	Enables initiation without heating; provides stable latex
Flexible coatings (VAE or VAc/BA copolymers)	APS or redox system	70–80	Dosage adjustment allows control of molecular weight and particle size

6.11 Technical Note: Discoloration in VAM-based Paints and the Role of Initiator Ions

6.11.1 Overview

In emulsion paints based on vinyl acetate monomer (VAM) or its copolymers, discoloration is a frequent challenge, especially in white or light-colored formulations. Under alkaline conditions and elevated temperatures, residual monomer and certain initiator residues undergo side reactions that yield colored by-products. The problem is particularly pronounced in hot climates or during extended storage.

6.11.2 Mechanism of Discoloration

1) Residual VAM hydrolyzes under alkaline conditions ($pH > 7$), forming acetic acid and acetaldehyde.
2) In the presence of ammonium ions, introduced either by ammonium persulfate (APS) or other ammonium-based additives, acetaldehyde reacts to form Schiff base–like compounds with a yellowish color.
3) Residual initiators such as peroxides or APS may also contribute to discoloration, either by generating reactive oxidative fragments or by releasing ammonium ions.
4) Residual monomer from insufficient initiator dosage or lack of chasing initiator leads to higher unreacted VAM, which later hydrolyzes to aldehydes, intensifying yellowing under alkaline conditions.
5) Excess initiator remaining in the binder can oxidize white pigments such as titanium dioxide, causing yellowing or loss of brightness even when residual monomer is minimal.
6) Discoloration accelerates in hot storage conditions, particularly in summer months.

6.11.3 Role of Initiator Choice

Ammonium persulfate (APS) introduces ammonium ions into the system and increases the risk of aldehyde-driven discoloration. Potassium persulfate (KPS) avoids ammonium-ion introduction and significantly reduces the risk of yellowing in white or pastel paints. Proper dosing is essential: too little initiator elevates residual monomer and in situ aldehyde formation; too much initiator leaves oxidative species that can attack pigments and cause yellowing.

6.11.4 Solutions for APS-based Systems

Where APS use is unavoidable, the following corrective actions are effective.

1) Post-add aldehyde scavengers such as sodium bisulfite, sodium metabisulfite, or suitable hydrazine derivatives.
2) Reduce residual monomer to below about 0.3% by redox finishing or vacuum stripping.
3) Avoid ammonium-based neutralizers; use sodium hydroxide or potassium hydroxide instead.
4) Store latex under cool, stable conditions to limit discoloration reactions.

6.11.5 Optimal Strategy

To mitigate yellowing in white and pastel systems, the prioritization of initiator and neutralizer choices is required. Formulators should avoid ammonium- and formaldehyde-related pathways and actively manage residual species that promote oxidative discoloration. The following actions are practical controls at the formulation level:

1) Prefer potassium persulfate (KPS) where formulation latitude permits, thereby eliminating ammonium-ion contribution.
2) Minimize residual monomer by maintaining adequate initiator during polymerization and applying an appropriate finishing ("chasing") initiator dose.
3) Avoid ammonium-based neutralizers and dispersants in white and pastel systems.
4) Monitor acetaldehyde; introduce suitable aldehyde scavengers if necessary.
5) Eliminate formaldehyde-releasing biocides, as formaldehyde can drive similar yellow-discoloration pathways.
6) Control residual oxidizing species—particularly peroxides/persulfates—to prevent oxidative discoloration of pigments.

6.12 Polymerization Onset Time

Initiators are selected for decomposition temperature, redox potential, reaction conditions, and the desired polymerization kinetics. One of the most critical parameters affected by initiator choice is the polymerization onset time.

Polymerization onset time is the interval between establishing reaction conditions—typically introduction of the initiator and application of heat—and the first evidence of polymer formation. Onset is usually indicated by a measurable temperature rise from the exotherm, an increase in viscosity, or the appearance of turbidity as particles are generated.

In emulsion polymerization, the onset time is a key indicator of reactivity and readiness. A rapid, well-defined onset reflects efficient initiator decomposition, proper emulsification, and minimal interference from inhibitors or oxygen. A delayed onset suggests issues that can compromise product quality or process control.

The principal factors that influence polymerization onset time include the following.

1) Type and dosage of initiator.
2) Presence of oxygen or monomer inhibitors.
3) Reactor temperature and mixing efficiency.
4) pH of the medium and the surfactant system used.

Controlling and monitoring the onset time allows the formulator to do the following.

1) Ensure consistent polymer nucleation and particle-size distribution.
2) Diagnose initiation problems or contamination.
3) Optimize reaction kinetics for improved process control.

6.12.1 Case Comparison Effect of Initiator System on Polymerization Onset Time

Table 6.4 Comparison of common initiator systems, their typical operating temperatures, and the corresponding polymerization onset times.

6.12.2 Interpretation

Thermal-only initiators such as ammonium persulfate (APS) or potassium persulfate (KPS) generally display longer onset times and are more vulnerable to

Table 6.4 Effect of initiator system on polymerization onset time.

Initiator system	Reaction temperature (°C)	Typical onset time (min)	Remarks
Ammonium Persulfate (APS)	75–85	8–15	Requires thermal activation; sensitive to oxygen; delayed onset at low temperature
APS + Sodium Bisulfite (Redox)	40–60	2–5	Rapid initiation at low temperature; suitable for temperature-sensitive monomers
Hydrogen Peroxide + Fe^{2+} (Fenton)	25–35	1–3	Very fast onset; suitable for cold-start systems; requires careful control
Potassium Persulfate (KPS)	75–85	10–20	Similar to APS; latency depends on monomer purity

inhibition caused by dissolved oxygen or residual stabilizers in monomers. Redox systems, such as APS combined with sodium bisulfite, initiate rapidly at lower temperatures and are particularly valuable for continuous or cold-polymerization processes. Fenton-type systems, based on hydrogen peroxide and ferrous ions, provide an almost immediate onset; however, they require strict control of both iron and peroxide concentrations to avoid uncontrolled or runaway reactions.

6.13 Fenton-type Redox Initiator Systems in Emulsion Polymerization

6.13.1 Overview

Fenton-type systems are a class of redox initiators that rely on hydrogen peroxide (H_2O_2) in the presence of a ferrous iron catalyst (Fe^{2+}) to generate radicals at room temperature or under mild heating conditions. These systems are widely used in emulsion polymerization, where rapid initiation at low temperature is required.

6.13.2 Mechanism

The radical generation mechanism is based on the reaction:

$$Fe^{2+} + H_2O_2 \rightarrow Fe^{3+} + OH^- + \cdot OH \tag{6.6}$$

The hydroxyl radical ($\cdot$OH) produced in this process is highly reactive and can immediately initiate polymerization.

6.13.3 Advantages

The main advantages of Fenton-type redox systems are the immediate onset of polymerization without the need for external heating, their effectiveness even at very low temperatures (0–30 °C), and their particular suitability for heat-sensitive formulations.

6.13.4 Limitations and Safety Considerations

The system is highly reactive and requires strict control to avoid instability. *Caution:* excess concentrations of Fe^{2+} or H_2O_2 may cause uncontrolled thermal runaway. Gelation or coagulum formation can occur if the system is not carefully balanced. The use of chelating agents such as EDTA can help moderate iron activity and improve process safety.

Table 6.5 Thermal decomposition of ammonium persulfate in different states.

State	Decomposition behavior
Dry solid APS	Stable at room temperature. Decomposition occurs significantly only above 120–130 °C, often closer to 150 °C, and produces ammonium bisulfate (NH_4HSO_4), with oxygen gas released.
Aqueous solution (emulsion polymerization)	Begins decomposing at approximately 60–65 °C. Optimal radical generation occurs between 70 and 85 °C. Water promotes faster and lower-temperature decomposition.

6.14 Thermal Decomposition of Ammonium Persulfate (APS)

When evaluating the decomposition of ammonium persulfate (APS), it is important to distinguish between its stability in the dry solid state and its decomposition behavior in aqueous solution during emulsion polymerization (Table 6.5). Table 6.5 summarizes the thermal decomposition behavior of ammonium persulfate in different physical states.

In emulsion polymerization, the decomposition of APS in water is fundamental for generating free radicals. The temperature range of 70–85 °C is considered optimal for efficient initiation and subsequent polymer growth. In contrast, APS in its dry solid form remains thermally stable until it is subjected to much higher temperatures.

6.15 Persulfate Initiators and the Role of pH in PVAc Polymerization

Persulfate salts, such as ammonium persulfate and potassium persulfate, are widely used initiators in emulsion polymerization because of their ability to produce highly reactive sulfate radicals ($SO_4{\cdot}^-$) in aqueous media. Their efficiency, however, is strongly influenced by the pH of the reaction system.

6.15.1 Optimum pH for Persulfate Efficiency

Persulfates decompose most efficiently at or near neutral pH (6.5–7.5). In this range, they generate sulfate radicals that initiate polymerization rapidly. Such conditions are advantageous in systems that do not require pH-sensitive stabilization.

6.15.2 Typical pH Conditions in PVAc and Copolymer Systems

In contrast, polyvinyl acetate (PVAc) and its copolymers are generally polymerized under acidic conditions, with a pH range of 4.0–5.0. Operating at this pH enhances latex stability and prevents hydrolysis of the polymer. Hydrolysis would otherwise release acetic acid, thereby compromising the integrity of the emulsion.

6.15.3 Tradeoffs Between Persulfate Efficiency and Polymer Stability

While persulfates show their highest decomposition efficiency at neutral pH, PVAc polymerization is normally carried out under acidic conditions. This creates an inherent tradeoff between initiator performance and system stability. At neutral pH, persulfates decompose quickly, providing abundant radicals, but the resulting reaction can be difficult to control, and the risk of hydrolysis is significant. Conversely, in acidic conditions, the efficiency of radical generation is somewhat reduced, yet latex stability and polymer quality are preserved.

6.15.3.1 Case Study: Neutral-pH APS Chase for Reducing Residual Monomer

A polyvinyl acetate dispersion polymerized at pH 4.5–5.0 exhibited high residual monomer content. Redox finishing and vacuum stripping were not available, and a conventional thermal persulfate chase at acidic pH reduced monomer levels only partially. To address this limitation, the latex was briefly adjusted to pH 6.7–6.9 and treated with 0.05–0.10 wt% ammonium persulfate (APS), calculated on monomer, at 68–72 °C for 20–30 min. After the chase, the pH was restored to 4.8–5.2 with acetic acid.

Rationale: At near-neutral pH, APS decomposition accelerates and generates both sulfate radicals ($SO_4^{\cdot -}$) and hydroxyl radicals (·OH), creating a high radical flux that efficiently consumes residual monomer. Although PVAc latexes are generally more stable under acidic conditions, limiting the near-neutral exposure to ≤40 min mitigated hydrolysis of acetate groups and preserved polymer properties.

Observation: Free monomer dropped significantly compared with a thermal chase conducted under acidic conditions. Hydrolysis and viscosity drift were negligible when the near-neutral stage was tightly time-limited. A slight increase in sulfate residue was detected, which may reduce water resistance unless compensated through formulation adjustments.

Lesson: The approach highlights the persulfate finishing tradeoff: greater radical efficiency at near-neutral pH versus greater polymer stability under acidic conditions. A neutral-pH APS chase can serve as an effective fallback to reduce residual monomer when stripping or redox systems are unavailable, provided exposure time and pH are carefully controlled (Table 6.6).

Table 6.6 demonstrates that neutral conditions favor initiator efficiency, while acidic conditions favor the stability of the polymer dispersion. For PVAc systems,

Table 6.6 Effect of pH on persulfate efficiency and PVAc system stability.

Factor	At neutral pH (~7.0)	At acidic pH (~4.5–5.0)
Persulfate efficiency	High decomposition rate; rapid radical generation	Slightly slower radical generation
Latex stability (PVAc)	Increased risk of hydrolysis	Improved latex stability
Polymerization control	Very fast; difficult to control exotherm	More controlled reaction
Risk of gelling	Higher risk	Lower risk

formulators generally accept a slightly slower initiation rate in order to maintain latex stability and reduce the risk of hydrolysis. In practice, this compromise allows the polymerization to proceed smoothly, balancing initiator performance with product quality.

6.16 Hydrogen Peroxide in Emulsion Polymerization

6.16.1 Overview

Hydrogen peroxide (H_2O_2) is extensively used in redox initiation systems for emulsion polymerization, particularly when low-temperature initiation is required. Acting as a strong oxidizing agent, it participates in systems such as H_2O_2/Fe^{2+} (Fenton-type) or H_2O_2/ascorbic acid, enabling radical generation under mild thermal conditions. This makes hydrogen peroxide one of the most versatile and widely adopted initiators in modern polymerization practice.

6.16.2 Type and Concentration Used

Commercial hydrogen peroxide is supplied as an aqueous solution, typically in concentrations of 30%, 35%, or 50% by weight. In industrial polymerization processes, the most common form is the 30% aqueous solution. In emulsion polymerization, hydrogen peroxide is normally employed at a level of 0.1–1.0% based on monomer weight. It is usually diluted further in the initiator feed to ensure safe and controlled addition during the polymerization process.

6.16.3 Inhibitor Content

Commercial grades of hydrogen peroxide may contain small quantities of stabilizers to minimize spontaneous decomposition. Common stabilizers include phosphoric acid, sodium stannate, and acetanilide. These additives do not interfere significantly with the course of polymerization. For applications requiring high-purity conditions, stabilized grades with minimal additives can be selected to ensure consistency and reliability.

6.16.4 Storage and Handling

Hydrogen peroxide must be stored in dark, vented polyethylene containers under controlled temperature conditions, ideally between 2 and 8 °C. It is essential to avoid contact with metals such as iron, copper, or manganese, as well as with heat, light, and organic matter, all of which can accelerate decomposition. Only clean, acid-washed plastic or glass vessels should be used for dilution and transfer to ensure safe handling and to prevent contamination.

All transfer lines and pumps should be made of stainless steel 304/316L or appropriate plastics (e.g., PTFE).

6.16.5 Safety Precautions

Hydrogen peroxide is a powerful oxidizer that can cause severe burns to the skin and eyes. Its decomposition releases oxygen gas, creating the risk of pressure buildup if improperly stored or handled. In emulsion polymerization, hydrogen peroxide should always be added slowly and in diluted form to prevent uncontrolled reactions. The use of proper protective equipment and strict adherence to safety procedures are mandatory when working with this initiator.

6.17 Formaldehyde-free Organic Redox Systems in Emulsion Polymerization

Formaldehyde-free redox systems are increasingly favored in emulsion polymerization, particularly in paints and adhesives designed to meet eco-label requirements. These systems avoid the use of formaldehyde-releasing reducing agents, such as sodium formaldehyde sulfoxylate (SFS) or Rongalite®, and instead employ naturally derived or environmentally friendly alternatives.

6.17.1 Preferred System TBHP/Ascorbic Acid

One of the most widely adopted formaldehyde-free redox systems uses tert-butyl hydroperoxide (TBHP) as the oxidant with ascorbic acid (vitamin C) as the reducing agent. It initiates efficiently under mild conditions and is well established in adhesive and coating applications. The system operates effectively at pH 3–6 (typically 4.5–5.5) and 25–45 °C, and is suitable for eco-focused paints, adhesives, and sealants. In addition to avoiding formaldehyde release, it offers low

Table 6.7 Key features of the TBHP/ascorbic acid system (formaldehyde-free).

Component/Parameter	Description
Peroxide	*Tert*-butyl hydroperoxide (TBHP)
Reducing agent	Ascorbic acid (vitamin C)
Optimal pH range	3–6 (best performance 4.5–5.5)
Operating temperature	25–45 °C
Application areas	Eco-focused paints, adhesives, sealants
Advantages (summary)	No formaldehyde emission; low discoloration risk; biodegradable, low-odor profile; efficient radical generation at mild temperatures

discoloration risk, a biodegradable, low-odor profile, and clean radical generation at mild temperatures (Table 6.7).

6.17.2 Other Formaldehyde-free Reducing Agents

Several alternative reducing agents can be paired with TBHP to achieve formaldehyde-free initiation. Their key features are summarized in Table 6.8

6.17.3 Redox Systems to Avoid

The following reducing agents are not formaldehyde-free and should be excluded from green-certified or low-VOC-compliant formulations. Both

Table 6.8 Alternative formaldehyde-free reducing agents in emulsion polymerization.

Reducing agent	Typical peroxide pairing	Key notes
Erythorbic acid	TBHP	Isomer of ascorbic acid; slightly more stable
Sodium ascorbate	TBHP	Buffered derivative of ascorbic acid; improved shelf stability
Sodium metabisulfite (SMBS)	TBHP	Low-cost option; may cause yellowing in some formulations
Thiourea[a]	TBHP	Effective, but must be evaluated carefully due to toxicity concerns

a) The use of thiourea requires careful safety assessment to ensure regulatory compliance.

Table 6.9 Redox reducing agents to exclude (formaldehyde-releasing).

Reducing agent	Reason for exclusion
Sodium formaldehyde sulfoxylate (SFS)	Releases formaldehyde during redox reaction
Rongalite® (hydroxymethanesulfinic acid derivatives)	Releases formaldehyde during redox reaction

Note: Rongalite® is a registered trademark.

release formaldehyde during the redox process and are therefore unsuitable for environmentally compliant applications (Table 6.9).

6.17.4 Summary

The TBHP/Ascorbic Acid system remains the best-in-class redox pair for formaldehyde-free emulsion polymerization. It combines effective low-temperature initiation with excellent film clarity and minimal discoloration. Alternative reducing agents, such as sodium ascorbate or erythorbic acid, provide flexibility in formulation design. Ultimately, the choice of system should balance performance, operating temperature, safety, and environmental compliance.

6.18 Auxiliary Ingredients in Emulsion Polymerization

In addition to monomers, water, surfactants, and initiators, emulsion polymerization systems frequently contain auxiliary ingredients. These components are used to control molecular weight, improve film properties, enhance colloidal stability, adjust pH, or extend storage life. The most important groups include chain transfer agents, surface-active chain transfer agents, coalescing aids, buffers, plasticizers, antimicrobial agents, antioxidants, defoamers, and rheology modifiers.

6.18.1 Chain Transfer Agents (CTAs)

Chain transfer agents regulate the molecular weight and distribution of the polymer by terminating growing chains and initiating new ones. Control over chain length is especially important in adhesives and coatings, where viscosity and film properties must be tuned for application performance. Proper CTA selection helps prevent excessively high molecular weights, enabling manageable viscosities and reliable film formation (Table 6.10).

Table 6.10 summarizes representative chain transfer agents and their primary roles in formulation.

Table 6.10 Representative chain transfer agents and their primary formulation role.

CTA	Primary role in formulation
n-Dodecyl mercaptan	Caps molecular weight to control viscosity; supports film formation
tert-Nonyl mercaptan	Caps molecular weight to control viscosity; supports film formation

6.18.2 Surface-active Chain Transfer Agents (Transurfs)

Surface-active chain transfer agents combine classical chain-transfer functionality with surfactant-like behavior. They can participate in micelle formation, enhance latex stability, and reduce reliance on separate surfactants. These agents are particularly valuable in high-solids or low-surfactant formulations, where conventional stabilization alone may be insufficient. By integrating chain-transfer and surface activity, they improve processability while controlling polymer molecular weight and narrowing the molecular-weight distribution.

A widely studied class comprises mercapto-functional surfactants that incorporate a thiol group (–SH) for chain transfer, together with a hydrophilic head group and hydrophobic tail for interfacial activity. Representative examples are summarized in Table 6.11

Table 6.11 Representative surface-active chain transfer agents (transurfs) and roles.

Transurf	Chemistry/Type	Primary role in emulsion polymerization
Sodium dodecyl mercaptosuccinate (SDMS)	Anionic mercapto-functional surfactant	Acts as a chain transfer agent and anionic stabilizer; widely used in vinyl acetate–acrylate copolymerizations
n-Dodecyl mercaptan ethoxylates	Nonionic mercapto-functional surfactants	Provide nonionic stabilization while regulating molecular weight in styrene–acrylic and VAc-based emulsions
Commercial transurfs (e.g., Transurf® series)	Proprietary mercapto-functional surfactant packages	Combine latex stabilization with molecular-weight control; enable higher-solids emulsions with reduced foaming and improved film properties

By unifying stabilization and chain-transfer functions, transurfs can simplify formulations, reduce free-surfactant content, and help balance colloidal stability with controlled polymer properties.

6.18.3 Coalescing Aids

Coalescing aids are temporary plasticizers that lower the minimum film-forming temperature (MFFT) of the latex. During film formation, they soften the polymer particles and promote their fusion into a continuous film. They are widely used in paints and adhesives when ambient-temperature film formation is required (Table 6.12).

6.18.4 Buffers

Buffers help maintain a stable pH during polymerization and storage. Because pH strongly affects initiator decomposition, surfactant ionization, and colloidal stability, buffering is essential in many formulations. After polymerization, the latex pH is adjusted according to the polymer type to optimize stability and downstream compatibility (Table 6.13).

Table 6.12 Representative coalescing aids and typical use.

Coalescent	Typical context/note
Texanol® (2,2,4-trimethyl-1,3-pentanediol monoisobutyrate)	Common in architectural coatings; effective MFFT reduction with moderate odor
Butyl Carbitol® (diethylene glycol monobutyl ether)	Used where slower evaporation is beneficial in coatings and adhesives
Dipropylene glycol *n*-butyl ether (DPnB)	Low-odor option; balanced volatility and effectiveness

Table 6.13 Typical buffers and pH targets.

System/polymer	Common buffer(s)	Typical pH window
PVAc and PVAc-based dispersions (persulfate-initiated)	Sodium bicarbonate	4.5–5.5 (limits hydrolysis of acetate groups)
Acrylic latexes	Phosphate, ammonium salts, citric acid (as acidic buffers)	7.5–8.5 after neutralization (storage stability, additive compatibility)

6.18.5 Plasticizers

Plasticizers increase film flexibility and lower the polymer glass transition temperature (Tg). They are indispensable in pressure-sensitive adhesives and flexible coatings where elasticity and cold-temperature performance are required (Table 6.14). Table 6.14 summarizes representative plasticizers and their primary use in latex formulations.

6.18.6 Antimicrobial Agents

Antimicrobial agents protect latexes against microbial contamination during storage and after application. Biocidal protection is essential in warm or humid environments to prevent spoilage, maintain rheology, and preserve film integrity (Table 6.15).

6.18.7 Antioxidants and UV Absorbers

To ensure long-term durability, antioxidants and UV absorbers are incorporated to mitigate oxidative and photo-oxidative degradation. This improves color retention, mechanical performance, and outdoor service life (Table 6.16).

6.18.8 Defoamers and Air Release Agents

Foam formation and entrained air are persistent issues in emulsion polymerization and in subsequent application steps. Defoamers suppress foam, while

Table 6.14 Representative plasticizers and primary use.

Plasticizer	Primary use/note
Dibutyl phthalate (DBP)	Legacy PSA and coatings plasticizer; strong Tg reduction
Diisononyl phthalate (DINP)	Widely used in flexible systems; a balance of permanence and compatibility
Triethyl citrate	Nonphthalate option; lower odor; used where regulatory profile is prioritized

Table 6.15 Common preservatives and notes.

Preservative	Typical note/use
Isothiazolinones (e.g., MIT, BIT)	Broad-spectrum in-can preservation; check regional regulatory limits
Bronopol	Often used in blends; supports control of bacteria in storage

Table 6.16 Representative stabilizers for long-term durability.

Additive	Function/examples	Typical note
Antioxidants	Hindered phenolics (e.g., BHT, BHA)	Inhibit oxidative chain scission; verify compatibility and potential odor
UV absorbers	Benzotriazoles, benzophenones	Protect against UV radiation; commonly used in exterior coatings

Table 6.17 Typical defoamer chemistries and selection notes.

Type	Selection note
Silicone-based defoamers	Highly efficient; monitor for potential cratering in films
Mineral-oil defoamers	Robust defoaming; consider impact on gloss and recoatability
Nonsilicone defoamers	Used when silicone sensitivity is a concern

air-release agents facilitate the removal of entrapped bubbles to prevent film defects (Table 6.17).

6.18.9 Rheology Modifiers

Rheology modifiers control viscosity, flow, and application behavior. They enable tailoring for brushing, rolling, or spraying, and they improve leveling, sag resistance, and handling characteristics (Table 6.18).

Table 6.18 Examples of rheology modifiers and typical roles.

Class	Examples	Typical role
Cellulosics	Hydroxyethylcellulose (HEC)	Builds viscosity; improves sag resistance; widely used in adhesives and paints
HEURs	Hydrophobically modified ethoxylated urethanes	Associative thickening; excellent flow/leveling balance in coatings
Acrylic thickeners	ASE/HASE types	pH-responsive thickening; good for spray and roll applications

6.19 Conclusion

Each auxiliary ingredient must be selected in view of the final application, compatibility with the polymer system, regulatory requirements, and specific performance targets. Used judiciously, these additives fine-tune emulsion-polymer formulations to deliver optimal film properties, improved processability, extended shelf life, and reliable end-use performance.

7

Process, Mechanism, and Variables in Vinyl Acetate-Based Emulsion Polymerization

7.1 Introduction

The emulsion polymerization of vinyl acetate (VAc) and its copolymers occupies a unique position in industrial polymer chemistry. Unlike styrene or acrylates, where classical micellar nucleation dominates, VAc systems rely on the protective colloid concept, with polyvinyl alcohol (PVOH) providing interfacial stabilization and enabling efficient particle formation. This distinction has not only shaped the scientific understanding of emulsion polymerization but also led to practical innovations in adhesive, coating, and paper-sizing technologies.

Over the past decades, the drive for improved performance—higher water resistance, better adhesion, lower volatile content—has shifted attention from simple homopolymers to carefully designed copolymers. Incorporating comonomers such as acrylates, VeoVa, ethylene, and maleates allows formulators to tailor glass-transition temperature, hydrophobicity, flexibility, and durability. As a result, VAc-based copolymer dispersions have become indispensable binders across diverse applications, ranging from architectural coatings to flexible packaging adhesives.

At the same time, the methods of emulsion polymerization—batch, semibatch, and continuous processes—have been refined to balance conversion, stability, and heat management. Process engineers and chemists must navigate a complex interplay of variables: initiator decomposition, monomer solubility, surfactant and colloid adsorption, particle size control, and reactor design. The optimization of these factors is crucial not only for latex performance but also for safe, economical, and environmentally compliant production at industrial scale.

This chapter explores the mechanisms, variables, and process optimization strategies for the emulsion polymerization of vinyl acetate and its copolymers. It begins with a detailed account of VAc homopolymerization, then extends to the principles of copolymerization and the influence of different process types.

By integrating mechanistic insight with practical considerations, the chapter provides both a conceptual foundation and a practical guide for scientists and engineers working in this field.

7.2 Overview of VAc Emulsion Polymerization

Vinyl acetate (VAc) is one of the most widely used monomers in emulsion polymerization due to its moderate reactivity, relatively low toxicity, and cost-effectiveness. Polymerization of VAc in aqueous media produces polyvinyl acetate (PVAc) latexes, which serve as the base for adhesives, sealants, nonwoven binders, and paper coatings.

Unlike many hydrophobic monomers, vinyl acetate is appreciably soluble in water, with a solubility of about 2.5% at room temperature. This partial solubility plays a decisive role in its polymerization mechanism, as it allows polymerization to proceed by homogeneous nucleation rather than relying on surfactant micelles.

7.2.1 Distinctive Features of VAc Polymerization

The emulsion polymerization of vinyl acetate (VAc) is most commonly conducted with polyvinyl alcohol (PVOH) as the primary stabilizer. During nucleation and growth, PVOH adsorbs on particle surfaces through its polymer backbone, while hydroxyl groups extend into the aqueous phase. This conformation supplies strong steric stabilization and hydrogen-bonding interactions that suppress droplet coalescence and prevent particle coagulation. Small amounts of low-foaming surfactant may be included to aid initial droplet dispersion or to fine-tune viscosity, but stabilization in VAc systems is dominated by the protective-colloid layer provided by PVOH.

A practical strength of the VAc process is the ability to reach relatively high solids (often ~55%) while maintaining manageable viscosity and good storage stability. Semicontinuous operation is preferred because the reaction is strongly exothermic, and particle size evolves during the feed. Controlled monomer addition and heat removal in semibatch mode stabilize temperature, narrow particle-size distribution, and support high solids without compromising colloidal stability.

This protective-colloid mechanism distinguishes VAc systems from classical styrene- or acrylate-based emulsions, where micellar nucleation driven by surfactant micelles is predominant. In VAc/PVOH systems, micelles are minimal or absent; nucleation occurs mainly by homogeneous pathways when aqueous-phase radicals add to dissolved monomer, with the growing oligomers captured and stabilized by adsorbed PVOH. Monomer droplets act primarily as reservoirs that replenish the dissolved fraction throughout the reaction, rather than as the principal loci of particle formation.

7.2.2 Prepolymerization State

Before initiation, the vinyl acetate emulsion comprises (i) a dissolved monomer fraction in the aqueous phase and (ii) dispersed monomer droplets stabilized by a protective colloid. This distribution governs the nucleation pathway and the subsequent particle-stabilization mechanism.

Aqueous phase: Vinyl acetate has an appreciable solubility in water (≈2.5% by weight at room temperature). This dissolved fraction is directly accessible to initiator-derived radicals formed in the aqueous phase. Polymerization, therefore, begins in solution; short oligomeric radicals form and then precipitate to create nuclei.

Monomer droplets: Most of the monomer remains as discrete droplets because water cannot dissolve more than about 2.5% vinyl acetate. These droplets are generated by agitation during charging and mixing and are stabilized primarily by polyvinyl alcohol adsorbed at the monomer–water interface (rather than by conventional surfactant micelles). The droplets do not nucleate particles directly; instead, they act as reservoirs that continuously replenish the dissolved fraction as polymerization proceeds.

Protective colloid layer: Polyvinyl alcohol provides colloidal stability by adsorbing strongly at both vinyl acetate droplet surfaces and the surfaces of newly formed polymer particles. Its amphiphilic character arises from hydrophilic hydroxyl groups (typical partially hydrolyzed grades are ~89% hydrolyzed) and residual hydrophobic acetate groups (~11%). This dual composition allows chains to align at the interface: acetate segments and the hydrophobic backbone associate with the droplet or particle surface, while hydroxyl groups extend into the aqueous phase, imparting steric stabilization and hydrogen-bonding interactions that prevent droplet coalescence and particle coagulation.

This protective-colloid mechanism distinguishes vinyl acetate systems from classical styrene or acrylate emulsion polymerizations, where surfactant micelles dominate nucleation. In vinyl acetate systems, micelles are absent or negligible; nucleation occurs homogeneously when radicals attack the dissolved monomer, while droplets serve primarily as reservoirs that sustain the dissolved fraction throughout the reaction (Figure 7.1).

7.2.3 Reaction Intervals

In vinyl acetate emulsion polymerization, the reaction typically progresses through three characteristic intervals governed by the availability of dissolved monomer and the presence (or depletion) of the droplet reservoir.

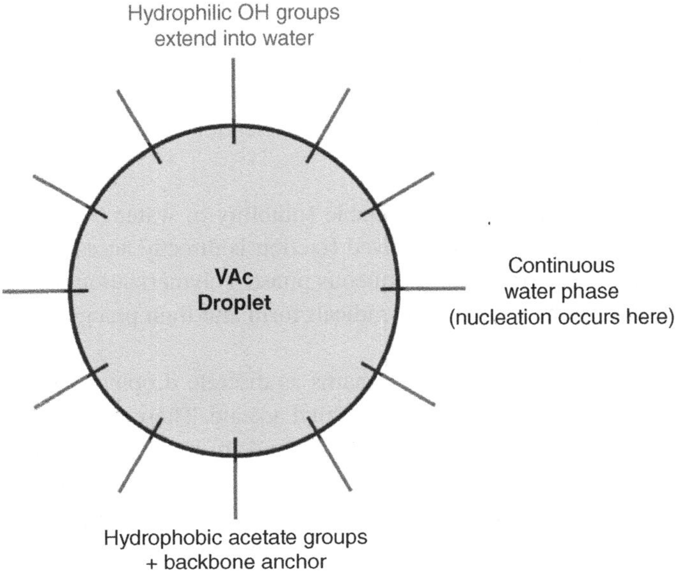

Figure 7.1 Protective colloid action of polyvinyl alcohol at the vinyl acetate droplet–water interface. Schematic representation of a vinyl acetate droplet acting as a monomer reservoir within the continuous aqueous phase. Polyvinyl-alcohol chains form a protective interfacial layer: hydrophobic acetate segments and the polymer backbone orient toward the dispersed monomer or particle surface, while hydrophilic hydroxyl groups extend into water, preventing coalescence and ensuring colloidal stability.

Interval I—Particle nucleation: Free radicals generated in the aqueous phase react with dissolved vinyl acetate. The resulting short oligomeric radicals precipitate to form primary nuclei, which are immediately stabilized by adsorbed polyvinyl alcohol. This homogeneous nucleation pathway is the dominant route for particle formation in vinyl acetate systems.

Interval II—Particle growth: As polymerization proceeds, monomer diffuses from the dispersed droplet reservoir into the growing particles, where conversion continues. Polyvinyl alcohol remains adsorbed at the particle surface, maintaining colloidal stability and preventing coagulation while particle size increases.

Interval III—Monomer depletion: When the droplet reservoir is exhausted, polymerization continues on the residual dissolved monomer. The growth rate declines as dissolved monomer is consumed and the system approaches completion, after which finishing steps may be applied to reduce residual monomer.

7.3 Key Influencing Factors in the Homopolymerization of Vinyl Acetate Monomer

The homopolymerization of vinyl acetate monomer (VAM) via emulsion polymerization is governed by a set of interdependent variables that determine initiation kinetics, particle-size distribution, monomer conversion, colloidal stability, and the performance characteristics of the resulting polyvinyl acetate (PVAc) latex. Although detailed experimental data appear in later sections, the following factors provide a strategic overview for process design and troubleshooting.

Protective colloid type and concentration: Polyvinyl alcohol is the primary stabilizer in VAM systems. Its degree of hydrolysis and molecular weight strongly influence particle size, viscosity, and film properties through adsorption at particle surfaces and the strength of the protective colloid layer.

Monomer solubility and monomer–water ratio: VAM solubility in water ($\approx$2.5% by weight at room temperature) underpins the homogeneous nucleation mechanism. The monomer-to-water ratio controls solids content, viscosity development, and heat release throughout the reaction.

Initiator type and dosage: Persulfates, redox systems, and hydrogen-peroxide pairs determine radical generation rates, molecular-weight distribution, and residual monomer. Selection balances rapid nucleation with controlled propagation and finishing.

Temperature profile: Temperature governs initiator decomposition and the balance between nucleation and growth. For persulfate initiation, 60–80 °C is typical, whereas suitable redox pairs allow operation at lower temperatures.

Agitation and mixing: Uniform mixing prevents localized overheating, ensures efficient dispersion of monomer droplets, and enhances mass transfer into growing particles. Agitation intensity influences droplet size and, consequently, polymerization kinetics.

pH of the reaction medium: VAM polymerizations are usually run under mildly acidic conditions (pH 4.0–5.0). This suppresses PVAc hydrolysis, maintains polyvinyl alcohol integrity, and supports reliable initiator decomposition.

Reactor addition strategy: Semicontinuous feeds of monomer and initiator provide efficient heat management, controlled particle growth, and consistent product quality. Batch operation is less common due to safety and stability constraints.

Electrolyte and impurity levels: Traces of multivalent ions or contaminants can destabilize the colloid, leading to coagulation. High-purity water and raw materials are critical to minimize these risks.

Solids level and conversion: Target solids (often 50–55%) affect viscosity, storage stability, and process efficiency. Achieving high conversion while maintaining manageable viscosity is a key industrial performance target.

A systematic understanding and coordinated optimization of these variables are essential to ensure reproducibility, scale-up fidelity, and consistent performance of VAM-based emulsions in industrial production.

7.4 Conclusion

The emulsion polymerization of vinyl acetate is distinct from systems dominated by surfactant micelles. Its mechanism is driven by the interplay between dissolved monomer in the aqueous phase, dispersed monomer droplets acting as reservoirs, protective colloids such as PVOH, initiators, and the operating conditions of the reactor. Homogeneous nucleation of the dissolved fraction, followed by stabilization through PVOH adsorption, defines the particle formation pathway.

By understanding and controlling the reaction intervals and the influencing factors outlined above, formulators can achieve reliable control of particle size distribution, molecular weight, latex stability, and the overall performance of PVAc dispersions.

7.5 Experimental Observations in VAM Polymerization

Experimental work on vinyl acetate emulsion polymerization has revealed a number of recurring patterns that highlight the delicate interplay between stabilizers, electrolytes, initiators, and reaction conditions. These observations not only provide mechanistic insight but also offer practical guidance for optimizing industrial formulations.

7.5.1 Effect of Protective Colloids

Increasing the concentration of polyvinyl alcohol (PVOH) accelerates the polymerization rate, with the rate showing an approximate three-fifths power dependence on PVOH concentration. High-molar-mass PVOH enhances the decomposition of potassium persulfate more effectively than low-molar-mass grades. This accelerated initiator decomposition increases the radical flux, thereby raising the rate of polymerization. The choice of PVOH grade is therefore a critical process variable, influencing both initiation kinetics and the stability of the resulting latex.

7.5.2 Influence of Ionic Strength

The addition of electrolytes increases ionic strength, screens electrostatic repulsion between particles, and facilitates coalescence, leading to larger particle sizes.

While useful in certain coating applications, this must be carefully balanced to avoid destabilization.

7.5.3 Temperature Effects

Higher polymerization temperatures reduce the average chain length and particle size of PVAc while increasing the number of sulfate end-groups. At elevated temperature, the decomposition constant (k_e) of persulfate increases, producing a higher flux of sulfate radicals ($SO_4{\cdot}-$). More frequent initiation and termination events shorten chains, and each new chain bears a sulfate-derived end group—hence the rise in sulfate end-groups.

A second, often overlooked consequence is the increased likelihood of generating radicals on the protective colloid itself. Sulfate radicals (or growing PVAc radicals via chain transfer) can abstract hydrogen from PVOH to form PVOH macroradicals. These PVOH· species can graft by adding to vinyl acetate or to particle-bound PVAc radicals, yielding covalently attached PVOH-g-PVAc. Grafting anchors PVOH to particle surfaces thickens the adsorbed steric layer and reduces desorption.

These two effects—smaller particles with larger total surface area and thicker, covalently anchored PVOH layers—combine to increase latex viscosity at higher polymerization temperatures. The apparent hydrodynamic volume rises due to stronger interparticle interactions and, at excessive grafting levels, possible bridging. Practically, modest grafting can improve wet tack and creep resistance, whereas excessive grafting may raise viscosity beyond handling limits or increase gel content.

7.5.4 Effect of Surfactant Concentration

In vinyl acetate homopolymerization, colloidal stabilization is provided mainly by poly(vinyl alcohol) (PVOH) rather than by conventional surfactants. Nevertheless, small amounts of emulsifiers are sometimes used during emulsification and seed formation, and their concentration can influence particle number and size.

When anionic surfactants are employed, they can assist in dispersing monomer droplets and, at higher levels, increase the number of nuclei formed. This may lead to smaller particle sizes and a modest increase in polymerization rate. However, excess anionic surfactant can be detrimental, since unbound surfactant remains in the latex and lowers the water resistance of the final film.

Nonionic surfactants contribute mainly through steric stabilization and can improve dispersion uniformity. They are less sensitive to pH and electrolytes than anionics but tend to broaden the particle-size distribution. In PVAc homopolymer systems, their effect on polymerization kinetics is limited, but they are sometimes combined with anionics in low levels to fine-tune dispersion stability. Nonionic surfactants can be used as auxiliaries to improve shelf life, freeze–thaw stability, and resistance to electrolytes.

Increasing surfactant concentration (when present) decreases particle size and increases particle number, though with diminishing returns at higher levels.

Overall, the role of surfactant concentration in PVAc homopolymerization is secondary compared to PVOH. The protective colloid dominates particle stabilization, while surfactants serve primarily to aid emulsification and provide auxiliary control over particle size distribution.

7.5.5 Effect of Initiator Decomposition and Feed Strategy

The decomposition of persulfate initiators such as ammonium or potassium persulfate is strongly temperature dependent, with the decomposition constant increasing exponentially as temperature rises. Under semicontinuous operation, the monomer feed rate directly influences the radical environment inside the reactor.

When monomer is introduced too quickly (monomer-flooded conditions), the local monomer concentration rises above the rate of radical generation. This causes polymer particles to swell excessively, increases the probability of secondary nucleation, and creates hot spots due to the exothermic heat of polymerization. Such conditions often lead to broader particle size distributions and difficulties in controlling conversion.

In contrast, a starved-feed strategy keeps the instantaneous monomer concentration below the consumption rate by radicals. This approach maintains polymerization within existing particles, minimizes secondary nucleation, and ensures smoother heat release. Importantly, it moderates the effective initiator decomposition rate, since radicals are consumed efficiently without excessive accumulation. The result is a more uniform particle population, better control of molecular weight, and lower residual monomer content.

Thus, initiator efficiency in vinyl acetate emulsion polymerization is not determined solely by the chemical choice of initiator but is critically linked to the operational design of the feed strategy.

7.5.6 Effect of Agitation (5-Ton PVAc Homopolymer Reactor)

In a 5-ton polyvinyl acetate homopolymerization reactor, agitation plays a decisive role in nucleation quality, feed uniformity, and long-term latex stability. During nucleation, which involves the initial charge and hold, operators typically apply a high agitation speed of approximately 400 rpm to establish strong circulation. This level of mixing ensures uniform distribution of vinyl acetate monomer, initiator, and protective colloid. If the agitation speed is insufficient, nucleation becomes irregular, resulting in broad particle-size distributions and occasional wall deposits. Conversely, excessively high speeds promote foaming and partial stripping of the protective colloid layer.

During the semibatch feeding stage, agitation is reduced to a steady yet assertive level, typically around 250 rpm. At this speed, incoming monomer or preemulsion disperses immediately upon entry, while the reactor temperature remains uniform. Agitation that is too low at this stage produces monomer-rich pockets and localized heating, whereas excessive agitation encourages foaming and destabilization of the colloid layer.

As viscosity increases toward the end of the feed and during conversion, polyvinyl acetate grades may reach viscosities in the range of hundreds of thousands of mPa s. In this regime, agitation is generally maintained within the assertive range or slightly reduced. At high viscosities, the geometry of the agitator rather than speed becomes the critical factor in ensuring bulk turnover and renewal of the cooled reactor wall. Increasing the rpm at this stage provides little heat-transfer benefit and instead introduces risks of air entrainment and foaming.

Since a single reactor is often required to handle both homopolymer and copolymer batches across a wide viscosity spectrum, the agitator must be capable of effective operation in both thin and highly viscous states. Broad-spectrum designs, such as EKATO's polymerization-duty impellers, are widely employed. These include multipurpose viscous impellers, such as the PARAVISC type, as well as anchor or helical ribbon impellers that maintain circulation and heat transfer at moderate speeds under very high viscosity conditions.

In practice, operators recognize effective agitation by specific cues. During nucleation, good agitation is indicated by a smooth temperature rise, the absence of foam surges, and steady conversion. During feeding, proper agitation is evident when the incoming feed disappears immediately upon contact, reactor temperature remains uniform, and the liquid surface appears calm but firm. At later stages, clean reactor walls, steady bulk turnover, and the absence of creeping foam confirm that agitation is appropriate.

In summary, agitation at high speed ($\approx$400 rpm) during nucleation promotes uniform particle formation. During feeding and late stages, assertive agitation at approximately 250 rpm provides steady mixing and preserves colloid stability. At elevated viscosities, further increases in rpm are not recommended; circulation and heat-transfer efficiency are best maintained through broad-spectrum agitator geometry rather than speed adjustment.

7.5.7 Summary of Experimental Observations

The individual effects of protective colloids, electrolytes, temperature, emulsifiers, and feed strategy on the emulsion polymerization of vinyl acetate can be complex when viewed in isolation. However, when presented in a structured way, these observations reveal the underlying mechanistic patterns that govern polymerization kinetics, particle size, and film properties.

Table 7.1 consolidates these experimental findings into a concise reference, linking each observation with its mechanistic basis and practical implication. This

Table 7.1 Summary of experimental observations in VAM polymerization.

Observation	Mechanistic interpretation	Industrial implication
Higher PVOH concentration increases the polymerization rate	Rate ∝ [PVOH]∧0.6; high-MW PVOH accelerates persulfate decomposition and increases radical flux	Select PVOH grade and dosage to balance rate acceleration with latex stability
Addition of electrolytes leads to larger particles	Ionic screening lowers repulsion, increasing coalescence probability	Adjust salt levels to target particle size while avoiding destabilization
Higher temperature decreases chain length and increases sulfate end-groups; viscosity also increases	Greater SO_4·− radical flux shortens chains; more PVOH radicals graft to PVAc; smaller particles plus grafted PVOH layers increase hydrodynamic volume	Use moderate temperature and initiator dosage to control grafting and viscosity
Higher emulsifier concentration increases the number of latex particles	Micellar nucleation is promoted, though diminishing returns occur at high levels	Optimize surfactant level to control particle size distribution and viscosity
Faster initiator decomposition occurs with surfactants and rapid monomer feed	Colloids and feed profile influence initiator efficiency	Couple initiator dosage with feed strategy; favor starved-feed to control heat release
PVOH versus surfactants as stabilizers	PVOH provides steric and grafted stabilization; surfactants provide micellar stabilization	PVOH latexes offer Newtonian flow, high wet tack, strong adhesion; grafting must be managed to prevent viscosity runaway
Agitation level influences nucleation and stability	High rpm (≈400) during nucleation ensures uniform VAM/initiator/PVOH distribution; assertive rpm (≈250) during feeding maintains circulation; agitator geometry governs heat transfer and stability at high viscosity	Maintain staged rpm control and use broad-spectrum agitator design to ensure reproducible nucleation, safe heat removal, and latex stability across viscosity range

format provides a quick overview for formulators and process engineers, while also highlighting the critical variables that require control in industrial practice.

7.6 Stability of PVAc Homopolymer Latexes

The stability of PVAc homopolymer latexes represents the most distinctive aspect of vinyl acetate emulsion polymerization. Unlike styrene or acrylate systems, where micelles and surfactants dominate stabilization, PVAc relies primarily on poly(vinyl alcohol) (PVOH) as a protective colloid. Stability is therefore governed by the interaction between PVOH, operating conditions, and reactor hydrodynamics.

7.6.1 Role of PVOH

PVOH irreversibly adsorbs at the polymer–water interface, providing steric stabilization. Radical attack during polymerization generates PVOH-g-PVAc chains, which strengthen anchoring. This dual role explains the Newtonian flow, high wet tack, and adhesion typical of PVOH-stabilized latexes.

7.6.2 Auxiliary Emulsifiers

Nonionic emulsifiers can extend stability under freeze–thaw or electrolyte stress, though their impact on particle size is modest. Anionic surfactants generate smaller, more uniform particles but increase radical flux and grafting, which may cause a viscosity rise. Blends of nonionic/anionic species are used selectively.

7.6.3 Temperature and Viscosity

Raising the temperature accelerates initiator decomposition, increases radical flux, and decreases particle size. The higher surface area demands more stabilizer and increases hydrodynamic volume through grafted PVOH, leading to higher viscosity. Excessive temperature risks flocculation or partial crosslinking.

7.6.4 Effect of Agitation

Agitation requirements in vinyl acetate emulsion polymerization are stage-dependent. During nucleation (initial charge and hold), a high agitation speed of approximately 400 rpm is applied to ensure uniform distribution of vinyl acetate monomer, initiator, and poly(vinyl alcohol) (PVOH), leading to a consistent particle number and narrow size distribution. During the semibatch feeding stage, agitation is reduced to an assertive level near 250 rpm so that the incoming monomer or preemulsion disperses immediately on entry and temperature remains uniform; excessive speeds at this stage risk foaming and partial stripping of the protective PVOH layer. As viscosity rises toward the end of feed and conversion, the agitation speed is held steady or slightly reduced. In this viscous regime, impeller geometry rather than additional speed becomes the dominant factor for maintaining bulk circulation and continual renewal of the cooled reactor wall. For reactors that produce both low- and high-viscosity grades, broad-spectrum impellers are preferred. Suitable options include anchor or helical-ribbon designs and polymerization-duty multipurpose viscous impellers (e.g., EKATO PARAVISC), which sustain circulation and heat transfer at moderate rotational speeds.

7.6.5 Reactor Hydrodynamics

Impeller and baffle design should prevent dead zones and excess shear at the liquid–air interface, where foaming and oxygen ingress threaten stability. Practical measures include controlled antifoam dosing, effective venting, and reliable clean-in-place routines.

7.6.6 Perspective

PVAc homopolymer stability is the outcome of synergy: PVOH anchoring, controlled temperature, staged agitation, and well-designed hydrodynamics. Together, these factors produce latexes with reproducible particle size, stable viscosity, and strong end-use performance.

7.6.7 Key Quality Parameters and Preventive Actions in PVAc Homopolymerization

The performance and stability of PVAc latexes depend not only on the intrinsic chemistry of vinyl acetate polymerization but also on the way operating parameters are controlled in practice. Even under semicontinuous conditions, variations in particle nucleation, temperature control, or agitation can lead to bimodal particle distributions, excessive viscosity growth, or early coagulation.

To ensure reproducibility at scale, it is essential to identify the most sensitive quality metrics and the preventive actions that mitigate risk. These include careful adjustment of solids content and stabilizer dosage, tight control of the seeding and feed protocols, and moderation of agitation and pH conditions during nucleation.

Table 7.2 summarizes the key quality parameters in PVAc homopolymerization, their measurement or root causes, and the corresponding preventive actions

Table 7.2 Key quality parameters and preventive actions in PVAc homopolymerization.

Metric/risk	Measurement or cause	Preventive action
Total particle surface area	Calculated from particle size distribution and solids content	Monitor routinely; adjust solids and stabilizer dosage to maintain stable conversion
Bimodal particle distribution	Arises from dual loci (early seeds vs. secondary particles)	Tighten seeding protocol; optimize feed window for unimodal growth
Early coagulation	Caused by high stirring, extreme T, or poor pH control	Use conservative agitation during nucleation; buffer pH; validate temperature profile

applied in industrial practice. This structured overview provides process engineers with a direct link between laboratory observations and plant-scale operational safeguards.

7.6.8 Operating Levers in PVAc Homopolymerization (Semicontinuous)

The semicontinuous process provides operators with a wide range of control points that directly influence conversion efficiency, particle size, viscosity, and final film properties. Unlike batch processes, where all parameters unfold passively once the reaction begins, the semicontinuous mode allows each domain—from monomer purity to agitation and feed profile—to be adjusted proactively.

A systematic view of these domains highlights how purity, thermal management, seeding protocols, growth kinetics, stabilization strategies, and feed tuning act as "operating levers" that determine whether a process remains robust or veers toward instability. For industrial practice, recognizing these levers is essential for scale-up and for consistent quality across different reactor capacities.

Table 7.3 summarizes the major domains in PVAc homopolymerization under semicontinuous operation. Each domain is linked to its key considerations and the practical levers typically used in manufacturing, providing a structured reference for process engineers.

7.6.9 Guidelines for Controlling Properties

The following practical guidelines for adjusting particle size, viscosity, and molecular weight in vinyl acetate emulsion polymerization are adapted from the authoritative work of Dr. Yildirim H. Erbil (2000). These principles provide a concise framework for process engineers to fine-tune latex characteristics under semicontinuous industrial conditions (Table 7.4).

Table 7.3 Major domains in PVAc homopolymerization under semicontinuous operation.

Domain	Key considerations	Practical levers
Monomer and inhibitors	Trace acetaldehyde lowers MW; hydroquinone-stabilized VAc is usable	Certify purity; adjust target MW for expected chain transfer
Thermal and pH	Strongly exothermic; hydrolysis risk outside pH window	Cooling capacity; buffer to pH ≈ 4.5–6.5; in-process temperature control
Initiation and seeding	Homogeneous nucleation; PVOH-stabilized seeds	Initiator dosing profile; PVOH grade/level
Growth and kinetics	Rate scales with particle number/surface area	Control nucleation window; manage particle population
Stabilization	PVOH primary; auxiliaries optional	Nonionic/anionic blends as needed; watch grafting/viscosity
Temperature and rheology	Higher T: smaller particles, more sulfate ends, higher viscosity	Optimize T and initiator; avoid excessive grafting/bridging
Mixing and hydrodynamics	Shear/foam can destabilize; design matters	Impeller/baffle optimization; antifoam; controlled agitation
Feed strategy	Starved-feed improves control; poor tuning broadens PSD	Match feed to radical kinetics; monitor residual monomer
Additives	Plasticizers/glycols tune film and freeze–thaw	Dose for end use; verify compatibility

Table 7.4 Summary of guidelines for controlling PVAc latex properties.

Property	To increase	To decrease
Particle size	Decrease emulsifier/colloid concentration Reduce initiator concentration and feed rate Lower polymerization temperature	Increase emulsifier/colloid concentration Increase initiator concentration Slow monomer feed rate Increase polymerization temperature Increase monomer-to-water ratio
Viscosity	Decrease particle size Increase emulsifier/colloid concentration Increase solids content Raise polymerization temperature	Opposite of the listed factors
Molecular weight	Avoid chain transfer agents Lower temperature Use redox initiators at low T Lower initiator concentration Increase monomer precharge (15–20%) Increase monomer feed rate	Add chain transfer agents Increase initiator concentration Increase polymerization temperature

Molecular weight control by initiator level, emulsifier concentration, or agitation speed is usually impractical in industrial settings. In the absence of chain transfer agents, vinyl acetate polymerization tends to yield high molecular weight polymers with significant branching.

7.7 Copolymerization of Vinyl Acetate

The copolymerization of vinyl acetate represents the logical extension of homopolymerization, enabling the design of binders and adhesives with properties that cannot be achieved by PVAc alone. While PVAc homopolymers provide excellent film formation and strong adhesion to porous substrates, they remain limited in water resistance, alkali stability, and long-term durability. By introducing comonomers with tailored polarity, steric effects, or reactive groups, these deficiencies can be systematically addressed.

A wide range of comonomers has been employed in practice. Acrylic esters lower the glass-transition temperature and impart flexibility; ethylene reduces brittleness and enhances adhesion to flexible substrates; vinyl versatate (VeoVa) introduces bulky, hydrophobic units that improve alkali and water resistance; and functional monomers such as acrylic acid or methacrylic acid create carboxyl sites that allow for further crosslinking. The particular combination and feed strategy define the copolymer's sequence distribution, particle morphology, and, ultimately, its application profile.

From a technological perspective, copolymerization expands the scope of vinyl acetate emulsions from paper and woodworking adhesives into high-performance applications such as flexible packaging, construction adhesives, coatings, and paints. This section, therefore, examines the mechanistic considerations, experimental observations, and industrial implications of copolymerizing vinyl acetate, providing a framework for understanding how each comonomer type modifies performance.

7.7.1 Emulsion Polymerization Context

As in the homopolymerization of vinyl acetate, the emulsion polymerization process plays a decisive role in shaping the outcome of copolymerization. Surfactants are typically used at concentrations above their critical micelle concentration (CMC), where they form micelles capable of solubilizing monomers. Before polymerization begins, large VAM droplets also exist in the aqueous medium; these droplets, stabilized by surfactant adsorption, act as monomer reservoirs. A smaller fraction of monomer partitions into micelles, where it is solubilized within the hydrophobic core.

It is important to distinguish between the two roles of surfactants. Emulsification stabilizes larger VAM droplets in water, preserving a dispersed two-phase system, while solubilization incorporates monomer into micelles at the molecular level, producing a one-phase environment for initiation. On the macroscopic scale, surfactants appear as droplet stabilizers; on the microscopic scale, they enable micellar solubilization.

When surfactant concentration falls below the CMC, micellar nucleation cannot account for particle formation. In such cases, homogeneous nucleation in the aqueous phase becomes dominant. Initiator radicals, generated by thermal decomposition, propagate with dissolved monomer until they form oligomeric radicals of sufficient chain length to precipitate as nuclei. These unstable nuclei undergo limited flocculation, which is arrested once surfactant molecules adsorb to their surfaces. Surfactants may originate from the aqueous phase or desorb from VAM droplets.

To further reinforce colloidal stability, formulators often introduce small amounts of functional monomers such as acrylic acid or methacrylic acid. These comonomers provide ionic stabilization, improve particle integrity during growth, and create reactive sites for subsequent crosslinking (Table 7.5). Table 7.5 compares the performance characteristics of PVAc homopolymers and VAc copolymers in VAM systems across key application and formulation parameters.

Table 7.5 Homopolymer vs. copolymer performance in VAM systems.

Aspect	PVAc homopolymer	VAc copolymers (acrylics, VeoVa, ethylene, AA, etc.)
Film formation	Excellent film-forming ability; low MFFT	Maintains film formation; adjustable via comonomer Tg
Adhesion	Strong to porous substrates (paper, wood)	Extended adhesion to plastics, metals, and coated surfaces
Water resistance	Poor; films swell or whiten in water	Greatly improved with hydrophobic comonomers (acrylates, VeoVa)
Alkali resistance	Low; sensitive to cementitious substrates	Enhanced through hydrophobic or bulky comonomers
Flexibility	Brittle at low Tg	Adjustable flexibility via soft acrylates or ethylene
Viscosity behavior	Newtonian; governed by PVOH stabilization	Wider range, influenced by surfactants and comonomer selection
Applications	Paper adhesives, woodworking, general-purpose binders	Flexible packaging, construction adhesives, paints, industrial coatings

7.8 Types of Emulsion Polymerization Processes

Although all emulsion polymerizations share the same fundamental intervals of particle nucleation, growth, and termination, the way these intervals are orchestrated in practice defines three distinct process types: batch, semicontinuous, and continuous polymerization. Each approach has been applied to vinyl acetate, but their industrial significance differs greatly. Understanding these modes is essential for appreciating why semicontinuous operation has become the industry standard for PVAc and its copolymers.

7.8.1 Semibatch (Semicontinuous) Process

The semicontinuous process combines the simplicity of batch charging with the control benefits of continuous feeding. In this approach, water, stabilizer, and a small initial charge of vinyl acetate monomer (VAM) and initiator are introduced first. The remaining monomer and initiator are then fed gradually over several hours, usually in the form of a preemulsion.

This gradual feeding strategy prevents uncontrolled exotherms, reduces the risk of gelation, and enables a narrower particle-size distribution compared to batch operation. By avoiding high instantaneous monomer concentrations, secondary nucleation is minimized, while heat evolution is spread over time and easily dissipated. The result is a latex of higher solids content, improved stability, and reproducible performance.

Another advantage is formulation flexibility. The delayed-feed mode allows comonomers, functional monomers (e.g., acrylic acid), or chain transfer agents to be introduced at specific stages, tailoring particle morphology and polymer architecture. Semicontinuous processes also adapt well to modern reactor control systems, ensuring consistent product quality and safe operation at commercial scales.

For these reasons, the semicontinuous process has become the dominant mode in the manufacture of PVAc homopolymers and copolymers. It strikes a practical balance between safety, control, and product quality, bridging the limitations of batch processing with the complexities of continuous operation.

7.8.2 Batch Process

In batch polymerization, all components are charged at once, and polymerization proceeds through the three intervals in an uncontrolled sequence. This often results in poor heat management, broader particle size distributions, and instability at high solids. As such, batch processes are now primarily used in laboratories for mechanistic studies rather than in commercial PVAc production.

7.8.3 Continuous Process

In continuous polymerization, reactants are fed steadily while product latex is withdrawn. Although this mode offers efficiency and high throughput, it is technically demanding. Maintaining constant particle size and preventing fouling over extended operation is challenging, which limits its use in PVAc production to specialized cases.

7.8.3.1 Comparison of the Three Emulsion Polymerization Processes

The three main approaches to emulsion polymerization—batch, semicontinuous, and continuous—share the same fundamental reaction intervals but differ greatly in how those intervals are managed in practice. These differences strongly influence process control, scalability, safety, and industrial applicability.

The batch process is simple and flexible, but it suffers from heat control and scalability issues. The semicontinuous process, now the industry standard for PVAc latex production, offers an excellent balance between control, stability, and product quality. Continuous processes, while highly efficient in theory, are technically demanding and therefore reserved for specific large-scale operations.

Table 7.6 provides a direct comparison of the three approaches.

7.8.4 Reaction Intervals in Semicontinuous Polymerization

The three-interval model—nucleation, particle growth, and monomer depletion—has long provided a useful framework for understanding emulsion polymerization. In the semicontinuous process, however, these intervals take on particular importance because they can be actively shaped by feed strategy. Unlike batch systems, where all monomer is present from the outset, and intervals follow one

Table 7.6 Summary of emulsion polymerization processes.

Criteria	Batch process	Semicontinuous process	Continuous process
Process flexibility	High	Moderate to high	Low
Process control	Moderate	High	Very high
Monomer addition mode	All at once	Gradual/Delayed	Constant
Heat control	Low	Moderate	High
Conversion per batch	Yes	Sometimes	Rarely
Scale-up flexibility	Limited	Excellent	Good (if steady-state)
Industrial adoption	Limited	Most common	Specific high-volume uses

another passively, semicontinuous operation allows the formulator to prolong or compress each stage. This flexibility translates directly into safer heat control, narrower particle-size distributions, higher solids stability, and reproducible product performance.

For vinyl acetate polymerization, this distinction is decisive. PVAc homopolymers and copolymers are highly exothermic and sensitive to stabilization limits. By metering monomer addition, operators prevent uncontrolled Interval II flooding and instead maintain a starved-feed regime that delivers the desired latex characteristics. In this sense, the semicontinuous process does not merely reproduce the classical three intervals—it transforms them into tools for process control.

Interval I: Micellar Nucleation (Particle Birth). Radicals generated in the aqueous phase react with dissolved vinyl acetate (~2.5% solubility in water). Oligomeric radicals precipitate as primary nuclei, which are rapidly stabilized by PVOH. Only a fraction of the initially charged monomer is present, so the number of nuclei is limited and controllable. The stabilizer concentration and seeding recipe determine the number and size of particles formed in this stage.

Interval II: Particle Growth (Reservoir-Controlled Growth). Once nuclei are established, monomer from the dispersed droplets diffuses steadily into growing particles. In a semicontinuous process, the rate of feed addition dictates whether conditions are monomer-flooded (feed > polymerization rate; droplets accumulate) or monomer-starved (feed < polymerization rate; particles remain unsaturated). Monomer-flooded conditions mimic classical Interval II, whereas monomer-starved conditions provide tighter control over temperature, molecular weight, and particle morphology. Industrial PVAc processes typically favor the starved approach.

Interval III: Monomer Depletion (Conversion and Finishing). As feed addition ceases and droplet reservoirs are exhausted, the reaction enters Interval III. Polymerization continues within the swollen particles using residual dissolved monomer. Conversion slows, viscosity rises, and final latex properties are determined. A chasing step (extra initiator) is often employed to reduce residual monomer content to below regulatory limits.

The ability to shape each interval through feed strategy is what makes the semicontinuous process so versatile. By extending nucleation (Interval I), controlling growth (Interval II), and carefully finishing (Interval III), producers achieve narrow particle size distributions, stable high-solids dispersions, and reproducible product performance (Figure 7.2).

7.8.5 Effects of Various Parameters in Copolymerization

The semicontinuous copolymerization of vinyl acetate with comonomers such as acrylates, VeoVa, or ethylene is influenced by the same basic variables that

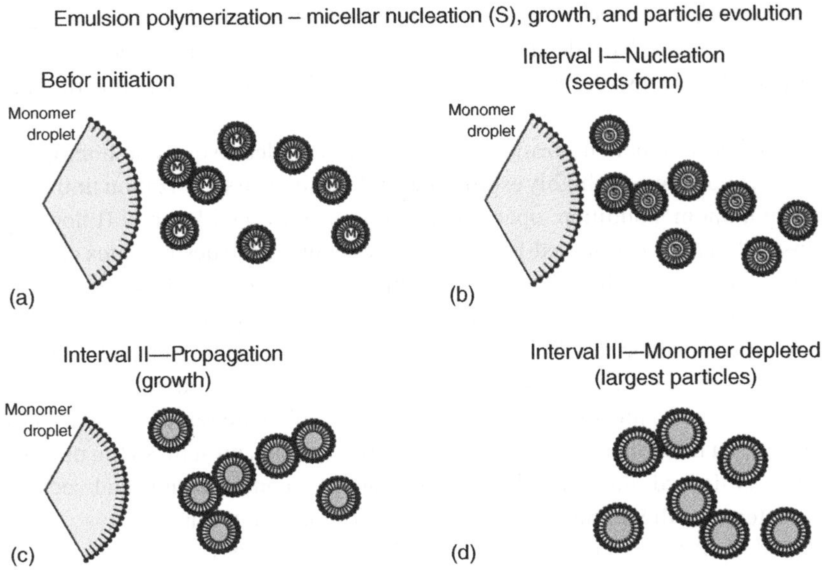

Figure 7.2 Emulsion polymerization via micellar nucleation and growth. (a) Before initiation: surfactant-stabilized micelles contain monomer (M). (b) Interval I–Nucleation: radicals form polymer seeds (S) inside micelles. (c) Interval II–Propagation: polymer cores enlarge as monomer diffuses from the droplet; micelles expand accordingly. (d) Interval III–Monomer Depleted: droplets are gone, and particles reach their largest size. Surfactant "pins" are shown with heads outside the interface and tails inward; pin length is constant across panels. M, monomer; S, seed. Not to scale.

govern homopolymerization, but their impact is magnified by the presence of hydrophobic or functional monomers. These comonomers introduce additional requirements for stabilization, heat removal, and conversion efficiency. Hydroxyethyl cellulose (HEC) is typically used as a protective colloid and thickener in these systems, often in combination with surfactants, since pure PVOH is insufficient to stabilize copolymer latexes containing large hydrophobic units.

Surfactants: In copolymerization, surfactants play a dominant role in stabilization. While homopolymer PVAc relies primarily on PVOH, copolymer latexes require blends of anionic and nonionic surfactants to manage the poor solubility of acrylates or VeoVa. HEC provides a steric contribution but must be supplemented by surfactants to prevent phase separation.

Monomer-to-water ratio: Hydrophobic comonomers have very low water solubility, unlike VAM. As a result, copolymerizations are more dependent on micellar nucleation. High monomer-to-water ratios increase droplet size and raise the risk of secondary nucleation if the feed is not carefully controlled.

Temperature: Although vinyl acetate has the highest heat of polymerization, copolymerizations with acrylates or VeoVa present greater challenges in heat management. Their low aqueous solubility and high propagation rates concentrate heat release in the particle phase, making exotherms sharper and more difficult to dissipate. Strict starved-feed regimes are therefore essential.

Agitation: Agitation must balance the dispersion of VAM, which is moderately soluble, with that of hydrophobic comonomers, which exist only as stabilized droplets. Effective mixing prevents monomer-rich zones and ensures uniform incorporation of all comonomers. Broad-spectrum impellers are again required to maintain circulation from low to high viscosity. Use the following rpm targets as standalone guidance for semicontinuous PVAc runs:

1) *Nucleation (initial charge and hold)*: High rpm ≈ 400.
   Purpose: Ensure uniform distribution of VAM, initiator, and HEC; clean start-up.
   Watch-outs: Foam surge, stripping of HEC if exceeded.
2) *Feed/growth stage*: Assertive rpm ≈ 250.
   Purpose: Disperse incoming VAM/preemulsion instantly; maintain temperature uniformity.
   Watch-outs: Too low speed leads to monomer-rich pockets and local heating. Too high speed causes foaming and destabilization of the colloid layer.
3) *Late stage (viscosity build-up)*: Keep rpm in the assertive range or slightly reduced. Agitator design, not higher speed, ensures bulk circulation and renewal of the internal jacket surface.

Conversion: Residual monomer control is more challenging in copolymerization, especially with bulky or hydrophobic species. Chasing steps with additional initiator or redox systems are often required to reduce residual levels to below regulatory limits.

Taken together, these factors highlight that while the same operating levers apply to both homopolymerization and copolymerization, the latter demands stricter process control and more sophisticated stabilization strategies. HEC, in synergy with optimized surfactant systems, plays a central role in ensuring colloidal stability and process reproducibility.

7.8.6 Comparison Between Homopolymerization and Copolymerization Processes

A direct comparison between PVAc homopolymerization and copolymerization clarifies why the latter requires different stabilization and control strategies. Table 7.7 summarizes the key differences across major process parameters and industrial outcomes.

Table 7.7 Comparison between homopolymerization and copolymerization processes.

Criteria	Homopolymerization (PVAc)	Copolymerization (VAc with acrylates, VeoVa, ethylene, etc.)
Stabilization	Primarily, PVOH provides steric stabilization	HEC with anionic/nonionic surfactants; PVOH insufficient
Nucleation mechanism	Largely homogeneous nucleation; some dispersed monomer droplets contribution	Hybrid nucleation (micellar + droplet); hydrophobic comonomers dominate micelle use
Heat control	Moderate exotherm, manageable with semicontinuous feed	Higher exotherm; strict starved-feed regimes essential
Agitation	Ensures VAM dispersion; less demanding at high viscosity	Critical to disperse VAM and hydrophobic comonomers simultaneously; broad-spectrum impellers required
Conversion/finishing	Chasing step reduces residual VAM to acceptable levels	Residual monomer control is more difficult; redox chasers often used in addition to vacuum stripping
Applications	Paper, wood adhesives, general-purpose binders	Exterior paints, flexible packaging adhesives, pressure-sensitive adhesives, construction coatings

7.8.7 Industrial Outcomes

Semicontinuous copolymerization of vinyl acetate has enabled the development of a broad family of dispersions.

1) VAc–BA copolymers for flexible coatings and pressure-sensitive adhesives.
2) VAc–VeoVa copolymers for exterior paints with improved alkali resistance and weatherability.
3) VAc–ethylene (VAE) copolymers, produced under pressure, for adhesives and construction binders with excellent toughness and water resistance.

Across all of these systems, the combination of HEC stabilization, surfactant micelles, and starved-feed addition makes the semicontinuous process indispensable. Without these features, stable latexes with the desired performance profile could not be achieved at a commercial scale.

These two effects—smaller particles with larger total surface area and thicker, covalently anchored HEC layers—combine to increase latex viscosity at higher polymerization temperatures. The apparent hydrodynamic volume rises due to stronger interparticle interactions and, at excessive grafting levels, possible bridging. Practically, modest grafting can improve wet tack and creep resistance, whereas excessive grafting may raise viscosity beyond handling limits or increase gel content.

7.9 Electrostatic Stabilization and Zeta Potential in Emulsion Polymers

To stabilize the polymer particles in the continuous aqueous phase, the particles must be sheathed by a transitional layer that prevents them from agglomerating. Two stabilization mechanisms can be distinguished:

a) *Electrostatic stabilization*: This stabilization arises from ionic groups with negative charges present at the surface of the polymer particles, which repel each other. Such electrostatic repulsion prevents the particles from coagulating. These negative ionic groups can originate from:
   1) Anionic surfactants (e.g., alkyl or aryl sulfates, sulfonates)
   2) Anionic end-groups (e.g., sulfates from persulfate initiators or anion-containing monomers such as carboxylic acids)

   The density of the negative charge on the polymer particles significantly increases their stability. For example, an anionic surfactant like alkylated di-sulfonated diphenyl oxide, containing two sulfonate groups, can double the charge density, offering excellent emulsion stability.
b) *Steric stabilization*: Steric stabilization arises from bulky groups at the surface of the polymer particles. These groups are typically:
   1) Polymeric surfactants
   2) Protective colloids such as hydroxyethyl cellulose (HEC) or polyvinyl alcohol (PVOH)

   In most latex systems, stabilization is achieved by a combination of both mechanisms. This is implemented by using a mixture of:
   1) Anionic and nonionic surfactants
   2) Hydrophilic polymers (e.g., HEC or PVOH) as protective colloids

   This dual mechanism ensures robust colloidal stability throughout the polymerization process and improves long-term stability of the latex.

7.10 Mechanism of Colloidal Stability of Emulsion Polymers Including PVAc Homopolymer

7.10.1 Electric Fields and Charge Interactions

Electric fields are created by electric charges, which come in two types: positive (+) and negative (−). Like charges repel each other, while opposite charges attract. These interactions produce forces that either push or pull other charges. Neutral objects, having no net charge, are not affected by electric fields.

The force between two charges is described by Coulomb's Law, which states that the magnitude of the force is directly proportional to the product of the two charges and inversely proportional to the square of the distance between them.

By convention, electric field lines are always drawn from positive to negative. These field lines represent a normalized force–force per unit charge—independent of any specific second charge. Thus, the electric field at any point in space specifies the force that would act on a hypothetical unit charge placed at that location.

7.10.2 Electric Potential and Colloidal Interactions

Electric potential, or voltage, is defined as the difference in electric potential energy per unit charge between two points in an electric field. It reflects the work required to move a charge against the electric field.

1) If a positive particle is moved farther away from a negative particle, more energy is required, and its electric potential energy increases.
2) Doubling the negative or positive charge increases the energy needed to move the opposing charge.

7.10.3 Implication for Colloidal Stability

Increasing the negative surface charge on polymer particles (e.g., by using a surfactant with multiple negative moieties) raises the electric potential. This intensifies the repulsive forces between particles, enhancing colloidal stability.

Conversely, adding positive electrolytes to the medium increases the attractive force to the negatively charged particles. This reduces the electric potential, weakens the repulsive forces, and promotes coalescence and coagulation.

7.10.4 Electric Double Layer and Electrostatic Stabilization

To prevent aggregation, colloidal stability is achieved by balancing attractive and repulsive forces. Negative surface charges on polymer particles influence ion distribution in the aqueous medium. Oppositely charged counter-ions (e.g., ammonium or potassium) are attracted to the surface, while co-ions (e.g., bicarbonate or acetate) are repelled. This results in an electric double layer, composed of:

1) A negatively charged polymer surface
2) A surrounding layer of neutralizing counter-ions (diffuse and Stern layers)

This unequal charge distribution creates an electric potential across the interface. Because of the net negative charge on the particle and the neutralizing positive charges in the solution, an electric field forms, resisting particle coagulation.

The double layer resembles a dipole—opposing charges separated by a distance. The stronger the field (i.e., higher charge density), the more energy is required for particles to overcome repulsion.

Adding anionic surfactants increases the negative surface charge, raising the potential barrier. Conversely, adding electrolytes (e.g., NaCl) compresses the electric double layer, lowers the potential, and increases coagulation risk.

7.10.5 Structure of the Electric Double Layer

1) *Stern layer*: Closest to the particle surface, composed of tightly bound counter-ions. It is immobile and moves with the particle.
2) *Diffuse layer*: Surrounds the Stern layer, with loosely bound counter-ions. It extends farther from the surface but still contributes to stabilization.

The Slipping Plane is the boundary of the liquid moving with the particle. The electric potential at this plane is called the Zeta Potential.

7.11 Zeta Potential and Colloidal Stability

Zeta potential (in mV) represents the potential difference between the dispersion medium and the charged particle layer. It is directly related to colloidal stability. Higher absolute values of zeta potential indicate stronger repulsion and greater stability.

7.11.1 Approximate Guide to Zeta Potential and Stability

Zeta potential (ζ) provides a convenient empirical indicator of colloidal stability in aqueous dispersions. Although the exact thresholds vary with system composition, ionic strength, and stabilizer type, approximate ranges have been widely reported in the colloid science literature. The magnitude of ζ reflects the strength of electrostatic repulsion between particles: low values correspond to coagulation risk, while high values indicate long-term stability.

Table 7.8 Approximate guide to zeta potential and stability in aqueous dispersions.

Zeta potential (mV)	Stability behavior
0 to ±5	Coagulation (unstable)
±10 to ±25	Incipient instability
±25 to ±40	Moderate stability
±40 to ±60	Good stability
±61 and greater	Excellent stability

Table 7.8 summarizes the approximate ranges of ζ-potential and their corresponding stability behaviors for emulsion polymers, including PVAc latexes.

As the zeta potential approaches zero, the activation energy for particle approach also approaches zero. Attractive interactions dominate, causing coagulation and formation of clumps (coagulum).

Zeta potential is a practical and predictive tool in formulation and quality control. For example:

1) Higher zeta potential = more stable dispersion
2) formulation Electrolyte addition (e.g., NaCl) = compresses the electric double layer, reduces repulsion, and facilitates aggregation

The measurement of zeta potential can be used routinely to control batch-to-batch consistency and optimize dispersion stability in design.

7.11.2 Zeta Potential and Emulsion Stability

The stability of emulsion polymers is strongly governed by electrostatic interactions at the particle–water interface. When particles acquire surface charge, counter-ions in the aqueous phase arrange themselves into a structured electrical double layer. The magnitude of the electrostatic potential at the slipping plane, referred to as the zeta potential (ζ), provides a practical measure of the repulsive forces between particles.

A high absolute value of ζ generally indicates strong electrostatic repulsion and resistance to flocculation, whereas a low value suggests weak repulsion and a tendency toward aggregation or coagulation. Ionic strength exerts a decisive influence: in low-salt systems, the diffuse double layer is extended and electrostatic repulsion is strong, while in high-salt systems, the diffuse layer is compressed, reducing $|\zeta|$ and thereby weakening stability.

Figure 7.3 schematically illustrates the electrical double layer and zeta potential under conditions of low and high ionic strength. Under low ionic strength, the

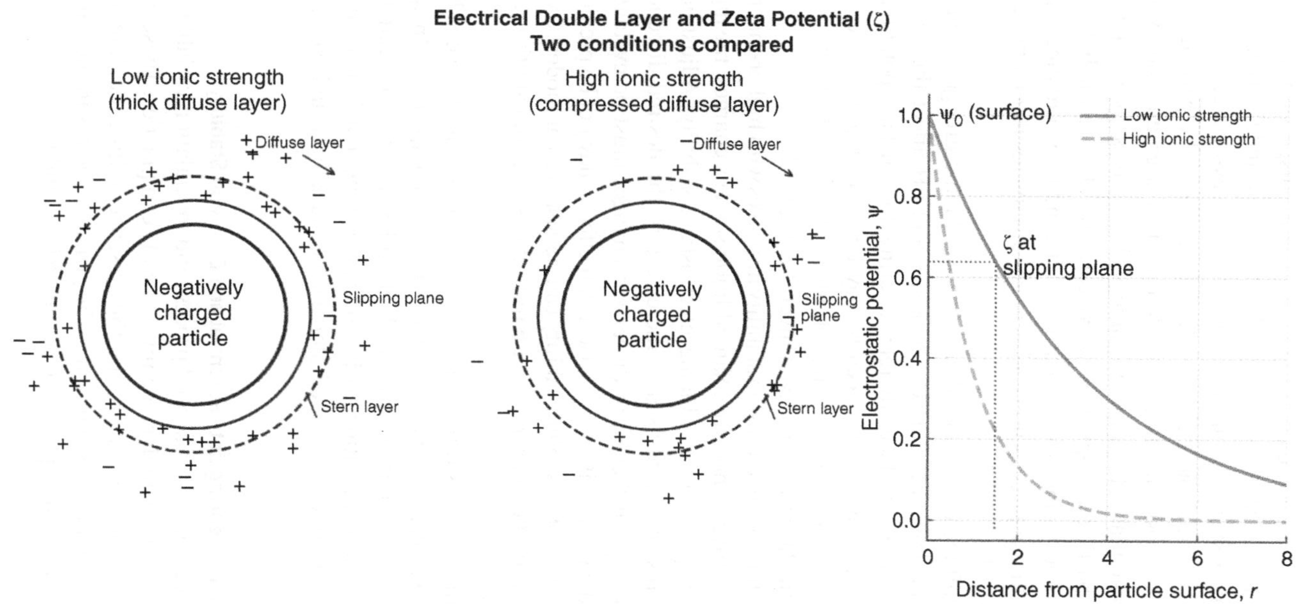

Figure 7.3 Electrical double layer and zeta potential under low and high ionic strength conditions. Schematic representation of the electrical double layer and zeta potential under low and high ionic strength conditions. The diagram shows the potential decay (ψ) with distance from the particle surface, indicating the Stern layer, diffuse layer, and slipping plane (ζ).

diffuse layer is relatively thick, and the zeta potential has a large magnitude, leading to strong repulsive forces that maintain particle dispersion. Under high ionic strength, the diffuse layer is compressed, ζ decreases, and the risk of flocculation increases unless steric stabilizers are present. The associated curve shows how the electrical potential (ψ) decays with distance from the particle surface: a steep drop occurs across the Stern layer, followed by a gradual decay in the diffuse layer. The potential at the slipping plane is identified as the zeta potential, while far from the particle ψ approaches zero, corresponding to the bulk liquid. A comparison of curves demonstrates that low-salt conditions produce a slower decay and thicker diffuse layer, while high-salt conditions result in a more rapid decay and a smaller |ζ|.

Several factors shift the curve or modify ζ. Increased ionic strength compresses the double layer and lowers ζ. Surface chemistry and pH affect the density of ionizable groups at the interface, thereby altering the surface potential and ζ. Adsorption of surfactants or dispersants may either increase or decrease ζ, depending on their charge characteristics. Multivalent ions such as Ca^{2+} or Al^{3+} strongly compress the layer and may even invert the sign of ζ.

From a practical standpoint, emulsion stability improves with higher-magnitude ζ values, whether positive or negative. Salt addition screens charges, raising conductivity but lowering |ζ|, thus reducing electrostatic stability. Adjustment of pH or surface chemistry can be used to increase ζ in the desired direction for a given latex chemistry. Surfactant or dispersant changes must be evaluated by re-measuring ζ and conductivity, since adsorption effects may not be predictable. In sterically stabilized systems, high stability can persist even at modest ζ values, and measurements must be interpreted with respect to steric contributions, solids content, and viscosity.

In practice, drifting ζ values during storage may indicate salt uptake, addition of biocides or coalescents, or depletion of surface-active components. Instability occurring at unchanged ζ values may instead reflect loss of steric protection, particle size growth, or contamination with multivalent ions. For reliable control, measurements of ζ are best considered alongside conductivity and particle size analysis.

7.11.2.1 Factors in Influencing Zeta Potential and Emulsion Stability

Beyond the magnitude of ζ itself, several process and formulation variables determine how stable a PVAc or copolymer emulsion will be in practice. These include ionic strength, pH, surfactant system, and the presence of functional monomers or protective colloids. Table 7.9 summarizes the principal factors influencing zeta potential and their practical implications for emulsion stability.

Table 7.9 Practical guidelines for interpreting and controlling zeta potential in emulsion polymers.

Factor	Effect on ζ and stability	Practical guidance
Magnitude of ζ	High |ζ| (positive or negative) indicates strong electrostatic repulsion and stable dispersions. Values near zero signal risk of flocculation	Maintain |ζ| sufficiently far from zero to ensure stability
Ionic strength (salt concentration)	Increased ionic strength compresses the diffuse layer, reducing |ζ|	Limit salt uptake during formulation and storage; monitor conductivity as an indirect indicator
pH and surface chemistry	Alter ionizable groups at the particle surface, shifting ζ more positive or negative	Adjust pH to the range that maximizes |ζ| for the specific latex chemistry
Surfactant and dispersant adsorption	Can increase or decrease surface charge depending on surfactant type	Reassess ζ after any change in surfactant grade or supplier
Multivalent ions (Ca^{2+}, Al^{3+})	Strongly compress the double layer; may reduce or invert ζ	Control water quality and fillers to avoid unintended multivalent ion contamination
Steric stabilizers (nonionic polymers, protective colloids)	Provide stability independent of ζ magnitude	Interpret ζ values in combination with steric contributions, solids level, and viscosity

8

PVAc-based Adhesives—From Homopolymers to Crosslinkable Copolymers Unlocking New Applications

8.1 Introduction

Polyvinyl acetate (PVAc) is one of the most important waterborne adhesive polymers, widely used in woodworking, packaging, paper converting, and textile applications. Its industrial success stems from low raw material cost, safe handling, strong adhesion to porous substrates, and reliable film-forming properties.

Two distinct families of PVAc-based adhesives exist. The first comprises homopolymer adhesives, often referred to as "white glues" or "wood glues," which dominate in applications such as furniture assembly, carton sealing, and general bonding of cellulosic materials. The second comprises copolymer adhesives, where vinyl acetate is polymerized with comonomers such as ethylene, acrylates, or vinyl versatate to overcome inherent weaknesses of the homopolymer, including susceptibility to creep, limited water resistance, and poor adhesion to nonporous substrates.

8.2 Homopolymer PVAc Adhesives

PVAc homopolymer adhesives represent the archetype of waterborne emulsion adhesives and are widely recognized in both industrial and consumer markets. They are often referred to as "white glues" or "wood glues," reflecting their dominance in furniture assembly, woodworking, carton sealing, paper converting, and school adhesives.

The performance of homopolymer PVAc adhesives is closely tied to the inherent characteristics of the polymer. The polymer provides excellent film-forming ability, rapid setting, and strong adhesion to porous substrates such as wood, paper, and textiles. These qualities, combined with safe handling and cost efficiency, have established PVAc homopolymers as one of the most commonly used adhesive systems worldwide.

However, PVAc homopolymers also display inherent weaknesses. Their hydrophilic character leads to limited resistance against water and moisture, while their relatively low glass transition temperature results in creep under sustained load. These limitations restrict their use in outdoor or high-stress structural applications.

Despite these drawbacks, PVAc homopolymer adhesives remain indispensable in many industries. Their formulation flexibility allows for significant property adjustment through the addition of plasticizers, stabilizers, fillers, and crosslinking agents. These strategies extend their range of applications and ensure that homopolymer-based systems remain competitive against more complex copolymer adhesives.

8.3 Additives in Homopolymer PVAc Adhesives

The formulation of PVAc homopolymer adhesives depends not only on the polymer backbone but also on a carefully selected set of additives. These ingredients determine the adhesive's viscosity, film formation, flexibility, tack, water resistance, and long-term durability. The following subsections present the principal additive groups used in PVAc homopolymer formulations.

8.3.1 Plasticizers

Plasticizers are incorporated into PVAc adhesives to reduce the glass transition temperature (Tg) of the polymer, thereby improving flexibility, softness, and tack. They can be introduced internally during polymerization or externally by postaddition. Internal plasticizers are covalently bound to the polymer structure and are permanent, while external plasticizers are physically blended and may gradually migrate out of the film (Table 8.1). Plasticizer dosage rarely exceeds 10% on

Table 8.1 Plasticizers used in PVAc homopolymer adhesives.

Plasticizer	Characteristics	Effect on adhesive properties
Dibutyl phthalate (DBP)	Highly efficient, widely used	Improves softness and tack; may increase setting time
Di-isobutyl phthalate	Lower cost, lower efficiency	Provides flexibility with reduced performance
Benzyl butyl phthalate	Low volatility, tougher films	Enhances water resistance and durability

Plasticizer dosage rarely exceeds 10% on polymer solids, since higher levels may impair cohesive strength and extend setting times.

polymer solids, since higher levels may impair cohesive strength and extend setting times. The selection of plasticizer significantly influences film flexibility, tack development, and setting behavior in PVAc systems, as summarized in Table 8.1.

8.3.2 Coalescing Agents

Coalescing agents are low-volatility solvents that temporarily soften latex particles, allowing them to fuse into a continuous film during drying. This function is particularly important in cold or humid environments, where incomplete coalescence could lead to brittle, cracked adhesive layers (Table 8.2).

8.3.3 Stabilizers and Protective Colloids

Stabilizers are essential in maintaining emulsion stability during polymerization, storage, and application. They prevent coagulation of latex particles and improve adhesive performance. Polyvinyl alcohol (PVOH) is the most widely used protective colloid in PVAc homopolymer adhesives. Its degree of hydrolysis critically influences adhesive properties: partially hydrolyzed grades offer improved colloidal stability and tack, while fully hydrolyzed grades enhance water resistance but reduce flow and wetting (Table 8.3).

8.3.4 Surfactants

In PVAc homopolymer adhesives, surfactants are generally of secondary importance compared to polyvinyl alcohol (PVOH), which functions as the principal stabilizer and locus for polymerization. Unlike systems such as styrene–butadiene or acrylates, where micelle nucleation is dominant, PVAc polymerization occurs primarily within protective colloid domains provided by PVOH. This makes surfactants less critical for particle nucleation and growth.

Table 8.2 Coalescing agents in PVAc homopolymer adhesives.

Coalescing agent	Key features	Application benefits
Butyl glycol acetate	Moderately volatile	Improves film continuity in cold conditions
Diethylene glycol monoethyl ether	High boiling point, strong plasticization	Reduces brittleness; increases flexibility
Texanol™	Industry standard	Enhances adhesion to semiporous substrates; improves durability

Typical use levels are 1–2% based on solids, balancing effective film formation with VOC and odor restrictions.

Table 8.3 Stabilizers and protective colloids in PVAc homopolymer adhesives.

Stabilizer	Function	Effect on adhesive properties
Polyvinyl alcohol (PVOH)	Steric stabilization, film formation	Enhances tack and strength; the degree of hydrolysis governs water resistance
Hydroxyethyl cellulose (HEC)	Thickening, stabilization	Improves viscosity and brushability
Casein	Natural stabilizer, film modifier	Enhances adhesion to wood and paper; increases rigidity

Stabilizers must be selected in harmony with plasticizers and surfactants to avoid incompatibility and phase separation.

Table 8.4 Limited use of surfactants in PVAc homopolymer adhesives.

Surfactant type	Example	Role in homopolymer systems
Anionic	Alkyl sulfates, alkylbenzene sulfonates	Occasionally used in small amounts to assist emulsification
Nonionic	Ethoxylated alcohols	Enhance freeze–thaw stability and substrate wetting
PVOH (protective colloid)	Primary stabilizer	Governs particle formation and stability; reduces reliance on surfactants

Nevertheless, small quantities of surfactants may be introduced for specific purposes, including improving wetting of certain substrates, enhancing freeze–thaw stability, and assisting in viscosity control. Because surfactants can migrate during film formation and reduce water resistance, their use in PVAc homopolymer adhesives is generally limited. A careful balance is necessary to ensure that performance is enhanced without compromising durability (Table 8.4).

8.3.5 Fillers and Extenders

Fillers are widely used in PVAc adhesives for cost control, rheology adjustment, and property modification. Moderate filler loading can improve viscosity, reduce film shrinkage, and provide dimensional stability. However, excessive filler reduces cohesive strength and weakens adhesion (Table 8.5).

Table 8.5 Fillers in PVAc homopolymer adhesives.

Filler	Function	Effect on properties
Calcium carbonate	Cost reduction, rheology control	Improves viscosity, reduces shrinkage
Kaolin clay	Reinforcement, opacity	Enhances stiffness, reduces tack
Talc	Film reinforcement	Improves dimensional stability and heat resistance

8.4 Homopolymerization Formulations and Procedures by Application

The following subsections present representative formulations and procedures for PVAc homopolymer emulsions tailored to different adhesive applications. All formulations are expressed relative to 100 parts by weight of vinyl acetate monomer. Polyvinyl alcohol (PVOH) serves as the protective colloid during polymerization, while surfactant levels are minimized except in carpet-backing grades, where high penetration is required. Initiation may be thermal or redox, and postpolymerization chases are employed to reduce residual monomer.

In practical adhesive formulations, several additional components are often incorporated to achieve the required balance of film formation, storage stability, and durability. Coalescent agents (typically 2–5 wt% on polymer solids) are added postpolymerization not only to reduce the effective film-formation temperature but also to ensure film integrity, preventing cracking or incomplete coalescence during drying. Plasticizers (5–15 wt%) enhance flexibility and adhesion to porous substrates, though their dosage must be balanced to avoid excessive creep. Biocides and preservatives (0.1–0.3 wt%) are essential for storage stability and to prevent microbial degradation during use.

Polyvinyl alcohol (PVOH) solution is often postadded to PVAc adhesive formulations at levels of 1–5 wt%. In this context, it does not influence nucleation but instead functions as a protective colloid and film-forming aid. PVOH improves the wet tack and adhesion to polar substrates (such as paper and carton) by hydrogen bonding, enhances viscosity control, and provides additional stability against coagulation and sedimentation during storage. Its hydroxyl functionality also imparts improved water retention and brushing properties in certain end uses.

These additives, while not explicitly listed in every formulation table, represent the standard toolbox for tailoring PVAc homopolymer emulsions to specific end-use requirements.

8.4.1 Woodworking Grade (Interior; Filler-free)

This subsection presents an interior woodworking adhesive based on a PVAc homopolymer binder stabilized predominantly with PVOH, designed for clean, filler-free formulations. The target is D2-class applications where water resistance demands are moderate and fast set, high wet tack, and clear dry film are prioritized over humidity durability. Key levers are latex solids (≈50–55%), molecular weight (viscosity window matched to application method), and minimal free surfactant to protect adhesion and water sensitivity. Storage pH should be held at 4.5–5.5 (Table 8.6).

8.4.2 Paper and Packaging Grade (Fast-set)

Designed for paper converting and light packaging lines where rapid handling strength is critical, this PVAc homopolymer grade targets fast wet-tack

Table 8.6 Polymerization recipe and process parameters for PVAc homopolymer binder (woodworking, interior; filler-free).

Stream	Component	pbw (per 100 pbw VAc)	Purpose
A—Initial charge	Deionized water	55.0	Vehicle
	PVOH (10% solution, 88% hydrolysis)	50.0	Protective colloid (≈5 pbw solids)
	Sodium acetate	0.30	Buffer
B—Monomer feed	Vinyl acetate	100.0	Monomer
C—Initiator feed	Ammonium persulfate	0.30	Thermal initiator
	Deionized water	5.0	Solvent
E—Redox chase	Hydrogen peroxide (35%)	0.15	Oxidant
	Ascorbic acid	0.08	Reductant
	$FeSO_4 \cdot 7H_2O$ (0.1% solution)	0.50	Catalyst

Procedure: The PVOH solution and buffer are charged into the reactor, and the system is heated to 70 °C under nitrogen. Monomer and initiator feeds are introduced simultaneously over three to 4 h under starved conditions. A short hold is applied until residual monomer stabilizes, after which the latex is cooled to 35–40 °C and subjected to a redox chase. The final latex typically exhibits 54–56% solids and a viscosity in the range of 1,500–3,000 cP, suitable for interior wood adhesives.

Table 8.7 Polymerization recipe and process parameters for PVAc homopolymer binder (paper and packaging, fast-set).

Stream	Component	pbw (per 100 pbw VAc)	Purpose
A—Initial charge	Deionized water	65.0	Vehicle
	PVOH (10% solution)	40.0	Protective colloid (≈4 pbw solids)
	Sodium acetate	0.25	Buffer
	Nonionic surfactant	0.15	Improves wetting on coated paper
B—Monomer feed	Vinyl acetate	100.0	Monomer
C—Initiator feed	Ammonium persulfate	0.35	Thermal initiator
	Deionized water	6.0	Solvent
D—Chain transfer	Isopropanol	0.30	Reduces MW for fast set
E—Redox chase	Hydrogen peroxide (35%)	0.20	Oxidant
	Ascorbic acid	0.10	Reductant
	$FeSO_4{\cdot}7H_2O$ (0.1% solution)	0.50	Catalyst

Procedure: The PVOH solution, buffer, and surfactant are charged to the reactor and heated to 62–68 °C under nitrogen. A small portion of vinyl acetate monomer and initiator is introduced to form the seed stage, ensuring nucleation of stable polymer particles. Once the seed latex is formed, the remainder of the vinyl acetate monomer and initiator are co-fed over 3 h, with isopropanol metered continuously to reduce molecular weight. After a short hold, the batch is cooled to 35–40 °C, and a redox chase is performed. The resulting latex typically has 50–52% solids, a viscosity of 600–1,200 cP, and fine particle size (~0.25–0.45 μm) suited to paper and packaging adhesives.

development and short open time to support high machine speeds. Typical uses include carton side seams, case and tray assembly, and paper-to-paper laminations under indoor (D2) conditions. Formulations emphasize clean, filler-free systems to maximize fiber tear while controlling penetration for good holdout on porous substrates (Table 8.7). Table 8.7 summarizes the polymerization recipe and key process parameters for the PVAc homopolymer binder used in paper and packaging fast-set applications.

8.4.3 Textile and Nonwoven Grade (Penetration; Soft Film)

This PVAc homopolymer grade is formulated for textile finishing and nonwoven bonding, where deep substrate penetration and a soft, flexible film are required.

Table 8.8 Polymerization recipe and process parameters for PVAc homopolymer binder (textile and nonwoven, penetration; soft film).

Stream	Component	pbw (per 100 pbw VAc)	Purpose
A—Initial charge	Deionized water	62.0	Vehicle
	PVOH (10% solution)	30.0	Protective colloid (≈3 pbw solids)
	Sodium acetate	0.25	Buffer
	Nonionic surfactant	0.20	Aids fiber wetting
B—Monomer feed	Vinyl acetate	100.0	Monomer
C—Initiator feed	Ammonium persulfate	0.30	Thermal initiator
	Deionized water	6.0	Solvent
D—Chain transfer	Isopropanol	0.20	Controls MW for flexible films
E—Redox chase	Hydrogen peroxide (35%)	0.18	Oxidant
	Ascorbic acid	0.09	Reductant
	$FeSO_4 \cdot 7H_2O$ (0.1% solution)	0.50	Catalyst

Procedure: The PVOH solution, buffer, and surfactant are charged and heated to 60–65 °C under nitrogen. A small portion of vinyl acetate and initiator is introduced to form the seed stage, establishing stable polymer nuclei. Once the seed latex is generated, the remainder of the vinyl acetate and initiator are co-fed over 3 h, with isopropanol introduced to regulate molecular weight. After holding for 30–45 min, a redox chase is carried out at 35–40 °C. The resulting latex typically shows 50–52% solids, a viscosity of 300–800 cP, and a particle size of 0.25–0.40 μm, which allows penetration into textile substrates and provides soft films.

It provides good hand feel, fiber encapsulation, and durable bonding under indoor service conditions. Applications include interlinings, nonwoven hygiene products, and textile laminations requiring softness and resilience without external plasticizers (Table 8.8).

8.4.4 Carpet Backing Grade (Low Viscosity; Higher Surfactant; Filler-free)

This PVAc homopolymer grade is developed for use in carpet backing applications, where low viscosity ensures adequate filler acceptance and uniform coating. A higher surfactant level is employed to provide stability during application and drying, while maintaining a filler-free baseline binder. The resulting adhesive film delivers good flexibility and adhesion to fibrous carpet substrates under interior conditions (Table 8.9).

Table 8.9 Polymerization recipe and process parameters for PVAc homopolymer binder (carpet backing; low viscosity, higher surfactant, filler-free).

Stream	Component	pbw (per 100 pbw VAc)	Purpose
A—Initial charge	Deionized water	68.0	Vehicle
	PVOH (10% solution)	25.0	Protective colloid (≈2.5 pbw solids)
	Sodium acetate	0.20	Buffer
	Nonionic surfactant (HLB 12–14)	0.60	Higher level of penetration
B—Monomer feed	Vinyl acetate	100.0	Monomer
C—Initiator feed	Ammonium persulfate	0.25	Thermal initiator
	Deionized water	7.0	Solvent
D—Chain transfer	Isopropanol	0.50	Reduces MW, enhances flow
E—Redox chase	Hydrogen peroxide (35%)	0.18	Oxidant
	Ascorbic acid	0.09	Reductant
	$FeSO_4 \cdot 7H_2O$ (0.1% solution)	0.50	Catalyst

Procedure: The PVOH solution, buffer, and surfactant are charged into the reactor and heated to 58–62 °C under nitrogen. A small initial charge of vinyl acetate and initiator is introduced to establish the seed stage, creating stable nuclei for particle growth. Once the seed latex has formed, the remainder of vinyl acetate and initiator are co-fed over 2.5–3 h, with isopropanol fed to control molecular weight and viscosity. Following a 20–30 min hold, the latex is cooled to 35–40 °C, and a redox chase is conducted. The final product has 50–52% solids, a viscosity of 300–700 cP, and fine particle size (~0.20–0.35 μm), providing deep penetration into carpet backings.

8.4.5 School and Consumer Grade (Low Odor; Low Residual Monomer)

This PVAc homopolymer grade is intended for school glues and consumer adhesives where safety, low odor, and very low residual monomer content are essential. The adhesive is formulated to provide clean handling, good clarity, and safe use in indoor environments, while maintaining sufficient bond strength for paper, light crafts, and general household applications (Table 8.10).

8.4.6 Carton-to-Carton (High-speed Machine Adhesive)

This grade is optimized for high-speed folding and gluing equipment where rapid tack development, excellent machinability, and low stringing are essential. The

Table 8.10 Polymerization recipe and process parameters for PVAc homopolymer binder (school & consumer; low odor, low residual monomer).

Stream	Component	pbw (per 100 pbw VAc)	Purpose
A—Initial charge	Deionized water	85.0	Vehicle
	PVOH (10% solution; 88–98% hydrolysis blend)	70.0	Protective colloid (≈7 pbw solids)
	Sodium acetate	0.25	Buffer
	Nonionic surfactant	0.05	Minimal
B—Monomer feed	Vinyl acetate	100.0	Monomer
C—Initiator feed	Hydrogen peroxide (35%)	0.20	Redox initiator
	Ascorbic acid	0.10	Reductant
	$FeSO_4·7H_2O$ (0.1% solution)	0.50	Catalyst
	Deionized water	8.0	Solvent
E—Additional chase	Redox system	as needed	Drive residual monomer <300 ppm

Procedure: The PVOH solution and buffer are charged, and the reactor is purged with nitrogen. A small initial dose of vinyl acetate and redox initiator is added at 25–35 °C to form the seed stage and generate stable nuclei. Once the seed latex is established, the remainder of the monomer is fed over 4 h with gentle control of temperature. After completion, an extended redox chase and optional vacuum stripping are used to reduce residual vinyl acetate to below 300–500 ppm The latex has 45–50% solids, a viscosity of 800–2,000 cP, and low odor, meeting safety and handling requirements for consumer adhesives.

formulation incorporates colloidal silica and a polyurethane rheology modifier to control viscosity and prevent nozzle clogging or tailing (Table 8.11).

8.4.7 PVAc Homopolymer Formulations

Polyvinyl acetate homopolymers represent the historical and commercial backbone of water-borne adhesives. Owing to their ease of manufacture, low cost, and inherent affinity for porous substrates, they have long served as the default choice for bonding paper, board, wood, and related cellulosic materials. While their resistance to water and heat is limited, judicious formulation with stabilizers, plasticizers, rheology modifiers, and tackifiers has enabled their use across a wide spectrum of everyday applications.

Table 8.11 Formulation for high-speed carton-to-carton PVAc homopolymer.

Ingredient	% (on total weight)	Function
De-ionized water	46.0	Vehicle
Vinyl acetate monomer	48.0	Monomer
Partially hydrolyzed PVOH	4.5	Protective colloid
Ammonium persulfate	0.2	Initiator
Sodium acetate	0.2	Buffer
Colloidal silica (25%)	1.0	Rheology modifier; reduces stringing
Rheovis PU 1211	0.5	Polyurethane thickener; pseudoplasticity
Biocide	0.1	Preservation

Procedure: A PVOH solution is first prepared by heating in water at 85–90 °C with stirring until fully dissolved, followed by cooling to ambient. Sodium acetate, colloidal silica, and biocide are then introduced into the cooled solution. After a brief seeding stage, vinyl acetate monomer and ammonium persulfate are gradually added under controlled stirring. Polymerization is maintained at 60–70 °C for 4–6 h. The latex is subsequently cooled to room temperature and filtered if required.

Formulation note: Colloidal silica serves as a rheology modifier and stringing reducer, critical for high-speed nozzle applications. It enhances shear-thinning behavior, improves wetting on carton surfaces, and ensures clean glue lines with minimal tailing or clogging. The inclusion of Rheovis PU 1211 imparts a pseudoplastic profile, preventing sagging and dripping during application. For best results, the pH is adjusted to approximately 6.5–7.5 prior to the addition of the rheology modifier, which should be incorporated slowly under low shear to avoid foaming.

The following section presents a series of representative formulations in which polyvinyl acetate homopolymer latexes are employed as the principal binder. These examples illustrate how relatively simple modifications allow tailoring of performance for specific end uses such as school glues, carton side-seam adhesives, high-tack label glues, wood assembly adhesives, paper-core winding, and bookbinding. Each formulation highlights both the advantages and the constraints of homopolymer systems, setting the stage for subsequent discussion on functional copolymerization with acrylic acid and methacrylic acid to overcome the intrinsic bottlenecks of polyvinyl acetate homopolymers (Tables 8.12–8.15).

8.4.8 The Bottleneck of PVAc Homopolymers and the Path Forward

Polyvinyl acetate homopolymers have demonstrated versatility across a range of adhesive applications, including woodworking, packaging, nonwoven bonding, carpet backing, and consumer glues. Despite these strengths, their inherent weaknesses—particularly in water resistance, alkali stability, and resistance to

Table 8.12 Polyvinyl acetate homopolymer adhesive (baseline formulation).

Ingredient	Parts by weight (solids)	Function
Polyvinyl acetate emulsion (55% solids, Mw ≈ 100,000)	100	Base polymer binder
Plasticizer (e.g., di-2-ethylhexyl adipate or triacetin)	5–10	Flexibility; improved film formation
Polyvinyl alcohol (88% DH, 30–50 k Mw), if not present in latex	2–3	Protective colloid; viscosity control
Defoamer (silicone-based)	0.2–0.5	Foam control
Preservative (isothiazolinone blend)	0.1–0.2	Microbial protection
Water	to 100	Balance; viscosity adjustment

Application: Paper bonding, school glue, carton side-seam. Fast set and good dry adhesion on porous substrates; limited water resistance (D1–D2).

Table 8.13 Polyvinyl acetate homopolymer adhesive for wood assembly.

Ingredient	Parts by weight (solids)	Function
Polyvinyl acetate emulsion (55% solids, Mw ≈ 100,000–150,000)	100	Base polymer binder
Polyvinyl alcohol (88–90% DH, 40–60 k Mw)	3–5	Protective colloid; creep resistance; viscosity control
Plasticizer (triacetin or di-2-ethylhexyl adipate)	5–8	Film flexibility and improved bond formation
Mineral filler (calcium carbonate, fine grade)	5–10	Cost reduction; improved heat resistance
Defoamer (silicone-based)	0.2–0.5	Foam suppression
Preservative (isothiazolinone blend)	0.1–0.2	Microbial protection
Water	to 100	Balance; viscosity adjustment

Application: Designed for interior wood bonding (furniture assembly, veneer lamination). Provides good initial tack and creep resistance when PVOH content is optimized, but durability remains limited to D2 class. For higher durability (D3/D4), copolymerization with acrylic acid or methacrylic acid and external crosslinkers is required.

creep under sustained load—form a significant bottleneck that restricts their use in higher performance or outdoor applications.

Industrial producers have successfully supplied homopolymer-based adhesives for regional markets, demonstrating the value of PVAc as a commodity adhesive

Table 8.14A Core winding (PVAc, no filler, with DYNOL 360) for improved early grab.

Ingredient	Parts by weight (solids or as supplied)	Function/rationale
Polyvinyl acetate emulsion (55% solids, medium Mw)	100	Base binder; intrinsic green tack + cohesive strength
DYNOL 360 (as supplied)	0.10–0.20	Dynamic wetting; boosts early "grab"; low foam
Plasticizer/coalescent (triacetin or DEHA)	3.5 (solids)	Lowers MFFT/Tg; accelerates early set
Polyvinyl alcohol (as solids, from 10% solution)	1.5	Stabilization; fiber anchoring; film strength
Rheology modifier (HEC, as solids)	0.35	Shear-thinning; full wet film; controls roll-off
Defoamer (silicone-based)	0.25 (as supplied)	Foam control during high-speed mixing and application
Preservative (isothiazolinone blend)	0.15 (as supplied)	Microbial protection
Water	q.s. to viscosity	Target 5,000–10,000 mPa s (RVT #4/20 rpm, 25 °C)

Application: Spiral/convolute paper cores (tissue, towel, textile). Typical solids ~50–55%. Wet coat 60–120 g/m^2 per ply; nip pressure to drive dewatering. Moisture resistance ≈ D2.

Procedure

1) Charge 70–80% of batch water at 20–25 °C; set 300–400 rpm (gentle vortex).
2) Add defoamer; mix 2–3 min. Add DYNOL 360; mix 10 min at 300–500 rpm.
3) Add emulsion slowly over 5–10 min while mixing at 400–600 rpm (avoid air).
4) Add plasticizer/coalescent; mix 5 min.
5) Add PVOH solution; mix 10 min.
6) Adjust viscosity with HEC to target; mix 20–30 min for full hydration.
7) Top with water to final viscosity; add preservative; mix 5 min.
8) Check pH 4.8–5.5; filter 200–300 μm; age 12–24 h.

platform. Yet, as manufacturing demands evolve toward more durable and versatile products, homopolymers alone cannot meet the performance expectations of industries such as furniture, construction, and specialty packaging.

The step-change in performance is achieved by introducing functionality into the polymer. One route is copolymerizing vinyl acetate with acrylic acid (AA) or methacrylic acid (MAA), which, even at modest incorporation levels (typically 1–5% by weight), provides carboxylic functionality that serves as reactive sites for crosslinking. However, AA or MAA are not strictly necessary in all cases, since certain internal crosslinkers are capable of forming self-condensed networks without additional acid functionality.

Table 8.14B Core winding (PVAc, with mineral filler, with DYNOL™ 360).

Ingredient	Parts by weight (solids or as supplied)	Function/rationale
Polyvinyl acetate emulsion (55% solids, medium Mw)	100	Base binder; cohesion and film formation
Ground calcium carbonate (GCC, d50 ≈ 3 μm)	12 (solids)	Cost reduction; higher effective solids; faster dewatering → faster tack
Dispersant for GCC (sodium polyacrylate, 40% active)	0.20 (solids)	Keeps filler dispersed; prevents flocculation
DYNOL 360 (as supplied)	0.10–0.20	Dynamic wetting; boosts early "grab"; low foam
Plasticizer/coalescent (triacetin)	2.5 (solids)	Early coalescence; maintains flexibility
Polyvinyl alcohol (as solids, from 10% solution)	1.0	Stabilization; fiber anchoring
Rheology modifier (HEC, as solids)	0.25	Viscosity and film build; adjust postfiller
Defoamer (silicone-based)	0.25 (as supplied)	Foam control
Preservative (isothiazolinone blend)	0.15 (as supplied)	Microbial protection
Water	q.s. to viscosity	Target 6,000–12,000 mPa s (RVT #4/20 rpm, 25 °C)

Application: Same as 8.14A but with faster drying/earlier grab at lower cost. Typical solids ~52–58%. Avoid filler >15 parts/100 PVAc solids to prevent embrittlement/powdering.

Procedure

1) Charge 60–70% of batch water at 20–25 °C; set 400–600 rpm.
2) Add defoamer and dispersant; mix 2–3 min. Add mineral filler; grind 10–15 min to smooth dispersion.
3) Dose DYNOL 360 into the grind; mix 5–10 min.
4) Add emulsion slowly over 5–10 min (avoid air).
5) Add plasticizer; add PVOH solution; mix 10 min.
6) Adjust viscosity with HEC; mix 20–30 min for full hydration.
7) Top with water to final viscosity; add preservative; mix 5 min.
8) Check pH 4.8–5.5; filter 200–300 μm; age 12–24 h.

Two principal approaches to crosslinking can be employed in PVAc-based systems:

1) *External crosslinkers (postaddition)*: Examples include aziridines and carbodiimides. These are introduced into the finished latex and react with hydroxyl or carboxyl groups in the polymer or stabilizer system. In PVAc–AA or PVAc–MAA copolymers, the pendant carboxyl groups act as reaction sites:

Table 8.15 Polyvinyl acetate homopolymer adhesive for bookbinding.

Ingredient	Parts by weight (solids)	Function
Polyvinyl acetate emulsion (55% solids, medium-to-high Mw)	100	Base polymer binder
Plasticizer (triacetin or dibutyl adipate)	5–8	Flexibility; reduces brittleness on the spine
Polyvinyl alcohol (88–90% DH, 40–60 k Mw)	2–4	Creep resistance; viscosity control
Mineral filler (calcium carbonate, fine grade)	5–10	Cost reduction; rheology adjustment
Defoamer (silicone-based)	0.2–0.5	Foam control
Preservative (isothiazolinone blend)	0.1–0.2	Microbial protection
Water	to 100	Balance; viscosity adjustment

Application: Used for bookbinding and perfect binding, where flexibility, film strength, and creep resistance are critical. Provides strong dry adhesion to paper and board but limited durability under moisture exposure.

aziridines undergo ring-opening with –COOH groups, while carbodiimides form stable *N*-acylurea linkages. The result is covalent network formation during drying, greatly enhancing water and solvent resistance. This approach provides curing on demand and is versatile, but requires careful handling and increases formulation cost.

2) *Internal crosslinkers (copolymerizable)*: Internal crosslinkers are comonomers incorporated directly into the polymer backbone during emulsion polymerization. In this book, acrylamide and *N*-methylolacrylamide (NMA) are excluded despite their historical effectiveness due to health and regulatory concerns associated with their toxicity and handling risks.

As a safer alternative, glycidyl methacrylate (GMA) is employed. GMA is copolymerized with vinyl acetate and, optionally, with acrylic acid (AA) or methacrylic acid (MAA), embedding reactive epoxide groups into the polymer backbone. During film formation, these epoxide groups undergo ring-opening reactions with hydroxyl groups in partially hydrolyzed polyvinyl alcohol (PVOH) and with carboxyl groups introduced by AA or MAA. The resulting covalent bonds generate an internally crosslinked network within the film.

The use of GMA supports one-component (1K) adhesive systems without pot-life limitation. Reactivity can be tuned through formulation to avoid premature gelation during polymerization or storage, providing a practical and safer pathway to internal crosslinking in PVAc-based adhesives.

8.4.8.1 Role of Tackifiers in PVAc-copolymerized with Acrylic Acid (AA)

The role of tackifier is to increase wet tack/initial grab and shorten set time.

In PVAc homopolymer, the effect is minor and does not fix water resistance or creep (resins mostly soften the film). In PVAc-AA/MAA, carboxyl functionality improves compatibility with tackifier emulsions and interphase wetting on coated/lacquered substrates; when paired with external crosslinkers such as carbodiimide (effective near pH 4.8–5.5), Zr complexes, or ZnO/$AlCl_3$ (acidic conditions)—the system retains fast tack while gaining humidity/heat resistance.

Performance improves with PVAc-AA (vs. neat PVAc) because AA increases polarity, enabling co-continuous or well-dispersed resin phases; neat PVAc is often just plasticized, leading to creep. AA raises surface energy, so tackifier-enhanced viscoelastic response translates into real fiber tear on coated boards. External crosslinkers react with –COOH from AA, building a network that locks in fast-grab benefits (durability). Tackifiers add wet-film body and reduce excessive penetration of adhesive into porous substrates, ensuring that a continuous film remains at the bond line. This shortens the set-to-fiber-tear time and improves bond reliability.

Typical level and handling. Tackifier emulsion (50% solids) 10–15% as supplied (up to 20% for very high grab).

To process: add after the latex under steady shear; add crosslinker last (gentle mix); monitor blocking/whitening at higher resin loadings.

Table 8.16 provides a comparative summary of internal and external crosslinkers used in PVAc-based adhesive systems, highlighting representative examples, reaction mechanisms, the influence of AA/MAA where relevant, and the main advantages and limitations of each strategy. For clarity, tackifiers are treated separately; detailed guidance on tackifier types, selection, and dosing is provided in the dedicated tackifier section later in this chapter.

8.4.9 Scope and Safety Note—Crosslinkers Used in This Book

To align with modern safety and compliance expectations, this book permits two categories of crosslinking strategies for PVAc-based adhesives: (A) external postaddition crosslinkers and (B) internal (copolymerizable) functionalities incorporated in the polymer backbone.

A) *External crosslinkers included in application formulations*: carbodiimides (CDI), aluminum chloride ($AlCl_3$), zirconium complexes (Zr), and zinc oxide (ZnO). These systems primarily target carboxyl functionality introduced via acrylic acid or methacrylic acid (AA/MAA) in PVAc-AA/MAA copolymers and are presented in 1K, production-practical recipes.

Table 8.16 Comparison of internal and external crosslinkers in PVAc adhesives.

Type of crosslinker	Examples (full names)	Reaction mechanism (concise)	Role of AA/MAA	Advantages	Limitations
External (postaddition)	Carbodiimides (aliphatic/aromatic); ~30% aqueous dispersions	*N*-acylurea formation with –COOH groups	Essential: –COOH sites enable efficient crosslinking (opt. pH ~4.8–5.5)	Strong humidity/heat resistance; 1K latency window; versatile	Cost; narrow pH range; gradual deactivation by hydrolysis
External (postaddition)	Zirconium complexes (e.g., zirconium ammonium carbonate; Zr chelates)	Coordination/chelation with $–COO^-$ → rapid wet-strength build	Beneficial: requires $–COOH/–COO^-$ for density/durability	Fast set; low odor; effective at low dosage	Sensitive to chelators/phosphates; pH-dependent; brittleness if overdosed
External (postaddition)	Zinc oxide (ZnO) dispersions	Lewis-acid zinc carboxylates (–COO–Zn–OOC–) formed during drying	Beneficial: more effective with higher AA/MAA	Cost-effective water resistance boost; simple dosing	Opacity/whitening; sedimentation risk; viscosity increase; sensitive to chelators
External (postaddition)	Aluminum chloride ($AlCl_3$) aqueous solutions	Lewis-acid aluminum carboxylates (–COO–Al–OOC–) at acidic pH	Beneficial: requires $–COOH/–COO^-$; best at pH ~3.5–4.5	Quick network build at low dose; supports fast set on paper/coated board	Corrosive (chloride); food-contact verification needed; sensitive to anions; risk of embrittlement
Internal (copolymerizable)	AAEM (acetoacetoxyethyl methacrylate)	Enolizable acetoacetate group reacts with dihydrazide/amine nucleophiles (if present)	Optional: AA/MAA not required; –COOH can aid adhesion	Provides latent acetoacetate sites; versatile for postformulation crosslinking	Requires external nucleophiles to activate; sensitive to pH and nucleophile balance
Internal (copolymerizable)	GMA (glycidyl methacrylate)	Methacrylate copolymerizes into backbone; epoxy ring opens with $–COOH/–OH/–NH_2$	Beneficial to Essential: –COOH greatly accelerates; PVOH –OH is also reactive	True 1K potential; strong covalent bonding; high water/solvent resistance	Raises Tg (higher MFFT); risk of premature gelation if overdosed; catalyst/pH control needed

Notes

1) Acrylamide, acrylamide-derived monomers (e.g., diacetone acrylamide, *N*-methylolacrylamide), and aziridines are excluded for health and safety reasons.
2) Preferred internal systems are acetoacetate (AAEM) and epoxy-functional (GMA) chemistries. AAEM provides internal acetoacetate sites that can later react with suitable nucleophiles. GMA exhibits dual functionality: its methacrylate group copolymerizes into the polymer backbone, while its epoxy group remains reactive for postpolymerization crosslinking with –COOH, –OH, or $–NH_2$ groups. This dual mechanism makes GMA especially valuable for durable 1K adhesives with strong interfacial bonding.
3) Crosslinking through PVOH –OH groups is not considered a principal pathway in this context. External crosslinkers act mainly via carboxyl groups introduced by AA/MAA.

B) *Internal (copolymerizable) crosslinkers permitted in polymer design and selected 1K applications*: (1) Glycidyl methacrylate (GMA), which is dual-functional—its methacrylate group copolymerizes into the backbone, while its epoxy group remains reactive postpolymerization for covalent bonding with –COOH/–OH/–NH_2 groups—enabling true 1K cure; and (2) Acetoacetoxyethyl methacrylate (AAEM), which introduces acetoacetate sites that can participate in condensation or Michael-type reactions with suitable nucleophiles. In this book's 1K focus, AAEM is discussed mechanistically and in polymer design notes; practical recipes that require added hydrazides/diamines for cure are classified as 2K and are not included in application sections.

Exclusions: Application recipes based on acrylamide or *N*-methylolacrylamide (formal-dehyde-releasing risk) are excluded. Aziridines (external) are excluded from formulas due to sensitization hazards and stringent handling requirements. Short background notes on these chemistries are retained for completeness and comparison only.

Operational boundaries: External systems are typically most effective at mildly acidic pH (≈3.5–5.5), depending on the crosslinker (e.g., $AlCl_3$ near 3.5–4.5; CDI around 4.8–5.5) and require AA/MAA for robust performance. GMA-modified polymers require careful control of epoxy content to avoid premature gelation and may increase Tg/MFFT; AAEM-modified polymers provide latent sites but generally require a reactive partner for full cure. Wherever a potential safety tradeoff exists, lower-hazard routes (CDI/Zr/ZnO/$AlCl_3$ or GMA-based 1K) are preferred.

Implementation note: All practical application formulations in subsequent sections use CDI, $AlCl_3$, Zr, or ZnO as external crosslinkers, and where internal crosslinking is employed, only GMA-based 1K designs are presented. AAEM is included in polymer design guidance and performance discussion, with 2K activation routes intentionally omitted from recipes (Table 8.17).

8.4.9.1 New Applications Enabled by Crosslinking and Functional Monomers

The incorporation of crosslinkable functionality into PVAc-based adhesives fundamentally extends their performance profile beyond the traditional domains of woodworking, carton sealing, and school glues. By overcoming the limitations in water resistance, creep strength, and alkali stability, crosslinked PVAc copolymers unlock new areas of application. Representative examples include:

Furniture and wood composites: Crosslinked PVAc copolymers provide improved water resistance and creep strength, enabling use in high-pressure laminates, edge banding, and moisture-resistant furniture adhesives.

Table 8.17 Optimal dosage ranges for external and internal crosslinkers.

Crosslinker	Form supplied/basis	Optimal dosage (active %)	pH window (use)	Activation/cure (concise)	Key cautions
Carbodiimide (CDI) dispersions	≈30% active aqueous dispersion; dosage expressed as % active on total polymer solids (TPS)	0.3–1.0% active on TPS (start ~0.5–0.7%)	≈4.8–5.5	Room-temp; reacts with –COOH to form *N*-acylureas; humidity-resistant network builds over 24–72 h	Hydrolysis over storage; avoid high pH; cost vs. performance tradeoff
Zirconium complexes (e.g., zirconium ammonium carbonate; Zr chelates)	Typically 20–25% metal complex in water; report as % active Zr complex on TPS	0.2–0.8% active on TPS (or 1–3% of a 25% product)	≈4.0–5.5	Rapid wet-strength via Zr–carboxylate coordination; develops on drying	Sensitive to phosphates/chelators; brittleness if overdosed
Zinc oxide (ZnO) dispersions	50–60% ZnO slurry; dosage as % ZnO on TPS	0.5–1.5% ZnO on TPS (start ~0.8–1.0%)	≈4.0–6.0	Forms zinc carboxylates during drying; boosts water resistance	Whitening/opacity ↑ with higher loadings; possible thickening/ sedimentation
Aluminum chloride ($AlCl_3$) solutions	5–10% aqueous $AlCl_3$; dosage as % $AlCl_3$ (anhydrous basis) on TPS	0.15–0.35% on TPS (start ~0.2–0.25%)	≈3.5–4.5	Fast ionic/coordination network with $-COO^-$; quick set on paper/board	Corrosive anions; verify food-contact; embrittlement if overdosed

(Continued)

Table 8.17 (Continued)

Crosslinker	Form supplied/basis	Optimal dosage (active %)	pH window (use)	Activation/cure (concise)	Key cautions
Glycidyl methacrylate (GMA)—internal	Comonomer in polymerization; dosage as wt% of total monomer (TM)	1.0–2.5% on TM (start ~1.5%); ensure AA/MAA 0.5–2% on TM for epoxy acceleration	Polymer design: pH during use typically 4.5–6.0	Dual function: methacrylate copolymerizes; epoxy ring opens with –COOH/–OH/–NH_2 postpolymerization for true 1K cure	Raises Tg/MFFT; risk of premature gelation if GMA too high; manage catalyst/pH
Acetoacetoxyethyl methacrylate (AAEM)—internal	Comonomer; dosage as wt% of total monomer (TM)	0.5–2.0% on TM (start ~1.0%)	Polymer design: storage/use pH typically 4.5–6.0	Provides acetoacetate sites; in 1K designs offer latent reactivity and adhesion benefits; full cure typically requires added nucleophiles (hydrazides/amines)	If activated with external nucleophiles, it becomes 2K; monitor yellowing with some amines; pH sensitivity

Notes
Dosages are expressed as % active relative to total polymer solids (TPS) for external crosslinkers, and as wt% of total monomer (TM) for internal comonomers.
Effective performance of CDI/Zr/ZnO/$AlCl_3$ generally requires AA/MAA in the PVAc copolymer to provide –COOH/–COO^- sites.
Starting points are suggested mid-range values; optimize for substrate, coat weight, and drying conditions.

Construction adhesives: Copolymers modified with AA or MAA, combined with external crosslinkers, deliver higher durability and alkali resistance. These properties support applications in flooring adhesives, tile adhesives, and joint sealants.

Packaging beyond paperboard: While homopolymers dominate carton gluing, crosslinked PVAc copolymers extend performance to coated, lacquered, and partially nonporous substrates, broadening utility in food and consumer goods packaging.

Specialty adhesives: In combination with carbodiimides, AA/MAA-modified PVAc systems can achieve levels of water and chemical resistance suitable for labeling under wet-fill conditions.

D3 and D4 performance classes: The introduction of functional monomers and crosslinkers enables PVAc adhesives to achieve classifications D3 and D4 according to EN 204/205 standards. D3 adhesives are suitable for interior applications exposed to running water or high humidity, such as kitchen furniture and interior joinery. D4 adhesives extend use to exterior applications exposed to weathering, such as windows, doors, and façade panels. Achieving these classifications with waterborne PVAc systems represents a major advance beyond the limitations of homopolymers.

Adhesion to hard-to-bond substrates: Functional PVAc copolymers extend compatibility to challenging substrates that are inaccessible to homopolymers. Examples include plastics such as surface-treated polyethylene.

Polypropylene, polyvinyl chloride, and polyester films; metals such as aluminum, galvanized steel, and tinplate, particularly when used in laminated packaging or construction panels; and coated or lacquered paperboards in high-performance packaging, where crosslinked copolymers combined with tackifiers deliver reliable adhesion.

8.5 PVAc–AA/MAA Copolymer Adhesives

Scope: This section covers adhesives based exclusively on polyvinyl acetate–co-acrylic or methacrylic acid) (PVAc–AA/MAA) systems, including tackifier use and external crosslinkers (CDI, $AlCl_3$, Zr, ZnO).

The transition from homopolymer polyvinyl acetate (PVAc) to copolymer systems represents the most important development in waterborne adhesive technology. While PVAc homopolymers provide low cost, good film formation, and excellent adhesion to porous substrates, they remain limited in resistance to water, alkali, and creep under load. These shortcomings restrict their use to commodity markets and prevent penetration into demanding sectors such as construction, high-performance packaging, and exterior wood bonding.

Copolymers of vinyl acetate with functional monomers overcome these limitations by introducing polar or reactive groups into the polymer chain. The incorporation of acrylic acid (AA) or methacrylic acid (MAA) provides carboxylic groups that enhance adhesion and enable crosslinking.

This section presents the principles of vinyl acetate copolymerization, followed by detailed formulations and processes for representative adhesive systems. Emphasis is placed on the relationship between comonomer selection, polymerization strategy, and final adhesive performance.

8.5.1 Principles of VAM Copolymerization

The copolymerization of vinyl acetate monomer (VAM) with functional comonomers represents a key strategy for enhancing the properties of waterborne adhesives. The selection of comonomers, such as acrylic acid (AA), methacrylic acid (MAA), introduces polar or reactive groups into the polymer backbone, thereby overcoming limitations of the PVAc homopolymer in water resistance, alkali stability, and adhesion to nonporous substrates.

Reactivity considerations: VAM is characterized by a moderate propagation rate and a strong tendency toward alternating copolymerization with more reactive monomers. Acrylic acid and methacrylic acid, due to their high reactivity and polar nature, readily copolymerize with VAM, albeit with low incorporation levels if used sparingly (1–5% by weight) of functional groups throughout the polymer chain.

Impact on polymer structure: Even small amounts of comonomer significantly affect the polymer's glass transition temperature (Tg), hydrophilicity, and hydrogen bonding capacity. Incorporation of AA or MAA introduces carboxyl groups that enable ionic interactions and covalent crosslinking. These modifications transform the adhesive film from a linear thermoplastic into a partially crosslinked network.

Formulation and process implication: Because AA/MAA are hydrophilic, careful control of emulsifier and protective colloid packages is required to prevent instability, and a slow, controlled comonomer feed is essential for uniform incorporation. pH discipline is critical: during emulsion polymerization, keep the reactor mildly acidic ($\approx$3.5–4.5) and do not exceed ~4.5. Any partial neutralization of AA (if used) should be limited to the aqueous pre-emulsion/side-feed with a volatile base, while the reactor itself remains acidic. After polymerization and cool-down, adjust the finished latex to pH ~4.8–5.2 for storage and subsequent formulating/crosslinking.

Performance outcomes: Through judicious copolymerization, PVAc-based adhesives can be tailored to achieve D3 or D4 water resistance ratings, improve adhesion to coated boards, plastics, and metals, enhance creep resistance and dimensional stability in wood adhesives, and provide wash durability and crosslinking capacity in textile binders.

8.5.2 VAM–AA Copolymers

Copolymerizing vinyl acetate (VAM) with acrylic acid (AA) at 1–5 pbw per 100 pbw VAM introduces pendant carboxyl groups that raise polarity, strengthen adhesion to cellulosics, and provide sites for external crosslinkers (CDI, $AlCl_3$, Zr, ZnO) (Table 8.18).

The representative formulation for a VAM–AA copolymer adhesive is shown in Table 8.18. Surfactant guardrails (AA system): avoid phosphate-containing surfactants if Zr crosslinkers will be used later; avoid cationics.

8.5.3 VAM–MAA Copolymers

Methacrylic acid (MAA) yields a slightly more hydrophobic copolymer than AA at similar feed levels (approximately 2–4 pbw typical), improving water and alkali resistance with a modest tradeoff in fiber-tear on highly polar substrates. Combined with external crosslinkers (CDI, $AlCl_3$, Zr, ZnO), D3 to D4 performance can be achieved (Table 8.19).

Surfactant guardrails (MAA system): prefer phosphate-free packages when Zr crosslinkers are planned; avoid cationics.

8.5.4 Comparative Note—VAM–AA vs. VAM–MAA Copolymers

Purpose. Provide a concise comparison between acrylic acid (AA) and methacrylic acid (MAA) as comonomers with vinyl acetate (VAM) for waterborne PVAc copolymer adhesives. Emphasis is on performance, process, pH/buffering, crosslinker synergy, and corrosion considerations (including chloride-bearing systems such as $AlCl_3$) (Table 8.20). Table 8.20 presents a side-by-side comparison of VAM–AA and VAM–MAA copolymer adhesive systems, highlighting differences in comonomer levels and adhesion performance.

8.5.4.1 Selection Guidance

Selection of comonomer systems for PVAc adhesives can be guided by a set of practical rules. Vinyl acetate–acrylic acid (VAM–AA) copolymers are favored when early wet grab and fiber-tear on coated board represent the primary performance indicators under moderate humidity conditions. In such cases, durability can be secured by pairing with carbodiimide (CDI) or zirconium (Zr) crosslinkers.

Table 8.18 Representative formulation for VAM–AA copolymer adhesive.

Stream	Component	pbw (per 100 pbw VAM)	Function
A—Initial charge	Deionized water	60.0	Medium
	PVOH (10% solution, 88% hydrolysis)	45.0	Protective colloid
	Anionic surfactant (e.g., SDS or DOSS, active)	0.20–0.50	Nucleation/particle size/stability
	Nonionic surfactant, optional (HLB 12–14, active)	0.10–0.30	Flow/defect control (keep low)
	Sodium acetate	0.30	Acetate buffer component
	Acetic acid (10% aq), q.s.	to pH 4.0–4.3	Set initial buffer pH
B—Monomer feed	Vinyl acetate	100.0	Principal monomer
	Acrylic acid	3.0	Functional comonomer (1–5 typical)
C—Initiator feed	Ammonium persulfate	0.35	Radical initiator
	Water	5.0	Solvent
D—Chain transfer	Isopropanol	1.0	Molecular-weight control (optional)
E—Redox chase	Hydrogen peroxide (35%)	0.20	Oxidant
	Ascorbic acid	0.10	Reductant
	Ferrous sulfate (0.1% solution)	0.50	Catalyst

Procedure **(Seeded, Starved-feed; Acetate Buffer; Surfactant-explicit)**

(1) *Reactor charge (aqueous)*: Add water, then dissolve the anionic surfactant and optional nonionic under agitation. Add the PVOH solution and mix to clarity.

(2) *Buffer set*: Add sodium acetate, then adjust with 10% acetic acid to pH 4.0–4.3. Dose a small, compatible defoamer if needed. Maintain a nitrogen blanket.

(3) *Heat-up and initiation*: Purge with nitrogen and heat to 65–75 °C. Start the ammonium persulfate feed to nucleate; hold 5–10 min after the exotherm stabilizes.

(4) *Seed shot*: Add approximately 2.5% of the premixed VAM/AA monomer blend; hold about 20 min to stabilize particle number.

(5) *Starved monomer feed*: Feed the remaining VAM/AA blend over 3–4 h under starved conditions. Maintain 65–75 °C and reactor pH 3.5–4.5; correct upward drift only with dilute acetic acid.

(6) *Chain transfer (optional)*: Meter isopropanol during the first half of the monomer feed to tune molecular weight/viscosity.

(7) *Finish and chase*: Hold 30–60 min at temperature after feed completion. Cool to 35–40 °C, then perform redox chase ($FeSO_4$ → ascorbic acid → H_2O_2) for 30–45 min to residual VAM ≤ 500 ppm.

(8) *Postadjust and finish*: Adjust the latex to pH 4.8–5.2, add preservative, verify solids/viscosity, filter 200–300 μm, and age 12–24 h before formulating.

Table 8.19 Representative formulation for VAM–MAA copolymer adhesive.

Stream	Component	pbw (per 100 pbw VAM)	Function
A—Initial charge	Deionized water	60.0	Medium
	PVOH (10% solution, 88% hydrolysis)	45.0	Protective colloid
	Anionic surfactant (e.g., SDS or DOSS, active)	0.20–0.50	Nucleation/particle size/stability
	Nonionic surfactant, optional (HLB 12–14, active)	0.10–0.30	Flow/defect control (keep low)
	Sodium acetate	0.30	Acetate buffer component
	Acetic acid (10% aq), q.s.	to pH 4.0–4.3	Set initial buffer pH
B—Monomer feed	Vinyl acetate	100.0	Principal monomer
	Methacrylic acid	2.5	Functional comonomer (≈2–4 typical)
C—Initiator feed	Ammonium persulfate	0.35	Radical initiator
	Water	5.0	Solvent
D—Chain transfer	Isopropanol	1.0	Molecular-weight control (optional)
E—Redox chase	Hydrogen peroxide (35%)	0.20	Oxidant
	Ascorbic acid	0.10	Reductant
	Ferrous sulfate (0.1% solution)	0.50	Catalyst

***Procedure* (Seeded, Starved-feed; Acetate Buffer; Surfactant-explicit)**

(1) *Reactor charge (aqueous)*: Add water, dissolve the anionic surfactant and optional nonionic under agitation, then add the PVOH solution to clarity.
(2) *Buffer set*: Add sodium acetate and adjust with 10% acetic acid to pH 4.0–4.3. Add minimal, compatible defoamer. Maintain a nitrogen blanket.
(3) *Heat-up and initiation*: Purge with nitrogen and heat to 65–75 °C. Start ammonium persulfate to nucleate; hold 5–10 min after the onset.
(4) *Seed shot*: Add approximately 2.5% of the premixed VAM/MAA monomer blend; hold about 20 min.
(5) *Starved monomer feed*: Feed the remaining VAM/MAA blend over 3–4 h under starved-feed conditions. Maintain 65–75 °C and pH 3.5–4.5; correct with dilute acetic acid only.
(6) *Chain transfer (optional)*: Meter isopropanol during the first half of the feed to tune molecular weight/viscosity.
(7) *Finish and chase*: Hold 30–60 min at temperature; cool to 35–40 °C; perform redox chase ($FeSO_4$ → ascorbic acid → H_2O_2) 30–45 min to residual VAM ≤ 500 ppm.
(8) *Post adjust and finish*: Adjust to pH 4.8–5.2; add preservative; QC; filter 200–300 μm; age 12–24 h.

Table 8.20 Side-by-side comparison of VAM–AA and VAM–MAA.

Dimension	VAM–AA	VAM–MAA
Typical comonomer level	AA: 1–5 pbw per 100 pbw VAM (2–3 common)	MAA: 2–4 pbw per 100 pbw VAM (≈2.5–3.0 common)
Adhesion to cellulosics (fiber-tear on coated board)	Stronger H-bonding; usually higher wet grab and earlier fiber-tear at equal dosage	Slightly lower than AA at equal mol%; may need tackifier optimization to match wet grab
Water and humidity resistance (uncrosslinked)	Improved over PVAc, but more hydrophilic than MAA at equal feed	More hydrophobic matrix; generally better baseline water uptake and creep resistance
Alkali resistance/hydrolysis stability	Good, but AA is more susceptible to saponification than MAA at equal feed	Superior alkali resistance; better long-term durability
Crosslinker efficacy (CDI, Zr, ZnO, $AlCl_3$)	Higher –COOH availability → high reactivity; excellent with CDI and Zr; ZnO/$AlCl_3$ also effective	Slightly fewer ionizable sites → sometimes needs ~10–20% higher comonomer or crosslinker to reach same network density
Tackifier synergy (rosin esters, terpene-phenolics)	Very good compatibility and interphase wetting; fast set on coated stock	Good; sometimes benefits from higher-polarity tackifiers or a small dose of superwetter to match AA's early set
Viscosity/process latitude	Broader resin acceptance; watch pH and surfactant level to avoid whitening	Slightly narrower, but yields smoother films at lower total surfactant
Polymer Tg/film character	Slightly lower Tg at equal mol% (more polar, softer at room temp)	Slightly higher Tg at equal mol% (more hydrophobic, firmer films)
EP pH discipline (reactor)	Maintain 3.5–4.5 throughout polymerization (acetate buffer + acetic acid). Do not exceed ~4.5	Same as AA: 3.5–4.5; do not exceed ~4.5
Postadjust pH (storage/formulating)	Adjust to 4.8–5.2 after polymerization	Adjust to 4.8–5.2 after polymerization

(Continued)

Table 8.20 (Continued)

Dimension	VAM–AA	VAM–MAA
Corrosion notes (including $AlCl_3$ use)	AA itself is a weak acid; normal EP metallurgy is acceptable. When using $AlCl_3$, the risk of *chloride corrosion* rises—use lined reactors, avoid carbon steel contact points, and keep pH 3.5–4.5 only just-in-time during addition	Same as AA; MAA systems also face chloride-induced corrosion with $AlCl_3$. Prefer Zr/CDI where sensitive equipment or food-contact constraints exist
Regulatory/odor	AA has sharper odor; ensure adequate venting	MAA has a lower odor intensity; often preferred in plants with strict odor limits
Cost/availability (regional)	Generally favorable and widely available	Often slightly higher cost; regional variability
Best-fit applications (baseline)	Labels, carton-to-carton (gloss/matte), window patching, bookbinding (with crosslinker)	Exterior-leaning packaging, alkaline exposure, construction adhesives (with crosslinker)

By contrast, vinyl acetate–methacrylic acid (VAM–MAA) copolymers are preferred where long-term water and alkali resistance dominates, such as in the transition from D3 to D4 classifications or in applications leaning toward exterior durability. Because VAM–MAA systems develop slower early grab, tackifier, and wetter selection can be adjusted to compensate.

When aluminum chloride ($AlCl_3$) is used for rapid ionic crosslinking, strict attention must be given to corrosion control. Stainless steel 316L or glass-lined surfaces should be employed, carbon steel should be avoided, and residence times at low pH must be minimized. Where corrosion constraints are severe, zirconium or CDI crosslinkers offer more suitable alternatives.

For both VAM–AA and VAM–MAA systems, the polymerization process is generally conducted at pH 3.5–4.5, followed by postadjustment to 4.8–5.2 using a sodium acetate/acetic acid buffer. A starved-feed strategy with approximately 2.5% seed latex is commonly applied to achieve controlled particle size.

Recommended starting dosages are as follows: acrylic acid (AA) at 1–5 parts per hundred monomer (phm) VAM, with 2–3 phm as the typical range; methacrylic acid (MAA) at 2–4 phm VAM. External crosslinkers may be introduced at CDI 1–3% (as a 30% dispersion), zirconium 0.1–0.5% (as metal), zinc oxide (ZnO) 0.5–2% (as solids), and aluminum chloride 0.05–0.20% as Al^{3+}, the latter to be added just-in-time with strict chloride management.

8.5.5 Process Safety Guide—Handling AA and MAA in VAM Copolymerization (with Corrosion Controls)

Scope: Practical checklist for safe handling of acrylic acid (AA) and methacrylic acid (MAA) during VAM emulsion polymerization and subsequent compounding. Includes corrosion notes (e.g., with $AlCl_3$), pH discipline, and sequencing (Table 8.21).

Table 8.21 Key materials and controls in handling AA and MAA and other ingredients in copolymerization with VAM.

Material	Primary hazards	Key controls (plant)
Vinyl acetate (VAM)	Highly flammable; eye/skin irritant; explosive vapors	Explosion-proof gear; inerting with care; bond/ground; LEV; keep <30 °C in storage; no ignition sources
Acrylic acid (AA)	Corrosive (skin/eyes); strong odor; polymerizes if uninhibited	Closed transfer; splash protection; cool storage; use inhibitor as supplied; add AA to water (not the reverse)
Methacrylic acid (MAA)	Corrosive; polymerizable; pungent odor	Same as AA; keep cool; do not remove inhibitor; avoid contamination
Acetic acid (10% aq)	Corrosive to eyes/skin	Dilute additions; face shield; add to water; keep away from hypochlorites
Sodium acetate	Low hazard (dust/irritant)	Solution prep in water; avoid dusting
Ammonium persulfate	Strong oxidizer; respiratory sensitizer	Segregate from organics; dedicated scoop; add as aqueous solution only
Hydrogen peroxide (35%)	Strong oxidizer; burns; decomposition/exotherm	Ventilated cabinet; dedicated lines; add last in redox; prevent metal contamination
Ascorbic acid/$FeSO_4$	Irritant; Fe salts stain/corrode	Add in sequence; avoid contact with peroxides until step
Isopropanol	Flammable	Bond/ground; keep away from oxidizers; small-metered feed
Zr complexes	Irritant; hydrolyze at high pH	Add last; pH 4.5–6.5; avoid phosphate/chelators
ZnO dispersions	Dust/respiratory (as powder); aquatic toxicity	Use premade dispersions; avoid powder handling when possible
$AlCl_3$ (aq)	Strongly acidic; chloride-induced corrosion	Glass-lined/316L; minimize low-pH residence; immediate water rinse; segregated hoses

8.5.5.1 Engineering and Administrative Controls

Safe operation of VAM–acid copolymerizations requires strict application of engineering and administrative measures. Closed charging and local exhaust ventilation should be installed at all addition points for acrylic acid (AA), methacrylic acid (MAA), and acetic acid. Reactors must be operated under a nitrogen blanket to minimize oxygen ingress, while storage tanks for acrylic monomers should retain minimal oxygen if the supplier's inhibitor requires it. Explosion protection includes the use of classified electrical equipment in vinyl acetate monomer (VAM) and isopropanol (IPA) areas, with bonding and grounding provided for all solvent lines. Interlocks should include high-temperature cut-offs, automatic monomer feed shutdown on high pressure or high pH, and initiator feed logic tied to monomer feed to ensure starved-feed control.

Materials of construction must be carefully selected. Stainless steel 316L or glass-lined surfaces are appropriate for acidic service, while carbon steel must be avoided in the presence of chloride sources such as aluminum chloride ($AlCl_3$). PTFE-lined hoses and gaskets are recommended for low-pH dosing. Relief systems must be in place, with off-gas routed to a scrubber and peroxide decomposition venting available.

Operating discipline requires maintaining reactor pH between 3.5 and 4.5 during polymerization. Any upward drift should be corrected only with diluted acetic acid. Acetate buffer should be prepared by dissolving sodium acetate in water and adjusting to pH 4.0–4.3 with 10% acetic acid. AA or MAA is added to the aqueous phase and never directly into neat VAM. For seeding, approximately 2.5% of the VAM–AA (or VAM–MAA) blend is charged, held for about 20 min, and followed by starved-feed addition over 3–4 h. Initiators and oxidizers are always added as aqueous solutions and must remain segregated from organic reducers until the redox chase step. The chase sequence proceeds with ferrous sulfate, followed by ascorbic acid, and finally hydrogen peroxide, under strict rate control with venting and cooling available. Post polymerization, the latex is adjusted to pH 4.8–5.2 only after cooldown and completion of the chase.

Personal protective equipment (PPE) must include chemical splash goggles and a face shield during handling of AA, MAA, and acetic acid. Acid-resistant gloves (nitrile or neoprene), long sleeves, and a chemical apron are required, with flame-resistant coats used in solvent handling zones. Respiratory protection is specified by the sites industrial hygiene assessment when ventilation is inadequate or during transfers. Antistatic footwear and grounding straps are mandatory in solvent areas.

Corrosion control is essential when using $AlCl_3$. Glass-lined or 316L stainless steel wetted parts should be employed, while carbon steel, copper alloys, and galvanized surfaces are unsuitable. Residence times at pH below 4.5 must be minimized, and $AlCl_3$ should be dosed just-in-time, followed by restoration of pH to about 5.0 if compatible with the formulation. Dedicated hoses and pumps for $AlCl_3$ should be clearly labeled and immediately flushed with water after use, according to a rinse-and-neutralize standard operating procedure. Periodic inspection of nozzles, valves, and gaskets for chloride stress corrosion cracking (CSCC) is mandatory.

8.5.5.2 Spill and Emergency Actions

Spill and emergency response procedures must be clearly established. For a VAM spill, ignition sources are eliminated, ventilation is maximized, and the solvent is absorbed with inert media for disposal according to solvent waste regulations. For AA, MAA, or acetic acid spills, the material should be contained and cautiously neutralized with sodium bicarbonate, with the resulting residue collected for disposal. In a peroxide incident, the area is evacuated, cooling is applied from a safe distance, decomposition is not confined, and the peroxide emergency protocol is consulted. For an exotherm or runaway reaction, all feeds are stopped, cooling is applied, venting is directed to a scrubber, and the reactor emergency procedure is followed.

Environmental management requires separate waste streams for acids, oxidizers, and solvents. Peroxides must never be mixed with organic waste. Acidic effluents must be neutralized to pH 6–8 before discharge, with monitoring and treatment of Zn or Al content if ZnO or $AlCl_3$ has been used. Inhibitors must not be stripped from AA or MAA storage, and supplier inhibitor levels as well as oxygen requirements must be maintained to prevent autopolymerization.

8.5.5.3 Final Safety Checklist

Before each batch, the following checklist is completed: all safety data sheets have been reviewed and required permits signed; eyewash and shower facilities are available within 10 m; reactor relief and scrubber systems are online; cooling and interlocks are operational. Personnel must be equipped with the correct PPE, and bonding and grounding are verified in solvent zones. The acetate buffer must be prepared and adjusted to pH 4.0–4.3; monomer preblend, initiator, and chase solutions must be freshly prepared. If $AlCl_3$ is employed, associated hardware must be isolated and flushed, with dedicated lines in place and corrosion inspection completed. Spill kits containing neutralizer, absorbents, and a peroxide kit must be staged before initiating the batch.

8.5.6 Three-Interval Copolymerization of Vinyl Acetate with Acrylic or Methacrylic Acid

The copolymerization of vinyl acetate (VAc) with small amounts of acrylic acid (AA) or methacrylic acid (MAA) provides an instructive case in emulsion polymerization where water-soluble monomers, protective colloids, and surfactants interact in a finely tuned balance. The acidic comonomers, though present in only a few percent, strongly influence nucleation, the distribution of functional groups, and ultimately the adhesion properties of the resulting latex. The process may be considered within the framework of three characteristic intervals.

8.5.6.1 Interval I: Nucleation and Particle Birth

Initiator radicals react with dissolved vinyl acetate and acrylic (or methacrylic) acid in the aqueous phase. Because AA/MAA is water-soluble, it rapidly incorporates into the earliest oligomers. Surfactant micelles and PVOH colloidal layers provide loci for particle nucleation. The soluble acid comonomer, therefore, enhances the number of primary nuclei compared with pure VAc systems.

8.5.6.2 Interval II: Particle Growth and Monomer Transport

As the reaction stabilizes, vinyl acetate diffuses continuously from the aqueous phase into growing particles, while AA/MAA partitions between the aqueous phase and the particle surface. This dynamic equilibrium often enriches the particle periphery with acid groups, supported by surfactant stabilization and PVOH adsorption. The cooperative action of surfactant and protective colloid maintains colloidal stability despite the increased polarity of the system.

8.5.6.3 Interval III: Monomer Depletion and Particle Maturation

At lower residual monomer concentrations, the system enters a diffusion-controlled regime. The remaining AA/MAA is scavenged preferentially at the particle interface, leading to a final shell-like enrichment of functional groups. This arrangement is particularly valuable in adhesives, where accessible acid groups can participate in hydrogen bonding or react with external crosslinkers such as carbodiimides or zirconium complexes. PVOH consolidates the surface, producing a robust hybrid layer of physical and chemical stabilization that extends into the dry film.

The evolution of monomer composition during copolymerization, particularly the depletion of vinyl acetate and the surface enrichment of AA/MAA across the three polymerization intervals, is illustrated in Figure 8.1.

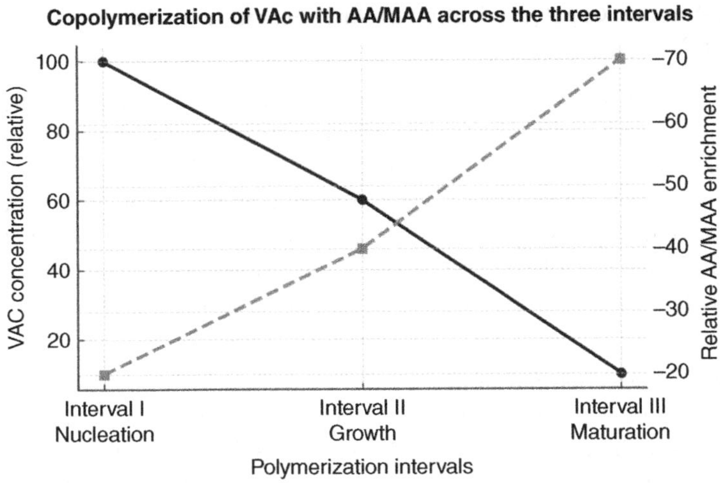

Figure 8.1 Copolymerization of vinyl acetate with acrylic/methacrylic acid across the three polymerization intervals. The solid line (•—•) represents the decreasing concentration of vinyl acetate in the system, while the dashed line (■– –■) represents the progressive enrichment of AA/MAA at the particle surface.

8.5.7 Selection of the Surfactant for the Copolymerization of VAM/AA or MAA

The choice of surfactant in the emulsion copolymerization of vinyl acetate with small amounts of acrylic acid (AA) or methacrylic acid (MAA) plays a decisive role in controlling particle nucleation, colloidal stability, and the final distribution of functional groups. Unlike pure polyvinyl acetate homopolymerizations, which can often be stabilized effectively by protective colloids such as polyvinyl alcohol, the incorporation of water-soluble acidic monomers increases the ionic demand of the system. This requires a carefully balanced surfactant system to ensure reproducible polymerization and stable dispersions.

In the initial nucleation interval, soluble AA or MAA tends to localize at the particle–water interface, where it interacts with both initiator radicals and surface-active agents. Conventional anionic surfactants, such as sodium lauryl sulfate or alkyl ether sulfates, provide electrostatic stabilization but may lead to excessive surface charge when combined with ionizable acid groups. Nonionic surfactants, particularly ethoxylated alcohols with HLB values between 13 and 15, are therefore frequently introduced to moderate electrostatic repulsion and provide steric stabilization. The combined use of anionic–nonionic blends allows for a more uniform particle size distribution and prevents coagulation at high conversion.

In aqueous polymer dispersions, the smallest particles are less stable than larger ones because of their higher surface energy. As a result, polymer or residual monomer molecules tend to migrate from the surface of these very small particles into the surrounding water phase. Over time, these molecules diffuse through the continuous phase and redeposit onto larger particles, which are more stable because of their lower surface energy. This molecular transfer leads to a gradual disappearance of the tiniest particles and a growth of the larger ones, broadening the particle size distribution and reducing latex stability.

Polyvinyl alcohol, when used as a stabilizer in VAc–AA or VAc–MAA formulations, plays a critical role in counteracting this process. By forming a strong, adsorbed protective layer around the polymer particles, it effectively creates a physical barrier that slows down the diffusion of molecules from small particles to large ones. This barrier action keeps the particle size distribution narrower, enhances the long-term stability of the latex, and improves the mechanical properties of the final adhesive film.

However, the protective action of PVOH alone is often insufficient when soluble acid monomers are introduced, since the latter alter the interfacial energy and may lead to particle flocculation unless supported by surfactant stabilization.

The surfactant system must also be selected with regard to the intended application. For adhesive grades where reactive acid groups must remain available for crosslinking, excessive use of strongly adsorbing anionic surfactants should be avoided, as these can block functional sites or compromise water resistance. In contrast, nonionic surfactants or polymerizable surfactants (surfmers) can be used to achieve partial covalent anchoring within the particle shell, thus reducing the level of free surfactant and enhancing film performance. The balance between mobility, colloidal stabilization, and reactivity determines the ultimate efficiency of the copolymer in practical applications (Table 8.22). Recommended surfactants commonly used in VAc–AA/MAA copolymerization and their functional roles are summarized in Table 8.22.

In PVAc–AA–carbodiimide (CDI) adhesives, minimizing free surfactant is critical for achieving high water resistance and durable bonding. Free surfactant refers to molecules that remain dissolved in the aqueous phase rather than adsorbed on particle surfaces or anchored in the polymer shell. Although rarely applied in plant practice, R&D laboratories may use the following methods to assess free surfactant:

Serum analysis: Centrifuge or ultrafilter the latex to separate polymer particles from the aqueous phase. Surface tension (values significantly below 72 mN/m indicate surfactant), conductivity (for ionic surfactants), and foam tests can provide a quick check of residual surfactant.

Chromatographic methods: HPLC, ion chromatography, or LC–MS can quantify surfactant species in the water phase with precision.

Table 8.22 Recommended surfactants for VAc–AA/MAA copolymerization.

Surfactant type	Typical reference values	Main benefits	Risks/limitations
Anionic (e.g., sodium lauryl sulfate, alkyl ether sulfates)	No HLB (CMC ~1–2 g/L; surface tension ~38–40 mN/m)	Strong electrostatic stabilization; effective particle nucleation	Excessive charge with AA/MAA; can reduce water resistance; residual surfactant migration
Nonionic (e.g., ethoxylated alcohols, ethoxylated alkylphenols)	HLB 12–15; cloud point 60–80 °C	Steric stabilization; moderates electrostatic repulsion; improves film clarity	Possible surfactant migration; reduced efficiency at high temperature
Anionic–nonionic blends	No true HLB (performance guided by CMC, cloud point, and blend ratio)	Balanced stabilization; narrower particle size distribution; higher tolerance to ionic stress	Complexity in optimization; potential surfactant desorption at high conversion
Poly(vinyl alcohol) (protective colloid)	—	Strong adsorption; enhances mechanical and freeze–thaw stability	Alone insufficient with soluble acids; needs supplemental surfactant
Polymerizable surfactants (surfmers)	HLB 10–14 (for nonionic part); anchoring confirmed by surface analysis	Covalent anchoring in particle shell; reduces free surfactant; enhances durability	May alter polymer Tg and surface polarity; higher cost

Note: The HLB concept is valid for nonionic surfactants and is widely used for selecting them. For ionic surfactants (anionic, cationic), hydrophilicity is determined by charge interactions rather than by HLB values, so parameters such as CMC, cloud point, surface tension, and electrolyte tolerance are more practical for selection. For blends of anionic and nonionic surfactants, performance is better predicted by these parameters rather than by a calculated HLB value.
Practical Note: Checking for Free Surfactant in PVAc–AA Systems.

Spectroscopic methods: NMR or IR may be used for specific functional surfactants, though less routine.

Indirect performance tests: Excessive water uptake, blocking, or poor wet adhesion in adhesive films often signal high free surfactant. In PVAc–AA–CDI systems, reduced water resistance after curing can indicate surfactant migration.

Formulators aiming for minimal free surfactant often rely on poly(vinyl alcohol) as the main stabilizer, using only a small amount of surfactant for particle nucleation. Polymerizable surfactants (surfmers) further reduce free surfactant by covalent anchoring into the polymer shell.

8.5.8 External Crosslinkers for VAM-AA/MAA Copolymer Adhesives

External crosslinkers can be formulated as one-component (1K) systems by incorporating them into the latex at storage pH 4.5–5.5. Under these mildly acidic conditions, the dispersion remains stable in-can; covalent network formation proceeds primarily during drying and mild thermal cure, when mobility increases, and local neutralization at the substrate interface may occur. This subsection provides concise guidance on the selection and use of water-dispersible carbodiimides in AA/MAA-functional VAM copolymers (e.g., VAM–AA and VAM–MAA).

8.5.8.1 Water-dispersible Carbodiimides (1K Use)

Mechanism and scope. Aliphatic carbodiimides react with carboxylic acids to form hydrolytically stable *N*-acylurea linkages. In 1K systems at pH 4.5–5.5, they remain usable in-can; network formation progresses during drying and mild heat (Table 8.23). Key formulation and processing parameters for carbodiimide (1K) crosslinking systems are summarized in Table 8.23.

Application Notes–Carbodiimide (CDI)

Scope: Adding carbodiimide (CDI) to PVAc–AA/MAA copolymers unlocks a range of durability classes; actual results depend on binder composition, pH, cure, and substrate. In practice, CDI enables D3-class performance at modest levels and, with tighter control and thermal assist, can reach D4-class. (Selection of tackifiers and wetting aids is addressed in the dedicated tackifier section later in this chapter.)

Use window: CDI 1.0–3.0% (as a 30% dispersion, ≈0.3–0.9% active) at pH 4.8–5.5. Add last, mix gently, and observe product-specific pot-life.

Table 8.23 Key parameters–carbodiimide (1K).

Use level (% on polymer solids)	1.0–2.0
Storage pH/in-can stability	pH 4.5–5.5; shelf-stable (weeks; minimize moisture ingress)
Application and cure	Apply as-is; dry 24–72 h or 40–60 °C for 30–90 min
Compatibility	Avoid amines; moderate alkalinity at application can accelerate

Procedure: Add carbodiimide slowly to the latex under low shear at pH 4.5–5.5, mix for 15 min, de-air, and screen. Apply and allow to dry; mild heat expedites cure. Prepare fresh batches periodically to account for gradual crosslinker consumption by moisture.

D3 (baseline): Binder: PVAc–AA (~2–3 pbw AA) or PVAc–MAA (~2.5–3 pbw).
CDI: 1.0–2.0% (as 30% dispersion).
Cure: ambient 24–48 h.
Outcome: D3-capable humidity/heat resistance on paper and coated board when free/mobile surfactant is controlled.
D4 (optimized): CDI: 2.0–3.0% (as 30% dispersion).
Matrix controls: minimize mobile plasticizer and free surfactant; consider slightly higher acid content or MAA-rich binders for network density.
Thermal assist: warm press 50–70 °C, 1–5 min, or postcure 50–60 °C, 12–24 h before testing.
Outcome: D4-level durability can be achieved, subject to verification by EN 204 (or local) testing.

Substrate Notes

Coated/lacquered cartons: maintain low free-surfactant to avoid interfacial bloom; ensure adequate coat weight and compression for rapid set.
Exterior/alkali-leaning uses: PVAc–MAA/CDI offers higher baseline hydrolytic stability; confirm creep and water class per specification.
Good practice: Hold storage/formulating pH 4.8–5.2; avoid strong bases; dose CDI last; confirm blocking/whitening risk at your solids/viscosity window. (For the effect of tackifiers and dynamic wetting agents on initial grab and set time, see the dedicated tackifier section later in this chapter.)

8.5.8.2 Aluminum Chloride ($AlCl_3$)

Mechanism and scope. Aluminum chloride functions as a Lewis acid, coordinating with carboxylate groups in AA/MAA-functional copolymers. This ionic crosslinking provides water resistance and cohesion at relatively low addition levels. However, $AlCl_3$ systems are salt-like and sensitive to hydrolysis, necessitating careful pH control and in-can stabilization (Table 8.24).

8.5.8.3 Zirconium Complexes (Zr)

Mechanism and scope. Zirconium salts and chelates (e.g., ammonium zirconium carbonate, zirconium acetate) form strong coordination complexes with carboxylate groups. Unlike $AlCl_3$, these complexes are more hydrolytically stable and can achieve D3/D4 water resistance, especially in MAA-rich copolymers (Table 8.25).

8.5.8.4 Zinc Oxide (ZnO)

Mechanism and scope: ZnO, introduced as fine dispersions, reacts with carboxylic acid groups to form zinc carboxylates that act as ionic crosslinks. Compared with $AlCl_3$, ZnO offers greater hydrolytic stability and moderate covalent/ionic character, but requires slightly higher dosages. It can also provide ancillary benefits such as mildew resistance and opacity (Table 8.26).

Table 8.24 Key parameters–$AlCl_3$ (1K).

Parameter	Typical range/notes
Use level (% on polymer solids)	0.5–1.5
Storage pH/in-can stability	pH 4.0–4.5; sensitive to hydrolysis above pH 5
Application and cure	Apply at acidic pH; dries under ambient conditions; heating not required
Compatibility	Avoid phosphate dispersants; precipitation possible with anionic thickeners

Procedure: $AlCl_3$ is introduced as an aqueous solution (≈5–10%) into the latex at pH 4.0–4.5 under gentle agitation. Avoid over-neutralization during formulation. The adhesive should be applied and dried without delay, since prolonged storage can reduce effectiveness due to slow hydrolysis.

Table 8.25 Key parameters–Zr crosslinkers (1K).

Parameter	Typical range/notes
Use level (% on polymer solids)	0.5–2.0
Storage pH/in-can stability	Stable at pH 4.5–5.5 (weeks–months, depending on supplier grade)
Application and cure	Apply as formulated; dry 24–48 h; mild heating accelerates
Compatibility	Sensitive to high phosphate load; nonionic thickeners preferred

Procedure: Add zirconium complex (e.g., 20–30% dispersion) to the latex at pH 4.5–5.5 with slow agitation. Avoid strong bases or polyphosphates, which can precipitate zirconium salts. De-air and apply to the substrate; curing proceeds during drying or under mild heating (40–60 °C).

Table 8.26 Key parameters–ZnO (1K).

Parameter	Typical range/notes
Use level (% on polymer solids)	1.0–3.0
Storage pH/in-can stability	pH 4.5–5.5; moderate shelf stability (several weeks)
Application and cure	Apply as-is; ambient cure; heat (50–70 °C) improves network density
Compatibility	Risk of viscosity drift with polyacrylate thickeners; avoid high phosphate dispersants

Procedure: ZnO is introduced as an aqueous slurry (≈20–30%) into the latex at pH 4.5–5.5 with low-to-moderate shear. Good dispersion is critical to prevent agglomeration. Formulations should be screened and de-aired before application.

8.6 External Crosslinker–Application Mapping

Each external crosslinker confers a distinct performance profile in VAM–AA/MAA adhesives. The following summary outlines the types of adhesives that can be most effectively produced using each crosslinker.

Carbodiimide (CDI)

Carbodiimides are the most versatile covalent crosslinkers, capable of achieving high durability. They are ideally suited for:

1) Wood adhesives requiring D3 and D4 classification under EN 204.
2) Exterior-exposed laminations where hydrolytic stability and creep resistance are essential.
3) Packaging adhesives for coated and lacquered cartons where moisture resistance and thermal set are demanded.

Aluminum Chloride ($AlCl_3$)

$AlCl_3$ provides economical ionic crosslinking with moderate water resistance. It is mainly applied in:

- Carton sealing adhesives for interior use (D2–D3 class).
- Paper-to-paper laminations where fast set and low cost are important.
- General-purpose packaging adhesives not requiring long-term hydrolytic stability.

Zirconium Complexes (Zr)

Zirconium salts and chelates deliver strong ionic coordination with excellent durability. They are widely used in:

1) Exterior wood adhesives requiring consistent D3/D4 classification.
2) High-humidity resistant laminations, such as for beverage cartons and specialty boards.
3) Industrial adhesives for substrates with alkaline tendencies, where hydrolytic stability is critical.

Zinc Oxide (ZnO)

ZnO provides moderate crosslinking with additional functional benefits. It is best suited for:

1) Moisture-resistant packaging adhesives in cost-sensitive applications (D2–D3).
2) White or pigmented laminating adhesives, where ZnO imparts opacity and color uniformity.
3) Adhesives with ancillary resistance properties, including mildew inhibition in paper and board applications.

8.6.1 Formulations and Applications (1K Crosslinkers)

This section provides ready-to-use formulations aligning each external crosslinker with its most suitable adhesive applications. All systems are designed as one-component (1K) adhesives formulated and stored at pH 4.5–5.5. Stabilization relies primarily on poly(vinyl alcohol) with minimal free surfactant; the crosslinker is dosed last. Where tackifier or dynamic wetting aid is mentioned, refer to the dedicated tackifier section for selection guidance. Target solids and viscosity can be adjusted by water and thickener to the coater or applicator in use.

8.6.2 D3 Wood Adhesive Based on PVAc–AA/Carbodiimide

This section presents a practical 1K pathway to EN 204 D3 performance using a PVAc–AA copolymer crosslinked with a carbodiimide (CDI) dispersion. The copolymer supplies a controlled level of carboxyl functionality (typically 0.5–1.5% AA on total monomer) for CDI to convert into *N*-acylurea links during drying, yielding durable wet strength without a two-pack mix. The adhesive is stabilized primarily by PVOH with minimal free surfactant to balance water resistance and wood wetting. CDI is postadded to the finished base at pH ~4.8–5.3 and mixed under low shear; storage pH is maintained at 4.5–5.5 to limit CDI hydrolysis while preserving latency.

Formulation targets typically include 50–55% solids and a mid-range viscosity suited to roller, bead, or brush application on softwoods and hardwoods used indoors with frequent short-term water exposure or prolonged high humidity. Open time and press time are tuned by solids, plasticizer level, and temperature; wood moisture of 8–12%, uniform spread, and firm clamping pressure are essential for consistent bondlines. Performance verification should cover EN 204 D3 conditioning, creep/heat resistance (e.g., WATT 91), and joint durability after cyclic humidity. Table 8.27 provides an illustrative starting point; dose and pH should be optimized for the specific PVAc–AA latex, wood species, and process conditions.

8.6.3 D4 Adhesive Room-temperature Cure Routes—1K vs. 2K

This section presents two production-ready routes for room-temperature (RT) curing VAM–AA/MAA adhesives. Route A is a one-component (1K) high-acid system that deliberately uses a dual crosslinker package (carbodiimide + zirconium) to exploit the elevated carboxyl content. Route B is a two-component (2K) system based on a hydroxyl-functional latex and polyisocyanate, combined with carbodiimide, mixed at use. Both routes avoid reliance on PVOH hydroxyls for cure.

Table 8.27 D3 wood adhesive (PVAc–AA/CDI; 1K)–illustrative starting point.

Ingredient	% by weight	Notes
Deionized water	9.0	Balance viscosity and solids
PVOH solution (10% active)	15.0	Protective colloid; degree of hydrolysis ≥88%
PVAc–AA latex (55% solids; 2–3 pbw AA)	60.0	Base copolymer; low free surfactant
Nonionic wetting aid (optional)	0.2	Only if needed; avoid strong anionics
Defoamer	0.2	Low-foam grade
Biocide	0.2	Per supplier
Thickener (HEC/HASE), as required	q.s.	Set to application viscosity
Carbodiimide dispersion (30% active)	2.0	≈0.6% active on solids; dose lasts at pH 4.8–5.2
Water	q.s. to 100	Adjust to 50–55% total solids

Procedure: Charge water and PVOH solution, then latex under gentle agitation at pH 4.8–5.2. Add defoamer, biocide, and optional wetting aid; adjust viscosity. Dose carbodiimide last with low shear, mix for 15 min, de-air, and screen. Cure 24–48 h at ambient or 40–60 °C for 30–60 min prior to testing (EN 204 D3).

8.6.3.1 One-component D4 Adhesive High-acid Route–Mandatory Dual Crosslinkers (CDI + Zr)

Design intent: Maximize –COOH content (AA/MAA) so that both carbodiimide (CDI) and zirconium complex (Zr) can simultaneously crosslink at RT. CDI forms covalent *N*-acylurea linkages with –COOH; Zr forms strong coordination complexes with carboxylates, increasing creep and hydrolysis resistance. The combination reduces ambiguity in D3/D4 outcomes while preserving 1K handling (Tables 8.28A–8.28C).

8.6.3.2 Two-component D4/Hydroxyl-Functional Route–CDI/Polyisocyanate (RT Cure)

Design intent: Build –OH directly into the polymer (e.g., HEA/HPMA 1.5–2.0 pbw) and pair with CDI for –COOH plus a dispersed aliphatic polyisocyanate as Part B. Mix at use. RT cure forms urethane/urea/amide linkages with high crosslink density, robustly achieving D4 when application variables are controlled (Tables 8.29A–8.29C).

Table 8.28A Design targets—1K D4 high-acid PVAc–MAA/CDI/Zr (RT cure).

Parameter	Mandatory target/notes
AA/MAA level (pbw on monomer)	5.0–6.0 (push upper bound; favor MAA for hydrolytic stability)
Protective colloid (PVOH)	10–16% of batch as solution (stabilization only; not relied upon for cure)
Free surfactant (total)	As low as compatible with stability; prefer nonionic traces; avoid strong anionics
Primary crosslinker (mandatory)	Carbodiimide dispersion 30%: 2.5–3.0% on batch (≈0.75–0.9% active on polymer solids)
Co-crosslinker (mandatory)	Zr complex 25%: 0.7–1.2% on batch (avoid phosphate dispersants)
Formulation/storage pH	4.8–5.2 (critical for CDI stability and latex integrity)
Cure profile	Room temperature; functional cure 24–72 h, depending on humidity and coat weight

Table 8.28B Controls to enable very high AA/MAA without instability 1K D4.

Control lever	Purpose/guidance
Seed and preemulsion strategy	Use robust PVOH-stabilized seed; meter AA/MAA starved to avoid aqueous-phase gelation
Initiation	Persulfate or redox at low ionic strength; consider ascorbate/Fe for low-temp starts
Buffering	Minimal; hold pH 4.8–5.2; avoid phosphate with Zr
Neutralization state	Keep AA/MAA mostly free acid during polymerization; minimize fixed alkali
Electrolytes	Keep ionic strength low; screen defoamer/dispersant for salt-in effects
Viscosity management	Target 50–55% solids; adjust water and HEC/HASE; avoid gel-point approach
Free surfactant control	Strip to minimum that maintains stability; verify no interfacial bloom

Table 8.28C Example-1K D4 high-acid PVAc–MAA/CDI/Zr (RT cure).

Ingredient	% by weight	Notes
Deionized water	7.5	Viscosity control
PVOH solution (10–12% active)	15.0	Protective colloid; stabilization only
PVAc–MAA latex (55% solids; 5.5–6.0 pbw MAA)	60.0	High-acid binder; low free surfactant
Nonionic defoamer	0.2	Low-foam
Biocide	0.2	Per supplier
Optional nonionic wetting aid	0.15	Only if needed for difficult substrates
Carbodiimide dispersion (30% active)	2.8	≈0.84% active; mandatory primary crosslinker
Zirconium complex (25% active)	1.0	Mandatory co-crosslinker; add after CDI
Thickener and water	q.s. to 100	Target 50–55% solids

Procedure: Make down at pH 4.8–5.2. Add CDI, mix for 10 min under low shear, then add Zr and mix an additional 10 min. De-air and screen. Apply at room temperature; allow 24–72 h cure depending on humidity and substrate. Verify D3/D4 per EN 204.

Table 8.29A 2K D4 Part A-PVAc–MAA–HEA latex with CDI (storage pH 4.8–5.2).

Ingredient	% by weight	Notes
Deionized water	8.0	Viscosity control
PVOH solution (10–12% active)	12.0	Stabilization only
PVAc–MAA–HEA latex (55% solids; 2.5–3.0 pbw MAA; 1.5–2.0 pbw HEA)	60.0	–OH built into polymer; low free surfactant
Nonionic defoamer	0.2	Low-foam
Biocide	0.2	Per supplier
Carbodiimide dispersion (30% active)	1.8	Target 0.6–0.9 eq CDI per eq –COOH
Thickener and water	q.s. to 100	Target 50–55% solids

Table 8.29B 2K D4 Part B-dispersed aliphatic polyisocyanate (RT cure).

Ingredient	% by weight	Notes
Polyisocyanate dispersion (HDI trimer/biuret, 60–75% solids)	100.0	Water-dispersible; aliphatic

Table 8.29C Mix-at-use recipes (RT cure).

Component	pbw	Notes
Part A (PVAc-MAA-HEA + CDI)	100.0	At 20–25 °C; pH 4.8–5.2
Part B (polyisocyanate dispersion)	2.5–4.0	Set NCO index 0.5–0.8 vs. polymer -OH + residual -COOH

Procedure: Bring both parts to 20–25 °C. Add Part B into Part A over 1–2 min under gentle agitation. Mix 10 min; de-air and screen. Apply and press at RT. Functional cure within 24–72 h at 23 °C. Minimize free surfactant and avoid amines/strong bases. Manage foaming risk by targeting -OH conversion as the dominant NCO sink.

8.6.4 Humidity-resistant Adhesives—(1K and 2K High-acid Variants)

To align with the agreed D4 design rules, the prior 1K Zr-only formula is re-scoped to D3. A D4-capable 1K variant is provided using the mandatory dual crosslinker package (CDI + Zr) and a high-acid binder. Both systems cure at room temperature.

8.6.4.1 D3 Humidity-resistant Adhesive-PVAc-AA/MAA/Zirconium Complex (1K)

A representative formulation of a one-component D3 humidity-resistant PVAc adhesive based on a zirconium complex crosslinking system, designed for room-temperature cure, is presented in Table 8.30.

Table 8.30 D3 humidity-resistant adhesive (Zr complex; 1K)—room-temperature cure.

Ingredient	% by weight	Notes
Deionized water	10.0	Balance viscosity
PVOH solution (10–12% active)	15.0	Protective colloid; stabilization only
PVAc-AA/MAA latex (55% solids; ~2.5–3.0 pbw acid)	60.0	Low free surfactant
Nonionic wetting aid (optional)	0.15	Avoid phosphates
Defoamer	0.2	Per supplier
Biocide	0.2	Per supplier
Zirconium complex (25% active)	2.0	≈0.5% active; dose lasts at pH 4.8–5.2
Thickener and water	q.s. to 100	Adjust to 50–55% solids

Procedure: At pH 4.8–5.2, blend water, PVOH, and latex; add auxiliaries. Dose zirconium complex last under slow agitation; avoid phosphate dispersants. De-air and screen. Cure at room temperature 24–72 h. Target: D3 humidity resistance.

8.6.4.2 D4 Humidity-resistant Adhesive 1K High-acid PVAc–MAA/CDI/Zr (RT Cure)

A high-acid, one-component D4 humidity-resistant PVAc adhesive formulation employing a dual crosslinking system based on carbodiimide and zirconium complex, designed for room-temperature cure, is presented in Table 8.31.

8.6.4.3 Note on Durability Classes (DIN EN 204)

The formulations presented above can be directly mapped onto the durability classes defined by DIN EN 204, which are widely used in the wood adhesive industry to benchmark water resistance. Pure PVAc homopolymer adhesives typically achieve only the D1 or D2 classes, which are suitable for interior applications with little or no water contact. This limitation reflects the inherent hydrolytic sensitivity of the polyvinyl acetate ester backbone. By copolymerizing with acrylic acid or methacrylic acid, the adhesive acquires carboxyl functionality that improves cohesion and water resistance, thereby lifting the performance to the D3 class. When a crosslinker such as carbodiimide is introduced into a D3-level copolymer system, ionic or covalent bridges reinforce the polymer network, allowing the adhesive to withstand exterior exposure

Table 8.31 D4 humidity-resistant adhesive (1K high-acid with mandatory CDI/Zr)–room-temperature cure.

Ingredient	% by weight	Notes
Deionized water	7.5	Viscosity control
PVOH solution (10–12% active)	15.0	Protective colloid; not relied upon for cure
PVAc–MAA latex (55% solids; 5.5–6.0 pbw MAA)	60.0	High-acid binder; very low free surfactant
Nonionic wetting aid (optional)	0.15	Only if needed; avoid phosphates
Defoamer	0.2	Per supplier
Biocide	0.2	Per supplier
Carbodiimide dispersion (30% active)	2.8	≈0.84% active; add first of the crosslinkers at pH 4.8–5.2
Zirconium complex (25% active)	1.0	Add after CDI; mandatory co-crosslinker
Thickener and water	q.s. to 100	Target 50–55% solids

Procedure: Make down at pH 4.8–5.2. Add CDI under low shear and mix for 10 min; then add Zr and mix for an additional 10 min. De-air and screen. Apply and press at room temperature; allow 24–72 h cure depending on humidity and coat weight. *Target: D4 humidity resistance* when glue-line thickness, low free surfactant, and high-acid binder are maintained. Verify per EN 204.

Table 8.32 Durability classes of PVAc-based adhesives (DIN EN 204).

Class	Description	Typical application
D1/D2	Homopolymer PVAc	Interior joinery, furniture, dry conditions
D3	VAM–AA or VAM–MAA copolymer	Kitchen joinery, window frames under cover
D4	Crosslinked copolymer (e.g., carbodiimide-modified)	Exterior load-bearing joints, outdoor wood structures

and even boiling water tests. This modification elevates the system into the D4 class, which represents the highest level of durability for PVAc-based adhesives (Table 8.32).

8.6.4.4 D2–D3 Carton Sealing—PVAc-AA/$AlCl_3$ (Economical)

This section provides an economical 1K route to D2–D3 performance for carton and case sealing based on a PVAc–AA copolymer activated with aluminum chloride ($AlCl_3$). Low-level acrylic acid in the binder (typically 0.5–1.5% on total monomer) supplies carboxylate sites that form aluminum carboxylate networks during drying at mildly acidic pH, delivering a fast set, good fiber tear, and humidity-tolerant bonds suitable for interior packaging (folding cartons, folding box board, and recycled board) where occasional moisture or high RH is expected but prolonged water exposure is not.

The system is true 1K: an aqueous $AlCl_3$ solution is postadded to the finished base at pH ~3.8–4.5 under low shear and used as-is. Target solids 48–55% and a low-to-mid viscosity tailored for wheel, nozzle, or pot application with low foam and clean machining. Keep plasticizer modest to preserve cuttability and minimize blocking. Verify adhesion on coated/varnished boards; highly lacquered or low-energy surfaces may require alternative crosslinkers or surface treatment.

For robust performance, avoid phosphate or strong chelating additives, maintain storage/use pH in the recommended window, and dose $AlCl_3$ conservatively (optimize around the starting point in Table 8.33). Observe food-contact compliance where relevant and handle $AlCl_3$ with corrosion precautions (HDPE/PP lines, minimal contact with carbon steel). Table 8.33 gives an illustrative starting formulation; fine-tune $AlCl_3$ level, pH, and coat weight to meet line speed, substrate, and D2–D3 durability targets.

8.6.4.5 Moisture-resistant Packaging—PVAc-AA/ZnO

This section describes a cost-efficient 1K route to moisture-resistant carton and case sealing using a PVAc–AA binder reinforced with zinc oxide (ZnO) dispersion. The acrylic acid content of the copolymer (typically 0.5–1.5% on total monomer) provides carboxyl functionality that forms zinc carboxylates during

Table 8.33 Interior packaging adhesive ($AlCl_3$; 1K).

Ingredient	% by weight	Notes
Deionized water	12.0	Balance viscosity
PVOH solution (8–10% active)	12.0	Colloidal stabilization
PVAc–AA latex (55% solids)	61.0	Base copolymer
Low-foaming wetting aid (optional)	0.2	For coated board
Defoamer	0.2	Per supplier
Biocide	0.2	Per supplier
$AlCl_3$ solution (10% active)	1.4	≈0.14% $AlCl_3$ on batch; add at pH 4.2–4.5
Water and thickener	q.s. to 100	Target 48–52% solids

Procedure: Formulate at pH 4.2–4.5. Add $AlCl_3$ solution toward the end under gentle agitation. Avoid prolonged storage to limit hydrolysis. Apply and dry under ambient conditions for D2–D3 service.

drying. The resulting ionic network improves wet strength and creep resistance under elevated humidity while retaining clean machining and good fiber tear for interior packaging environments.

The system remains single-component: ZnO is postadded to the finished base as a stabilized aqueous dispersion. A practical starting point is solids 48–55% and a low-to-mid application viscosity compatible with wheel, nozzle, or pot application. Maintain pH in the mildly acidic range (approximately 4.2–5.5) to promote carboxylate formation and preserve dispersion stability. Keep plasticizer levels modest to avoid blocking and to maintain cuttability on high-speed lines.

For robust performance and storage stability, use well-milled ZnO dispersions, agitate before use, and avoid strong chelating additives or high-phosphate packages that can suppress zinc–carboxylate formation. Watch for thickening over time and potential whitening at higher ZnO loadings; adjust defoamer and wetting aid judiciously to maintain clean laydown. Table 8.34 provides an illustrative starting formulation; optimize ZnO level, pH, and coat weight to meet the required line speed, substrate, and humidity-tolerance targets (Table 8.34).

8.6.4.6 Coated/Lacquered Cartons—High-grab PVAc–AA/CDI

This section outlines a high-grab, single-component packaging adhesive based on a PVAc–AA copolymer crosslinked with carbodiimide (CDI). The acrylic acid level (typically 0.5–1.5% on total monomer) provides carboxyl functionality for CDI to form *N*-acylurea linkages during drying, delivering humidity-resistant bonds while maintaining fast wet tack. The system is suited to coated or lacquered cartons where early green strength and clean machining are critical.

Table 8.34 Moisture-resistant packaging adhesive (ZnO; 1K).

Ingredient	% by weight	Notes
Deionized water	10.0	Balance viscosity
PVOH solution (10% active)	14.0	Protective colloid
PVAc–AA latex (55% solids)	60.0	Low free surfactant
Defoamer	0.2	Per supplier
Biocide	0.2	Per supplier
ZnO slurry (30% active, fine grade)	2.5	≈0.75% active on solids; dose lasts at pH 4.8–5.2
White pigment extender (optional)	1.0	Opacity and color uniformity
Thickener and water	q.s. to 100	Adjust to 50–54% solids

Procedure: Blend at pH 4.8–5.2. Add ZnO slurry with moderate shear to ensure good dispersion. De-air and screen. Ambient cure; optional 50–60 °C warm press improves crosslink density.

Formulation guidance: target 50–55% solids with a low-to-mid application viscosity for wheel or nozzle application. Postadd CDI at pH ~4.8–5.3 under low shear, and maintain storage pH at 4.5–5.5 to limit CDI hydrolysis while preserving latency. Where line speed demands exceptional wet-grab, a small amount of a compatible, low-migration tackifier dispersion may be employed (optional), balancing open time and set without compromising cuttability.

Table 8.35 High-grab packaging adhesive (CDI; 1K)—tackifier optional.

Ingredient	% by weight	Notes
Deionized water	10.0	Balance viscosity
PVOH solution (10–12% active)	16.0	Surface consolidation
PVAc–AA latex (55% solids)	56.0	Low free surfactant
Dynamic wetting aid (nonionic, low-foam)	0.15	Only as needed for lacquered surfaces
Defoamer	0.2	Per supplier
Biocide	0.2	Per supplier
Carbodiimide dispersion (30% active)	2.5	≈0.75% active; dose lasts at pH 4.8–5.2
Optional tackifier dispersion	2.0	Select per tackifier section; evaluate blocking/whitening
Thickener and water	q.s. to 100	Target 50–54% solids

Procedure: Formulate at pH 4.8–5.2. Add CDI last under low shear. If a tackifier is used, confirm compatibility and blocking resistance.

Substrate considerations: coated or varnished boards can show reduced surface energy; verify cleanliness and surface energy, and, if feasible, raise it by corona or plasma treatment. Avoid strong chelators and high-phosphate packages that may interfere with adhesion. Optimize CDI level, pH, and coat weight using the starting point in Table 8.35, and validate performance under expected humidity and handling conditions (e.g., creep and fiber-tear on coated stocks).

Table 8.36 VAM–AA copolymer adhesive with ZnO/carbodiimide (1K) for lacquered carton.

Raw material	Parts by weight (pbw)	Function
VAM–AA copolymer dispersion (55% solids, Tg ≈ +5 to +10 °C; low-surfactant grade preferred)	100.0	Base polymer binder
ZnO aqueous dispersion (40–50% solids, fine grade)	1.0–1.5 (on polymer solids)	1K ionic adhesion promoter; improves wet tack on lacquer
Carbodiimide crosslinker (water-dispersible, latent at pH 4.5–5.5)	1.0–2.0 (on polymer solids)	1K external crosslinker; boosts wet strength and peel
Polyvinyl alcohol solution (10% aq., 88–90% DH)	1.5–2.5	Colloidal stabilizer; keep low to minimize migration
Plasticizer (triacetin or citrate type)	2.0–4.0	Balances brittleness; controls open time
Very-low-wetting aid (only if needed)	0.05–0.10	Laydown on slick lacquers; keep minimal to avoid migration
Silicone defoamer	0.3–0.5	Foam control
Preservative (isothiazolinone blend)	0.2–0.3	Biocidal protection
Deionized water	Adjust to 50–55% solids	Viscosity adjustment

Incorporation Procedure

1) Charge VAM–AA dispersion under moderate stirring (pH 4.5–5.5).
2) Add ZnO dispersion slowly under shear; mix for 10–15 min.
3) Add carbodiimide; avoid alkaline drift (keep pH ≤5.5). Mix 15 min.
4) Add PVOH solution and plasticizer; mix until uniform.
5) If wetting is insufficient, add minimal wetting aid (0.05–0.10%).
6) Add defoamer and preservative; adjust solids/viscosity with water.

Application and Performance

Designed for carton-to-lacquered carton. Cold-applied at 18–30 °C; mild warming (30–40 °C) is optional for high-speed lines. The combined ZnO + carbodiimide 1K system improves interfacial anchoring and internal cohesion, yielding stronger peel and higher fiber integrity. Target coat weight 3.5–5.0 g/m^2 wet; forced dry 40–60 °C improves early strength. Minimize total surfactant to reduce migration and blocking.

Table 8.37 VAM–AA copolymer adhesive with amino silane for glass, metal, and mineral substrates.

Raw material	Parts by weight (pbw)	Function
VAM–AA copolymer dispersion (55% solids, Tg ≈ +8 to +12 °C; low-surfactant grade)	100.0	Base polymer binder
Amino silane (3-aminopropyltrimethoxysilane, supplied as a 50% aqueous solution)	0.5–1.0 (on polymer solids)	1K adhesion promoter; hydrolyzes to silanols that bond to oxide surfaces
Polyvinyl alcohol solution (10% aq., 88–90% DH)	2.0–3.0	Colloidal stabilization; viscosity control
Plasticizer (citrate or adipate type)	3.0–5.0	Flexibility; reduces brittleness
Silicone defoamer	0.3–0.5	Foam control during mixing/application
Preservative (isothiazolinone blend)	0.2–0.3	Biocidal protection
Deionized water	Adjust to 50–55% solids	Viscosity adjustment

Incorporation Procedure

1) Charge VAM–AA dispersion to the vessel under moderate agitation.
2) Add amino silane slowly as a 50% aqueous solution; maintain pH 4.5–5.5 to keep silane latent until application.
3) Introduce PVOH and plasticizer sequentially; stir until uniform.
4) Add defoamer and preservative; adjust with water to the final viscosity/solids.

Application

Designed for glass, metal, and mineral substrates where strong bonding to oxide surfaces is required. Cold-applied at 18–30 °C; coat weight 3.5–5.0 g/m^2 wet. The amino silane hydrolyzes during film formation, producing silanol groups that bond to hydroxylated surfaces, while its amino groups interact with the copolymer matrix, creating durable interfacial adhesion.

General Notes for All Formulations

Maintain storage pH between 4.8 and 5.2 for 1K stability. Dose crosslinkers last under low shear. Minimize free surfactant and mobile plasticizer to improve water class and creep resistance. Confirm final pH, viscosity, and solids prior to scale-up, and verify performance to EN 204 (D2–D4) or local standards (Tables 8.36 and 8.37).

8.7 Tackifiers in PVAc-based Adhesives

Among the many additives explored in emulsion adhesives, tackifiers hold a special position. These low-molecular-weight resins are widely used in

Table 8.38 Effectiveness of tackifiers in PVAc adhesives.

Adhesive type	Role of tackifiers	Practical outcome
PVAc homopolymer	Minor surface wetting; limited compatibility; acts mainly as an inert resin phase	Marginal benefit; effect not durable; little impact on initial grab
PVAc–AA/MAA copolymers	Act as a plasticizing resin phase within the adhesive film; weak polar interactions (hydrogen bonding, polarity effects) with carboxyl groups improve miscibility	Significant tack improvement with strong initial grab; supports pressure-sensitive and packaging adhesives

Summary: The main mechanism of tackifiers in PVAc-based systems is physical, as they act as a plasticizing resin phase within the adhesive film. This effect provides body, wet tack, and initial grab. Tackifiers offer minimal benefit in pure PVAc homopolymer adhesives, but substantial performance gains are achieved in PVAc–AA/MAA copolymers, where polarity-driven interactions improve compatibility and allow effective use in pressure-sensitive and packaging applications.

pressure-sensitive, packaging, and paper adhesives to improve wet tack and peel adhesion. However, their relevance depends strongly on the chemistry of the polymer binder.

In PVAc homopolymer adhesives, tackifiers have only a trivial effect. The vinyl acetate backbone lacks functional groups capable of strong interaction with tackifier molecules, and the improvement in adhesion or durability is marginal at best.

The situation changes markedly when PVAc is copolymerized with functional monomers such as acrylic acid (AA) or methacrylic acid (MAA). The carboxyl groups introduced by these comonomers provide polar sites that interact favorably with tackifiers, greatly improving compatibility, adhesion strength, and film performance. When further combined with external crosslinkers such as carbodiimides, the system achieves not only strong initial tack but also long-term cohesion and water resistance.

This section examines the ineffectiveness of tackifiers in PVAc homopolymers, their true role in PVAc–AA copolymer adhesives, and their synergy with carbodiimide crosslinkers. It concludes with representative formulations that highlight their practical applications in modern adhesive systems (Table 8.38).

8.7.1 Suitable Tackifiers for PVAc-AA

The incorporation of tackifiers into PVAc–AA adhesive formulations is essential to enhance initial tack, peel strength, and adhesion to low-energy or porous substrates. Unlike pure PVAc, where tackifiers contribute little, the presence of

Table 8.39 Widely used tackifiers for PVAc-AA adhesives.

Tackifier type	Main examples	Key benefits in PVAc-AA	Application fit
Rosin esters	Glycerol ester of gum rosin, Pentaerythritol ester of rosin	Excellent adhesion to polar substrates (paper, board), improves tack and film stickiness, compatible with PVAc-AA emulsions	Paper and board adhesives, laminating adhesives
Terpene phenolic resins	Terpene-phenolic blends (from α-pinene or β-pinene with phenols)	Strong tack at low temperatures, high shear resistance, effective in PSAs	Pressure-sensitive adhesives (PSAs), high-performance packaging
Hydrogenated rosin esters	Hydrogenated glycerol ester of rosin	Light color, better oxidation resistance, improved storage stability, suitable for transparent films	Clear film adhesives, laminates, premium labels
Aromatic hydrocarbon resins (C9)	Styrene-modified aromatic resins	Cost-effective, good adhesion to difficult surfaces, enhances cohesive strength	General-purpose adhesives, carton sealing, cost-sensitive packaging
Aliphatic hydrocarbon resins (C5)	Polyterpene resins, C5 petroleum resins	Good initial tack, fast set speed, compatibility with PVAc-AA in hot-melt emulsions	Fast-tack adhesives, high-speed converting
Dimerized rosins	Dimerized rosin esters	High molecular weight, boosts cohesion, useful in formulations needing water and heat resistance	Heat-resistant adhesives, specialty laminates

acrylic acid provides anchoring sites and improved compatibility, making the system receptive to a wide range of tackifying resins. The selection of the tackifier must balance compatibility, film clarity, softening point, and polarity to match the end-use requirements (Table 8.39).

8.7.2 Optimizing the Use of Tackifiers in PVAc-based Adhesives

The use of tackifiers in polyvinyl acetate (PVAc) homopolymer adhesives has only a minor influence on performance. Since the polymer backbone lacks functional groups for strong interaction, tackifiers in this case act mainly as an inert resin phase. Improvements in initial tack, wet grab, or bond strength are therefore marginal, and the effect is not durable when blended into pure PVAc emulsions.

A more effective approach is to copolymerize PVAc with functional monomers such as acrylic acid (AA) or methacrylic acid (MAA). The introduction of carboxylic groups along the polymer chain does not create covalent anchoring sites for tackifiers, but it does enhance polarity and compatibility. This allows tackifiers to disperse more effectively within the adhesive matrix, functioning as a plasticizing resin phase that increases body, wet tack, and initial grab. As a result, PVAc–AA/MAA copolymer adhesives with tackifiers show marked improvements in adhesion, particularly on paper, carton, and other porous substrates where fast setting and fiber-tear strength are required.

For more demanding applications, however, the presence of tackifier and acid comonomer alone may not suffice. The introduction of carbodiimides or other crosslinkers provides an additional performance dimension. These crosslinkers react with pendant acid groups in the copolymer, generating covalent bonds that enhance cohesive strength, water resistance, and thermal durability. The result is an adhesive system that can be tailored to both conventional packaging needs and high-performance lamination or film-bonding applications.

Accordingly, the following sections first present formulations based on PVAc–AA copolymers combined with tackifiers, representing practical one-component adhesive systems. The discussion then progresses to advanced formulations where tackifiers are combined with carbodiimides or other crosslinkers, thereby expanding the scope of applications to substrates and environments where enhanced durability is critical.

8.7.3 Formulations of PVAc-AA with Tackifier and Crosslinker (Carbodiimide)

For corona-treated biaxially oriented polypropylene (BOPP) and polyethylene terephthalate (PET), tackified PVAc-AA can achieve reliable adhesion in many standard packaging and print-lamination conditions. Carbodiimide is not strictly mandatory for all treated BOPP/PET work; however, it becomes highly advantageous when durability requirements increase, such as exposure to high humidity, hot-fill or pasteurization cycles, oil/grease contact, or elevated service temperatures. Carbodiimide reacts with the pendant carboxylic acid groups in the PVAc-AA backbone, forming covalent crosslinks that reduce water sensitivity and increase cohesive strength and heat resistance while maintaining a one-component workflow at pH 4.5–5.5.

Table 8.40A PVAc–AA/polar tackifier (No CDI) for treated BOPP/PET—standard duty.

Ingredient	% by weight	Notes
Deionized water	12.0	Balance viscosity; final 50–52% solids
PVOH solution (10–12% active)	10.0	Protective colloid; minimal free surfactant
PVAc–AA latex (55% solids; AA 1.5–2.0%)	60.0	Low free surfactant; clarity-optimized
Polar tackifier emulsion (50% solids; rosin-ester or terpene-phenolic)	6.0–10.0	3–5% active; improves wet tack/initial grab and optics vs. HC
Wetting aid (low level, nonionic)	0.2	Promote uniform laydown; avoid excess
Defoamer	0.2	Per supplier
Biocide	0.2	Per supplier
pH adjust (e.g., ammonium acetate)	q.s. to pH 4.8–5.2	Maintain 1K stability; avoid high pH

Process notes: Treat films ≥38–40 dyn/cm; coat 2–3 g/m^2 dry; nip 0.2–0.4 MPa; dwell 24 h at 23 °C. Target viscosity 600–900 cP (25 °C). Optical clarity prioritized; heat/water resistance limited vs. CDI grade.

Polyethylene (PE) presents an additional challenge due to its low surface energy. For untreated PE, crosslinker addition alone cannot overcome poor wetting and the lack of chemical interaction; surface treatment (e.g., corona or flame) or a dedicated primer is typically required. For treated PE (≥40 dyn/cm), PVAc-AA formulations that combine a tackifier with carbodiimide and a silane adhesion promoter can provide practical bonding to paper or carton, particularly in non-sterilized, ambient-service laminations (Tables 8.40A, 8.40B, and 8.41).

8.7.4 Treated PE: Do You Need a Carbodiimide?

For corona- or flame-treated polyethylene (PE), a tackified PVAc–AA adhesive can provide practical bonding for paper- and carton-lamination when surface energy is sufficiently raised. Carbodiimide is not categorically mandatory for treated PE; its necessity depends on durability expectations. Use carbodiimide when exposure to humidity, elevated temperature, plasticizer/oil contact, or long open/press times is expected. For ambient, dry-service laminations with adequate treatment, a non-crosslinked system can be suitable.

Table 8.40B PVAc–AA/CDI/polar tackifier for treated BOPP/PET–performance grade.

Ingredient	% by weight	Notes
Deionized water	10.5–11.5	Balance viscosity; target 50–52% solids
PVOH solution (10–12% active)	8.5–9.5	Protective colloid; minimize free surfactant
PVAc–AA latex (55% solids; AA 1.8–2.3%)	60.0	AA window supports CDI crosslinking; clarity-optimized latex
Polar tackifier emulsion (50% solids; rosin-ester or terpene-phenolic)	6.0–8.0	3–4% active; preserves clarity and initial grab
Carbodiimide dispersion (25% active)	2.0–3.2	≈0.5–0.8% active; add last at pH 4.8–5.2; 1K latent
Wetting aid (low level, nonionic)	0.2	Uniform laydown
Defoamer	0.2	Per supplier
Biocide	0.2	Per supplier
pH adjust (e.g., ammonium acetate)	q.s. to pH 4.8–5.2	Maintain latency; avoid amine surfactants

Process notes: Add CDI immediately before use; gently mix to avoid micro-gelation. Coat 2–3 g/m^2 dry; calendared nip 0.3–0.5 MPa; condition 24–48 h at 23–25 °C. Expect improved wet/heat resistance, reduced blocking, and clean optics.

Table 8.41 Primer/adhesive for treated PE-to-carton laminations (with silane and carbodiimide).

Ingredient	% by weight
Deionized water	12.0
PVAc-AA copolymer emulsion (55% solids)	60.0
Vinyl-silane emulsion (≈40% solids)	5.0
Rosin ester tackifier dispersion (50%)	10.0
Carbodiimide dispersion (30%)	3.0
Polyvinyl alcohol (10% solution)	6.0
Wetting agent (low-foam)	0.5
Coalescent (e.g., butyl carbitol)	3.0
Defoamer	0.3
Preservative	0.2

Formulation note: Apply to PE treated to ≥40 dyn/cm. Use as a primer (thin coat, dry, then bond) or as a direct adhesive for ambient-service laminations. For untreated PE, surface treatment is strongly recommended; otherwise, consider alternate chemistries or specialized primers outside the scope of PVAc-AA systems.

Table 8.42 Adhesive for treated PE-to-carton lamination (standard duty; with carbodiimide).

Ingredient	% by weight
Deionized water	11.6
PVAc–AA copolymer emulsion (55% solids)	66.0
Rosin ester tackifier dispersion (50%)	10.0
Vinyl/alkoxy silane emulsion (≈40% solids)	2.0
Polyvinyl alcohol (10% solution)	6.0
Carbodiimide dispersion (25% active)	2.8
Coalescent (e.g., butyl carbitol)	1.0
Wetting agent (low-foam)	0.2
Defoamer	0.2
Preservative	0.2

Formulation notes: Maintain pH 4.8–5.2 (e.g., ammonium acetate) to keep carbodiimide latent; avoid amine-bearing additives. Use on PE treated to ≥42 dyn/cm (ideal 44–46 dyn/cm). Apply thin, ensure sufficient wetting, and laminate promptly after treatment. Suitable for ambient, dry-service laminations with improved durability vs. no-CDI grade.

Process: Add CDI last under gentle stirring to avoid micro-gelation. Target viscosity 700–1,100 cP (25 °C). Typical coat weight 2–3 g/m^2 dry. Set the roller pressure (nip) to 0.3–0.5 MPa (≈44–73 psi) to ensure intimate contact and squeeze out air without flooding the joint. After lamination, lay the laminate flat and let it condition for 24–48 h at room temperature (23–25 °C) before testing or packing to allow water to evaporate, the PVAc film to finish forming, and the carbodiimide (CDI) crosslinking to develop full strength and durability (Table 8.43).

Recommended treatment levels: aim for ≥42 dyn/cm (preferred 44–46 dyn/cm). Verify immediately before coating, as PE surfaces can decay (age) within hours to days. A small dose of a silane adhesion promoter improves interfacial strength on treated PE and can be used with or without carbodiimide. Maintain pH 4.5–5.5 to preserve one-component latency when carbodiimide is present (Table 8.42).

By contrast, BOPP and PET films, once corona-treated, typically reach stable surface energy levels that remain adequate for lamination for longer periods. Adhesion with tackified PVAc-AA emulsions is therefore more reliable, and crosslinkers are only required for demanding conditions such as heat or humidity exposure. For PE, however, silane promoters and sometimes carbodiimides are strongly recommended even when the surface has been treated, in order to ensure durable adhesion (Table 8.44).

Table 8.43 Adhesive/primer for treated PE (high duty; with carbodiimide).

Ingredient	% by weight
Deionized water	11.0
PVAc–AA copolymer emulsion (55% solids)	60.0
Terpene-phenolic tackifier emulsion (50%)	12.0
Carbodiimide dispersion (30%)	3.0
Vinyl/alkoxy silane emulsion (≈40% solids)	3.0
Polyvinyl alcohol (10% solution)	6.0
Coalescent (e.g., butyl carbitol)	3.5
Wetting agent (low-foam)	0.5
Defoamer	0.3
Preservative	0.2

Formulation note: Enhanced water and heat resistance after 24–72 h ambient cure. Use when humidity or light thermal load is expected. Can be run as a primer (thin coat, dry, then bond with PVAc–AA adhesive) or as the adhesive itself for single-pass lamination, depending on machine setup.

Procedure: Charge water, dispersant/wetting agent, and defoamer; slowly add silane emulsion under moderate shear and mix 10 min; add PVAc–AA emulsion; premix tackifier emulsion separately, then add; add PVOH solution and coalescent; for high-duty grade, add carbodiimide last with gentle mixing; adjust pH to 4.8–5.2; filter 200–300 μm; age 12–24 h before production trials. Target coat weight 3–6 g/m^2 (dry).

Why Is Treated PE Handled Differently from Treated BOPP/PET?

Although both polyethylene (PE) and polypropylene (PP/BOPP) have low surface energies, treated PE remains more difficult to bond with PVAc-based adhesives. The reason is twofold. First, PE has an extremely inert, wax-like backbone with very low polarity. Corona or flame treatment temporarily raises the surface energy by generating oxidized groups, but these functional groups are unstable and decay quickly through hydrophobic recovery. This makes adhesion highly variable unless a coupling agent, such as a silane, is introduced. Second, PE often contains slip and anti-block additives that migrate to the surface after treatment, reducing wettability and interfering with adhesive bonding.

8.7.5 Formulation for Food Packaging/Multilayer Laminations (Paper/Film/Aluminum Foil)

Food packaging laminations often combine paper, polymer films, and aluminum foil to create multilayer structures with barrier properties against moisture, oxygen, and grease. Adhesives for such laminates must provide strong initial bonding, maintain integrity under humidity and heat, and resist chemical attack from oils and food components. PVAc–AA copolymers supply the necessary polarity and reactive sites, while tackifiers ensure rapid wet grab on porous paper or board substrates. Carbodiimide crosslinkers react with the acid groups in

Table 8.44 Formulation for carton and paper packaging under humid/cold-chain conditions (PVAc–AA/CDI/hydrogenated rosin-ester).

Ingredient	% by weight	Notes
Deionized water	10.8	Balance viscosity; target 50–52% solids
PVOH solution (10–12% active)	9.0	Protective colloid; minimize free surfactant
PVAc–AA latex (55% solids; AA 1.8–2.2%)	61.5	Base polymer; low free surfactant; paper/carton grade
Hydrogenated rosin-ester tackifier dispersion (50% solids)	6.0	≈3% active; boosts initial grab while maintaining humidity resistance
Carbodiimide dispersion (25% active)	3.2	≈0.8% active; add LAST at pH 4.8–5.2; 1K latent
Colloidal silica (30–40% dispersion)	0.4	Controls cold-flow; improves beading control on carton
PU rheology modifier	0.2	Builds body; reduces tailing/stringing on fast lines
Propylene glycol (optional, antifreeze/handling)	1.0	Use only if required for low-temp handling; verify open time
Coalescent (e.g., butyl carbitol)	0.3	Keep low to protect humidity resistance
Wetting agent (low-foam, nonionic)	0.3	Uniform laydown; avoid amine-bearing surfactants
Defoamer	0.2	Per supplier
Preservative	0.1	Per supplier
pH adjust (e.g., ammonium acetate)	q.s. to pH 4.8–5.2	Maintain CDI latency; avoid ammonia/strong bases

Process and Application Notes

The recommended mixing sequence begins with the addition of water, followed by polyvinyl alcohol (PVOH), and then the latex. The tackifier should be introduced slowly to ensure proper dispersion. The pH is then adjusted to 4.8–5.2, after which the rheology modifier and silica are incorporated. Finally, carbodiimide (CDI) is added under gentle stirring in order to avoid micro-gelation.

For application, the adhesive is applied at 3–5 g/m^2 (dry) on carton. The roller pressure (nip) should be set to 0.3–0.5 MPa, while the line speed is adjusted according to the capability of the equipment. In cold-room environments, both the adhesive and substrates should be preconditioned to at least 15–18 °C whenever possible to ensure proper film formation and adhesion.

Curing and conditioning are typically carried out by holding the laminated structure for 24–48 h at 23–25 °C and 50–60% relative humidity. Longer dwell times are recommended if the laminate is stored in cold or humid environments.

Quality control targets include a viscosity of 800–1,200 cP at 25 °C (measured with a Brookfield viscometer, with spindle and speed specified if required). Bond strength should yield fiber-tear on carton after the specified conditioning period. Blocking resistance must be confirmed by the absence of unacceptable blocking after a 24 h stack test at 40 °C. Finally, humidity and cold-chain durability are verified by maintaining adhesion after 24 h at 4–8 °C and 80–90% relative humidity, followed by a recovery test.

Table 8.45 Formulation for food packaging laminations (paper/film/aluminum foil) (PVAc–AA/CDI/hydrogenated rosin-ester/epoxy-silane).

Ingredient	% by weight	Notes
Deionized water	q.s. (≈10–12)	Balance viscosity; target 50–52% solids
PVOH solution (10–12% active)	8.5–9.5	Protective colloid; minimize free surfactant
PVAc–AA latex (55% solids; AA 1.8–2.2%)	61.0	Base polymer; paper/film/foil grade; low odor
Hydrogenated rosin-ester tackifier dispersion (50% solids)	4.0–6.0	→ 2–3% active; low-odor grade for food packaging; boosts initial grab
Carbodiimide dispersion (25% active)	3.2	→ ≈ 0.8% active; add LAST at pH 4.8–5.2; avoid amines
Epoxy-functional silane emulsion (≈40–50% solids)	1.5–2.0	→ 0.6–0.8% active; improves adhesion to treated film and aluminum oxide
Wetting agent (low-foam, nonionic)	0.2	Uniform laydown; food-contact suitable
Defoamer (food-contact suitable)	0.2	Per supplier; low silicone bleed
Preservative	0.1	Per supplier; verify food-contact status/limits
Coalescent	0–0.3	Use only if needed; choose low-odor, compliant grade
pH adjust (e.g., ammonium acetate)	q.s. to pH 4.8–5.2	Maintain CDI latency; avoid ammonia/strong bases

Process Notes

Suitable substrates include paper and carton, treated polyolefin and polyester films (PE, BOPP, PET) with a surface energy of at least 42 dyn/cm, and aluminum foil that is clean and free of oil with a stable oxide surface.

The recommended mixing order is as follows: add water, then polyvinyl alcohol (PVOH), followed by the PVAc–AA latex. Introduce the tackifier slowly to ensure proper dispersion. Adjust the pH to 4.8–5.2, add the silane coupling agent, and finally add carbodiimide (CDI) under gentle stirring. Avoid amine-bearing additives to minimize side reactions and maintain stability.

For application, apply 2–4 g/m^2 (dry). Set the roller pressure (nip) to 0.3–0.5 MPa and condition the laminate for 24–48 h at 23–25 °C. When laminating foil, ensure a uniform laydown to prevent tunnel defects.

Quality-control targets include a viscosity of 800–1,200 cP at 25 °C, a clear, defect-free film appearance, fiber-tear on paper, and cohesive peel on treated films and foil after the specified conditioning period.

Food-contact Compliance Note

All grades should be selected for their suitability under the relevant regional food-contact regulations (for example, EU 10/2011 and FDA 21 CFR, where applicable). Confirm specific migration limits, residual monomer levels, and additive status with suppliers. Specify hydrogenated rosin-ester and epoxy-silane in low-odor, food-grade versions.

the copolymer, enhancing water and heat resistance. To improve adhesion to metallic and polar film surfaces, an epoxy-functional silane adhesion promoter can be introduced, reinforcing interfacial bonding—especially with aluminum foil and corona-treated polymer films. The tackifier is used at low, polar levels (e.g., hydrogenated rosin-ester), it boosts initial grab and wet-film body and helps control excessive penetration at the bond line, enabling faster set without compromising clarity or barrier performance (Table 8.45).

8.7.6 Formulations for Labels and Self-adhesive Constructions

Label substrates span porous papers and boards, as well as polar inorganics (glass, aluminum) used in wet-service applications such as beer and wine bottles. Silane adhesion promoters are not required for paper-to-paper or board labels. For glass/metal wet-service labels, prefer **epoxy- or methacryloxy-functional silanes** to promote interfacial bonding and moisture resistance. (Avoid

Table 8.46 Formulation for general-purpose paper labels (no silane).

Ingredient	% by weight	Notes
Deionized water	10.5	Balance viscosity; target 50–52% solids
PVAc–AA copolymer emulsion (55% solids)	60.0	Base polymer; PVOH-dominant stabilization; low free surfactant
Terpene-phenolic or hydrogenated rosin-ester tackifier emulsion (50%)	10.0–14.0	Prefer polar tackifiers; avoid hydrocarbon tackifiers for compatibility
Carbodiimide dispersion (30%)	2.0–2.7	≈0.6–0.8% active; add LAST at pH 4.8–5.2; avoid amine additives
Polyvinyl alcohol (10% solution)	7.0	Protective colloid; improves body and label lay
Coalescent (e.g., butyl carbitol)	0.5–1.0	Keep low to protect humidity resistance
Plasticizer (e.g., triethylene glycol)	0.5–1.0	Boosts low-temperature tack; verify blocking
Wetting agent (low-foam, nonionic)	0.2–0.3	Uniform laydown; minimize foam
Defoamer	0.2–0.3	Per supplier
Preservative	0.2	Per supplier

Application scope: Paper-to-paper/carton labels for ambient or refrigerated service; strong initial tack and fast set. For film-lined boards, verify surface energy and, if needed, use a vinyl- or epoxy-silane **primer coat** separately. For glass/metal wet-service labels (e.g., ice-bucket), formulate a variant by adding **epoxy- or methacryloxy-silane** (≈0.6–0.8% active) and consider reducing tackifier toward the low end while increasing CDI toward ≈0.8% active (Table 8.47).

amino-silanes in systems crosslinked with carbodiimide, as amines can interfere with CDI latency.)

Accordingly, a general-purpose grade without silane is provided below for paper labels; a wet-service variant for glass/metal may be formulated by adding an epoxy- or methacryloxy-silane (≈0.6–0.8% active) and adjusting tackifier/CDI balance (Table 8.46).

8.7.7 Formulation for Bookbinding and Hardcover Manufacturing

Bookbinding and hardcover manufacturing adhesives require a delicate balance between strong initial tack for page hold, creep resistance under load, and flexibility to withstand repeated opening and handling. Conventional PVAc homopolymer adhesives may suffer from brittleness or loss of adhesion over time, particularly under humid conditions. A PVAc–AA copolymer backbone improves compatibility with tackifiers and allows crosslinking with carbodiimide to enhance durability. The inclusion of a tackifier ensures rapid set and fiber tearing strength, while a plasticizer provides flexibility to prevent cracking along the spine. Partial replacement of polymer with starch or polyvinyl alcohol (PVOH) helps to control viscosity, reduce cost, and improve penetration into paper fibers. The following formulation is representative of a durable one-component adhesive system for bookbinding and hardcover applications (Table 8.48).

8.7.8 Formulation for Wood Veneer and Edge Gluing (Nonstructural Interior)

Adhesives for wood veneer and edge gluing in nonstructural interior applications must combine fast tack to hold pieces in position during pressing with sufficient creep resistance to prevent joint failure over time. Conventional PVAc adhesives can be prone to slippage or cold flow, especially under variable humidity. A PVAc–AA copolymer backbone enhances the interaction with tackifiers for improved wet grab, while carbodiimide crosslinking strengthens cohesion and water resistance. Plasticizers are included to balance flexibility and prevent embrittlement, and PVOH helps stabilize the emulsion and improve film strength. The following formulation provides a practical balance for veneer pressing and edge bonding under ambient interior service conditions (Table 8.49).

8.7.9 Formulation for Tissue and Hygiene Laminations

Tissue and hygiene laminations, such as two-ply tissue, paper towels, and sanitary products, require adhesives that provide reliable bonding while maintaining softness, flexibility, and absorbency. Conventional PVAc systems may stiffen the

Table 8.47 Formulation for glass/metal labels (wet-service; epoxy-silane recommended with zirconium crosslinker).

Ingredient	% by weight	Notes
Deionized water	10.2	Balance viscosity; target 50–52% solids
PVOH solution (10–12% active)	7.0	Protective colloid; minimizes free surfactant
PVAc–AA latex (55% solids; AA 1.8–2.2%)	61.0	Base polymer; label grade; low odor
Terpene-phenolic tackifier emulsion (50% solids)	8.0–10.0	→ 4–5% active; boosts initial grab; choose low-odor grade
Zirconium crosslinker dispersion (ZrO_3 15–20%)	2.0–3.0	≈0.5–0.8% ZrO_2 on polymer solids; add last at pH 4.8–5.2; avoid amines/strong chelators
Epoxy-functional silane emulsion (≈40–50% solids)	1.5–2.0	→ 0.6–0.8% active; promotes adhesion to glass/metal oxides
Wetting agent (low-foam, nonionic)	0.2–0.3	Uniform laydown; food-contact suitable if required
Defoamer	0.2–0.3	Per supplier; low silicone bleed
Plasticizer (e.g., triethylene glycol)	0.3–0.7	Low-temp tack; verify blocking
Coalescent (low-odor)	0–0.5	Use only if needed; humidity tolerance first
Preservative	0.2	Per supplier
pH adjust (e.g., ammonium acetate)	q.s. to pH 4.8–5.2	Maintain zirconium crosslinker stability

Use and Performance Notes

Substrates: Clean, oil-free glass and metal; ensure oxide surfaces are uncontaminated. For wet-service performance, pair an epoxy- or methacryloxy-functional silane with the zirconium crosslinker. Avoid amino-silanes in the same mix.

Application: Apply 3–5 g/m^2 (dry). Set the roller pressure (nip) to 0.3–0.5 MPa. Condition for 24–48 h at 23–25 °C. Validate ice-bucket resistance (0–4 °C water, 60–120 min) and refrigerated-condensation performance.

Process order: Add water, then PVOH, then latex. Add the tackifier slowly. Adjust pH to 4.8–5.2. Add the silane. Add the zirconium crosslinker last with gentle mixing. Avoid amine-bearing additives and strong chelators (e.g., phosphate, citrate, oxalate).

If an amino-silane must be used

Amines can complex zirconium and compromise crosslinking. If an amino-silane is mandated, either (i) apply it as a separate primer coat and keep the zirconium crosslinker in the adhesive, or (ii) omit the zirconium crosslinker and rely on the amino-silane with a slightly higher AA level—accepting reduced wet/heat resistance relative to zirconium-crosslinked systems.

Table 8.48 Formulation for bookbinding and hardcover manufacturing (PVAc–AA/CDI/hydrogenated rosin-ester).

Ingredient	% by weight	Notes
Deionized water	q.s. (≈10–12)	Balance viscosity; target 50–52% solids
PVAc–AA copolymer emulsion (55% solids)	60.0	Base polymer; PVOH-dominant stabilization; low free surfactant
Hydrogenated rosin-ester tackifier emulsion (50%)	6.0–8.0	→ 3–4% active; boosts initial grab/body; low-odor grade
Carbodiimide dispersion (30% active)	2.0–2.7	→ 0.6–0.8% active; add LAST at pH 4.8–5.2; avoid amines
Polyvinyl alcohol (10% solution)	7.5–8.5	Protective colloid; improves wet-film body; limits excessive penetration
Pregelatinized starch dispersion (20%)	5.0–7.0	Adds body and lay-flat; avoid embrittlement at high levels
Plasticizer (DEGDB/TEGDB)	2.0–3.0	Low-odor dibenzoate plasticizer; alternative: TEG 1–2% for softer feel
Coalescent (low-odor)	0–0.5	Use only if needed; protects block/humidity resistance
Wetting agent (low-foam, nonionic)	0.2–0.3	Uniform laydown; avoid amine-bearing types
Defoamer	0.2–0.3	Per supplier
Preservative	0.2	Per supplier
pH adjust (e.g., ammonium acetate)	q.s. to pH 4.8–5.2	Maintain CDI latency; avoid ammonia/strong bases

Process and Application Notes

Mixing order: Add water, then PVOH, then PVAc–AA latex. Add the tackifier slowly. Adjust pH to 4.8–5.2. Add the starch dispersion, then the plasticizer, then the coalescent (if used). Add CDI last with gentle stirring to avoid micro-gelation.

Application: Apply 3–6 g/m^2 (dry) on covers and cases (higher on the spine if needed). Set roller pressure (nip) or press according to line requirements. Condition for 24–48 h at 23–25 °C before testing or packing.

Targets: Viscosity 800–1,200 cP (25 °C); strong initial grab; good lay-flat; minimal blocking at 40 °C.

QA Checks

Quality assurance should confirm page-pull and flex-crack resistance at the spine after the specified conditioning period. Blocking and creep must be evaluated by a 24-h stack test at 40 °C. Where storage or distribution occurs under refrigeration, an optional refrigerated-adhesion check should be performed to ensure bond integrity.

Advisory

Hydrocarbon tackifiers should be avoided due to compatibility issues and odor; instead, select hydrogenated rosin-ester or low-odor terpene-phenolic grades. Coalescent content should be kept low, since excessive levels increase blocking and odor. Carbodiimide (CDI) must be added last at pH 4.8–5.2, and amine-bearing additives should be avoided.

Table 8.49 Formulation for wood veneer and edge gluing (nonstructural interior) (PVAc–AA/CDI; low-tackifier, low-coalescent for creep/water balance).

Ingredient	% by weight	Notes
Deionized water	q.s. (≈9–11)	Balance viscosity; avoid over-dilution
PVAc–AA copolymer emulsion (55% solids)	62.0–64.0	Base polymer; PVOH-dominant stabilization; minimize free surfactant
Hydrogenated rosin-ester tackifier emulsion (50%)	4.0–6.0	→ 2–3% active; optional for initial grab; higher levels increase creep/blocking
Carbodiimide dispersion (30% active)	2.3–2.7	→ 0.7–0.8% active; add LAST at pH 4.8–5.2; avoid amines
Polyvinyl alcohol (10% solution)	7.5–8.5	Protective colloid; improves wet-film body and green strength
Pregelatinized starch dispersion (20%) (optional)	0–4.0	1% active typical; adds body/anti-slip; excessive use can embrittle
Plasticizer (DEGDB/TEGDB)	1.0–2.0	Low-odor dibenzoate; alternative: TEG 0.5–1.0%
Coalescent (low-odor)	0–0.5	Keep minimal to protect humidity resistance and reduce creep
Wetting agent (low-foam, nonionic)	0.2–0.3	Uniform wetting; avoid amine-bearing types
Defoamer	0.2–0.3	Per supplier
Preservative	0.2	Per supplier
pH adjust (e.g., ammonium acetate)	q.s. to pH 4.8–5.2	Maintain CDI latency; avoid ammonia/strong bases

Process and Application Notes

Substrates: Interior wood veneer and edge joints; moisture content 8–12%. Surfaces are clean and well-mated.

Application: Bead/roll/brush; 150–250 g/m^2 wet (joint-line coverage). Clamp/press per line; avoid squeeze-out puddling.

Conditioning: 24–48 h @ 23–25 °C before strength testing; longer for dense woods.

Targets: Viscosity 800–1,200 cP (25 °C); strong initial grab without stringing; balanced creep and D2–D3 moisture resistance (nonstructural).

Cautions

Hydrocarbon tackifiers should be avoided because they often show limited compatibility with PVAc/PVOH systems at mildly acidic pH and can plasticize the adhesive film, lowering cohesive strength and increasing creep and blocking. Odor and yellowing risk are additional concerns.

Tackifier dosage should be kept low—typically 2–3% on total formulation solids—or omitted altogether when clamp times and line conditions permit. While tackifiers can raise initial tack, excessive levels depress the effective glass-transition temperature and minimum film-formation temperature, leading to softer films, greater migration potential, and higher creep under load.

Dibutyl phthalate (DBP) must not be used due to regulatory and toxicological concerns, as well as odor and migration risk. Dibenzoate plasticizers (for example, diethylene glycol dibenzoate or triethylene glycol dibenzoate) or triethylene glycol (TEG) are preferred options, offering lower odor profiles and improved regulatory acceptance while maintaining film formation.

Carbodiimide (CDI) should be added last. Primary and secondary amines will react with CDI to form ureas, consuming the crosslinker and reducing its availability to couple carboxyl groups on the polymer; this diminishes final water resistance. Maintain pH 4.8–5.2 during CDI addition and avoid amine-bearing additives in the let-down.

tissue or lose strength in high-humidity environments. Incorporating acrylic acid into the PVAc backbone improves compatibility with tackifiers and enables carbodiimide crosslinking, which imparts resistance to moisture and dimensional changes. A tackifier ensures fast wet tack to hold delicate substrates during

Table 8.50 Formulation for tissue and hygiene laminations (PVAc–AA/CDI; low-odor, low-VOC; softener optional and nonamine).

Ingredient	% by weight	Notes
Deionized water	q.s. (≈10–12)	Balance viscosity; target 50–52% solids
PVAc–AA copolymer emulsion (55% solids)	62.0	Base polymer; PVOH-dominant stabilization; minimal free surfactant
Hydrogenated rosin-ester tackifier emulsion (50%)	6.0–8.0	→ 3–4% active; lower odor than rosin-phenolic; improves initial grab/body
Carbodiimide dispersion (30%)	2.3–2.7	→ 0.7–0.8% active; add LAST at pH 4.8–5.2; avoid amine-bearing additives
Polyvinyl alcohol (10% solution)	7.0–8.0	Protective colloid; improves wet-film body; limits excessive penetration
Plasticizer (PEG-400)	1.0–2.0	Keeps film flexible without odor; verify absorbency impact
Coalescent (low-odor)	0–0.5	Use only if needed; higher levels raise odor/blocking
Softener (optional, nonionic—e.g., fatty acid PEG ester)	0–1.0	Add only if handfeel is critical; avoid cationic quats in CDI systems
Wetting agent (low-foam, nonionic)	0.2–0.3	Uniform laydown with minimal foam
Defoamer	0.2–0.3	Per supplier
Preservative	0.2	Per supplier
pH adjust (e.g., ammonium acetate)	q.s. to pH 4.8–5.2	Maintain CDI latency; avoid strong bases

Process and Application Notes—Tissue and Hygiene Laminations

Substrates: Tissue and hygiene webs; confirm absorbency targets and basis weight.

Mixing order: Add water, then PVOH, then latex. Add the tackifier slowly. Adjust pH to 4.8–5.2. Add the plasticizer and, if required, the coalescent. Add the softener only if handfeel is a KPI. Add CDI last with gentle stirring to avoid micro-gelation.

Application and conditioning: Apply 1–3 g/m^2 (dry). Set the roller pressure (nip) according to line requirements. Condition for 24–48 h at 23–25 °C before testing or packing.

QC checks: Viscosity 600–1,000 cP (25 °C); balance bond strength vs. softness; blocking after 24 h at 40 °C; odor/migration checks.

high-speed converting, while plasticizers help preserve flexibility and hand-feel. The following formulation represents a balanced system suitable for tissue and hygiene laminations, delivering durability in humid bathrooms and kitchens without compromising softness (Table 8.50).

Remark: Why softener is optional: In tissue/hygiene laminations, the adhesive should not noticeably stiffen or harshen the feel of the web. A small amount of softener can improve handfeel and reduce film brittleness, but it can also lower bond strength, increase blocking, and raise migration/odor risks. Because not all lines need extra softness, the softener is optional and used only when handfeel is a key performance indicator (KPI). Avoid cationic quaternary ammonium softeners in this CDI-crosslinked system due to charge incompatibility and potential performance tradeoffs; prefer nonamine, nonionic softeners if required.

8.7.10 Formulation for Window Patching Adhesives

Window patching adhesives are used to bond transparent polymer films such as PET, BOPP, or PVC onto cartons for display and protective purposes. These applications require adhesives with excellent clarity, high initial tack, and resistance to whitening under humidity or temperature fluctuations. PVAc–AA copolymers provide polarity and compatibility with tackifiers, while carbodiimide crosslinkers improve water and heat resistance. To further strengthen bonding to films, a vinyl silane adhesion promoter may be included. Tackifiers enhance wet grab and set speed, ensuring precise film positioning at high machine speeds. The following formulation is representative of a one-component adhesive system suitable for window patching applications in folding cartons and specialty packaging (Table 8.51).

8.7.11 Formulation for Specialty Industrial Laminations

Specialty industrial laminations cover paper-to-foil, paper-to-metallized film, high-barrier films, coated boards, and specialty papers. Key requirements include high cohesive strength, resistance to water and chemicals, thermal stability, and durable adhesion to polar and semipolar substrates.

PVAc–AA copolymers supply pendant acid groups that react with carbodiimides (CDI) to raise water and creep resistance. Tackifiers contribute primarily through a physical mechanism as a plasticizing resin phase to accelerate wet grab and early set—use low, polar types to preserve durability. For interfacial adhesion, prefer **epoxy- or methacryloxy-functional silanes** with CDI systems (they do not consume CDI). Avoid amino-silanes in the same mix with CDI, as amines can react with CDI and reduce crosslink density. Vinyl silanes are generally less effective

Table 8.51 Formulation for window patching adhesives (PVAc–AA/CDI/polar tackifier; clarity-optimized).

Ingredient	% by weight	Notes
Deionized water	q.s. (≈10–12)	Balance viscosity; target 50–52% solids
PVAc–AA copolymer emulsion (55% solids)	61.0	Base polymer; PVOH-dominant stabilization; low free surfactant
Hydrogenated rosin-ester or terpene-phenolic tackifier emulsion (50%)	6.0–8.0	→ 3–4% active; prefer hydrogenated rosin-ester for optics/odor; avoid HC tackifiers
Carbodiimide dispersion (30% active)	2.3–2.7	→ 0.6–0.8% active; add LAST at pH 4.8–5.2; avoid amines
Epoxy-functional silane emulsion (≈40–50% solids)	1.5–2.0	→ 0.6–0.8% active; improves adhesion to treated films (BOPP/PET/PVC) and carton
Polyvinyl alcohol (10% solution)	7.0–8.0	Protective colloid; builds body; limits excessive penetration
Coalescent (low-odor)	0–0.5	Use only if needed; high levels increase blocking/odor
Plasticizer (triethylene glycol)	1.0–2.0	Flexibility at low temp; confirm optics/blocking
Wetting agent (low-foam, nonionic)	0.2–0.3	Uniform laydown; avoid amine-bearing surfactants
Defoamer	0.2–0.3	Per supplier; low silicone bleed
Preservative	0.2	Per supplier
pH adjust (e.g., ammonium acetate)	q.s. to pH 4.8–5.2	Maintain CDI latency; avoid strong bases

Process and Application Notes for Carton and Clear Films (Treated BOPP, PET, PVC)

Substrates: Carton with clear films (treated BOPP, PET, PVC). Verify surface energy ≥42 dyn/cm.

Mixing order: Add water, then PVOH, then latex. Add the tackifier slowly. Adjust pH to 4.8–5.2. Add the silane. Add the plasticizer and, if required, the coalescent. Add CDI last with gentle stirring.

Application and conditioning: Apply 2–4 g/m^2 (dry). Set the roller pressure (nip) to 0.3–0.5 MPa. Condition for 24–48 h at 23–25 °C before testing or packing.

QC checks: Viscosity 700–1,100 cP (25 °C); verify optical clarity (haze), blocking after 24 h at 40 °C, and peel vs. fiber-tear balance.

in these waterborne systems; if used, apply them as a separate primer on difficult polyolefin/barrier films.

This adhesive family is intended for light- to medium-duty industrial laminations. For heavy structural or high solvent/heat exposure, use reactive PU or epoxy systems (Table 8.52).

Table 8.52 Formulation for specialty industrial laminations (PVAc–AA/CDI/polar tackifier; silane-enabled).

Ingredient	% by weight	Notes
Deionized water	q.s. (≈9–11)	Balance viscosity; target 50–52% solids
PVAc–AA copolymer emulsion (55% solids)	60.0–62.0	Base polymer; PVOH-dominant stabilization; minimal free surfactant
Hydrogenated rosin-ester or terpene-phenolic tackifier emulsion (50%)	6.0–8.0	→3–4% active; polar, low-odor grades; avoid hydrocarbon tackifiers
Carbodiimide dispersion (30% active)	2.0–3.0	→0.6–0.9% active; add LAST at pH 4.8–5.2; avoid amine additives
Epoxy- or methacryloxy-silane emulsion (≈40–50% solids)	1.5–2.0	→0.6–0.8% active; promotes adhesion to metal oxides & metallized films
Polyvinyl alcohol (10% solution)	7.5–8.5	Protective colloid; improves wet-film body; limits excessive penetration
Plasticizer (triethylene glycol or dibenzoate)	1.0–2.0	Flexibility without excessive blocking/odor; validate for end use
Coalescent (low-odor)	0–0.5	Use only if needed; higher levels reduce heat/water resistance
Wetting agent (low-foam, nonionic)	0.2–0.3	Uniform laydown; avoid amine-bearing surfactants
Defoamer	0.2–0.3	Per supplier; low silicone bleed
Preservative	0.2	Per supplier
pH adjust (e.g., ammonium acetate)	q.s. to pH 4.8–5.2	Maintain CDI latency; avoid strong bases

Silane Selection Guide by Substrate

Aluminum foil and metallized films (aluminum-oxide surface): use an epoxy-functional silane at 0.6–0.8% active.

Glass and ceramic labels: use an epoxy-functional silane at 0.6–0.8% active.

Treated PET and BOPP: an epoxy-functional silane can assist; confirm surface energy ≥42 dyn/cm. For clarity-critical PET, keep the tackifier near 2–3% active.

Treated PE and barrier polyolefin laminates containing EVOH/PE: EVOH (ethylene–vinyl alcohol copolymer) is a highly polar, high-oxygen-barrier layer often co-extruded with PE. When bonding to such structures, prefer a separate primer step based on vinyl- or epoxy-functional silane applied to the low-energy surface. Use the main adhesive without silane added in-mix to protect carbodiimide crosslinker performance.

Process and Application Notes

Mixing order: Add water, then PVOH, then latex. Add the tackifier slowly. Adjust pH to 4.8–5.2. Add the silane. Add carbodiimide last with gentle stirring.

Application and conditioning: Apply 2–4 g/m^2 (dry) for film/foil and 3–5 g/m^2 (dry) for paper/board. Set the roller pressure (nip) to 0.3–0.5 MPa. Condition for 24–48 h at 23–25 °C.

QC checks: Viscosity 800–1,200 cP at 25 °C; verify clarity/haze for film laminates; blocking after 24 h at 40 °C; peel vs. fiber-tear balance; perform chemical spot tests (water, ethanol, surfactant) if relevant.

8.7.12 Formulation for Remoistenable and Gumming Applications

Remoistenable (water-activated) adhesives must dry to a nontacky film and rapidly regain tack upon contact with water. To balance reactivation with storage stability, the crosslink density must be kept low and the resin phase chosen for easy re-wetting. PVAc–AA improves compatibility with polar tackifiers and enables limited carbodiimide (CDI) crosslinking for block resistance, but CDI should be held to a very low level to avoid suppressing reactivation. Humectants (polyols) and PVOH aid flexibility and redispersibility (Table 8.53).

8.7.13 Formulation for Carton-to-Lacquered and Coated Carton (Gloss and Matte)

Gloss and matte coated cartons are low-porosity, lower-energy substrates. Adhesives must wet quickly, set fast, and maintain bond strength with minimal fiber penetration. A PVAc–AA copolymer provides polarity; a small amount of a polar tackifier adds wet-film body and initial grab. Carbodiimide (CDI) crosslinking improves moisture/heat resistance. For interfacial adhesion to lacquered/coated layers, prefer epoxy- or methacryloxy-functional silane; avoid vinyl silane in-mix with CDI. Coalescent levels should be kept minimal to protect optics and block resistance (Table 8.54).

8.7.14 Summary Table of PVAc/AA/Tackifier/Crosslinker Adhesive Applications

This table summarizes the eleven main application areas where PVAc/AA/tackifier/crosslinker adhesive systems are applied, together with their key performance features and benefits (Table 8.55).

8.7.15 Trade Name Tables—Tackifiers and Crosslinkers

Representative commercial tackifiers, including trade names, suppliers, and their corresponding chemistries, are summarized in Table 8.56.

Representative commercial sources and trade names of zinc oxide (ZnO), including typical grades and formats, are summarized in Table 8.60.

Table 8.53 Formulation for remoistenable and gumming adhesives (PVAc–AA, low-CDI, polar tackifier).

Ingredient	% by weight	Notes
Deionized water	q.s. (≈15–18)	Balance viscosity; target 48–52% solids
PVAc–AA copolymer emulsion (55% solids)	56.0	Base polymer; PVOH-dominant stabilization; minimal free surfactant
Hydrogenated rosin-ester tackifier emulsion (50%)	4.0–6.0	→ 2–3% active; improves wet grab after re-wet; avoid hydrocarbon tackifiers
Carbodiimide dispersion (30% active)	0.8–1.3	→ ~0.24–0.39% active; ultra-low to prevent blocking yet allow reactivation; add LAST at pH 4.8–5.2
Polyvinyl alcohol (10% solution)	9.0–10.0	Improves redispersibility and film flexibility; limits excessive penetration
Humectant plasticizer (glycerol or sorbitol)	3.0–4.0	Promotes rapid re-wet; verify blocking and curl
Wetting agent (low-foam, nonionic)	0.2–0.3	Uniform laydown; avoid amine-bearing types
Defoamer	0.2–0.3	Per supplier
Preservative	0.2	Per supplier
pH adjust (e.g., ammonium acetate)	q.s. to pH 4.8–5.2	Maintain CDI latency; avoid strong bases

Process and Performance Notes

The recommended coat weight (dry) is 2–5 g/m^2, selected according to substrate porosity to ensure adequate wetting and film continuity without excess squeeze-out. After application, allow the film to dry to a nonblocking state under ambient conditions before stacking or further converting.

For the reactivation test, wet the adhesive with neutral water (pH approximately 7) and allow a contact time of 2–5 seconds, then evaluate immediate tack and adhesion. Blocking performance is assessed by a 24-h stack test at 40 °C to confirm that sheets do not adhere undesirably under mild load and heat.

If reactivation is sluggish, reduce the carbodiimide (CDI) level toward approximately 0.2–0.3% active and/or reduce the tackifier toward 2% active; a slight increase in humectant may also be used to promote faster water uptake and wet-grab.

If blocking occurs in stacks (for example, sheets stick together after storage), increase CDI toward approximately 0.35–0.4% active, reduce humectant to raise the film's softening threshold, and consider a light dusting powder or the use of a release liner between layers during stacking.

Optional Variant (Faster Reactivation)

For faster reactivation, use the tackifier at 2–3% active and CDI at approximately 0.2% active (about 0.7% of a 30% dispersion). This variant provides quicker wet-grab but reduces block resistance; it is best reserved for envelopes and stamps that experience gentle stacking conditions.

Table 8.54 Carton-to-lacquered and coated carton (gloss and matte) (PVAc–AA/CDI/polar tackifier; epoxy-silane enabled).

Ingredient	% by weight	Notes
Deionized water	q.s. (≈9–11)	Balance viscosity; target 50–52% solids
PVAc–AA copolymer emulsion (55% solids)	61.0–62.0	Base polymer; PVOH-dominant stabilization; minimal free surfactant
Hydrogenated rosin-ester or terpene-phenolic tackifier emulsion (50%)	6.0–8.0	→ 3–4% active; boosts initial grab; prefer hydrogenated rosin-ester for optics/odor
Carbodiimide dispersion (30% active)	2.3–2.7	→ 0.7–0.8% active; add LAST at pH 4.8–5.2; avoid amine-bearing additives
Epoxy-functional (or methacryloxy-) silane emulsion (≈40–50% solids)	1.5–2.0	→ 0.6–0.8% active; promotes adhesion to lacquer/coat surfaces
Polyvinyl alcohol (10% solution)	7.0–8.0	Protective colloid; builds body; limits excessive penetration
Coalescent (low-odor)	0–0.5	Use only if needed; high levels increase blocking/odor
Plasticizer (triethylene glycol)	1.0–2.0	Flexibility at low temp; verify blocking/optics
Wetting agent (low-foam, nonionic)	0.2–0.3	Uniform laydown; avoid amines
Defoamer	0.2–0.3	Per supplier
Preservative	0.2	Per supplier
pH adjust (e.g., ammonium acetate)	q.s. to pH 4.8–5.2	Maintain CDI latency; avoid strong bases

Process and QA Notes—Coated Cartons (Gloss and Matte)

Suitable substrates are gloss and matte-coated cartons. Verify a surface energy of at least 38–40 dyn/cm for gloss grades and at least 36–38 dyn/cm for matte grades. Ensure that all surfaces are clean and free of dust before coating or lamination.

The recommended mixing order is: add water, then polyvinyl alcohol (PVOH), followed by the latex. Introduce the tackifier slowly to ensure proper dispersion. Adjust the pH to 4.8–5.2, add the silane, then add the plasticizer and, if required, the coalescent. Add carbodiimide (CDI) last under gentle stirring.

For application, apply 2–4 g/m^2 (dry). Set the roller pressure (nip) to 0.3–0.5 MPa. Condition for 24–48 h at 23–25 °C before testing or packing to ensure film development and bond maturation.

Quality-control checks include viscosity of 700–1,100 cP at 25 °C, confirmation of optical clarity with low haze, block resistance after a 24-h stack at 40 °C, and an appropriate balance between peel and fiber-tear on coated surfaces.

Table 8.55 Summary of PVAc/AA/tackifier/crosslinker adhesive applications.

Application	Key requirements	Benefits of PVAc/AA/ tackifier/crosslinker
Carton and paper packaging (humid/cold chain)	High wet tack, resistance to condensation, cold storage durability	Fast set, strong wet grab, moisture-resistant crosslinked bonds
Film-to-carton laminations (treated PE, BOPP, PET)	Adhesion to low-energy films, humidity, and grease resistance	Carbodiimide + silane improves film bonding, durable laminates
Food packaging laminations (paper/film/foil)	Moisture, oil, and heat resistance, adhesion to foil	Crosslinker improves barrier performance, amino silane boosts foil adhesion
Labels and self-adhesive constructions	Aggressive tack, refrigeration/ ice-bucket resistance	High wet grab, carbodiimide ensures water resistance; silane for glass/metal
Bookbinding and hardcover manufacturing	Creep resistance, flexibility, page pull strength	Tackifier gives fast set; plasticizer + crosslinker improve durability
Wood veneer and edge gluing (nonstructural)	Fast green strength, resistance to creep in humidity	Tackifier aids holding, carbodiimide adds water resistance
Tissue and hygiene laminations	Bonding without stiffness, humidity tolerance	Soft films maintained, crosslinked bonds resist humidity
Window patching adhesives	Clarity, tack to PET/BOPP/PVC films, no whitening	Silane improves film adhesion, carbodiimide prevents whitening
Specialty industrial laminations	Adhesion to foils, coated boards, barrier films	Dual silanes (amino/vinyl) enhance adhesion; durable crosslinked matrix
Remoistenable and gumming applications	Dry film integrity, reactivation on moistening	Crosslinker prevents blocking; tackifier enables quick re-tack
Carton-to-lacquered and coated carton (gloss/matte)	Low porosity, smooth surfaces, gloss/matte coatings	Silane aids adhesion; carbodiimide + tackifier ensures durable bonds

Table 8.56 Tackifiers – trade names and suppliers.

Supplier	Trade name (series)	Chemistry/type	Notes
ExxonMobil	Escorez™	Hydrocarbon resin (C5/C9; hydrogenated grades available)	Wide grade range; PSA and packaging; compatibility charts available
Eastman	Regalite™	Hydrogenated hydrocarbon resin (HCR)	Light color, low odor; good heat stability
Eastman	Eastotac™	Aliphatic hydrocarbon resin (C5)	Good tack; PSA and hot-melt
Eastman	Kristalex™	Styrenic resin (C9)	Clarity; adhesion to polar substrates
Eastman	Piccotac™	C5/C9 hydrocarbon resin	General-purpose tackifier series
Eastman	Plastolyn™	Aromatic-modified hydrocarbon resin	Hot-melt and solventborne uses
Kraton	SYLVATAC™/ SYLVALITE™	Rosin esters	Paper/packaging; good polarity for PVAc-based systems
Kraton	SYLVARES™	Hydrocarbon/modified resins	Broad portfolio for PSA/HMAs
DRT	Dercolyte™	Terpene/ terpene-styrene	High tack; good odor profile
Neville Chemical	NEVTAC®	C5/C9 hydrocarbon resin	Established in HMAs and PSAs
Arakawa	ARKON™ (P/M series)	Hydrogenated hydrocarbon resin	Good UV/heat stability; light color
Yasuhara Chemical	Clearon™	Hydrogenated terpene resin	Light color; good adhesion to polar substrates

Table 8.57 Carbodiimide (CDI) crosslinkers—trade names and suppliers.

Supplier	Trade name/grade	Format	Notes
Nisshinbo Chemical	CARBODILITE® (e.g., V-02, V-02-L2, E-02)	Water-dispersible CDI	Common for aqueous adhesives; 1K use at pH ~4.8–5.2
Advancion	ZOLDINE™ XL-29SE (ex-UCARLINK®)	Water-dispersible polyfunctional CDI	Used in coatings/adhesives; check supplier RTU solids
LANXESS	Stabaxol® grades	Polymeric carbodiimide (various forms)	Hydrolysis stabilizer; some grades are suitable in waterborne systems

Table 8.58 Aluminum chloride ($AlCl_3$)—representative sources (commodity).

Supplier	Trade name/grade	Format	Notes
Merck/ MilliporeSigma	Aluminum chloride solution; Aluminum chloride hexahydrate EMPROVE® Essential	Solution/solid (hexahydrate)	Specify assay and impurities for adhesive grade
American Elements	Aluminum Chloride Solution	Solution	Industrial supply; confirm concentration and metals
Thermo Fisher (industrial/ chemical supply)	Aluminum chloride, various grades	Solid/solution	Commodity Lewis acid; specify concentration and water content

Table 8.59 Zirconium crosslinkers—trade names and suppliers.

Supplier	Trade name/grade	Chemistry	Notes
Luxfer MEL Technologies	Bacote® 20; Bacote® M/XL	Ammonium zirconium carbonate (AZC)/zirconium complexes	Aqueous Zr sources; widely used in paper/coatings/adhesives
Luxfer MEL Technologies	Zirconium acetate	Zr acetate complex	Alternative Zr source; check pH compatibility
Dorf Ketal	Tyzor® Zr grades (e.g., Tyzor® 217, ZEC, 212/TEAZ)	Zirconium chelates/complexes	Aqueous/organic grades; verify with phosphate-containing systems

Table 8.60 Zinc oxide (ZnO)–representative trade names and sources.

Supplier	Trade name/grade	Format	Notes
Zochem	Zochem Grade 102/102T; 112 USP	Powder	Fine grades for dispersion; confirm treated/untreated surface
EverZinc (incl. G.H. Chemicals)	CR-series (e.g., CR-41)	Powder	Industrial/USP grades; verify particle size
Zinc Nacional/Legacy brands	KADOX® (e.g., 911)—legacy reference	Powder	Legacy branding still appears; match to current equivalents

8.8 Internal Crosslinkers–Introduction

This section introduces internal (copolymerizable) crosslinkers for PVAc-based adhesives, focusing on chemistries that can be implemented in 1K systems and deliver durable performance without on-site mixing. Internal crosslinkers are polymerized into the backbone and provide latent reactivity for bond formation during film formation and drying. Compared with external, postadded crosslinkers (e.g., CDI, $AlCl_3$, Zr, ZnO), internal systems can offer lower migration, tighter control of crosslink density, and strong interfacial coupling when properly designed.

Two internal systems are emphasized: (i) glycidyl methacrylate (GMA), a dual-functional monomer whose methacrylate group copolymerizes with VAc and acrylates while the pendant epoxy group remains reactive in the dried film; and (ii) acetoacetoxyethyl methacrylate (AAEM), which is likewise dual-functional—its methacrylate group copolymerizes into the polymer backbone, while the pendant acetoacetate group remains available postpolymerization for condensation with dihydrazides (acylhydrazone formation) or for Michael-type addition with suitable amines/diamines, and can participate in acid/base-assisted interactions that enhance cohesion and adhesion.

In 1K PVAc–AA systems containing GMA, the epoxy group reacts primarily with carboxyl functionality (-COOH) introduced by acrylic or methacrylic acid during and after drying, building a moisture-resistant network. Under the mildly acidic pH and temperatures used here, epoxy–hydroxyl reactions (e.g., with PVOH or cellulose -OH) are negligible and are not relied upon; amine-rich additives are avoided to preserve storage stability.

AAEM provides latent acetoacetate sites that contribute to toughness and adhesion; full network development typically requires a nucleophilic co-reactant (e.g., adipic dihydrazide or diamines). Where such co-reactants are added at use, the formulation becomes 2K.

Design considerations include comonomer level, catalytic/acid sites, latex stabilization, and target Tg/MFFT. Typical starting levels are GMA ≈ 1.0–2.5% on total monomer and AAEM ≈ 0.5–2.0% on total monomer; small AA/MAA additions (≈0.5–2.0% on TM) can accelerate epoxy reactions (GMA) and improve substrate bonding. Processing must prevent premature gelation (especially with higher GMA), and storage pH should remain in a mildly acidic window to preserve latency and colloidal stability.

Safety and scope: For health and regulatory reasons, acrylamide and *N*-methylol-acrylamide systems are excluded from practical recipes, and aziridines are not used as crosslinkers in this chapter. Example formulations that follow conform to the 1K intent unless explicitly noted otherwise.

8.8.1 PVAc–AA–GMA (1K)–Internal Crosslinker System

This section restarts from first principles and treats a single internal system: PVAc modified with glycidyl methacrylate (GMA) and acrylic acid (AA). The goal is a clear, production-practical 1K adhesive where the pendant epoxy of GMA primarily reacts with –COOH from AA during/after drying to build moisture-resistant bonding. Including AA is important: it accelerates epoxy ring-opening and provides covalent junctions that outperform epoxy–alcohol routes under our mildly acidic conditions.

Scope: polymer design, a concise semicontinuous procedure, then application-specific blending guidance. No fixed solids targets are imposed here; water is adjusted as needed for heat balance and viscosity (Table 8.61).

8.8.2 PVAc–AA–AAEM (1K)–Internal Crosslinker System

This section treats a single internal 1K system based on PVAc modified with acetoacetoxyethyl methacrylate (AAEM) and acrylic acid (AA). AAEM is dual-functional: its methacrylate group copolymerizes into the backbone, and its pendant acetoacetate remains latent in the dried film. Including AA is important because it improves substrate bonding and can facilitate limited acid-assisted reactions that strengthen the adhesive matrix without resorting to a two-component cure. In strictly 1K designs, AAEM enhances adhesion and toughness; high wet-strength targets generally require external crosslinkers or deferred 2K activation.

Table 8.61 Monomer design and formulation based on PVAc-AA-GMA (1K).

Monomer/package	Typical level	Notes
Vinyl acetate (VAc)	Balance	Main monomer
Acrylic acid (AA)	0.8–1.2% on total monomer (TM)	Essential for epoxy–COOH cure and substrate bonding
Glycidyl methacrylate (GMA)	1.2–1.8% on TM	Dual-functional; keep within range to avoid gel risk
Protective colloid	PVOH (10–12% solution), as needed	Favor PVOH-dominant stabilization; minimize free surfactant
Surfactant (seed)	Anionic, 0.20–0.35% (as product)	Use only for nucleation; keep migration low
Buffer	$NaHCO_3$ 0.08–0.12%	Maintain a mildly acidic pH after neutralization

Procedure: PVAc–AA–GMA (1K)—Semicontinuous seeded emulsion polymerization

Scope: This procedure describes the preparation of a polyvinyl acetate latex containing acrylic acid (AA) and glycidyl methacrylate (GMA) by seeded semicontinuous emulsion polymerization. In this context, “latex” denotes the resulting aqueous polymer dispersion stabilized predominantly by polyvinyl alcohol (PVOH).

Abbreviations: APS, ammonium (or potassium) persulfate; TBHP, t-butyl hydroperoxide; SMBS, sodium metabisulfite.

Procedure (Step-by-Step)

1) *Prepare the reactor*: Charge approximately 60% of the total deionized water, one-third of the PVOH solution, and the buffer. De-air, blanket with nitrogen, and heat to 70–72 °C under moderate agitation.
2) *Form the seed*: Dose 2–3% of the total vinyl acetate together with a small portion of the APS solution (≈5% aqueous) to initiate. Hold 20–30 min to establish a stable seed at about 0.5–1.0% solids.
3) *Prepare the preemulsion*: Combine the remaining vinyl acetate, AA, and GMA with the remaining water, the PVOH solution, and the anionic surfactant. Maintain 20–25 °C with gentle shear during preemulsification.
4) *Feed under starved conditions*: Begin parallel feeds of the monomer preemulsion and the APS solution over approximately 3 h while holding 72–76 °C. Adjust initiator flow to control residual monomer at low levels.
5) *Complete the polymerization*: Hold 30–45 min at 76–78 °C. Apply a redox chase using small alternating additions of TBHP (≈5% aqueous) and SMBS (≈5% aqueous) until residual monomer is low.
6) *Finish the batch*: Cool to ≈45 °C. Strip residual vinyl acetate if required. Adjust pH to 4.8–5.2, add defoamer, filter through 100–150 μm, and package.

Blending and Auxiliaries (PVAc–AA–GMA Latex)

Order of addition: Add deionized water, then the PVOH solution if used, then the PVAc–AA–GMA latex. Introduce the wetting aid, the defoamer, and the preservative last, while maintaining low-shear mixing. Maintain storage and use pH at 4.8–5.3 to preserve latency and limit premature consumption of epoxy functionality. Avoid amine-rich auxiliaries because they catalyze epoxy ring-opening and may cause viscosity growth or gel risk during storage. Avoid strong chelators and phosphate packages because they disturb pH and ionic balance, can interfere with interfacial bonding, and may leave migratory residues that reduce fiber tear and moisture resistance. Verify the compatibility of any new additive at the pilot scale before implementation. Under the mildly acidic conditions used here, epoxy–hydroxyl reactions with PVOH or cellulose are negligible and are not relied upon (Tables 8.62–8.65).

Table 8.62 Application (A)–coated/lacquered cartons (high grab, interior packaging).

Parameter	Guidance
Cure driver	Epoxy–COOH during/after drying (dominant)
Application method	Wheel or nozzle; clean machining; low foam
Coat weight (wet)	Tune to line speed/substrate (document actual values during trials)
pH window (use)	≈4.8–5.3
Set/handling	Fast wet tack; short-to-moderate press; validate fiber tear on coated stocks
Cautions	Highly lacquered/low-energy surfaces may need corona/plasma or a different system

Table 8.63 Application (B)–moisture-tolerant carton sealing (interior).

Parameter	Guidance
Cure driver	Epoxy–COOH; verify humidity creep at RH ≥ 80%
Application method	Wheel/nozzle; keep plasticizer modest to preserve cuttability
Coat weight (wet)	Establish via plant trials (document)
pH window (use)	≈4.8–5.3
Set/handling	Moderate open time; verify blocking and hot slip on stacks
Cautions	Avoid phosphate/chelators; maintain storage pH; monitor whitening on glossy stocks

Table 8.64 Application (C)–interior wood adhesive.

Parameter	Guidance
Cure driver	Epoxy–COOH; epoxy–OH is considered negligible under these conditions
Application method	Bead/brush; clamp pressure per species; wood moisture 8–12%
Open/press time	Tune via plasticizer and temperature (document bands during trials)
Performance	Potential D2–upper D2; D3 requires verification and may favor external CDI routes (covered elsewhere)
Cautions	Avoid amines; maintain pH; verify creep (WATT-91 if pursued)

Table 8.65 Key levers for PVAc–AA–GMA (1K).

Parameter	Guidance
Increase early wet strength	Raise AA within range; optimize GMA toward the upper end; ensure pH 4.8–5.2 at use
Improve machining/cuttability	Moderate plasticizer; defoamer in two portions; control coat weight
Reduce risk of gelation	Keep GMA ≤ ~1.8% on TM; avoid excessive temperature; control inhibitor and feed rates
Surface energy issues	Clean substrates; consider corona/plasma; otherwise, evaluate a different crosslinker system

Notes
Under the mildly acidic pH and temperatures used here, epoxy–OH reactions are negligible and are not relied upon.
AA is intentionally incorporated to catalyze epoxy ring-opening and build covalent junctions during drying.
Document actual plant solids, viscosity, and coat weight; the above provides composition and process envelopes only.

Scope: monomer design guidance, a concise semicontinuous procedure, production blending notes, and application-specific guidance. No fixed solids targets are imposed; water is adjusted according to heat balance and viscosity needs (Table 8.66).

8.8.3 Carbodiimide Augmentation of Internal 1K Systems–Carboxyl Functionality Balance and New Applications

This section explains how the postaddition of a carbodiimide (CDI) dispersion to PVAc–AA–GMA or PVAc–AA–AAEM latexes widens the application window while keeping a one-component workflow. CDI reacts with carboxyl groups (from AA) to form *N*-acylurea linkages during drying and early service, improving humidity and heat resistance. Under the mildly acidic conditions used here, CDI develops benefits over 24–74 h rather than providing instant wet strength.

8.8.3.1 Carboxyl Functionality Balance and AA Optimization (Constant AA Basis)

When GMA is present, both CDI and the pendant epoxy compete for the same carboxyl (–COOH) sites introduced by acrylic acid. For a constant AA level (as a percent of total monomer), too few –COOH sites can slow epoxy conversion or

Table 8.66 Monomer design and formulation based on PVAc–AA–AAEM (1K).

Monomer/package	Typical level	Notes
Vinyl acetate (VAc)	Balance	Main monomer
Acrylic acid (AA)	0.5–1.0% on total monomer (TM)	Improves bonding; supports latent reactions
Acetoacetoxyethyl methacrylate (AAEM)	0.8–1.5% on TM	Latent acetoacetate sites for cohesion/adhesion
Protective colloid	PVOH (10–12% solution), as needed	Favor PVOH-dominant stabilization; minimize free surfactant
Surfactant (seed)	Anionic, 0.25–0.35% (as product)	Use only for nucleation; limit migration
Buffer	$NaHCO_3$ 0.08–0.12%	Maintain a mildly acidic pH after neutralization

Procedure: PVAc–AA–AAEM (1K)—Semicontinuous seeded emulsion polymerization

Scope: This procedure describes the preparation of a polyvinyl acetate latex containing acrylic acid (AA) and acetoacetoxyethyl methacrylate (AAEM) by seeded semicontinuous emulsion polymerization. Here, "latex" denotes the resulting aqueous polymer dispersion stabilized predominantly by polyvinyl alcohol (PVOH).

Abbreviations: APS, ammonium (or potassium) persulfate; TBHP, t-butyl hydroperoxide; SMBS, sodium metabisulfite.

Procedure (Step-by-Step)

1) *Prepare the reactor*: Charge ~60% of the total deionized water, one-third of the PVOH solution, and the buffer. De-air, blanket with nitrogen, and heat to 70–72 °C under moderate agitation.
2) *Form the seed*: Dose 2–3% of the total vinyl acetate together with a small portion of the APS solution (≈5% aqueous) to initiate. Hold 20–30 min to establish a stable seed at about 0.5–1.0% solids.
3) *Prepare the pre-emulsion*: Combine the remaining vinyl acetate, AA, and AAEM with the remaining water, the PVOH solution, and the anionic surfactant. Maintain 20–25 °C with gentle shear during pre-emulsification.
4) *Feed under starved conditions*: Begin parallel feeds of the monomer pre-emulsion and the APS solution over ~3 h while holding 72–76 °C. Adjust initiator flow to keep residual monomer low.
5) *Complete the polymerization*: Hold 30–45 min at 76–78 °C. Apply a redox chase using small alternating additions of TBHP (≈5% aqueous) and SMBS (≈5% aqueous) until residual monomer is low.
6) *Finish the batch*: Cool to ≈45 °C. Strip residual vinyl acetate if required. Adjust pH to 4.8–5.2, add defoamer, filter through 100–150 μm, and package.

Blending and Auxiliaries (PVAc–AA–AAEM Latex)

Order of addition: Add deionized water, then the PVOH solution if used, then the PVAc–AA–AAEM latex. Introduce the wetting aid, the defoamer, and the preservative last, while maintaining low-shear mixing. Maintain storage and use pH at 4.8–5.3 to preserve the latency of the acetoacetate functionality. Avoid amine-rich auxiliaries and strong bases, which can slowly consume acetoacetate groups in storage and increase viscosity. Limit strong chelators and phosphate packages because they disturb ionic balance and may reduce adhesion by leaving migratory residues. Verify the compatibility of any new additive at the pilot scale before implementation (Tables 8.67–8.70).

Table 8.67 Application (A)–paper converting and labels (moderate humidity).

Parameter	Guidance
Primary effect	Latent acetoacetate sites increase cohesion and adhesion without a two-component cure
Method	Wheel or nozzle with clean machining and modest coat weight
pH during use	Approximately four point eight to five point three
Handling	Moderate open time; verify blocking and peel after 24–72 h
Cautions	Limit amines and strong bases; minimize chelators and phosphate packages

Table 8.68 Application (B)–carton sealing (interior, moisture-tolerant).

Parameter	Guidance
Primary effect	Toughness and adhesion improvements; humidity tolerance increases vs. PVAc base
Method	Wheel or nozzle; maintain clean laydown and a modest plasticizer for cuttability
pH during use	Approximately four point eight to five point three
Handling	Moderate press; document coat weight during trials
Cautions	If high wet strength is required, evaluate external crosslinkers in separate sections

Table 8.69 Application (C)–laminations with fibrous substrates (moderate service).

Parameter	Guidance
Primary effect	Improved interfacial strength and toughness; limited 1K network build
Method	Roller application with controlled coat weight and low foam
pH during use	Approximately four point eight to five point three
Handling	Moderate open time; verify creep under elevated humidity
Cautions	For demanding moisture exposure, consider systems presented elsewhere

Table 8.70 Key levers for PVAc–AA–AAEM (1K).

Objective	Adjustments
Increase adhesion and toughness	Move AAEM toward the upper end of the range and optimize PVOH level
Improve humidity tolerance	Optimize AA within range, control pH, and consider coat-weight increase
Maintain storage stability	Keep pH mildly acidic; avoid amines and strong bases; limit reactive auxiliaries
Machining and cuttability	Use defoamer in two portions and moderate plasticizer levels

Notes

AAEM is treated here as a 1K latent functionality that enhances adhesion and toughness; it does not typically achieve D3-level wet strength without additional crosslinking routes.

Acrylic acid is intentionally incorporated to improve bonding and to support limited acid-assisted reactions in the film.

Document plant-specific viscosity, solids, and coat weight; the above provides composition and process envelopes.

Table 8.71 PVAc–AA–GMA with CDI–dosage, AA target, and unlocked applications.

CDI dosage (active % on polymer solids)	AA target (on total monomer)	What this enables	Notes and cautions
≈0.30–0.50	≈1.0% (range 0.8–1.2%)	Coated and lacquered folding cartons facing higher humidity or short condensation exposure	Improves humidity creep while preserving clean machining; verify blocking on glossy stocks; monitor latency at pH ≈ 5
≈0.50–0.70	≈1.0–1.2%	Interior wood joints on a D3 pathway and moisture-tolerant case/carton sealing	Raises heat/creep resistance vs. GMA alone; confirm press/open time bands; track cure over 24–72 h
≈0.60–0.80	≈1.2% (consider +0.2 pp on TM if GMA near 1.8%)	Premium packaging and labels exposed to chilled chain and bathroom humidity	Balance dose with pH control to limit viscosity growth in storage; verify food-contact compliance where relevant

Table 8.72 PVAc–AA–AAEM with CDI–dosage, AA target, and unlocked applications.

CDI dosage (active % on polymer solids)	AA target (on total monomer)	What this enables	Notes and cautions
≈0.30–0.50	≈0.8% (range 0.5–1.0%)	Paper converting and laminations needing improved humidity tolerance over PVAc–AA–AAEM alone	Observe cure development over one to three days; document peel and humidity creep after conditioning
≈0.50–0.70	≈0.8–1.0%	Interior packaging with intermittent moisture or high relative humidity, including chilled-chain secondary packaging	Check blocking and hot slip; maintain use pH near 5; monitor film stiffness
≈0.60–0.80	≈1.0%	Labels and specialty cartons requiring both toughness and wet resistance	Higher CDI can stiffen films; monitor cuttability and fiber tear; verify regulatory constraints as applicable

Notes

Dosages are expressed on an active CDI basis relative to polymer solids; AA targets are expressed on total monomer.

Select the minimum CDI dose that meets durability targets to preserve latency and viscosity stability.

Maintain mildly acidic pH in storage and use; avoid amine-rich auxiliaries, strong chelators, and high-phosphate packages that can compromise cure or stability.

leave CDI under-utilized. Therefore, AA must be optimized jointly with CDI and GMA to balance site demand and maintain storage stability.

Recommended starting points: for PVAc–AA–GMA with GMA at approximately 1.2–1.8% on total monomer, target AA near 1.0% on total monomer (adjust within 0.8–1.2%). If CDI exceeds about 0.6% active on polymer solids, consider increasing AA by approximately +0.2 percentage points on total monomer or reducing GMA toward the lower end of its range.

For PVAc–AA–AAEM with AAEM at approximately 0.8–1.5% on total monomer, target AA near 0.8–1.0% on total monomer when CDI is used at mid-to-upper doses. Maintain storage and use pH at approximately 4.8–5.3 (Tables 8.71 and 8.72).

9

VAM-based Copolymers with Acrylates and VeoVa 10 in Paints and Coatings

9.1 Introduction

Vinyl acetate copolymerization with acrylates and versatic acid vinyl esters (VeoVa 10) has become an essential strategy for developing waterborne binders with enhanced flexibility, durability, and resistance to environmental stresses. Whereas the previous chapter dealt extensively with adhesives based on polyvinyl acetate, acrylic acid, carbodiimide, and tackifier modifications, the present chapter focuses primarily on paints and coatings where copolymer design is the central determinant of performance.

By varying the ratio of vinyl acetate to acrylates, formulators can balance hardness and flexibility while improving resistance to water, alkali, and outdoor exposure. Incorporation of VeoVa 10 monomers further enhances hydrophobicity and film integrity, leading to binders that meet the stringent demands of exterior coatings and roof protection systems. The addition of small amounts of functional monomers such as acrylic acid introduces sites for crosslinking or ionic stabilization, enabling self-dispersibility and compatibility with external crosslinkers when required.

This chapter examines the synthesis principles, structure–property relationships, and formulation aspects of these copolymers. Representative procedures and formulations are presented to demonstrate how copolymer choice influences critical paint properties such as scrub resistance, dirt pick-up, and weather durability. Although the main emphasis is on VAM-based systems, comparisons with styrene–acrylic and pure acrylic binders are also included where these provide practical insight, particularly in roof coatings and exterior architectural applications.

9.2 Chemistry and Mechanism of Copolymerization

The behavior of vinyl acetate in copolymerization is strongly influenced by the relative reactivity of the comonomers. Vinyl acetate typically exhibits low self-propagation ($r_1 < 0.1$ in most systems), which means it prefers to react with more active comonomers rather than adding to itself. The magnitude of the comonomer reactivity ratio (r_2) determines whether the chain becomes dominated by that comonomer or whether a more balanced, random structure is formed (Table 9.1).

The consistently low r_1 values confirm that vinyl acetate is a poor homopolymerizer and tends to incorporate other monomers into the chain. Acrylates, with their high r_2 values, dominate propagation and create gradient or acrylate-rich copolymers that are soft and flexible. Methyl methacrylate, with intermediate r_2, produces harder, higher-T_g copolymers due to partial self-propagation. In contrast, VeoVa 10 shows r-values close to unity with vinyl acetate, leading to nearly ideal random copolymerization. This randomness contributes to uniform film morphology and superior durability in coatings.

Table 9.1 Reactivity ratios in vinyl acetate copolymerizations.

Monomer pair	r_1 (VAM)	r_2 (Comonomer)	Polymer behavior
VAM/butyl acrylate	0.03	8.1	r_1 very low: vinyl acetate units preferentially add acrylate. r_2 very high: butyl acrylate strongly propagates itself. The result is acrylate-dominated sequences forming a soft gradient copolymer.
VAM/2-ethylhexyl acrylate	0.04	6.0–7.5	r_1 very low: VAM strongly favors cross-propagation with acrylate. r_2 high: acrylate propagates itself readily. The outcome is highly flexible acrylate-rich copolymers with very low T_g.
VAM/methyl methacrylate	0.05	2.0–2.5	r_1 low: VAM prefers MMA. r_2 moderate: MMA adds both to itself and to VAM. This produces harder copolymers, somewhat blocky, with higher T_g and increased rigidity.
VAM/VeoVa 10	0.99	0.92	Both r_1 and r_2 are near unity: each monomer has almost equal probability of reacting with itself or with the other. This produces a near-random distribution, giving coherent films with low water uptake and excellent alkali/UV resistance.

The presence of small amounts of acrylic acid or methacrylic acid (typically 1–5% of the total monomer) plays a distinct mechanistic role. The ionizable groups introduce charge stabilization and auto-dispersibility, reducing the need for free surfactant in the latex. In addition, they provide anchoring sites for external crosslinkers such as carbodiimides or zirconium complexes. However, overuse of acid monomers may lead to excessive water sensitivity or gel formation during polymerization.

Mechanistically, vinyl acetate participates through a head-to-tail radical addition pathway with relatively low propagation rate constants. In semibatch processes, the copolymer composition drifts according to the feed ratios, with acrylates consumed preferentially early in the reaction. Careful control of monomer feed and pH is therefore necessary to achieve uniform particle composition and to avoid excessive residual acrylates. VeoVa 10, being less water-soluble, partitions predominantly into the monomer droplets and feeds more steadily into the growing chains, ensuring consistent incorporation.

9.3 Structure–Property Relationships

Vinyl acetate copolymers display properties that are strongly determined by comonomer type and feed ratio. The balance between vinyl acetate, acrylates, and VeoVa 10 establishes the glass transition temperature (T_g), the hydrophobic or hydrophilic character, and the resulting film durability.

Glass transition temperature (T_g) and flexibility: Increasing acrylate content lowers T_g, yielding softer, more flexible films suitable for interior paints and roof coatings. Conversely, incorporation of methyl methacrylate or lower acrylate levels increases T_g, producing harder, more block-resistant films.

Hydrophobicity and water resistance: Butyl acrylate improves flexibility but does not strongly enhance water resistance. VeoVa 10, with its bulky hydrophobic side chain, markedly reduces water uptake, dirt pick-up, and alkali attack, making it the preferred choice for exterior applications.

Adhesion and durability: Acrylates provide strong adhesion to both porous and nonporous substrates, while VeoVa 10 enhances resistance to hydrolysis. Small amounts of acrylic acid introduce polarity and ionic stabilization, improving dispersion stability and crosslinking potential.

Weathering and scrub resistance: Copolymers rich in VeoVa 10 or incorporating crosslinkable sites show superior UV resistance and film integrity over long exposures. Addition of acrylic acid enables the use of external crosslinkers, which improve scrub resistance in interior paints and water resistance in roof coatings (Table 9.2).

Table 9.2 Structure–property relationships of VAM-based copolymers.

Copolymer system	T_g trend	Hydrophobicity	Adhesion characteristics	Durability/resistance
VAc–BA (soft, flexible)	Low T_g (−10 to 0 °C)	Moderate	Good adhesion, porous substrates	Flexible but moderate water resistance
VAc–2EHA (very soft)	Very low T_g (−20 °C)	Low–moderate	Excellent flexibility	Poor water/alkali resistance
VAc–MMA (hard)	High T_g (+20 to +30 °C)	Moderate	Good hardness, lower flexibility	Better block and scrub resistance
VAc–VeoVa 10 (balanced)	Mid-range T_g (0 to +10 °C)	High	Good adhesion + hydrolytic stability	Excellent water, alkali, and UV resistance
VAc–BA–AA (functional)	Low T_g (−5 to 0 °C)	Moderate	Adhesion + ionic stabilization	Enhanced scrub resistance with crosslinkers

9.4 VAc–Acrylate Copolymers

9.4.1 Overview

Vinyl acetate–acrylate copolymers form the basis of many interior and exterior architectural paints. The acrylate component lowers the glass transition temperature, providing flexibility and good adhesion, while vinyl acetate contributes cost efficiency and film formation. The balance between hardness and flexibility can be tuned by selecting different acrylates such as butyl acrylate, 2-ethylhexyl acrylate, or methyl methacrylate.

9.4.2 Copolymerization Process (5,000 kg batch, ~55% NV)

A semibatch emulsion polymerization process is used for preparing VAc–BA copolymer latex. The following tables represent a typical recipe for a 5,000 kg batch at approximately 55% nonvolatile content. Stabilization is achieved primarily by hydroxy ethyl cellulose (HEC) with minimized free surfactant. pH is maintained at 4.8–5.2 during polymerization, with postneutralization if required (Tables 9.3–9.7).

Procedure (Semibatch, Pre-emulsified; Seed Nucleation, Persulfate Initiation, Redox Finish)

Table 9.3 Seed/charge.

Component	Mass (kg)
Deionized water	700.0
HEC solution (10% active)	500.0
Nonionic surfactant (100% active)	3.0
Anionic surfactant (100% active)	2.0
Defoamer	1.5
Sodium bicarbonate	5.0

Table 9.4 Pre-emulsion.

Component	Mass (kg)
Deionized water	1, 750.0
Vinyl acetate	1, 700.0
Butyl acrylate	940.0
Nonionic surfactant (100% active)	3.0
Anionic surfactant (100% active)	2.0

Table 9.5 Initiators and catalysts.

Component	Mass (kg)
Ammonium persulfate	8.0
t-Butyl hydroperoxide (70%)	4.0
Sodium formaldehyde sulfoxylate	3.0
Fe catalyst solution	0.2

Table 9.6 Post-additions.

Component	Mass (kg)
Rinse water	380.0
Ammonia (10%)	15.0
Defoamer	1.5
Biocide (20%)	5.0

Table 9.7 Totals check.

Parameter	Value
Total batch size	5,000 kg
Nonvolatile content	≈55%
Residual monomer (after finish)	<0.1%
Particle size	120–220 nm

1) Charge the reactor with the seed materials. Heat to 72–75 °C under nitrogen purge with agitation at 1.5–2.0 m/s tip speed.
2) Prepare the pre-emulsion by dispersing surfactants in water and slowly adding vinyl acetate and butyl acrylate under high shear until homogeneous.
3) Add 10% of the APS initiator to the reactor; hold until a slight exotherm confirms seed formation.
4) Begin simultaneous feeds of pre-emulsion and the remaining APS solution over 3.0–3.5 hours, maintaining 75 ± 2 °C. Control the exotherm by cooling.
5) After feeds, hold 30–45 min at 75 °C. Reduce the temperature to 65–70 °C and add *t*-BHP and SFS alternately until residual monomer <0.3%.
6) Cool to 35–40 °C. Add rinse water, adjust pH with ammonia if required, and add post-additives. Filter before storage.
7) *QC targets*: NV 54–56%, viscosity 500–2,000 mPa s, particle size 120–220 nm, residual monomer <0.1%.

9.4.2.1 Representative Paint Formulations

The following formulations illustrate the use of VAc–BA copolymer latex in interior architectural paints. The binder provides a good balance of film formation, flexibility, and adhesion (Tables 9.8 and 9.9).

Table 9.8 Washable matt interior paint.

Component	% by weight
Water	35.0
Dispersant	0.3
Defoamer	0.3
TiO_2 (rutile)	20.0
Calcium carbonate	15.0
VAc–BA copolymer latex (55%)	27.0
Thickener (HEC)	1.2
Biocide	0.2
Coalescent	1.0

Table 9.9 Semigloss interior paint.

Component	% by weight
Water	30.0
Dispersant	0.4
Defoamer	0.3
TiO_2 (rutile)	18.0
Calcium carbonate	12.0
VAc–BA copolymer latex (55%)	35.0
Coalescent	1.5
Thickener (HASE)	2.0
Biocide	0.3

9.5 VAc–VeoVa 10 Copolymers

9.5.1 Overview

Copolymerization of vinyl acetate with versatic acid vinyl esters (VeoVa 10) produces binders of high hydrophobicity and excellent resistance to water, alkali, and ultraviolet light.

Durability on alkaline mineral substrates requires sufficient hydrophobic shielding of the acetate ester in polyvinyl acetate copolymers. VeoVa 10 (vinyl neodecanoate) provides this protection by introducing sterically hindered, hydrolysis-resistant ester units and lowering water uptake (the "umbrella effect").

For VAc–VeoVa 10 copolymers intended for exterior paints, VeoVa 10 should be held at not less than 20% on total monomer. The practical optimum is 20–25%: below 20%, alkali resistance is typically inadequate; above approximately 25%, further gains in alkali and water resistance plateau in the VAc–VeoVa 10 binary system.

When VeoVa 10 levels greater than 25% are desired for additional property tuning, improvements in exterior durability are generally realized only when a third comonomer is introduced (for example, an acrylate) to form a VAc–VeoVa 10–(meth)acrylate terpolymer. In such cases, feed strategy should maintain a homogeneous microstructure, and acid-functional comonomers should be kept low (no more than approximately 0.5–1.0% on total monomer) to preserve alkali resistance.

Design guidance (VAc–VeoVa 10 binders for exterior paints)

1) *VeoVa 10 on total monomer*: 20–25% (optimum window for alkali and water resistance).

2) *Vinyl acetate (VAc) balance*: Adjust the VAc content to achieve the target T_g/MFFT and film hardness, and confirm adequate early water resistance and blocking performance.
3) *Optional acrylate terpolymer*: Use for additional toughness and improved durability; maintain low – COOH levels and homogeneous feeds.

9.5.2 Copolymerization Process (5,000 kg Batch, ~55% NV)

A semibatch process is used to prepare VAc–VeoVa 10 copolymer latex with balanced flexibility and superior durability. The following tables represent a typical recipe for a 5,000 kg batch. Special care is required during pre-emulsion preparation due to the hydrophobic nature of VeoVa 10 (Tables 9.10–9.14).

Procedure (Semibatch; Notes Specific to VeoVa 10)

1) Charge the reactor with the seed materials and heat to 72–75 °C under a nitrogen sweep. Ensure the condenser and cooling system are fully operational.
2) Prepare the pre-emulsion by dispersing surfactants in water, then slowly adding vinyl acetate and VeoVa 10 under high shear until stable. Extended homogenization may be needed to prevent phase separation.

Table 9.10 Seed/charge.

Component	Mass (kg)
Deionized water	700.0
HEC solution (10% active)	500.0
Nonionic surfactant (100% active)	3.0
Anionic surfactant (100% active)	2.0
Defoamer	1.5
Sodium bicarbonate	5.0

Table 9.11 Pre-emulsion.

Component	Mass (kg)
Deionized water	1,750.0
Vinyl acetate	2,050.0
VeoVa 10	520.0
Nonionic surfactant (100% active)	3.0
Anionic surfactant (100% active)	2.0

Table 9.12 Initiators and catalysts.

Component	Mass (kg)
Ammonium persulfate	8.0
t-Butyl hydroperoxide (70%)	4.0
Sodium formaldehyde sulfoxylate	3.0
Fe catalyst solution	0.2

Table 9.13 Post-additions.

Component	Mass (kg)
Rinse water	380.0
Ammonia (10%)	15.0
Defoamer	1.5
Biocide (20%)	5.0

Table 9.14 Totals check.

Parameter	Value
Total batch size	5,000 kg
Nonvolatile content	≈55%
Residual monomer (after finish)	<0.1%
Particle size	150–250 nm

3) Add 10% of the APS initiator solution to the seed; wait for a small exotherm confirming nucleation.
4) Begin simultaneous feeds of the pre-emulsion and the remaining APS solution over 3.5–4.0 hours, maintaining 75 ± 2 °C. Control the exotherm by active cooling.
5) After feeds are complete, hold at 75 °C for 45 min, then lower the temperature to 65–70 °C. Add *t*-BHP and SFS alternately in small shots until residual monomer is <0.3%.
6) Cool to 35–40 °C. Add rinse water, adjust pH with ammonia (if required), and add post-additives. Filter before storage.

7) QC targets: NV 54–56%, viscosity 600–2,000 mPa s, particle size 150–250 nm, residual monomer <0.1%.

9.5.3 Representative Paint Formulations

The following formulations illustrate the use of VAc–VeoVa 10 copolymer latex in exterior architectural coatings and roof paints. The binder provides excellent weathering and alkali resistance (Tables 9.15 and 9.16).

Table 9.15 Exterior matt paint.

Component	% by weight
Water	32.0
Dispersant	0.4
Defoamer	0.3
TiO_2 (rutile)	20.0
Calcium carbonate	12.0
Talc	8.0
VAc–VeoVa 10 copolymer latex (55%)	25.0
Coalescent	1.3
Thickener (HASE)	0.8
Biocide	0.2

Table 9.16 Roof coating formulation.

Component	% by weight
Water	28.0
Dispersant	0.5
Defoamer	0.4
TiO_2 (rutile)	10.0
Aluminum silicate	15.0
Calcium carbonate	10.0
VAc–VeoVa 10 copolymer latex (55%)	33.0
Elastomeric thickener	2.0
Coalescent	1.0
Biocide	0.3

9.6 VAc–Acrylate–AA Terpolymers

9.6.1 Overview

Terpolymers of vinyl acetate, acrylates, and small amounts of acrylic acid (AA) combine the flexibility of acrylates with the hydrophobic film-forming ability of vinyl acetate, while introducing functional carboxyl groups. The acrylic acid contributes ionic stabilization, improves colloidal stability, and enables chemical crosslinking with carbodiimides, zirconium complexes, or aziridines. These binders are highly valued in paints where enhanced scrub resistance, adhesion, and water resistance are critical.

9.6.2 Copolymerization Process (5,000 kg Batch, ~55% NV)

The following recipe represents a semibatch polymerization of a VAc–BA–AA terpolymer (62:36:2 by weight ratio). The batch size is 5,000 kg at approximately 55% nonvolatile content (Tables 9.17–9.21).

Table 9.17 Seed/charge.

Component	Mass (kg)
Deionized water	700.0
HEC solution (10% active)	500.0
Nonionic surfactant (100% active)	3.0
Anionic surfactant (100% active)	2.0
Defoamer	1.5
Sodium bicarbonate	5.0

Table 9.18 Pre-emulsion.

Component	Mass (kg)
Deionized water	1,780.0
Vinyl acetate	1,610.0
Butyl acrylate	940.0
Acrylic acid	50.0
Nonionic surfactant (100% active)	3.0
Anionic surfactant (100% active)	2.0

Table 9.19 Initiators and catalysts.

Component	Mass (kg)
Ammonium persulfate	8.0
t-Butyl hydroperoxide (70%)	4.0
Sodium formaldehyde sulfoxylate	3.0
Fe catalyst solution	0.2

Table 9.20 Post-additions.

Component	Mass (kg)
Rinse water	410.0
Ammonia (10%)	15.0
Defoamer	1.5
Biocide (20%)	5.0

Table 9.21 Totals check.

Parameter	Value
Total batch size	5,000 kg
Nonvolatile content	≈55%
Residual monomer (after finish)	<0.1%
Particle size	130–220 nm

Procedure (Semibatch; Terpolymer with Small AA)

1) Charge the reactor with the seed phase and heat to 70–73 °C under nitrogen purge. Maintain pH at 4.8–5.2 to limit premature homopolymerization of acrylic acid.
2) Prepare the pre-emulsion by dispersing surfactants in water, then slowly adding vinyl acetate, butyl acrylate, and acrylic acid under high shear until homogeneous.
3) Add 10% of the APS initiator to the seed. Confirm nucleation with a slight exotherm (2–4 °C rise).

4) Begin feeds of pre-emulsion and APS simultaneously over 3.0–3.5 hours at 74–76 °C. Avoid over-neutralizing acrylic acid during this stage to prevent gelation.
5) After feeds are complete, hold 30–45 min. Lower the temperature to 65–70 °C and add *t*-BHP and SFS alternately until residual monomer <0.3%.
6) Cool to 35–40 °C. Add rinse water, adjust pH to 7.5–8.0 if crosslinker compatibility is required, and add postadditives. Filter before storage.
7) *QC targets*: NV 54–56%, viscosity 800–2,500 mPa s, particle size 130–220 nm, residual monomer <0.1%, acid value consistent with ~2% AA incorporation.

9.6.3 Representative Paint Formulations

The following formulations illustrate the use of VAc–BA–AA terpolymer latex in scrub-resistant and premium interior paints. The carboxyl functionality allows effective crosslinking with carbodiimides or zirconium complexes to further enhance resistance (Tables 9.22 and 9.23).

9.7 Comparative Analysis with Styrene–Acrylic and Pure Acrylic Binders

Although vinyl acetate copolymers dominate many architectural coatings, styrene–acrylic and pure acrylic binders play an important role where maximum

Table 9.22 Scrub-resistant interior paint.

Component	% by weight
Water	30.0
Dispersant	0.4
Defoamer	0.3
TiO_2 (rutile)	22.0
Calcium carbonate	15.0
VAc–BA–AA terpolymer latex (55%)	28.0
Coalescent	1.0
Thickener (HEC)	2.0
Biocide	0.3
Carbodiimide crosslinker (dispersion)	1.0

Table 9.23 Premium washable paint.

Component	% by weight
Water	32.0
Dispersant	0.3
Defoamer	0.3
TiO_2 (rutile)	20.0
Aluminum silicate	10.0
Calcium carbonate	12.0
VAc–BA–AA terpolymer latex (55%)	24.0
Coalescent	1.2
Thickener (HASE)	2.0
Biocide	0.2
Zirconium crosslinker (25% active)	0.8

weather and alkali resistance are required. Within the VAM family, functionalized terpolymers containing small amounts of acrylic acid (VAc–acrylate–AA) enable crosslinking and elevated scrub/water resistance. A comparative view helps determine when to select each binder platform.

9.7.1 Styrene–Acrylic Binders

Styrene–acrylic copolymers balance hardness and chemical resistance by incorporating styrene, which raises T_g and improves block resistance, with acrylates that provide flexibility. They are widely used in exterior paints, roof coatings, and traffic markings. Compared with VAc–acrylate copolymers, they generally offer better alkali and weather resistance, though they may yellow under prolonged UV exposure.

9.7.2 Pure Acrylic Binders

Pure acrylic latexes, based on combinations of butyl acrylate, methyl methacrylate, and related monomers, set the benchmark for long-term durability. They provide superior UV stability, water resistance, and dirt pick-up resistance compared with both VAc–acrylate and VAc–VeoVa 10 systems, at the expense of higher raw material costs and potential need for coalescents at ambient temperature.

9.7.3 VAc–Acrylate–AA Terpolymers

Introducing approximately 1–3% acrylic acid into VAc–acrylate copolymers adds carboxyl functionality that improves colloidal stability and provides anchor sites for postcrosslinking (e.g., carbodiimide, zirconium, aziridines). When crosslinked appropriately, these terpolymers achieve notable gains in scrub resistance, chemical spot resistance, and water resistance while preserving film formation and flexibility. Excessive acid levels, however, may increase water uptake or risk gelation if pH is not properly controlled during polymerization. Typical uses include premium interior paints (kitchen/bath), primers, and coatings demanding elevated scrub.

Maintaining the correct pH window requires the use of suitable buffer systems. Bicarbonate salts are commonly employed to stabilize persulfate-initiated reactions in the mildly acidic range (pH 4.5–5.5). Phosphate buffers may also be used where tighter control is needed, although excessive buffering capacity should be avoided to prevent inhibition of radical generation. These measures ensure that carboxyl groups remain partially ionized to aid colloidal stability while avoiding over-neutralization that could destabilize the latex or increase water sensitivity. Common buffer systems are summarized in Table 9.24, while Table 9.25 summarizes comparative performance of common binder classes.

Decision guidance: Use VAc–acrylate for balanced cost and film formation, VAc–VeoVa 10 for exterior durability and alkali resistance, VAc–acrylate–AA when crosslinker-enabled scrub/water upgrades are needed, and pure acrylic or styrene–acrylic where maximum outdoor durability is mandatory.

Table 9.24 Common buffer systems for VAc–Acrylate–AA polymerizations.

Buffer system	Effective pH range	Notes
Sodium bicarbonate ($NaHCO_3$)	4.5–5.5	Commonly used in persulfate-initiated systems; maintains mild acidity; prevents rapid pH drift.
Phosphate buffer (NaH_2PO_4/Na_2HPO_4)	5.0–7.0	Provides tighter pH control; should be used at low concentration; excessive buffering may slow radical generation.
Citrate buffer	3.0–6.2	Occasionally used where a lower pH is required; may interact with metal ions if present.
Acetate buffer (CH_3COOH/CH_3COONa)	3.5–5.5	Useful for stabilizing acidic conditions; less common due to the volatility of acetic acid.

Table 9.25 Comparative performance of common binder classes.

Binder type	Strengths	Limitations	Typical applications
VAc–acrylate copolymers	Good flexibility, cost-effective, tunable T_g	Moderate water/alkali resistance vs. premium acrylics	Interior paints, general-purpose coatings
VAc–VeoVa 10 copolymers	High hydrophobicity, excellent alkali and UV resistance	Higher cost than VAc–acrylate; lower flexibility vs. BA-rich systems	Exterior paints, roof coatings
VAc–acrylate–AA terpolymers	Functional sites for crosslinking; elevated scrub and water resistance when crosslinked; good adhesion	Can be more water-sensitive without crosslinker; AA requires careful pH control	Premium interior paints (kitchen/bath), primers, upgrade path for scrubbables
Styrene–acrylic copolymers	Good hardness, alkali resistance, weather durability	Potential yellowing; higher cost than VAc systems	Exterior paints, roof coatings, traffic markings
Pure acrylic copolymers	Top-tier durability, UV stability, dirt pick-up resistance	High cost; may require coalescents at ambient temps	Premium exterior paints, high-performance coatings

9.8 Application Formulations and Procedures

The practical value of VAc-based copolymers lies in their use as binders for architectural paints and roof coatings. The following formulations and procedures demonstrate how VAc–acrylate, VAc–VeoVa 10, and VAc–acrylate–AA terpolymers are applied in different classes of coatings. Each example is expressed as a representative laboratory or plant-scale paint formulation, with notes on application performance.

9.8.1 Interior Architectural Coatings

For interior paints, VAc–acrylate copolymers provide a cost-balanced binder with good adhesion and film formation. When acrylic acid functionality is incorporated and combined with a crosslinker, superior scrub resistance can be achieved (Tables 9.26 and 9.27).

Table 9.26 Washable matt interior paint (VAc–BA binder).

Component	% by weight
Water	35.0
Dispersant	0.3
Defoamer	0.3
TiO_2 (rutile)	20.0
Calcium carbonate	15.0
VAc–BA latex (55%)	27.0
Thickener (HEC)	1.2
Coalescent	1.0
Biocide	0.2

Table 9.27 Premium scrub-resistant interior paint (VAc–BA–AA terpolymer + carbodiimide).

Component	% by weight
Water	30.0
Dispersant	0.4
Defoamer	0.3
TiO_2 (rutile)	22.0
Calcium carbonate	15.0
VAc–BA–AA terpolymer latex (55%)	28.0
Thickener (HEC)	2.0
Coalescent	1.0
Biocide	0.3
Carbodiimide crosslinker (dispersion)	1.0

9.8.2 Exterior Architectural Paints

Copolymerization of vinyl acetate with versatic acid vinyl esters (VeoVa 10) produces binders of high hydrophobicity and excellent resistance to water, alkali, and ultraviolet light. The bulky branched structure of VeoVa 10 contributes to a more random copolymer sequence and reduces water sensitivity compared with vinyl acetate–acrylate copolymers. These copolymers are particularly suited for exterior architectural coatings and roof protection systems (Table 9.28).

Table 9.28 Exterior matt paint (VAc–VeoVa 10 binder).

Component	% by weight
Water	32.0
Dispersant	0.4
Defoamer	0.3
TiO_2 (rutile)	20.0
Calcium carbonate	12.0
Talc	8.0
VAc–VeoVa 10 latex (55%)	25.0
Coalescent	1.3
Thickener (HASE)	0.8
Biocide	0.2

9.8.3 Roof Coatings

Roof coatings demand maximum hydrophobicity, dirt pick-up resistance (DPUR), and above all, elasticity to withstand thermal movement and substrate stresses. VAc–VeoVa 10 copolymers are often combined with soft acrylates such as butyl acrylate to lower T_g and impart elongation capability. Terpolymers with small amounts of acrylic acid can be crosslinked to further improve water resistance without compromising elasticity (Table 9.29).

Table 9.29 Elastic roof coating formulation (VAc–VeoVa 10–BA binder).

Component	% by weight
Water	26.0
Dispersant	0.5
Defoamer	0.4
TiO_2 (rutile)	8.0
Aluminum silicate	15.0
Calcium carbonate	8.0
VAc–VeoVa 10–BA latex (55%)	36.0
Elastomeric thickener (ASE/HASE)	3.0
Coalescent	2.0
Biocide	0.3
UV absorber/HALS (optional)	0.8

These formulations illustrate how VAc-based copolymers and terpolymers can be tailored to meet the needs of different applications. For roof coatings, the binder must combine hydrophobicity with high elasticity; this is achieved by incorporating soft acrylates into VAc–VeoVa 10 systems. Crosslinking strategies, pigment volume concentration, and thickener choice are adjusted depending on whether the target is an economical interior paint, a scrub-resistant premium coating, a weather-resistant exterior, or a waterproof elastic roof system.

9.9 Summary and Practical Guidelines

Vinyl acetate copolymers with acrylates and VeoVa 10 offer a wide spectrum of performance tailored for architectural paints and roof coatings. Acrylate comonomers reduce glass transition temperature and impart flexibility, while VeoVa 10 contributes hydrophobicity, alkali resistance, and enhanced weather durability. Incorporation of small amounts of acrylic acid introduces functional groups for stabilization and crosslinking, enabling further performance upgrades.

Compared with alternative binder systems, VAc–acrylate copolymers are most cost-effective for general-purpose interior paints. VAc–VeoVa 10 copolymers excel in exterior coatings where resistance to water, alkali, and ultraviolet light is paramount. Terpolymers containing acrylic acid, when combined with appropriate crosslinkers such as carbodiimide or zirconium salts, offer significant gains in scrub resistance and chemical spot resistance for premium interior and specialty coatings. Styrene–acrylic and pure acrylic binders remain benchmarks for maximum durability, but at a higher raw material cost.

Practical guidelines for selecting and formulating VAc-based copolymer binders include the following:

1) Use VAc–acrylate copolymers for economical interior paints where flexibility and adhesion are essential but water resistance demands are moderate.
2) Select VAc–VeoVa 10 copolymers for exterior coatings and roof coatings where hydrophobicity, alkali resistance, and long-term durability are required.
3) Employ VAc–acrylate–AA terpolymers in premium interior paints and primers when elevated scrub and water resistance are necessary. Activate their potential with external crosslinkers.
4) Adjust pigment volume concentration and thickener type depending on target performance (e.g., high PVC for matt paints, lower PVC with coalescent for scrub-resistant coatings).
5) For roof coatings, ensure binder elasticity by incorporating soft acrylates such as butyl acrylate (BA) or 2-ethylhexyl acrylate (2-EHA) into

VAc–VeoVa 10 copolymer systems. These comonomers reduce the glass transition temperature, imparting flexibility necessary to withstand thermal cycling and mechanical stress. To maximize outdoor durability, the binder is typically combined with light-stabilization additives. While ultraviolet absorbers protect by filtering harmful radiation, hindered amine light stabilizers (HALS) act through a radical-scavenging mechanism that interrupts photo-oxidative degradation. Common examples include Tinuvin 770 or Tinuvin 292. The synergistic use of soft acrylates with VeoVa 10 for elasticity, and UV absorbers or HALS for stabilization, yields roof coating binders with superior weather resistance, crack resistance, and long-term film integrity.

6) *Maintain pH control during polymerization and storage*: 4.8–5.2 as-made for stability, adjust to 7.5–8.0 only if required for crosslinker compatibility or application needs.

By understanding the structure–property relationships and practical formulation strategies, formulators can exploit the versatility of VAc-based copolymers to design coatings that meet both performance and economic requirements across interior, exterior, and roofing applications.

9.10 Polymerizable Surfactants—Cross-reference and Selection for Coating Copolymers

This section provides coating-specific selection guidance without repeating the full surfmer list. For candidate chemistries and trade names, see Table 4.10 (Selected commercial surfmers and their suitable copolymer systems). For mechanism, order of addition, dosing strategy, and pH control, refer to Chapter 4 (Polymerizable Surfactants in PVAc copolymer).

Selection focuses on minimizing free-surfactant migration, lowering water whitening and alkali sensitivity, and improving wet adhesion. Use AA/MAA (≈0.5–3.0%) to reinforce colloidal stability and provide crosslinking sites (CDI or Zr), while maintaining pH ≈ 4.5–5.5.

Selection guidance by binder/application

1) *VAc–BA (interior scrub paints)*: Anionic reactive sulfate/sulfonate surfmer; 0.3–0.6% on monomer. Aim: scrub resistance, low water whitening; keep conventional free surfactant minimal.
2) *VAc–2-EHA (elastomeric/roof)*: Reactive phosphate ester at low dose (± small nonionic macromer); 0.2–0.5%. Aim: wet adhesion, alkali/water resistance; monitor persulfate kinetics with phosphate buffers.

3) *VAc–VeoVa 10 (exterior alkali/UV)*: Reactive phosphate ester (±macromer); 0.2–0.6%. Aim: reduce leachable surfactant, enhance alkali resistance and cleanability; pair with UV absorber/HALS.
4) *VAc–BA–VeoVa 10 (balanced exterior)*: Blend of anionic reactive + nonionic macromer; 0.3–0.7%. Aim: uniform film, balanced elasticity/durability; optimize EO length to avoid water sensitivity.
5) *VAc–MMA (harder primers/topcoats)*: Nonionic polymerizable surfactant (e.g., Sipomer® COPS-1); 0.3–0.6%. Aim: clarity and low migration in higher-T_g systems; add modest chain-transfer if molecular-weight growth is observed.

For additional performance, a polymerizable (reactive) surfactant can be used. Covalently anchoring the surfactant to the particle surface reduces the amount of free, migratory surfactant in the dried film. This typically lowers water uptake and surfactant leaching, improves wet adhesion and alkali resistance, and enhances resistance to blocking and dirt-pickup—especially valuable for exterior VAM/VeoVa 10 and VAM/acrylic systems. Reactive surfactants can also improve electrolyte tolerance during polymerization and reduce postfilm foaming and sensitivity.

Incorporating a small fraction of acrylic acid (AA) or methacrylic acid (MAA) (≈0.5–3% of total monomer) introduces carboxyl functionality that strengthens colloidal stabilization and provides anchoring sites for postcrosslinking (e.g., carbodiimide or zirconium complexes). AA tends to bolster dispersion stability and wet adhesion but may increase water sensitivity if overdosed. MAA contributes similar crosslinkability with a slightly higher hardness/higher T_g bias, supporting block, scrub, and chemical resistance in more demanding coatings. In all cases, maintain appropriate pH control during polymerization (typically mildly acidic) to avoid gelation or excessive water uptake.

9.11 Selection Guide for VAM-based Copolymer Systems

This flowchart helps in selecting the appropriate VAM-based copolymer system based on application and performance requirements. Start from the top and follow the decision path to the recommended system (Figure 9.1).

9.12 Selection Guide for VAM-based Copolymer Systems (Enhanced)

This enhanced flowchart includes performance notes and relative cost indicators for each copolymer system (Figure 9.2).

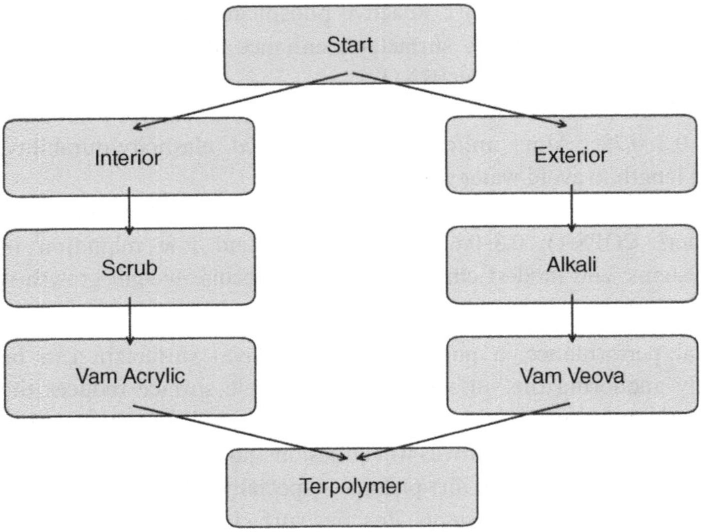

Figure 9.1 Selection guide for VAM-based copolymer systems. This figure presents a simplified decision flow for selecting VAM-based copolymer systems according to application category. It distinguishes between interior and exterior uses and indicates the general selection path leading to VAM/Acrylic, VAM/VeoVa 10, or their terpolymer variants based on performance demands.

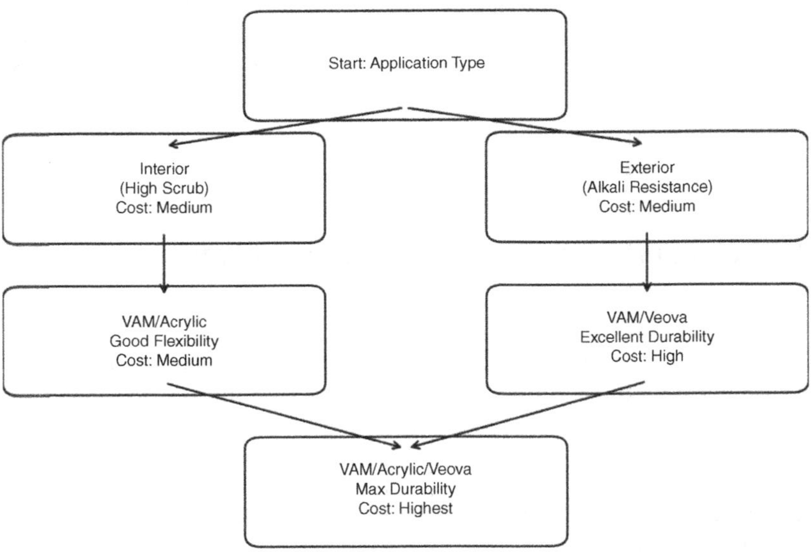

Figure 9.2 Enhanced selection guide for VAM-based copolymer systems. Explanation: This enhanced flowchart provides a structured decision guide incorporating performance attributes and relative cost levels for each system. It emphasizes when to choose VAM/acrylic for scrub resistance, VAM/VeoVa 10 for exterior alkali resistance, or terpolymers for maximum durability.

9.13 Surfactant and Colloid Compatibility

The selection and distribution of surfactants and protective colloids significantly influence latex stability, rheology, and ultimately the performance of VAM-based copolymer systems. These interactions are governed by the polymer's surface characteristics, which in turn depend on copolymer composition and architecture. For a detailed discussion of surfactant types, mechanisms, and guidelines on selecting the appropriate surfactant for emulsion polymerization systems, the reader is referred to Chapter 4.

9.13.1 Influence of Polymer Structure on Surfactant Adsorption

The incorporation of hydrophobic comonomers such as VeoVa 10 or long-chain acrylates modifies the surface energy and polarity of latex particles. This affects both the adsorption strength and distribution of surfactants across the particle interface: PVAc/VeoVa 10 copolymers present a less polar surface, reducing the affinity for strongly anionic surfactants; nonionic surfactants with ethoxylated chains typically provide better steric stabilization. Acrylate-rich copolymers, by contrast, exhibit higher polarity and hydrogen-bonding capacity and are best stabilized using anionic or mixed systems, promoting electrosteric stabilization.

9.13.2 Interaction with Protective Colloids

Protective colloids such as polyvinyl alcohol (PVOH) and hydroxyethyl cellulose (HEC) provide steric stabilization and viscosity control. The degree of hydrolysis and molecular weight of PVOH strongly influence adsorption efficiency. PVAc-rich and VeoVa 10-modified systems readily adsorb PVOH and HEC, whereas acrylate-dominant copolymers require higher colloid levels or synergistic surfactant combinations to achieve similar stability.

9.13.3 Effect on Viscosity and Application Rheology

Surfactant and colloid compatibility also affects rheology: Well-adsorbed colloids promote higher low-shear viscosity for sag resistance, while free surfactant in the serum phase can reduce viscosity and increase foaming. Excessive PVOH may lead to thickening and poor leveling. Careful optimization ensures controlled flow for brush, roller, or spray application.

9.13.4 Implications for Formulation Design

The interplay between monomer selection, surfactant system, and colloid compatibility provides a versatile toolkit for customizing adhesive and coating performance. By understanding these relationships, formulators can tailor systems for high-shear stability, freeze–thaw resistance, open time control, and optimized application behavior.

9.14 Soft Coalescents for VAM-based Copolymers

Soft coalescents play a pivotal role in lowering the minimum film formation temperature (MFFT), enhancing flexibility, and providing smooth, continuous films. Their use is particularly important in pressure-sensitive adhesives, laminating adhesives, and flexible coatings based on VAM/BA, VAM/2-EHA, or VAM/VeoVa 10 systems. The choice of coalescent determines not only the softness and flow characteristics but also the evaporation rate, environmental profile, and compatibility with crosslinkers (Table 9.30).

9.15 Structure–Property Relationship in VAM Copolymers

The performance advantages of vinyl acetate copolymers arise from the ability to tailor chemical structure to meet specific property requirements. Incorporating different comonomers modifies glass transition temperature (T_g), flexibility, water resistance, and durability. This section highlights the impact of copolymer composition on these critical attributes.

9.15.1 Glass Transition Temperature (T_g)

The glass transition temperature is a key determinant of flexibility, tack, and film formation behavior. Homopolymeric polyvinyl acetate exhibits a T_g in the range of +28 to +32 °C, which limits its suitability in low-temperature applications because of brittleness. Copolymerization with soft monomers allows

Table 9.30 Soft coalescents for VAM-based copolymers.

Coalescent name	Chemical type	Function/remark
Butyl carbitol (diethylene glycol monobutyl ether)	Glycol ether	Softens film, slows evaporation, improves flow and leveling
Texanol (2,2,4-trimethyl-1,3-pentanediol monoisobutyrate)	Ester–alcohol blend	Balances softening and drying time, widely used

Table 9.31 Effect of comonomer type on glass transition temperature and film characteristics.

Copolymer type	Approx. T_g (°C)	Film behavior
PVAc homopolymer	+28 to +32	Brittle at low temperatures
VAM/butyl acrylate	+10 to −5	Flexible, soft, and tacky
VAM/2-ethylhexyl acrylate	−10 to −20	Very soft; high adhesion and peel strength
VAM/VeoVa 10 10	+15 to +25	Tough, water-resistant; moderate flexibility

precise T_g adjustment, enabling improved performance in diverse applications (Table 9.31).

The incorporation of soft acrylates substantially lowers the T_g, facilitating film formation at ambient temperature and increasing tack and flexibility. These properties are particularly advantageous in pressure-sensitive adhesives and flexible coatings. In contrast, VeoVa 10 introduces a balance between moderate T_g and hydrophobic character, yielding tough, dimensionally stable films with superior weather resistance. This makes VeoVa 10-containing copolymers highly suitable for exterior architectural coatings and construction adhesives.

9.15.2 Water Resistance, Alkali Stability, and Durability

Water resistance and durability are essential performance criteria for coatings and adhesives intended for outdoor exposure or alkaline substrates such as cementitious materials. Homopolymeric polyvinyl acetate suffers from high water sensitivity because of its polar acetate groups and limited hydrolytic stability, leading to chalking and loss of adhesion over time.

Copolymerization significantly improves resistance characteristics. Incorporating hydrophobic comonomers such as VeoVa 10 or long-chain acrylates reduces water uptake and mitigates hydrolysis. These modifications not only enhance resistance to moisture penetration but also confer better alkali stability, extending the service life of architectural coatings and construction adhesives (Table 9.32).

The presence of hydrophobic comonomers reduces the accessibility of water and alkaline ions to the polymer backbone. VeoVa 10-modified copolymers exhibit superior performance due to their bulky, branched structure, which shields ester linkages from hydrolytic attack. Acrylate-containing systems offer improved flexibility and moderate water resistance but may require further modification or crosslinking for exterior durability.

Table 9.32 Influence of comonomer selection on water and alkali resistance.

Copolymer type	Water/alkali resistance	Relative durability
PVAc homopolymer	Poor	Low
VAM/butyl acrylate	Moderate	Moderate
VAM/2-ethylhexyl acrylate	Moderate	Moderate
VAM/VeoVa 10	Excellent	High

9.16 Copolymerization Without HEC-Process Advantages and Formulation Strategy

The manufacturing of vinyl acetate–acrylic (VAc/AC) and vinyl acetate–Versatic ester (VAM/VeoVa 10) copolymers is carried out without the inclusion of hydroxyethyl cellulose (HEC) or other protective colloids in the polymerization stage. Polymerization stability is ensured through a blend of anionic and nonionic emulsifiers, combined with starved-feed semicontinuous conditions, seed latex, and functional monomers such as acrylic acid (AA). This approach eliminates the negative impact of hydrophilic colloids on water resistance and alkali durability, which are critical for exterior and roof coating applications.

9.16.1 Why HEC Is Not Used in Polymerization

HEC is designed as a rheology modifier for paint formulations and not as a polymerization stabilizer. Adding HEC during polymerization introduces hydroxyl-rich material that increases serum viscosity, disrupts particle nucleation, and compromises water and alkali resistance of the final latex film. Furthermore, HEC can trap residual surfactants and salts, leading to higher water uptake and reduced weathering performance. Instead, polymer colloidal stability is achieved through optimized surfactant systems and minor levels of AA (0.3–0.8% on monomer basis) to provide electrostatic stabilization and pigment-binding sites.

9.16.2 Benefits of Copolymerization Without HEC

Eliminating HEC from the reactor improves the quality and durability of copolymer dispersions in several ways:

1) Enhanced water resistance and reduced swelling of dried films.
2) Improved alkali resistance on cementitious substrates.
3) Lower grit and cleaner particle size distribution under starved-feed conditions.
4) Easier crosslinking in specialized systems (e.g., CDI or aziridine for acidic terpolymer topcoats) without risk of reactive site interference.

9.16.3 Role of HEC in Coating Formulations

While HEC is excluded from polymerization, it plays an important role in the paint formulation stage as a rheology modifier. HEC provides thickening, sag resistance, and good flow in waterborne paints and coatings. It is typically introduced as an aqueous slurry to ensure full hydration before pigment dispersion and binder addition. Alternative or supplementary rheology modifiers include hydrophobically modified alkali-swellable emulsions (HASE) and hydrophobically modified ethoxylated urethanes (HEUR), which improve ICI viscosity control and flow (Table 9.33).

Summary

HEC is preferably excluded from the polymerization stage for vinyl-acrylic and VAM/VeoVa 10 copolymer systems, as its presence may reduce water resistance and exterior durability. Its use is therefore generally confined to the paint formulation stage for rheology control, typically in combination with associative thickeners to achieve balanced flow, sag resistance, and application performance.

Table 9.33 Practical use of associative and cellulosic thickeners in paint formulations.

Parameter	Typical range	Application notes	Targets
HEC slurry concentration	2.0–3.0% (w/w)	Prepare preslurry in 3–5× water; disperse under high shear; hydrate 20–30 min	Avoid clumping
Dosage (on total paint)	0.15–0.30%	Adjust after grind for KU control; combine with HASE/HEUR for ICI balance	KU 90–110
HASE (acid form)	0.3–0.6%	Add in let-down under moderate shear	Boost low-shear viscosity
HEUR	0.2–0.5%	Add late in let-down; tune for flow/leveling	ICI per spec (e.g., 1.0–2.0 P)

9.16.4 Applications and Copolymerization Without HEC

9.16.4.1 Best Applications for HEC-Free Binders

HEC-free, surfactant-stabilized binders offer superior water and alkali resistance. Recommended uses include:

1) Kitchens and bathrooms (high humidity, frequent washing).
2) Exterior masonry and concrete (rain, UV, alkaline substrates).
3) Roof coatings (elastomeric/reflective; ponded water, thermal cycling).
4) Hygienic and food-processing areas (frequent cleaning, chemical exposure).

9.16.4.2 Why Performance Improves Without HEC

Eliminating HEC from the polymerization stage reduces the hydrophilicity of the dried film and improves durability:

1) Lower water absorption; better alkali resistance on cementitious substrates.
2) Cleaner particle size distribution; reduced grit under starved-feed conditions.
3) Compatibility with crosslinkers in specialized acidic systems (e.g., VAM/VeoVa 10 CDI topcoats) (Table 9.34).

Process Highlights (Starved-Feed with In-Situ Seed)

1) Charge water + 40% of emulsifiers; purge N_2; heat to 80–85 °C.
2) Seed stage (pre-emulsion kick): add 5–10% of total monomer (styrene/VAc-rich), emulsifier, and initiator; polymerize for 15–20 min until exotherm subsides.
3) Starved co-feeds (3–4 h): remaining pre-emulsion and initiator; keep residual monomer <1%.
4) Hold 30 min; redox chase (APS/SFS) to residual <500 ppm.
5) Cool; adjust storage pH 8.0–8.5; filter 100–150 μm (Table 9.35).

9.16.4.3 Surfactant Selection and Dosage

An optimized anionic/nonionic blend provides electrostatic and steric stabilization without HEC. Typical total surfactant (active) is ~0.9–1.2% on total monomer. Use low-foam grades and verify freeze–thaw stability where required.

Formulation Note (Paint Stage): HEC is applied only during paint formulation as a rheology modifier. Use an aqueous slurry at 2.0–3.0% w/w, delivering 0.15–0.30% HEC on total paint. Combine with HASE (0.3–0.6%) or HEUR (0.2–0.5%) as needed to achieve KU 90–110 and specified ICI.

Table 9.34 Vinyl–acrylic copolymerization without HEC (semicontinuous, 5,000 kg).

Component	Amount (kg)	Function	Notes
Deionized water	q.s. to 5,000	Continuous phase	Initial charge ~60% of total
Anionic surfactant (active)	28	Primary emulsifier	SDS/sulfosuccinate/phosphate ester
Nonionic surfactant (active)	18	Co-emulsifier	Fatty alcohol ethoxylate, HLB 13–14
In-situ seed (pre-emulsion kick)	5–10% of total monomer	Nucleation control	Prepare in reactor before main feed
Vinyl acetate	1,750	Main monomer	—
Butyl acrylate (BA)	690	Soft monomer	Tg control, flexibility
Methyl methacrylate (MMA)	105	Hard monomer	Gloss/hardness
Acrylic acid (AA)	25	Functional monomer	Adhesion, pigment binding
Initiator (APS, 5% aq.)	16	Radical source	Co-feed with monomers
Reducing agent (SFS, 5% aq.)	10	Redox pair	Postfeed chase
Defoamer (process aid)	8	Foam control	Silicone-free grade

Table 9.35 Practical note: why in-situ seed is preferred.

Guideline	Reason
Prepare seed in the same reactor	Avoids cost/logistics of purchased seed; ensures chemistry match.
Polymerize 5–10% of monomer first	Improves PSD control; reduces secondary nucleation.
Increase emulsifier if skipping seed	Compensates for lack of particle templates; avoids grit formation.

9.16.4.4 Kitchen and Bathroom Paint Formulations Using HEC-Free Copolymer Binder

These formulations are designed for high-humidity environments such as kitchens and bathrooms, where washability, scrub resistance, and low water uptake are essential. Binders are vinyl–acrylic copolymers produced without HEC in the polymerization stage, relying on surfactant stabilization for improved water and alkali resistance. HEC and associative thickeners (HASE) are applied only at the paint formulation stage to achieve desired rheology targets (Tables 9.36 and 9.37).

Procedure (Both Formulations)

1) Preslurry HEC in 3–5× water at 2.0–3.0% solids; hydrate 20–30 min.
2) Charge water, dispersant, wetting agent, and half the defoamer into the mixing vessel; begin high-speed agitation.
3) Add pigments and extenders under high shear; grind to Hegman ≥6.
4) Add vinyl–acrylic binder under moderate shear; mix 15–20 min.
5) Incorporate HEC slurry and HASE thickener; adjust viscosity (KU 90–110, ICI per gloss target).
6) Add coalescent, remaining defoamer, and preservatives; adjust pH to 8.0–8.5.
7) Filter through a 150 μm mesh; package in lined containers.

Table 9.36 Standard washable matte paint using HEC-free binder (1,000 kg batch).

Component	Amount (kg)	Function	Notes
Deionized water	q.s. to 1,000	Carrier	Initial charge ~60%
Dispersant (polycarboxylate)	5.0	Pigment dispersion	Alkali-resistant
Nonionic wetting agent	2.0	Wetting	Improves pigment wetting
Defoamer	1.5	Air release	Add in two stages
TiO_2 (rutile)	160	Opacity	Chloride-process grade
Calcium carbonate	380	Extender	High brightness
Talc (platy)	30	Extender	Improves stain resistance
Vinyl–acrylic binder (50% solids) HEC-free	220	Film former	HEC-free latex
HEC slurry (2.5%)	500	Rheology	Delivers ~12.5 kg dry HEC (~0.25%)
HASE (acid form)	4.0	Associative thickener	Boosts low-shear viscosity
Coalescent (low VOC)	6.0	Film formation aid	Select plasticizer-type
Biocide (in-can + film)	As required	Preservation	Per regulatory guidelines

Table 9.37 Premium satin paint using HEC-free binder (1,000 kg batch).

Component	Amount (kg)	Function	Notes
Deionized water	q.s. to 1,000	Carrier	Initial charge ~55%
Dispersant (polycarboxylate)	5.5	Pigment dispersion	Alkali-resistant
Nonionic wetting agent	2.5	Wetting	Improves gloss uniformity
Defoamer	1.8	Air release	Silicone-free
TiO_2 (rutile)	180	Opacity	High-gloss grade
Calcium carbonate	300	Extender	Gloss control
Kaolin (fine)	40	Extender	Improves scrub resistance
Vinyl–acrylic binder (50% solids)	260	Film former	HEC-free latex
HEC-free			
HEC slurry (2.5%)	500	Rheology	Delivers ~12.5 kg dry HEC (~0.25%)
HASE (acid form)	5.0	Associative thickener	Improves flow and leveling
Coalescent (low VOC)	7.0	Film formation aid	Plasticizer type
Biocide (in-can + film)	As required	Preservation	Per regulatory guidelines

Performance Targets

1) *Washability*: ≥1,000 scrub cycles (per ASTM D2486).
2) *Gloss*: Matte (5–10 at 60°) for A; Satin (25–35 at 60°) for B.
3) *Water resistance*: No whitening after 1 hour immersion.
4) *pH stability*: 8.0–8.5; storage viscosity stable for 6 months.
5) *Fungus/mildew resistance*: Pass ASTM D3273 after 4 weeks.

9.17 VAM/VeoVa 10 Terpolymer Binder–Semicontinuous Emulsion Polymerization (No HEC)

This formulation describes a vinyl acetate–Versatic ester–acrylate terpolymer latex manufactured without any hydroxyethyl cellulose (HEC) or protective colloids. Colloidal stability is achieved by an optimized anionic/nonionic surfactant package, an in-situ seed, and starved-feed co-addition of pre-emulsified

Table 9.38 5,000 kg batch—VAM/VeoVa 10 terpolymer binder (solids ≈ 50–52%)—HEC-free.

Component	Amount (kg)	Function	Notes
Deionized water	q.s. to 5,000	Continuous phase	Initial charge ≈60% of the total water
Anionic surfactant (active)	18.2	Primary emulsifier	E.g., sulfosuccinate/phosphate ester
Nonionic surfactant (active)	11.7	Co-emulsifier	E.g., fatty alcohol EO (HLB 13–14)
pH adjusters (as required)	—	Polymerization pH	Set nucleation pH 4.5–5.0
In-situ seed (pre-emulsion kick)	5–10% of total monomer	Nucleation control	With 2,500 kg monomer, seed ≈ 125–250 kg
Vinyl acetate (VAM)	1,596	Main monomer	—
Butyl acrylate (BA)	616	Soft monomer	T_g control, flexibility
VeoVa 10™ (versatic vinyl ester)	195	Hydrophobe/durability	Select grade (e.g., VeoVa 10 10/11)
Methyl methacrylate (MMA)	78	Hard monomer	Gloss/hardness; optional
Acrylic acid (AA)	15	Functional monomer	0.5 wt% on total monomer (typical 0.4–0.6%)
Initiator—APS (5% aq.)	10.4	Radical source	Co-feed with monomer pre-emulsion
Reducer—SFS (5% aq.)	6.5	Redox chase	Residual monomer reduction
Defoamer (process aid)	5.2	Foam control	Low-silicone/compatible

monomers with initiator. Acrylic acid (AA) provides electrostatic stabilization and pigment-binding sites (Table 9.38).

Procedure (Starved-Feed, 80–85 °C)

1) Charge reactor with water and ~40% of total emulsifiers; adjust pH to 4.5–5.0. N_2 purge; heat to 80–85 °C.
2) *In-situ seed*: add 5–10% of the pre-emulsified monomers with initiator; hold 15–20 min to subside exotherm.
3) *Starved-feed (3–4 h)*: co-feed the remaining pre-emulsion (VAM/BA/VeoVa 10/MMA/AA) and APS to keep residual monomer <1%.

4) Postfeed hold 30 min, then redox chase with SFS/APS; cool, filter, and adjust pH if required.
5) Cool to 25–30 °C; set storage pH 7.8–8.5; filter 100–150 μm.

Quality Targets

Solids: 50–52%; Viscosity: 500–1,500 mPa s (Brookfield, 25 °C); Particle size: 120–220 nm; Grit (325 mesh): ≤100 ppm; T_g (Fox estimate): −5 to +5 °C (tune via BA/MMA/VeoVa 10 ratio).

Formulation and Use Notes

1) No HEC (and no protective colloids) are used in polymerization. Rheology for paints/coatings is set later with HEC/HASE/HEUR.
2) VeoVa 10 improves hydrophobicity and alkali/water resistance; AA provides adhesion and pigment-binding sites.
3) Surfactant totals (active) typically 0.9–1.2% on total monomer; use low-foam grades and verify freeze–thaw stability if required.

9.18 Closure-VAM-based Copolymers and Terpolymers without HEC

The use of VAM-based copolymers and terpolymers without hydroxyethyl cellulose (HEC) highlights the movement toward latexes that are inherently stable and performance-driven. By carefully balancing surfactant packages and controlling PVOH protection, these systems achieve dispersion stability without relying on external rheology modifiers that may otherwise increase water sensitivity. This approach creates cleaner formulations, more predictable viscosity profiles, and broader compatibility with downstream additives.

Within this framework, the functional role of acrylic acid (AA) becomes central. Even in small amounts, AA imparts anionic stabilization and introduces carboxylic sites that act as reactive handles for postpolymerization modification. These sites can be strategically activated through external crosslinkers, and carbodiimides (CDIs) are among the most effective options. By reacting with carboxyl groups to form acyl-urea linkages, CDIs provide permanent, water-resistant bonds that elevate the durability of films well beyond the baseline performance of uncrosslinked emulsions.

The result is a versatile design platform where VAM/acrylate/VeoVa 10 terpolymers can be tuned for hardness, flexibility, and hydrophobic balance, while AA and CDI chemistry supply the latent potential for long-term resistance and strength. Far from being a constraint, the absence of HEC enables a more direct route to functionalized binders: formulations that are lighter, more resilient, and capable of meeting both architectural and industrial performance standards. This marks a decisive step toward the next generation of waterborne technologies, in which monomer selection and crosslinking strategy are seamlessly aligned with application demands.

10

Advanced PVAc–PVOH Adhesives: Tackification, Reactive PVOH, and Crosslinking

10.1 Introduction

Polyvinyl acetate (PVAc) and polyvinyl alcohol (PVOH)-based adhesives have long been established as versatile systems in packaging, woodworking, and laminating applications. While homopolymer PVAc emulsions provide fast setting speed, good adhesion to porous substrates, and cost efficiency, they often fall short when more demanding performance criteria are required. Market needs such as high-speed packaging lines, adhesion to difficult substrates like coated boards and plastic films, and enhanced resistance to moisture drive the continuous modification of these adhesive systems.

The inherent polarity of PVAc contributes to excellent adhesion on hydrophilic surfaces, but its sensitivity to water, moderate cohesive strength, and limited heat resistance necessitate further improvement. PVOH, commonly present as a protective colloid in these systems, can be chemically tailored to participate in crosslinking or functionalization, thereby enhancing water resistance and durability. Tackifiers, plasticizers, crosslinkers, and hybrid dispersions all provide complementary strategies to adapt the adhesive to specific applications.

Recent developments have emphasized balancing reactivity with stability during storage, as many high-performance additives introduce risks of premature gelation or loss of shelf life. At the same time, industry regulations and environmental considerations increasingly restrict the use of solvent-based adhesives, further positioning waterborne PVAc/PVOH systems as an attractive, sustainable alternative.

This chapter examines advanced modification strategies for PVAc and PVAc–AA adhesives, covering tackification, plasticization, and crosslinking technologies. It also highlights functional PVOH derivatives, hybrid adhesive systems with other dispersions, and application-specific adjustments. Practical guidelines are provided to assist formulators in selecting the appropriate modification path for targeted performance, while maintaining the essential balance between processability, storage stability, and end-use requirements.

10.2 Tackified PVOH—Concept, Benefits, and Formulations

Tackified PVOH refers to a PVOH solution precompounded with a compatible tackifier to elevate initial tack (green strength) and accelerate setting in PVAc-based adhesives. The tackifier is incorporated as an aqueous dispersion and associates with the PVOH phase, increasing wet-tack response and film cohesion without requiring high levels of free surfactant. Specific tackifier chemistries and trade names are discussed in Chapter 11; they are not repeated here.

Benefits of tackified PVOH include higher initial grab and faster stick build, improved fiber tear on paper substrates and better hold on films when paired with PVAc–AA, tunable rheology (shear-thinning) for cleaner application, and reduced surfactant migration relative to late addition of free tackifier. Care is required to balance water resistance, since excess tackifier may increase water uptake unless a suitable crosslinking strategy is used. Maintain storage pH at 4.5–5.5 and avoid amine-bearing surfactants when zirconium or carbodiimide crosslinkers are planned (Table 10.1).

Process (concentrate): Charge water, add the hot PVOH solution with mixing, cool to below 40 °C, add the tackifier dispersion slowly under moderate shear, then

Table 10.1 Tackified PVOH concentrate (≈40% solids) for blending into PVAc/PVAc–AA adhesives.

Ingredient	% by weight	Notes
Deionized water	18.0	Balance viscosity; cool the PVOH solution below 40 °C before adding tackifier.
PVOH solution (20% active)	50.0	88–98% hydrolyzed grade; prepare separately until fully dissolved.
Tackifier dispersion (≈55% solids)	30.0	Chapter 8 selection; compatible anionic or nonionic dispersion.
Nonionic wetting aid (optional)	0.3	Avoid phosphates if CDI/Zr crosslinking is expected.
Defoamer	0.3	Per supplier recommendation.
Biocide/preservative	0.4	Per supplier.
Total	100.0	

Table 10.2 PVAc–AA adhesive with in situ tackified PVOH (t-PVOH) (general purpose).

Ingredient	% by weight	Notes
Deionized water	12.5	Viscosity control.
PVOH solution (10–12% active)	20.0	Protective colloid; adds cohesive strength.
PVAc–AA latex (≈55% solids)	60.0	Low free surfactant; AA provides crosslinkable sites.
Tackifier dispersion (≈55% solids)	6.0	Chapter 11 selection; dose for desired wet tack.
Nonionic wetting aid (optional)	0.2	Keep low to minimize migration.
Defoamer	0.2	Per supplier.
Biocide/preservative	0.1	Per supplier.
Total	99.0	Adjust water to reach 100 and the target viscosity.

add the wetting aid, defoamer, and preservative. Target viscosity as required for downstream blending (Table 10.2).

Process (in situ): Add water, introduce the PVOH solution, add the PVAc–AA latex slowly, add the tackifier dispersion slowly, then add the wetting aid, defoamer, and biocide. Avoid localized high tackifier concentration. Storage pH 4.5–5.5.

Recommended checks: solids content 50–55 wt%, viscosity according to application, pH within the specified window, 180° peel (ASTM D903), shear and humidity creep, and blocking resistance for foil or film laminations.

10.3 Incorporating t-PVOH into PVAc/PVAc–AA Systems

The incorporation of tackified PVOH (t-PVOH) into PVAc and PVAc–AA emulsions provides a practical route to enhance wet tack, machine grab, and open-time control. Because the PVOH phase governs stabilization and film coalescence, embedding tackifiers directly into PVOH ensures more efficient distribution at particle surfaces and in the adhesive film. When compounding t-PVOH concentrates into latex systems, the formulator must balance the ratio of polymer solids, protective colloid, and tackifier dispersion to maintain storage stability at pH 4.5–5.5. This is particularly critical in one-component (1K) systems designed to accept latent crosslinkers without premature reaction.

Table 10.3 Representative formulations of PVAc–AA and PVAc adhesives with tackified PVOH.

Ingredient	PVAc–AA system (%)	PVAc homopolymer system (%)
Deionized water	10.0	12.0
Tackified PVOH concentrate	15.0	15.0
PVAc–AA latex (55% solids)	60.0	—
PVAc latex (55% solids)	—	63.0
Carbodiimide dispersion (30% active)	1.5	— not applicable
Zirconium complex (25% active)	0.5–1.0 (alternative to CDI)	0.5–1.0 (optional)
Aluminum chloride solution	0.1–0.3 (caution with pH)	0.1–0.3 (optional, limited effect)
Defoamer	0.2	0.2
Biocide	0.2	0.2
Other additives (wetting aid, etc.)	q.s.	q.s.

Notes: In PVAc–AA systems, carbodiimide is functional and improves durability through reaction with carboxyl groups. In neat PVAc systems, carbodiimide is ineffective, and zirconium or aluminum salts only provide minor reinforcement. Formulators should prioritize tackifier optimization in these cases. Maintain storage pH at 4.5–5.5 for latent crosslinker latency. Discussion of formal performance classifications (such as D3/D4) will be introduced in later sections under application-specific contexts.

Crosslinker choice depends strongly on the polymer composition. In PVAc–AA systems, carbodiimide crosslinkers are highly effective because they react with carboxylic acid groups introduced by acrylic acid comonomer. In neat PVAc homopolymer, however, carbodiimides offer no meaningful advantage since hydroxyl groups from residual PVOH stabilizer do not provide sufficient reactivity. In such cases, formulators may rely on zirconium complexes or aluminum salts only for marginal benefits, and the emphasis remains on tackification rather than chemical crosslinking. The following table outlines representative formulations illustrating the integration of t-PVOH into PVAc–AA and PVAc systems (Table 10.3).

10.4 Crosslinking Technologies

Crosslinking is essential for upgrading the performance of PVAc/PVOH adhesives, where high water resistance, cohesive strength, and durability are required.

In advanced formulations, two crosslinkers stand out as the most reliable and widely applied: carbodiimide (CDI) and zirconium complexes (Zr). These agents enable 1K adhesive systems with enhanced performance when used under controlled pH conditions. Other crosslinkers, such as aluminum salts and zinc oxide, were discussed in detail in Chapter 8 and are therefore not covered again here. The focus of this chapter is limited to CDI and Zr as the principal crosslinkers for advanced PVAc/PVOH adhesives.

10.4.1 Carbodiimide Crosslinkers

Carbodiimides are highly effective in PVAc–AA systems, where they react with carboxylic acid groups to form stable N-acyl urea linkages. This chemical reaction significantly reduces water sensitivity and enhances wet strength. They are not effective in neat PVAc homopolymer systems, since residual hydroxyl groups from PVOH stabilizers do not provide sufficient functionality. CDI is typically used at 1–2% on latex solids. Maintaining a storage pH of 4.5–5.5 ensures latency in 1K systems. If required, the pH can be adjusted closer to neutrality at the point of use to trigger faster crosslinking. Mechanistic details and extended discussion are available in Chapter 8.

10.4.2 Zirconium Complexes

Zirconium complexes are versatile and effective crosslinkers in both PVAc homopolymers stabilized with PVOH and PVAc–AA copolymers. They coordinate with hydroxyl and carboxyl groups, providing strong water resistance and improved film durability. Zr complexes are used at 0.5–1.0% on solids and require careful pH control to maintain storage latency. The recommended pH range for stable 1K systems is 4.5–5.5. Zirconium complexes provide reliable performance without the sensitivity issues associated with aziridines.

Note: For more details on aluminum salts, zinc oxide, and other crosslinking agents, refer to the dedicated sections in Chapter 8 (Table 10.4).

Table 10.4 Principal crosslinkers for advanced PVAc/PVOH adhesives.

Crosslinker	Effective systems	Key considerations
Carbodiimide (CDI)	PVAc–AA copolymers	Requires –COOH groups; 1–2% dosage; latency at pH 4.5–5.5
Zirconium complexes	PVAc homopolymer (PVOH-stabilized) and PVAc–AA	Broad applicability; 0.5–1.0% dosage; reliable at pH 4.5–5.5

10.5 Reactive PVOH and Functional Additives

Polyvinyl alcohol (PVOH) in its conventional form serves primarily as a protective colloid and stabilizer in PVAc adhesives. The hydroxyl groups on native PVOH contribute to hydrogen bonding but are only weakly reactive in covalent crosslinking. To enhance performance, PVOH can be chemically modified to introduce functional groups capable of forming covalent bonds with crosslinkers or substrates. These derivatives are referred to as reactive PVOH. Unlike tackified PVOH discussed earlier, which relies on physical blending with tackifiers, reactive PVOH incorporates chemically bound groups that unlock new crosslinking pathways or adhesion mechanisms.

10.5.1 Acetoacetylated PVOH

In acetoacetylated PVOH, a portion of the hydroxyl groups is esterified with acetoacetyl moieties. These diketone structures react readily with carbodiimides, polyamines, or multifunctional nucleophiles to form covalent crosslinks. This significantly improves water resistance and wet strength. Such grades are used in wood adhesives meeting D3 and D4 classifications and in coatings requiring high durability.

10.5.2 Silane-modified PVOH

Silane-modified PVOH contains alkoxysilane groups grafted onto the polymer backbone. Upon hydrolysis, these groups form silanols that condense to create siloxane networks or bond directly with inorganic surfaces such as glass or metal oxides. This modification improves adhesion to films and nonporous substrates, making it useful in laminating adhesives and coatings for metallized boards or plastics.

10.5.3 Carboxylated PVOH

Carboxylated PVOH introduces pendant carboxylic acid groups by esterification with anhydrides such as succinic anhydride. These groups interact strongly with zirconium complexes, aluminum salts, or carbodiimides, enabling improved ionic and covalent crosslinking. The result is enhanced water resistance, adhesion to polar substrates, and controlled colloidal behavior (Table 10.5).

Reactive PVOH grades extend the functional scope of PVAc adhesives beyond what conventional PVOH can achieve. They provide chemical handles for efficient crosslinking or direct bonding to inorganic surfaces, enabling higher performance in demanding applications. For mechanistic pathways of these modifications, refer back to the discussions in Chapter 8.

Table 10.5 Main types of reactive PVOH and their applications.

Modification	Reactive groups introduced	Crosslinking/adhesion pathway	Typical applications
Acetoacetylated PVOH	Diketone (acetoacetyl)	Reacts with CDI, amines → covalent crosslinks	Wood adhesives (D3/D4), durable coatings
Silane-modified PVOH	Trialkoxysilane	Forms siloxane bonds; adheres to inorganic surfaces	Film lamination, glass/metal adhesion
Carboxylated PVOH	Carboxyl (–COOH)	Coordinates with Zr, Al; reacts with CDI	Polar substrates, water-resistant adhesives

Table 10.6 Reactive PVOH commercial grades.

Modification type	Trade names/series	Supplier
Silane-modified PVOH	KURARAY POVAL™ R series	Kuraray
Carboxylated PVOH	GOHSENX™ T series; KURARAY POVAL™ K series	Mitsubishi Chemical/ Soarus (USA); Kuraray
Acetoacetylated PVOH	GOHSENX™ Z series	Mitsubishi Chemical/ Soarus (USA)

Note: These trade names correspond to true reactive PVOH grades. Exceval™ (Kuraray) is an ethylene–vinyl alcohol copolymer (EPVOH) and should not be classified as reactive PVOH.

10.5.4 Reactive PVOH–Commercial Trade Names

Reactive PVOH refers to chemically modified grades of PVOH that introduce functional groups beyond the native hydroxyls. These functional groups enable covalent crosslinking or adhesion to inorganic substrates. Below is a list of commercial trade names corresponding to reactive PVOH categories (Table 10.6).

10.5.4.1 Reactive PVOH–Functional Moieties and Crosslinking Pathways

Reactive PVOH refers to PVOH that has been chemically modified to introduce functional groups beyond its native hydroxyl groups. These modifications give PVOH the ability to participate in covalent crosslinking reactions. The table below summarizes the major functional moieties, their chemical nature, and the compounds or crosslinkers with which they form covalent bonds (Table 10.7).

Table 10.7 Functional moieties in reactive PVOH and their crosslinking partners.

Reactive PVOH type	Functional moiety introduced	Reactive pathway	Typical crosslinkers/ partners
Acetoacetylated PVOH (e.g., GOHSENX™ Z series)	Acetoacetyl group (–O–CO–CH_2–CO–CH_3; diketone)	Nucleophilic attack on diketone carbonyls	Carbodiimide (CDI), multifunctional amines, hydrazides → covalent N-acyl urea or imine linkages
Silane-modified PVOH (e.g., Kuraray Poval™ R series)	Alkoxysilane (–Si$(OR)_3$)	Hydrolysis to silanol (–Si–OH) followed by condensation	Inorganic surfaces (glass, metals, oxides); self-condensation to siloxane networks
Carboxylated/ PVOH (e.g., GOHSENX™ T series)	Carboxyl groups (–COOH)	Acid–base and covalent condensation reactions	Carbodiimide (–COOH + CDI → N-acyl urea); zirconium complexes; aluminum salts

Summary: Reactive PVOH grades differ fundamentally from conventional PVOH. By incorporating diketone, silane, or carboxyl moieties, they create new covalent pathways for crosslinking. This enables compatibility with crosslinkers such as CDI and Zr complexes, or with inorganic substrates and comonomers.

10.6 Hybrid Adhesive Systems

Hybrid adhesive systems in the context of PVAc-based technologies are defined as formulations in which the main polymer binder, typically PVAc–AA, is combined with a modified PVOH phase. Two principal approaches exist: (Finch, 1992) physical tackification of PVOH, yielding tackified PVOH (t-PVOH), and (Hassan & Peppas, 2000) chemical modification of PVOH to produce reactive PVOH. These hybrid systems combine the film-forming and adhesion properties of PVAc–AA with the functional contributions of PVOH, thereby achieving performance levels unattainable by either polymer alone. Crosslinkers such as CDI or zirconium complexes (Zr) are often employed to reinforce reactive PVOH hybrids through covalent or coordination bonding.

10.6.1 Tackified PVOH/PVAc–AA Hybrids

In tackified PVOH (t-PVOH) hybrids, tackifiers such as rosin esters or terpene phenolic resins are predispersed into the PVOH phase before compounding with PVAc–AA. The resulting system delivers enhanced wet tack, faster set speed, and improved grab on high-speed packaging lines. While no covalent bonds are introduced, the physical hybridization of tackifiers within the PVOH domain ensures more efficient energy dissipation and adhesion under dynamic conditions. Such systems are particularly suited for carton sealing and paper converting applications where immediate bond strength is critical.

10.6.2 Reactive PVOH/PVAc–AA/CDI Hybrids

Reactive PVOH carrying acetoacetyl or carboxyl groups can be hybridized with PVAc–AA latex and crosslinked using carbodiimide. This system exploits the –COOH groups of the copolymer together with the reactive moieties on PVOH, producing a dense covalent network. The result is high water resistance and durability, enabling adhesives to achieve D3/D4 classifications for wood bonding. Formulations typically maintain storage pH between 4.5 and 5.5 to preserve CDI latency in 1K systems.

10.6.3 Reactive PVOH/PVAc–AA/Zr Hybrids

Zirconium complexes coordinate effectively with hydroxyl and carboxyl groups, making them highly suitable for hybrid systems based on reactive PVOH and PVAc–AA. When combined with silane-modified or carboxylated PVOH, Zr crosslinkers reinforce adhesion to challenging substrates such as PET, BOPP, and metallized boards. These hybrids maintain good storage stability at pH 4.5–5.5 and provide a balanced combination of water resistance, substrate versatility, and mechanical durability. Applications include film laminating adhesives and high-performance packaging formulations (Table 10.8).

Table 10.8 Hybrid adhesive systems based on tackified and reactive PVOH with PVAc–AA.

Hybrid type	PVOH modification	Crosslinker	Key properties	Typical applications
Tackified hybrid	t-PVOH (rosin ester, terpene phenolic)	None	Enhanced wet tack, fast set, improved grab	Carton sealing, paper converting
Reactive hybrid (CDI)	Acetoacetylated or carboxylated PVOH	Carbodiimide	Covalent crosslinks; high water resistance; durability	Wood adhesives (D3/D4), structural packaging
Reactive hybrid (Zr)	Silane-modified or carboxylated PVOH	Zirconium complexes	Coordination crosslinks; adhesion to films; balanced durability	Film laminating (PET, BOPP), metallized carton

Hybrid adhesive systems based on PVAc–AA and modified PVOH provide formulators with a flexible platform to address diverse applications. Tackified hybrids prioritize immediate adhesion and machine performance, while reactive hybrids extend performance into water resistance and difficult substrate bonding through CDI or Zr crosslinking. Together, they represent the advanced frontier of PVAc/PVOH adhesive technology.

10.7 Application-specific Modifications

Application performance of PVAc–AA adhesives can be optimized by targeted hybridization with tackified or reactive PVOH. Each application imposes unique demands on bond strength, water resistance, flexibility, and adhesion to different substrates. By selecting the appropriate PVOH modification and crosslinker, formulators can tune the adhesive system to meet industry standards such as EN 204 classifications for wood adhesives or the specific requirements of film laminating and carton sealing operations.

10.7.1 Carton Sealing (High-speed Machine Tack)

Carton sealing adhesives require immediate grab on high-speed packaging lines, where bond formation occurs within fractions of a second. Tackified PVOH hybrids with PVAc–AA are the preferred approach, as they deliver strong wet tack and fast setting without the need for chemical crosslinking. Rosin ester or terpene phenolic tackifiers dispersed into PVOH ensure efficient machine performance, while PVAc–AA provides the structural film. These formulations emphasize speed and handling performance rather than long-term water resistance.

10.7.2 Wood Adhesives (D3/D4 Classification)

Wood bonding imposes higher durability requirements, particularly under humid or wet conditions. Reactive PVOH hybrids with PVAc–AA and CDI are designed to achieve D3 and D4 classifications according to EN 204. Acetoacetylated or carboxylated PVOH introduces covalent crosslinking pathways that, when activated by CDI, yield strong wet strength and resistance to water penetration. Maintaining pH between 4.5 and 5.5 during storage ensures CDI latency, allowing 1K systems to be used effectively in industrial wood applications.

10.7.3 Film Laminating (Window Patching, BOPP, PET)

Film laminating requires adhesion to nonporous substrates such as BOPP, PET, or metallized films. Reactive PVOH hybrids combined with PVAc–AA and Zr

Table 10.9 Application-specific hybrid PVAc/PVOH adhesive systems.

Application	Hybrid system	Crosslinker	Key performance features	Standard/ requirement
Carton sealing	t-PVOH + PVAc–AA	None	High wet tack; fast set; machine grab	Packaging line speed; no water resistance requirement
Wood adhesives	Reactive PVOH (acetoacetylated/ carboxylated) + PVAc–AA	CDI	Covalent crosslinks; high water resistance; durability	EN 204 D3/D4 classification
Film laminating	Reactive PVOH (silane-modified/ carboxylated) + PVAc–AA	Zr	Coordination crosslinks; adhesion to PET, BOPP, metallized boards	Window patching and flexible packaging performance

complexes are particularly effective. Silane-modified or carboxylated PVOH provides functional groups for coordination with Zr, enabling strong bonds to plastics and metal oxides. This approach enhances water resistance and durability while preserving the processing advantages of PVAc–AA. Such systems are widely used in window patching, flexible packaging, and laminates for premium cartons (Table 10.9).

These application-specific hybrids demonstrate the versatility of combining PVAc–AA with tackified or reactive PVOH. By tailoring the modification of PVOH and selecting CDI or Zr crosslinkers where appropriate, adhesives can be designed to address diverse industrial needs ranging from rapid carton sealing to durable wood bonding and high-performance film laminating.

10.8 Formulations for Hybrid PVAc/PVOH Adhesives

The following formulations cover the three target applications using PVAc–AA as the base polymer, with either tackified PVOH (t-PVOH) or reactive PVOH. Crosslinkers are limited to CDI or Zr where applicable. All tables are boxed; prose precedes each table. Aziridines are excluded. Minimize free surfactant and avoid amine-bearing surfactants in crosslinkable systems. Store 1K systems at pH 4.5–5.5.

10.8.1 Carton Sealing–t-PVOH/PVAc-AA (No Crosslinker)

For high-speed packaging lines, a tackified PVOH hybrid maximizes wet tack and machine grab without chemical crosslinking. Predisperse the tackifier into the PVOH phase (t-PVOH concentrate), then compound with PVAc–AA (Tables 10.10A and 10.10B). Target Brookfield viscosity 3,000–6,000 mPa s (RV, spindle 4/20 rpm, 25 °C).

Preparation: Charge water, add t-PVOH concentrate under moderate shear, then add PVAc–AA slowly to avoid shock. Add wetting aid, defoamer, and preservative; adjust viscosity as needed. Store at pH 4.5–5.5.

Preparation: Dissolve PVOH completely (80–90 °C), cool to 35–40 °C, then add tackifier dispersion slowly under shear. Add defoamer and preservative. Keep pH 4.5–5.5 for compatibility with downstream 1K systems.

Table 10.10A Tackified PVOH/PVAc-AA carton sealing adhesive.

Ingredient	% by weight	Notes
Deionized water	12.6	Balance viscosity; DI only
t-PVOH concentrate	12.0	Pretackified PVOH
PVAc–AA latex (55% solids)	75.0	Low free surfactant grade
Low-foam wetting aid (nonionic)	0.2	Avoid amine-bearing types
Defoamer	0.2	Per supplier
Preservative/biocide	0.2	Per supplier
Rheology modifier (optional)	–	Adjust to viscosity target

Table 10.10B Tackified PVOH concentrate (for carton sealing).

Ingredient	% by weight	Notes
Deionized water	40.0	Heat to dissolve PVOH
PVOH solution (10–12% active)	35.0	Partially hydrolyzed grade
Tackifier dispersion (rosin ester, 50% solids)	24.0	Fine-particle dispersion
Defoamer	0.2	Per supplier
Preservative/biocide	0.3	Per supplier
Optional plasticizer (e.g., glycerol)	0.5	Open-time tuning

Table 10.11 Reactive PVOH/PVAc–AA/CDI wood adhesive D3 grade.

Ingredient	% by weight	Notes
Deionized water	9.2	Balance viscosity
Reactive PVOH solution (10% active; acetoacetylated or carboxylated)	20.0	Reactive PVOH grade
PVAc–AA latex (55% solids)	68.3	Low free surfactant grade
Carbodiimide dispersion (30% active)	1.5	1.5% on total; do not exceed 2%
Defoamer	0.2	Per supplier
Preservative/biocide	0.2	Per supplier
Low-foam wetting aid (nonionic)	0.2	Avoid amine-bearing types
Rheology modifier (optional)	—	Adjust to press viscosity

Preparation: Premix water and reactive PVOH solution; add PVAc–AA under gentle shear. Add CDI dispersion last at <30 °C. Defoamer, preservative, and wetting aid follow. Target viscosity 5,000–10,000 mPa s, depending on the spreader. Store at pH 4.5–5.5. Press guidance (typical): 0.5–1.0 MPa, 20–25 °C, 20–40 min; postcure improves wet strength.

10.8.2 Wood Adhesives D3 Grade—Reactive PVOH/PVAc–AA/CDI

This hybrid uses acetoacetylated or carboxylated PVOH with PVAc–AA and CDI. The CDI reacts with -COOH (from AA) and with reactive PVOH moieties to form covalent linkages. Store at pH 4.5–5.5 to maintain latency (Table 10.11).

10.8.3 Film Laminating—Reactive PVOH (Silane)/PVAc–AA/Zr

For surface-treated PET, BOPP, and metallized boards, silane-modified PVOH with PVAc–AA crosslinked by Zr provides strong film adhesion and moisture durability. Maintain pH 4.5–5.5 for storage stability and activate near use if required (Table 10.12).

Quality Control Targets (Typical)

Solids (wt%): 50–55. pH (storage): 4.5–5.5. Viscosity: see application above. Appearance: smooth, free of gels/skins. Freeze–thaw: ≥3 cycles recommended. Residual monomer: per regulatory limits.

Table 10.12 Reactive PVOH (silane)/PVAc–AA/Zr laminating adhesive.

Ingredient	% by weight	Notes
Deionized water	12.6	Balance viscosity
Silane-modified PVOH solution (10% active)	16.0	Silanol-forming PVOH grade
PVAc–AA latex (55% solids)	70.0	Low free surfactant grade
Zirconium complex (25% active)	0.7	Typical 0.5–1.0% on solids
Low-foam wetting aid (nonionic)	0.3	Avoid amine-bearing types
Defoamer	0.2	Per supplier
Preservative/biocide	0.2	Per supplier

Preparation: Combine water and silane-modified PVOH solution; add PVAc–AA under moderate shear. Dose Zr complex slowly while mixing; avoid localized high concentration. Add wetting aid, defoamer, and preservative. Target viscosity 2,500–5,000 mPa s for gravure/roll coating. For non-corona-treated films, surface preparation improves bonding.

10.9 Expanded Formulations for Hybrid PVAc/PVOH Adhesives

This section enriches the application coverage with additional formulations. All systems use PVAc–AA as the primary binder, combined with either tackified PVOH (t-PVOH) or reactive PVOH, and crosslinked where specified with CDI or zirconium complexes (Zr). Storage pH for 1K latency: 4.5–5.5. Avoid amine-bearing surfactants; minimize free surfactant to support water resistance.

10.9.1 Wood Adhesive—D3 Classification (Reactive PVOH/PVAc–AA/CDI, 1K)

A single-component, CDI-crosslinkable hybrid designed to meet EN 204 D3 requirements. Reactive PVOH carries acetoacetyl or carboxyl functionality; CDI reacts with -COOH in PVAc–AA and with reactive PVOH moieties (Table 10.13).

10.9.2 Wood Adhesive—D4 Classification (Reactive PVOH/PVAc–AA/CDI, 1K)

For D4 durability, increase reactive PVOH proportion and CDI within safe limits. Maintain pH 4.5–5.5 for storage latency. Ensure low free surfactant and adequate press times (Table 10.14).

Table 10.13 Reactive PVOH/PVAc–AA/CDI–D3 wood adhesive.

Ingredient	% by weight	Notes
Deionized water	8.8	Balance viscosity
Reactive PVOH solution (10% active; acetoacetylated or carboxylated)	22.0	Reactive PVOH grade
PVAc–AA latex (55% solids)	67.3	Low free surfactant grade
Carbodiimide dispersion (30% active)	1.5	Do not exceed 2% on total
Defoamer	0.2	Per supplier
Preservative/biocide	0.2	Per supplier
Low-foam wetting aid (nonionic)	0.2	Avoid amine types

Preparation: Premix water and reactive PVOH; add PVAc–AA slowly under moderate shear. Dose CDI last at <30 °C. Target viscosity 5,000–9,000 mPa s. Press guide: 0.5–1.0 MPa, 20–25 °C, 20–40 min; postcure 24 h.

Table 10.14 Reactive PVOH/PVAc–AA/CDI–D4 wood adhesive.

Ingredient	% by weight	Notes
Deionized water	7.9	Balance viscosity
Reactive PVOH solution (12% active; acetoacetylated)	25.0	Higher reactive handle density
PVAc–AA latex (55% solids)	65.1	Low free surfactant grade
Carbodiimide dispersion (30% active)	1.8	Upper end of CDI window
Defoamer	0.2	Per supplier
Preservative/biocide	0.2	Per supplier
Low-foam wetting aid (nonionic)	0.2	Avoid amine types

Preparation: As in D3, but verify pot stability over 4–8 weeks at 25–30 °C. Target viscosity 6,000–10,000 mPa s. Press guide: 0.6–1.0 MPa, 20–25 °C, 25–45 min; postcure recommended.

10.9.3 Carton-to-Lacquered Carton–Reactive PVOH (Silane)/PVAc–AA/Zr

For coated or lacquered boards, a silane-modified PVOH hybrid with Zr crosslinking improves adhesion to low-porosity, low-energy surfaces often encountered in premium packaging (Table 10.15).

Table 10.15 Silane-modified PVOH/PVAc–AA/Zr–lacquered carton adhesive.

Ingredient	% by weight	Notes
Deionized water	11.8	Balance viscosity
Silane-modified PVOH solution (10% active)	18.0	Silanol-forming grade
PVAc–AA latex (55% solids)	68.8	Low free surfactant grade
Zirconium complex (25% active)	0.7	0.5–1.0% typical
Low-foam wetting aid (nonionic)	0.3	Acetylenic diol type
Defoamer	0.2	Per supplier
Preservative/biocide	0.2	Per supplier

Preparation: Combine water and silane-modified PVOH; add PVAc–AA under moderate shear. Dose Zr slowly. Target viscosity 3,000–6,000 mPa s for wheel/roll application.

10.9.4 Window Patching–PET Film to Carton (Reactive PVOH/PVAc–AA/Zr)

For PET window patching, use silane-modified or carboxylated reactive PVOH with Zr to promote bonding to polyester and metallized films. The formulation below assumes surface-treated PET; untreated PET will be addressed separately (Table 10.16).

Table 10.16 PET window patching adhesive.

Ingredient	% by weight	Notes
Deionized water	12.4	Balance viscosity
Reactive PVOH solution (silane-modified, 10% active)	16.0	Silanol-forming grade
PVAc–AA latex (55% solids)	70.0	Low free surfactant grade
Zirconium complex (25% active)	0.7	0.5–1.0% typical
Low-foam wetting aid (nonionic)	0.3	Improves wetting on films
Defoamer	0.2	Per supplier
Preservative/biocide	0.2	Per supplier

Preparation: As in Table 10.15. Target viscosity 2,500–5,000 mPa s for wheel/gravure coaters. For non-treated PET, corona or plasma pretreatment is recommended (detailed later).

Table 10.17 BOPP window patching adhesive.

Ingredient	% by weight	Notes
Deionized water	12.1	Balance viscosity
Reactive PVOH solution (carboxylated or silane-modified, 10% active)	16.5	Select for film type
PVAc–AA latex (55% solids)	69.9	Low free surfactant grade
Zirconium complex (25% active)	0.8	0.5–1.0% typical
Low-foam wetting aid (nonionic)	0.5	Higher level for BOPP
Defoamer	0.2	Per supplier
Preservative/biocide	0.2	Per supplier

Preparation: As in Table 10.15, Target viscosity 2,500–5,000 mPa s. Recommend surface energy ≥38 dyn/cm; we will detail untreated options later.

10.9.5 Window Patching—BOPP Film to Carton (Reactive PVOH/PVAc-AA/Zr)

For BOPP window patching, increase wetting aid slightly and ensure fine dispersion quality. Surface treatment is generally needed; untreated cases will be addressed later (Table 10.17).

10.9.6 High-grab Window Patching—t-PVOH/PVAc-AA (No Crosslinker)

Where immediate machine grab is paramount and water resistance demands are moderate, a t-PVOH hybrid without a crosslinker can be effective for patching on coated boards (Table 10.18).

General notes and QC targets

Solids: typically, 50–55 wt%. pH (storage): 4.5–5.5 for 1K latency. Viscosity: adjust per machine (see each subsection). Appearance: smooth, free of gels/skins. Residual monomer: per regulatory limits. Freeze–thaw stability: ≥3 cycles recommended. Where adhesion to films is marginal, surface energy ≥38 dyn/cm is advised; detailed strategies for untreated films will follow.

10.9.7 Special Section: Treated vs. Untreated Films (PET, BOPP)

This section provides formulations and practical guidance distinguishing surface-treated films (corona/flame/plasma; ≥38 dyn/cm) from untreated films

Table 10.18 t-PVOH/PVAc–AA high-grab patching adhesive.

Ingredient	% by weight	Notes
Deionized water	12.0	Balance viscosity
t-PVOH concentrate (rosin ester, see 10.7A.2)	14.0	Pretackified PVOH
PVAc–AA latex (55% solids)	73.3	Low free surfactant grade
Low-foam wetting aid (nonionic)	0.3	Acetylenic diol type
Defoamer	0.2	Per supplier
Preservative/biocide	0.2	Per supplier

Preparation: Target viscosity 3,000–6,000 mPa s for high-speed lines. If film resistance is insufficient, migrate to reactive PVOH + Zr systems above.

(≤32 dyn/cm). All systems retain the chapter's constraints: PVAc–AA base, hybridized with PVOH (tackified or reactive as applicable), and crosslinked only with zirconium complexes for film laminating (no aziridines). Minimize free surfactant; avoid amine-bearing surfactants in crosslinkable systems. Store 1K systems at pH 4.5–5.5. The untreated variants increase functional PVOH content, Zr dosage, and wetting strategy to compensate for low surface energy.

10.9.7.1 Treated PET Film-carton (Reactive PVOH/PVAc–AA/Zr)

Baseline for corona/plasma-treated PET (≥38 dyn/cm). Silane-modified PVOH supplies silanol functionality; Zr complexes provide coordination crosslinks. Designed for window patching and laminating lines (Table 10.19).

10.9.7.2 Untreated PET Film-carton (Reactive PVOH/PVAc–AA/Zr)

For PET without pretreatment (≤32 dyn/cm). Increase silane-PVOH and Zr, and strengthen wetting to compensate for low surface energy. Whenever possible, add inline corona treatment; otherwise, use this reinforced chemistry (Table 10.20).

10.9.7.3 Treated BOPP Film-carton (Reactive PVOH/PVAc–AA/Zr)

For corona-treated BOPP (≥38 dyn/cm). Carboxylated or silane-modified PVOH can be used. Slightly higher wetting aid than the PET baseline (Table 10.21).

10.9.7.4 Untreated BOPP Film-carton (Reactive PVOH/PVAc–AA/Zr)

For low-energy, untreated BOPP (≈30 dyn/cm). Increase functional PVOH, Zr, and wetting aid. If possible, introduce an inline corona; otherwise, validate adhesion with standardized peel and shear tests (Table 10.22).

Table 10.19 Treated PET-carton adhesive.

Ingredient	% by weight	Notes
Deionized water	12.6	Balance viscosity
Silane-modified PVOH solution (10% active)	16.0	Silanol-forming grade
PVAc–AA latex (55% solids)	70.0	Low free surfactant grade
Zirconium complex (25% active)	0.7	Typical 0.5–1.0% on solids
Low-foam wetting aid (nonionic)	0.3	Acetylenic-diol type
Defoamer	0.2	Per supplier
Preservative/biocide	0.2	Per supplier

Target viscosity: 2,500–5,000 mPa s (25 °C). Surface energy target: ≥38 dyn/cm. Application: wheel/gravure coaters; nip pressure as recommended by machine OEM.

Table 10.20 Untreated PET-carton adhesive.

Ingredient	% by weight	Notes
Deionized water	11.0	Balance viscosity
Silane-modified PVOH solution (12% active)	20.0	Higher functional PVOH
PVAc–AA latex (55% solids)	67.5	Low free surfactant grade
Zirconium complex (25% active)	1.0	Upper end for low-energy films
Low-foam wetting aid (nonionic)	0.8	Higher level improves wetting
Defoamer	0.3	Per supplier
Preservative/biocide	0.4	Per supplier

Target viscosity: 3,000–6,000 mPa s. Aim for maximum attainable surface energy; if offline treatment is impossible, verify bond via peel testing under humidity. Storage pH: 4.5–5.5.

Table 10.21 Treated BOPP-carton adhesive.

Ingredient	% by weight	Notes
Deionized water	12.0	Balance viscosity
Reactive PVOH solution (carboxylated or silane-modified, 10% active)	17.0	Select per substrate
PVAc–AA latex (55% solids)	69.0	Low free surfactant grade
Zirconium complex (25% active)	0.8	0.5–1.0% typical
Low-foam wetting aid (nonionic)	0.6	Improved wetting on BOPP
Defoamer	0.3	Per supplier
Preservative/biocide	0.3	Per supplier

Target viscosity: 2,500–5,000 mPa s. Surface energy target: ≥38 dyn/cm. Check optical clarity and blocking resistance on stacked sheets.

Table 10.22 Untreated BOPP-carton adhesive.

Ingredient	% by weight	Notes
Deionized water	10.8	Balance viscosity
Reactive PVOH solution (carboxylated or silane-modified, 12% active)	21.0	Higher functional PVOH
PVAc–AA latex (55% solids)	66.2	Low free surfactant grade
Zirconium complex (25% active)	1.0	Upper end for low-energy films
Low-foam wetting aid (nonionic)	1.2	Elevated for BOPP wetting
Defoamer	0.4	Per supplier
Preservative/biocide	0.4	Per supplier

Target viscosity: 3,000–6,000 mPa s. Validate adhesion and blocking performance; monitor any slip additive transfer from BOPP affecting bond. Storage pH: 4.5–5.5.

Process and QC Notes

Order of addition: water, reactive PVOH solution, PVAc–AA, zirconium complex (slow dosing), wetting aid, defoamer, and preservative. Avoid localized high zirconium concentration.

For untreated films, maximize line cleanliness to reduce slip-agent contamination; use fresh film where possible. Recommended tests include 180° peel (ASTM D903), shear-onset under humidity, and blocking at the stack. Always verify surface energy; if feasible, apply corona or plasma treatment to raise the surface tension to ≥38 dyn/cm.

10.9.8 Expanded Polystyrene (EPS) Adhesive Formulations

Extruded polystyrene (XPS) and expanded polystyrene (EPS) are widely used as thermal insulation materials in building and packaging applications. Their low surface energy, hydrophobicity, and porous structure present significant challenges for adhesive bonding. Conventional PVAc-based systems typically fail to achieve durable adhesion due to poor wetting and insufficient penetration into the substrate.

To address these limitations, specialized waterborne formulations are developed that combine polymer dispersions with tailored tackifiers, reactive PVOH, crosslinkers, and surface-active agents. The binder selection must balance high initial tack with controlled film formation, while maintaining compatibility with the polystyrene surface. For structural or insulation panel lamination, resistance to creep, dimensional stability under thermal cycling, and low water absorption are also essential performance requirements.

Formulations for EPS adhesives in this work are primarily based on PVAc–AA copolymers, which provide enhanced substrate interaction through the introduction of carboxyl functionality. Crosslinkers, including carbodiimides or zirconium complexes, may be employed to further improve heat and water resistance.

In addition, application-specific rheology modifiers and defoamers ensure uniform coating and penetration without collapse of the polystyrene structure.

This section presents practical formulations and process guidelines designed for bonding EPS in industrial and construction applications. Emphasis is placed on balancing adhesive performance with ease of processing, environmental compliance, and cost-effectiveness.

10.9.8.1 Polystyrene Foams (EPS/XPS) Adhesives—Notes

EPS and XPS are solvent-sensitive; use waterborne PVAc–AA systems only. XPS surfaces are lower energy and smoother than EPS; they generally require stronger wetting and/or reinforced reactive PVOH + Zr. For each formulation below, target solids 50–55 wt% unless noted; viscosity by application method (Tables 10.23–10.28).

Table 10.23 EPS to paperboard (packaging)—t-PVOH/PVAc–AA (no crosslinker).

Ingredient	% by weight	Notes
Deionized water	12.8	Balance viscosity
t-PVOH concentrate (PVOH 10–12% + rosin ester dispersion)	14.0	Pretackified PVOH
PVAc–AA latex (55% solids)	72.5	Low free surfactant grade
Low-foam wetting aid (nonionic)	0.3	Acetylenic-diol type
Defoamer	0.2	Per supplier
Preservative/biocide	0.2	Per supplier

Target viscosity: 3,000–6,000 mPa s (wheel/roller). Fast set and high grab; no crosslinker required.

Table 10.24 EPS to gypsum board—reactive PVOH (carboxyl/acetoacetyl)/PVAc–AA/CDI.

Ingredient	% by weight	Notes
Deionized water	8.8	Balance viscosity
Reactive PVOH solution (10–12% active; acetoacetylated or carboxylated)	18.0	Reactive PVOH grade
PVAc–AA latex (55% solids)	70.8	Low free surfactant grade
Carbodiimide dispersion (30% active)	1.8	1.0–2.0% as-is; add last, <30 °C
Defoamer	0.3	Per supplier
Preservative/biocide	0.3	Per supplier

Target viscosity: 5,000–9,000 mPa s (trowel/bead). Elevated wet resistance; strong gypsum adhesion.

Table 10.25 EPS to aluminum foil facer—reactive PVOH (silane)/PVAc–AA/Zr.

Ingredient	% by weight	Notes
Deionized water	12.0	Balance viscosity
Reactive PVOH solution (10% active; silane-modified)	16.0	Silanol-forming PVOH
PVAc–AA latex (55% solids)	70.4	Low free surfactant grade
Zirconium complex (25% active)	0.8	0.5–1.0% as-is; dose slowly
Low-foam wetting aid (nonionic)	0.3	Improves foil wetting
Defoamer	0.25	Per supplier
Preservative/biocide	0.25	Per supplier

Target viscosity: 2,500–5,000 mPa s (roll/gravure). Good wetting on foil; coordination network via Zr.

Table 10.26 EPS to EPS lamination—PVAc–AA/t-PVOH (optional).

Ingredient	% by weight	Notes
Deionized water	9.5	Balance viscosity
t-PVOH concentrate (optional)	8.0	Use for faster grab if needed
PVAc–AA latex (55% solids)	82.1	Higher solids for fast set
Defoamer	0.2	Per supplier
Preservative/biocide	0.2	Per supplier

Target viscosity: 3,000–6,000 mPa s. Avoid excessive water; press lightly to prevent foam crush.

Table 10.27 XPS to paperboard—reactive PVOH (carboxyl/silane)/PVAc–AA/Zr (low-energy surface).

Ingredient	% by weight	Notes
Deionized water	10.8	Balance viscosity
Reactive PVOH solution (12% active; carboxylated or silane)	20.0	Higher functional PVOH for XPS
PVAc–AA latex (55% solids)	66.9	Low free surfactant grade
Zirconium complex (25% active)	0.8	0.5–1.0% as-is
Low-foam wetting aid (nonionic)	1.0	Elevated for low-energy XPS
Defoamer	0.25	Per supplier
Preservative/biocide	0.25	Per supplier

Target viscosity: 3,000–6,000 mPa s. Mechanical scuffing of XPS surface improves bonding.

Table 10.28 XPS to aluminum foil facer—reactive PVOH (silane)/PVAc–AA/Zr (reinforced).

Ingredient	% by weight	Notes
Deionized water	10.2	Balance viscosity
Reactive PVOH solution (12% active; silane-modified)	20.0	Higher functional loading
PVAc–AA latex (55% solids)	67.7	Low free surfactant grade
Zirconium complex (25% active)	1.0	Upper end for foil + XPS
Low-foam wetting aid (nonionic)	0.6	For foil wetting
Defoamer	0.25	Per supplier
Preservative/biocide	0.25	Per supplier

Target viscosity: 2,500–5,000 mPa s. Validate peel under humidity and temperature cycling.

10.10 Process and QC Notes (Common to all EPS/XPS Formulations)

Order of addition: water, PVOH solution (thermoplastic PVOH or reactive PVOH), PVAc–AA (slow addition), zirconium complex (slow dosing), carbodiimide (if used; added last, below 30 °C), wetting aid, defoamer, and biocide. Avoid the use of amine-bearing surfactants. Storage pH should be maintained in the range of 4.5–5.5.

Recommended quality control checks include determination of solids content (50–55 wt%), viscosity according to the intended application, pH stability within the specified window, and adhesion performance by 180° peel (ASTM D903). Additional evaluations should be carried out for shear strength, humidity creep, and blocking resistance, particularly when used with foil or film substrates.

10.11 Summary and Practical Guidelines

This section consolidates the core principles and hands-on guidance established throughout Chapter 10. It is organized to help the formulator select between tackified PVOH and reactive PVOH hybrids with PVAc–AA, and to apply CDI or Zr crosslinking where appropriate. Troubleshooting tips are provided for rapid iteration on the bench.

Strategy selection depends on application speed, substrate class, and durability targets. The following table summarizes which hybrid path to choose and why (Table 10.29).

Table 10.29 Strategy selection for PVAc–AA/PVOH hybrids.

Use case	Preferred hybrid path	Crosslinker	Why this works
High-speed carton sealing	t-PVOH + PVAc–AA	None	Maximum wet tack and fast set without latency risks
Wood bonding (D3/D4)	Reactive PVOH (acetoacetylated/ carboxylated) + PVAc–AA	CDI	Covalent network (–COOH + CDI) for wet strength
Film laminating/ window patching	Reactive PVOH (silane/ carboxylated) + PVAc–AA	Zr	Coordination bonding to PET/BOPP/metallized boards
Carton to lacquered carton	Reactive PVOH (silane) + PVAc–AA	Zr	Improved adhesion to low-porosity coated boards

Table 10.30 Do's and don'ts for PVAc–AA/PVOH hybrid adhesives.

Do's	Don'ts
Store 1K CDI/Zr systems at pH 4.5–5.5 to preserve latency	Don't use CDI with neat PVAc (no benefit)
Minimize free surfactant; rely on PVOH for stabilization where possible	Don't include aziridines (excluded for sensitivity)
Predisperse tackifier into PVOH (t-PVOH) for machine tack gains	Don't rely on residual PVOH –OH for crosslinking; its effect is trivial
Add CDI last at <30 °C; dose Zr slowly to avoid local gelation	Don't over-alkalize storage (neutral/ alkaline pH kills latency and stability)
Match PVOH grade and modification to the substrate (silane for films; acetoacetyl/carboxyl for D-class wood)	Don't overload wetting aid—risk of blocking or plasticization on films
Target surface energy ≥38 dyn/cm for films; treat surfaces when feasible	

Before formulating, review the following do's and don'ts to avoid common pitfalls (Table 10.30). These points reflect the constraints set in this chapter (no aziridines; PVAc–AA backbone; 1K latency at pH 4.5–5.5).

Use the following checklist when scaling or transferring formulations to the plant (Table 10.31).

Table 10.31 Formulator's checklist.

Item	Target/guidance
Solids (wt%)	50–55 (adjust to machine)
Viscosity (25 °C)	Carton/patching: 2,500–6,000 mPa s; Wood: 5,000–10,000 mPa s
pH (storage)	4.5–5.5 (1K latency)
Order of addition	Water → PVOH (tackified/reactive) → PVAc–AA → Crosslinker → Additives
Surface energy	≥38 dyn/cm for PET/BOPP; treat if lower
Freeze–thaw stability	≥3 cycles (as applicable)
Residual monomer	Within regulatory limits

Table 10.32 Troubleshooting quick fixes.

Symptom	Likely cause	Quick fix
Low bond to PET/BOPP	Insufficient surface energy; low silane-PVOH or Zr	Treat film; raise silane-PVOH 2–4 pts; Zr to 1.0%
Blocking in stack	Excess wetting aid/plasticizer; soft film	Reduce wetting aid by 0.2–0.3 pts; increase PVAc–AA solids
Poor wet tack on carton	Tackifier not integrated in PVOH; low t-PVOH	Use t-PVOH concentrate; add 2–4 pts tackifier (as dispersion)
Viscosity drift in storage	pH too high; premature crosslinking	Hold pH 4.5–5.5; add crosslinker at end under cooling
Weak wet strength (wood)	CDI underdosed; reactive PVOH too low	Increase CDI to 1.8–2.0%; raise reactive PVOH 2–3 pts

Common issues and rapid fixes are summarized below for bench-top iteration (Table 10.32).

11

Vinyl Acetate–Ethylene (VAE) Copolymers–Applications in Adhesives and Coatings

11.1 Introduction

Vinyl acetate–ethylene (VAE) copolymers form a distinct class of water-borne binders within the broader family of vinyl acetate dispersions. Their development was motivated by the limitations of conventional polyvinyl acetate (PVAc) in applications requiring flexibility, low-temperature film formation, and resistance to cracking. By introducing ethylene as a comonomer, the glass transition temperature (Tg) of the polymer is reduced, which lowers the minimum film-forming temperature (MFFT) and enables film formation at or below zero degrees Celsius. This characteristic allows formulators to prepare coalescent-free systems, a significant advantage in cold-climate applications and in coatings designed for low-VOC compliance.

The influence of ethylene content on performance is fundamental. At low ethylene levels, the dispersion behaves similarly to PVAc, with relatively high stiffness and limited elongation. As ethylene content increases, the polymer chains become more mobile, and the resulting films exhibit greater elasticity, meaning they can deform under stress without immediate fracture. When this elasticity is balanced by sufficient cohesive strength—the internal resistance of the film arising from chain entanglement, intermolecular interactions, and the limited crystallinity of ethylene domains—the material becomes tough. Toughness in this context refers to the ability of the adhesive or coating film to absorb energy before failure, resisting cracking, peeling, or brittle fracture during use.

This balance of elasticity and cohesive strength explains why VAE copolymers are highly valued in both adhesives and coatings. They provide immediate tack and flexibility in adhesives for paper, packaging, and textiles, while in coatings they allow smooth, crack-free films that form reliably even at low ambient temperatures. Although water and alkali resistance are lower than in acrylate or VeoVa-modified systems, the cost-effectiveness, environmental profile, and versatility of VAE dispersions have established them as indispensable binders in numerous applications.

The following sections outline the copolymerization process of VAE, the specialized reactor requirements compared with other vinyl acetate–based copolymers, and the properties and application areas that define VAE technology. Comparative assessments are also included to guide formulators in selecting VAE where its balance of performance and economy is most advantageous.

11.2 Copolymerization Process of VAE

Vinyl acetate–ethylene (VAE) copolymers are produced by semibatch emulsion polymerization under elevated pressure. The incorporation of ethylene distinguishes VAE from other vinyl acetate copolymers because ethylene is a gaseous comonomer with low solubility in water. To achieve the desired copolymer composition, the polymerization must be carried out in pressure-rated reactors that allow continuous dosing and dissolution of ethylene into the aqueous phase. Typical operating pressures range from forty to one hundred bar, depending on the target ethylene content and plant design.

The process follows the principles of semibatch emulsion polymerization. A preemulsion of vinyl acetate and any optional comonomers is gradually fed into the reactor along with initiator solution, while ethylene is continuously supplied under pressure. Persulfates or redox systems are commonly employed as initiators, adapted for high-pressure conditions. The starved-feed strategy helps maintain low residual monomer content, narrow particle size distribution, and uniform incorporation of ethylene into the growing polymer chains.

Because ethylene incorporation is sensitive to solubility and pressure, process control focuses on balancing ethylene uptake, monomer feed rate, and heat removal. Adequate agitation and gas dispersion equipment are required to maintain ethylene concentration in the aqueous phase and avoid composition drift. The resulting latex typically contains particles in the range of 100–250 nm, stabilized by protective colloids such as polyvinyl alcohol or by surfactant systems.

The safety requirements for VAE polymerization exceed those of conventional vinyl acetate systems. In addition to standard emulsion polymerization precautions, pressure vessels must be equipped with reinforced seals, pressure relief systems, ethylene monitoring and dosing controls, and emergency venting capacity. The combination of flammable ethylene and vinyl acetate under pressure demands rigorous process design and operator training to ensure safe and reliable procedures.

11.2.1 Copolymerization of Vinyl Acetate–Ethylene (VAE)–Representative Formulation and Procedure (5,000 kg Batch)

Scope: This section provides a representative VAE formulation sized to 5,000 kg latex and a complete, stepwise procedure suitable for semicontinuous operation with preemulsification and ethylene pressure control (Table 11.1).

11.2.1.1 Preparation of HEC Solution (10% Active)

Charge a clean make-up tank with deionized water equal to 450 kg. Under vigorous vortex, slowly slurry in 50 kg HEC; mix until fully hydrated and lump-free. Allow adequate time for full dissolution before polymerization day. Screen through 80–100 mesh into the reactor feed tank.

11.2.1.2 Reactor Preparation and Inerting

Verify pressure rating, relief devices, rupture disc, and interlocks. Confirm oxygen is excluded from the reactor and feed lines. Close the reactor, then perform three

Table 11.1 Copolymerization of vinyl acetate–ethylene (VAE)–5,000 kg latex (≈55% total solids).

Component	Specification/form	kg
Deionized water (total)[a]	Process + preemulsion + chases	1,503
Hydroxyethyl cellulose (HEC) solution	10% active (contains 50 kg HEC)	500
Vinyl acetate (VAc) monomer	Inhibited, ≤10 ppm HQ/MEHQ	2,413
Ethylene (gas)[b]	Polymer-bound uptake target	385
Anionic surfactant (e.g., SLES, 30% active)	Feed-split seed/main; product	50
Nonionic surfactant (e.g., AE10–AE20, 100%)	EO-rich, low-foam	25
Ammonium persulfate (APS) solution	10% active (≈9.65 kg APS solids)	96.5
Sodium bicarbonate (buffer)	Technical grade	2.5
t-Butyl hydroperoxide (tBHP)	70% solution (postchase redox)	2.5
Sodium formaldehyde sulfoxylate (SFS) solution	20% active (postchase redox)	12.5
Defoamer	Mineral/silicone, nonionic	5
In-process seed water (startup)	For seed generation	150

Totals: 5,000 kg (≈55% solids). Target polymer composition (by polymer): ≈86 wt% polyvinyl acetate / ≈14 wt% ethylene of total monomer.

a) The listed 1,503 kg water is the net water in addition to the water already present in the listed solutions (e.g., HEC 10%, APS 10%, SFS 20%).

b) Ethylene mass is the estimated polymer-bound uptake required to reach ≈14 wt% ethylene in polymer solids. Actual uptake depends on pressure, temperature, agitation, and VAc feed profile; monitor by mass balance and off-gas analysis.

vacuum–nitrogen cycles to ≤ 1 vol% O_2 in headspace. Heat the jacket to standby at 60–65 °C.

11.2.1.3 Preemulsion Preparation-Offline

Prepare a VAc preemulsion by dispersing VAc into water with all anionic and most nonionic surfactants, and 10–30% of the HEC solution. Practical ratio: 1.2–1.8 parts water: 1.0 part VAc (by mass). Target droplet size ~1–10 μm and viscosity <300 mPa s to ensure reliable pumping. Keep under an N_2 blanket if available.

11.2.1.4 Seed Charge and Nucleation

Charge 150 kg of deionized water (seed water), 5–10 kg of the HEC solution, and a small fraction of the surfactants (no more than ~20% of the total planned surfactant). Start agitation at high shear (tip speed ≥4–6 m s^{-1}). Adjust pH to 4.8–5.2 with dilute acetic acid or sodium bicarbonate as needed. Bring to 65 °C.

From the preemulsion prepared offline, add an amount that provides 20–40 kg VAc equivalent to create the seed phase (recommendation: Continue agitation to ensure uniform dispersion.

Note. In this O/W system, VAc is the emulsified "oil." Ethylene is not emulsified.

11.2.1.5 Pressurization with Ethylene for Initiation

Introduce 30 kg of ethylene under pressure; it dissolves in the reaction medium. Allow 10–15 min at 65 °C for dissolution/equilibration. Maintain reactor pressure at 30–35 bar during initiation and subsequent feeds.

11.2.1.6 Initiation and Main Feed

Start APS 10% solution feed (total APS solids ≈ 9.65 kg, i.e., 0.25–0.45 wt% on total monomer). Maintain 65 ± 2 °C and keep ethylene pressure at 30–35 bar. Keep pH 4.8–5.2; dose buffer as needed. Use jacket cooling and feed-rate modulation to control exotherm.

Begin the preemulsion feed using the remainder of the same pre-emulsified VAc. Ramp the feed over 3.5–5 h to control heat release and maintain starved-feed conditions. Continue to hold pressure with ethylene feed as needed.

11.2.1.7 End of Monomer Feed and Conversion Push

After completing the VAc and ethylene feed, hold 30–60 min at temperature and pressure to raise conversion. Monitor residual monomer via online GC or calibrated off-gas; aim APS consumption to plateau.

11.2.1.8 Redox Chase and Finishing

Reduce temperature to 55–60 °C. Add tBHP (70%) in small aliquots and SFS (20%) as co-feed to consume residual monomer to ≤500 ppm VAc. Maintain agitation and 10–15 bar ethylene for 20–40 min if needed to stabilize the composition, then depressurize cautiously to flare or recovery.

11.2.1.9 Cooling, Monomer Stripping (Optional), and Let-down

Cool to <35 °C. If lower residual monomer is required, apply vacuum/steam stripping with condenser and carbon trap. Add defoamer as needed. Adjust solids to ≈55% with deionized water. Screen through 80–100 mesh to the finished-goods tank.

11.2.1.10 Preservation and Final pH

Add in-can preservative according to supplier's dosage. Adjust pH to 4.8–5.0 for storage. Record viscosity and particle-size data.

11.2.1.11 Safety and Engineering Controls

Ethylene is highly flammable; verify Ex-rated equipment, lower-explosive limit (LEL) monitoring, and oxygen <1 vol% at initiation. Confirm pressure relief and interlocks prior to feed. Handle VAc under closed transfer with appropriate inhibitor levels; avoid copper/brass contact. Provide adequate heat-removal capacity for the VAc exotherm; size the jacket/coil duty for the worst-case feed ramp. Route vent and depressurization gases to a flare or recovery system with knock-out and activated-carbon polishing. Use splash shields and PPE when handling initiators and peroxides.

11.2.1.12 Quality Control (In-process and Final)

Measure pH (target 4.8–5.2), temperature, and pressure continuously, and agitator load as a proxy for viscosity rise. Track conversion by gravimetric solids, residual VAc via GC, and off-gas mass balance for ethylene uptake. At finish, verify total solids (≈55%), viscosity (Brookfield, 20–25 °C; specify spindle/rpm), particle size (D50 typically 0.2–0.5 μm, method per ISO 13320/DLS equivalent), pH 4.8–5.0, and residual monomer per internal spec. Record film MFFT and Tg if required.

11.2.1.13 Three Intervals in Emulsion Polymerization of VAE

This section summarizes the classic three intervals of emulsion polymerization and adapts them to vinyl acetate–ethylene (VAE) systems using a single pre-emulsified VAc feed with ethylene held by pressure control. Use this as a plant-practical guide to heat release, particle development, and control levers during the batch (Table 11.2).

Interval I: Nucleation (Seed to ~10–15% Solids) Radicals from the aqueous phase enter micelles and small VAc droplets, creating particles. In VAE, ethylene

Table 11.2 Practical map of the three intervals (VAE, semibatch).

Interval	What is happening	Typical range (solids/time)	Heat release and risks	Primary controls
I—Nucleation	Micelles and small droplets form particles; particle number (Np) set. Ethylene dissolves into VAc-rich loci.	Seed: 10–15% solids; time: ~0.5–1 h[a)]	High heat; insufficient surfactant leads to grit/coagulum.	Initiator start-rate; fine preemulsion; hold ethylene pressure; high shear; pH 4.8–5.2
II—Particle growth with reservoir	Particle number fixed; particles grow fed by the preemulsion reservoir; composition steered by E-pressure and VAc feed.	~10–45% solids; bulk of feed time (≈2–4 h)	High but flatter heat profile; runaway risk if feed is not starved	Starved preemulsion feed; jacket duty; ethylene 30–35 bar; initiator feed; defoamer management
III—No droplet reservoir	Preemulsion droplets consumed; growth limited by monomer diffusion; viscosity rises; composition more sensitive.	≥45% solids to end (≈last 0.5–1 h + hold)	Heat falls; autoacceleration possible; gel/coagulum risk if stabilization weak	Trim feeds; hold temperature; maintain pressure; small late initiator shot if needed; reduce shear if foam

a) The nucleation phase is considered complete when the reactor heat removal demand (jacket duty) and the ethylene make-up required to maintain reactor pressure become steady, typically corresponding to approximately 10–15% solids.

dissolves rapidly into VAc-rich loci; composition control is mainly via pressure at this stage. Particle number is set here by surfactant package, shear, and the early initiator rate.

Seed Window Rationale Why 0.5–1.0 h for VAE (vs. other VAc copolymers). Seeding proceeds faster in VAE systems because dissolved ethylene at 65 °C and 30–35 bar plasticizes the newly formed polymer phase (nascent), which accelerates nucleation and early particle stabilization compared with VAc–VeoVa or VAc–acrylate systems. With a fine VAc preemulsion and APS at pH 4.8–5.2, nucleation typically completes within 20–40 min; an extra 10–20 min lets jacket duty and ethylene uptake stabilize, fixing particle number and avoiding secondary nucleation. Extending much beyond ≈1 h tends to waste ethylene, broaden PSD, and raise end-batch viscosity and coagulum; stopping before ≈0.5 h risks too few particles and instability. Hence, the 0.5–1.0 h seed window to ~10–15% solids, then transition to the starved preemulsion feed.

Plant Signs: fast temperature rise as feeds begin; milky, low-viscosity latex; foam sensitive. Controls: keep preemulsion fine (~1–10 μm), maintain pH 4.8–5.2, start initiator steadily, and hold ethylene at 30–35 bar.

Interval II: Particle Growth with Reservoir (~10–45% Solids) Particle number is fixed; particles grow at a rate set by the starved preemulsion feed and radical flux. Heat release is near maximum but flatter than in Interval I. Ethylene uptake remains strong; pressure control keeps composition near target as the reactor consumes E.

Plant signs: stable, high jacket duty; viscosity climbs modestly; droplet count falls as feed proceeds. Controls: keep the VAc preemulsion starved (3.5–5 h ramp), use feed-forward cooling, maintain ethylene pressure (30–35 bar), and meter initiator to sustain steady conversion. Manage foam to avoid air entrainment.

Interval III: No Droplet Reservoir (≥ 45% Solids to End) The preemulsion droplet reservoir is depleted. Growth is limited by monomer diffusion within polymer particles. Ethylene uptake drops as viscosity and polymer content rise; composition control via pressure becomes less responsive. Heat generation declines, but low mobility can cause local autoacceleration if control is poor.

Plant signs: falling heat load; thicker latex; possible rise in torque; droplets absent. Controls: taper the preemulsion tail, maintain temperature and pressure, consider a small late initiator shot to push conversion, and avoid excessive shear if foam persists. Prepare for strip/finish (residual monomer control).

Practical Notes for VAE

1) Ethylene uptake is highest in Intervals I–II and tapers in III; hold pressure rather than target a fixed mass after initiation.
2) As the reservoir vanishes (Interval III), both E and VA compositions are more sensitive to pressure and temperature; keep $65 \pm 2\,°C$ and avoid swings.
3) Poor stabilization in Interval I often shows up as end-batch coagulum; maintain pH 4.8–5.2 and the planned surfactant levels.

11.2.1.14 Formulation and Process Remarks

The HEC level corresponds to ≈1.0 wt% on total batch (50 kg HEC solids in 5,000 kg), substituting for traditional PVOH while keeping free surfactant modest. The surfactant package is deliberately conservative to minimize free surfactant in the final latex; it may be tuned after pilot runs to optimize particle size and mechanical stability. Ethylene composition is governed primarily by pressure, temperature, and feed profile; maintaining 30–35 bar at 65 °C with robust mixing typically yields ≈12–16 wt% ethylene in the polymer. If higher ethylene content is desired, increase pressure in small increments and verify heat-removal capacity before changing the feed profile.

11.3 Reactor Requirements and Comparison

The production of vinyl acetate–ethylene (VAE) copolymers requires dedicated high-pressure reactors. These reactors differ fundamentally from the atmospheric or mild-pressure vessels used for PVAc homopolymerization or VAc–acrylate/VeoVa copolymerization. Because ethylene is a gaseous comonomer with limited solubility, its incorporation into the polymer backbone can only be achieved by maintaining elevated pressures, typically between forty and one hundred bar. This requirement dictates a reactor design capable of safe and efficient gas–liquid dispersion under polymerization conditions.

A VAE reactor must be constructed to withstand high pressures and must include specialized components: gas-dispersion impellers or spargers to introduce ethylene into the aqueous phase; reinforced seals and gaskets; pressure monitoring

Table 11.3 Comparison of reactors for VAE and other VAc copolymers.

Feature	VAE copolymers	PVAc homopolymer	VAc–BA/ VAc–VeoVa copolymers
Operating pressure	40–100 bar	Atmospheric	Atmospheric or ≤2 bar
Reactor design	Pressure-rated vessel with gas-handling systems, reinforced seals	Standard emulsion polymerization reactor	Standard semibatch reactor
Monomer feeds	Vinyl acetate + gaseous ethylene (plus optional comonomers)	Vinyl acetate only	Vinyl acetate + acrylates or VeoVa
Gas handling	Continuous ethylene dosing and dispersion equipment required	Not required	Not required
Initiator systems	Persulfates or redox systems adapted to high pressure	Persulfates or redox	Persulfates or redox
Cooling requirements	High, due to exothermic reaction under pressure	Moderate	Moderate
Safety considerations	Pressure relief, explosion protection, gas monitoring	Standard venting and controls	Standard venting and controls
Interchangeability	Dedicated VAE reactor required	Conventional plant suitable	Conventional plant suitable

and relief systems; and automated dosing controls to regulate ethylene uptake during the reaction. The design must also account for the high heat of polymerization, requiring efficient cooling capacity to maintain reaction stability under starved-feed conditions.

By contrast, reactors for PVAc or VAc–BA/VeoVa copolymers are generally operated at atmospheric pressure and rely only on liquid monomer feeds. These systems do not require the same level of reinforcement or gas-handling infrastructure. While the fundamental chemistry of emulsion polymerization is shared across these processes, the engineering requirements for VAE production are considerably more complex, making it unsuitable to adapt a standard PVAc reactor for VAE manufacture (Table 11.3).

This comparison emphasizes that while the underlying emulsion polymerization chemistry is similar, the engineering demands of VAE manufacture require reactors specifically designed for high-pressure operation. Producers considering entry into VAE production must therefore plan for dedicated reactor capacity rather than relying on retrofitting existing PVAc facilities.

11.4 Properties of VAE Dispersions

The properties of vinyl acetate–ethylene (VAE) dispersions are governed primarily by the ratio of vinyl acetate to ethylene. Ethylene, introduced as a flexible comonomer, reduces the glass transition temperature (Tg) of the polymer and lowers the minimum film-forming temperature (MFFT). As a result, VAE binders are capable of forming continuous films even below zero degrees Celsius, making them particularly suitable for cold-climate coatings and adhesives that must perform under ambient conditions without added coalescents.

At a very low ethylene content, the polymer exhibits properties similar to PVAc homopolymer: relatively high stiffness, higher Tg, and limited elongation. As ethylene incorporation increases, the polymer becomes more elastic and flexible. However, a balance must be maintained; very high ethylene contents reduce cohesive strength, leading to soft, weak films. In practice, most commercial VAE dispersions contain between 10 and 30% ethylene, providing an optimal balance between film flexibility and cohesive strength.

Compared with other vinyl acetate–based binders, VAE dispersions combine good adhesion, film-forming ability, and cost-effectiveness. Although their water and alkali resistance is lower than that of acrylate or VeoVa-modified copolymers, their environmental advantages and versatility ensure broad application. Table 11.4 summarizes the key performance characteristics of VAE dispersions in relation to PVAc, acrylate copolymers, and VeoVa-based copolymers.

Table 11.4 Comparative properties of VAE and other VAc-based binders.

Property	PVAc homopolymer	VAE copolymer	Acrylate copolymer	VAc–VeoVa copolymer
Glass transition temperature (Tg)	28–32 °C	−5 to +15 °C (depending on ethylene content)	−20 to +20 °C (adjustable)	0 to +20 °C
Minimum film-forming temperature (MFFT)	15–20 °C	0 to −5 °C	Below 0 °C	5–10 °C
Flexibility/ elasticity	Low	Moderate to high	High	Moderate
Cohesive strength	High but brittle	Balanced (depending on ethylene)	Moderate to high	High
Water resistance	Poor	Moderate	High	High
Alkali resistance	Poor	Moderate	High	High
VOC requirement (coalescent need)	Yes	No (coalescent-free)	No	No
Typical applications	Paper, wood adhesives	Adhesives, interior paints, cold-climate coatings	Exterior coatings, high-durability adhesives	Exterior coatings, masonry paints

This comparison illustrates that VAE dispersions occupy an intermediate position between PVAc and acrylate or VeoVa copolymers. Their strength lies in combining low-temperature film formation and elasticity with cost-effectiveness, making them particularly valuable in adhesives and in interior and cold-climate coatings.

11.5 VAE in Adhesives

Vinyl acetate–ethylene (VAE) dispersions are widely employed as adhesive binders because they combine low film-forming temperature, flexibility, and good adhesion to paper and board at competitive cost. Their inherent elasticity, arising from the ethylene segments, gives better toughness than PVAc homopolymers, while their ability to form films at ambient or sub-ambient conditions allows fast setting without coalescent additives. VAE adhesives are used in packaging, lamination, paper converting, labeling, and wood bonding under D1–D2 service conditions.

Table 11.5 Representative VAE carton sealing adhesive (1,000 kg batch).

Component	Parts by weight	Function	Notes
VAE dispersion (55% solids; Tg ≈ 0 °C; MFFT ≈ 0 °C)	550.0	Main binder	Provides flexibility and quick tack
Polyvinyl alcohol solution (10% solids)	50.0	Protective colloid	Improves stability and adhesion to paper
Plasticizer (phthalate-free)	20.0	Flexibility	Optional for high-speed machines
Defoamer (low silicone)	2.0	Foam control	Add during mixing
Preservative/biocide	1.0	In-can preservation	Per supplier
Deionized water	Balance to 1,000	Diluent	Adjust viscosity to 3,000–4,000 mPa s

They are also applied in textile finishing and nonwoven bonding, where softness and strength are required (Table 11.5).

This formulation provides rapid setting and strong adhesion on paperboard surfaces. The film is flexible and clear, suitable for folding cartons and general packaging. Water resistance is limited; applications are restricted to dry conditions (D1–D2 service) (Table 11.6).

This adhesive is designed for paper-to-paper or paper-to-board lamination. The inclusion of a tackifier improves wet tack and running speed, facilitating high-output lamination machines. The formulation balances adhesion with film flexibility, but its resistance to water and heat is limited compared with PVAc–AA or acrylate systems.

Table 11.6 Representative VAE paper lamination adhesive (1,000 kg batch).

Component	Parts by weight	Function	Notes
VAE dispersion (55% solids; Tg ≈ 0 °C)	600.0	Main binder	Provides flexibility and adhesion
Polyvinyl alcohol solution (10% solids)	30.0	Protective colloid	Enhances wetting of paper surface
Tackifier dispersion (rosin ester, 50% solids)	50.0	Tack improvement	Improves lamination speed
Defoamer (low silicone)	2.0	Foam control	Add during mixing
Preservative/biocide	1.0	In-can preservation	Per supplier
Deionized water	Balance to 1,000	Diluent	Viscosity 2,000–3,000 mPa s; solids ≈ 50%

Table 11.7 Representative VAE labeling adhesive (1,000 kg batch).

Component	Parts by weight	Function	Notes
VAE dispersion (55% solids; Tg ≈ 0 °C; MFFT ≈ 0 °C)	600.0	Main binder	Provides film formation and adhesion
Polyvinyl alcohol solution (10% solids)	30.0	Protective colloid	Improves clarity and adhesion to paper
Plasticizer (phthalate-free)	15.0	Flexibility	Maintains open time and label slip control
Defoamer (low silicone)	2.0	Foam control	Add during mixing
Preservative/biocide	1.0	In-can preservation	Per supplier
Deionized water	Balance to 1,000	Diluent	Adjust viscosity to 2,500–3,500 mPa s; solids ≈ 50%

Performance characteristics: VAE adhesives provide fast setting and good flexibility, which makes them particularly effective for packaging and paper converting operations. They deliver immediate tack, clean machining, and resistance to cracking in folded or creased areas. Their limitations lie in water and alkali resistance, which remain moderate compared with acrylate-based or AA-modified PVAc systems. For D1 and D2 indoor uses they are reliable and economical, but for higher water resistance they require modification with acrylic acid or external crosslinkers, which is discussed in a later section (Table 11.7).

This adhesive is formulated for labeling applications such as wet-glue labels on paper or carton substrates. It provides high initial tack, good clarity on transparent films, and sufficient open time for high-speed bottling and labeling lines. The formulation is designed for room-temperature application and delivers flexible, transparent bonds, though water resistance remains moderate.

11.6 VAE in Coatings

Vinyl acetate–ethylene (VAE) dispersions are widely used in architectural coatings where low-temperature film formation, application ease, and cost effectiveness are prioritized over maximum chemical resistance. Ethylene lowers the glass transition temperature and minimum film-forming temperature, enabling coalescent-free formulations that comply with stringent VOC targets. VAE binders are particularly suited to interior matt paints, plasters, and primers, and

Table 11.8 Representative VAE interior matt paint (1,000 kg batch).

Component	Amount (kg)	Function	Notes
Deionized water	320	Continuous phase	Adjust viscosity
Dispersant (sodium polyacrylate)	5	Pigment wetting	pH 8–9
TiO_2 (rutile)	200	Opacity	High hiding power
Calcium carbonate (2–5 μm)	300	Extender	Balance cost and scrub
VAE dispersion (55% solids; Tg ≈ 0 °C; MFFT ≈ 0 °C)	130	Main binder	Coalescent-free film formation
HEUR thickener	10	Rheology control	HEC-free option
Defoamer (low silicone)	2	Foam control	Add in grind and let-down
Biocide/preservative	3	Preservation	Per supplier guidance
Coalescent	—	—	Not required with VAE

they are frequently selected in cold-climate markets where reliable film formation at low ambient temperature is essential.

While water and alkali resistance are moderate compared with pure acrylate or VeoVa-modified systems, carefully balanced formulations deliver durable films for interior service. The following examples provide representative starting points; pigment volume concentration, rheology modifiers, and additives should be optimized to local raw materials and application methods (Table 11.8).

This formulation yields a smooth, low-odor interior matt paint suitable for brush, roller, or spray application. The coalescent-free profile assists with VOC compliance and reduces sensitivity to application temperature. Scrub resistance is adequate for standard interior use; for higher scrub, reduce PVC and optimize thickener balance (Table 11.9).

The primer/sealer is designed for reliable film formation at low ambient temperature and for sealing porous substrates prior to topcoating. The balance of extenders and HEUR ensures good leveling and sandability. Where stain resistance or mildew resistance is required, zinc oxide or appropriate additives may be included subject to local regulations.

Performance characteristics: VAE coatings exhibit easy application, rapid dry to touch, and reliable film formation at low temperatures without added coalescent. Films are flexible and resist cracking on minor substrate movement. Water and alkali resistance are moderate; exterior durability is limited compared with acrylate or VeoVa-modified systems and should be considered when specifying for outdoor exposure (Table 11.10).

Table 11.9 Representative VAE cold-climate primer/sealer (1,000 kg batch).

Component	Amount (kg)	Function	Notes
Deionized water	350	Continuous phase	Viscosity adjustment
Dispersant (sodium polyacrylate)	5	Pigment wetting	pH 8–9
TiO_2 (rutile)	120	Opacity	Primer hiding
Calcium carbonate (fine)	250	Extender	Controls porosity
Talc (10–15 μm)	60	Sheen control	Improves sanding
VAE dispersion (55% solids; MFFT ≈ 0 °C)	150	Main binder	Cold-weather film formation
Zinc oxide (optional)	10	Stain block/mildew aid	As regulations allow
HEUR thickener	8	Rheology	Sag/leveling balance
Defoamer (low silicone)	2	Foam control	—
Biocide/preservative	3	Preservation	—

Table 11.10 Representative VAE exterior masonry paint (1,000 kg batch).

Component	Amount (kg)	Function	Notes
Deionized water	280	Continuous phase	Adjust viscosity
Dispersant (sodium polyacrylate)	6	Pigment wetting	pH 8–9
TiO_2 (rutile)	180	Opacity	Primary hiding pigment
Calcium carbonate (medium particle size)	320	Extender	Provides body and cost balance
Talc (10 μm)	40	Weathering aid	Improves flatting and durability
Silica (micronized)	25	Matting/scrub aid	Improves surface strength
VAE dispersion (55% solids; Tg ≈ 0 °C)	130	Main binder	Film formation at low temperature; moderate durability
Coalescent	5	Film formation aid	Optional; enhances early film integrity
HEUR thickener	10	Rheology	Balance sag and leveling
Defoamer (low silicone)	2	Foam control	—
Biocide/preservative	4	Preservation	In-can and dry film protection

This formulation illustrates the use of VAE as a binder in an exterior masonry paint. It provides good application properties, opacity, and flexibility, making it suitable for mild climates or cost-sensitive markets. However, its resistance to water, alkali, and long-term weathering is moderate. Compared with pure acrylate or VeoVa-modified binders, VAE films are less resistant to chalking, efflorescence, and alkali attack. For demanding exterior exposures, acrylate or VAc–VeoVa copolymers remain the preferred choice.

11.7 Comparative Assessment

Where VAE is preferred vs. PVAc, acrylates, or VeoVa; cost–performance balance; eco-label compliance.

This section provides a concise, decision-oriented comparison of four binder families used throughout this chapter: PVAc homopolymers, VAE (vinyl acetate–ethylene), pure acrylates (e.g., BA/MMA/EA systems), and VAc–VeoVa copolymers. Emphasis is placed on end-use performance, cost positioning, and practical compliance with major eco-labels via low-VOC and low-residual strategies (Table 11.11).

Eco-label compliance is typically enabled by (i) low-VOC design (coalescent-lean or coalescent-free, ammonia-free), (ii) low residual monomer (strip/finish to spec), (iii) APEO-free surfactants/low SVOC package, and (iv) biocide stewardship. Always verify current criteria for the specific scheme.

11.7.1 Selection Guidance

When to choose VAE: VAE offers a compelling middle ground where PVAc is too brittle or moisture-sensitive and pure acrylates are over-engineered or cost-prohibitive. Ethylene lowers Tg and MFFT without heavy plasticizer/coalescent demand, delivering high peel on films and good MVTR control for paper/packaging. Natural fit: film-to-paper/carton laminations, window patching, flexible packaging adhesives, and interior coatings requiring balanced block/water resistance without exterior-grade UV.

When PVAc suffices: For interior wood assembly, paper converting, and general-purpose carton where humidity exposure is modest and cost is critical, PVAc remains the most economical choice. Crosslinkable PVAc-AA variants (with appropriate external crosslinkers) can partially bridge performance gaps; flexibility at low temperature still favors VAE.

When acrylates are preferred: For exterior durability, UV stability, and low-energy substrate adhesion, pure acrylates lead. They support broad Tg engineering and

Table 11.11 Comparative assessment of PVAc, VAE, acrylates, and VAc–VeoVa (decision-oriented).

Criterion	PVAc (homopolymer)	VAE (VAc–ethylene)	Acrylates (pure)	VAc–VeoVa
Typical Tg/MFFT window	Tg ≈ +28 to +33 °C; MFFT typically ≥ 15–20 °C	Tg tunable to ~0…+15 °C by ethylene; low MFFT achievable	Broad Tg design (−20…+40 °C); MFFT tailored via coalescent	Tg lowered by VeoVa (neo-ester); MFFT reduced without plasticizer
Water resistance/ MVTR	Moderate; sensitive to hydrolysis/alkali	Better than PVAc; good MVTR for paper/packaging	Very good to excellent; best for exterior exposure	Improved hydrolysis resistance vs. PVAc; good alkali resistance
Block/heat resistance	Fair; softening near service temperatures	Fair–good; depends on ethylene level (lower Tg → softer)	Good–excellent with higher-Tg grades	Good block/heat with balanced VeoVa level
Adhesion—polar substrates	Strong (paper/ wood/PVOH-sized films)	Strong; improved flexibility and peel	Strong; wide substrate latitude	Strong; improved on chalky/alkaline mineral surfaces
Adhesion—low-energy films	Limited without primers/ treatments	Better than PVAc; still benefits from corona/primers	Best (with proper surfactants and wetting aids)	Improved vs. PVAc; may require adhesion promoters
Flexibility/ toughness	Moderate; can be brittle at low T	Good peel/impact from ethylene content	Tunable; soft grades excel in flexibility	Balanced; retains cohesion with alkali resistance
Exterior durability (UV)	Poor–fair; interior applications	Fair for sheltered use; not for long-term exterior	Best-in-class; exterior paints and roof coatings	Good for certain exterior masonry/ topcoats (check chalking)
Odor/residual monomer	Low odor possible at low-ppm VAM; easy to strip	Low odor feasible; manage dissolved ethylene off-gas	Low odor feasible; BA/ MMA/EA control required	Typically low odor; VeoVa has a characteristic neo-ester note
Cost positioning (resin)	$ (lowest)	$$ (low-mid)	$$$ (mid-high)	$$–$$$ (mid; VeoVa raises cost)
Eco-label pathway	Straightforward with low VOC/ low VAM ppm	Straightforward; ensure residuals/ethylene venting	Achievable; watch total VOC and residual acrylates	Achievable; manage neo-ester odor and residual monomer
Typical sweet-spot uses	Wood joinery, paper converting, carton laminations	Flexible packaging, film-to-paper, cold-flex adhesives	Exterior paints, roof coatings, low-energy film adhesion	Alkali-resistant coatings, masonry paints, durable interiors

maintain alkali/hydrolysis resistance, making them the first choice for architectural exterior paints, roof coatings, and plastic-rich assemblies.

When VAc–VeoVa shines: Incorporating VeoVa introduces a neo-ester that enhances hydrolysis and alkali resistance while easing MFFT—ideal for masonry and durable interior topcoats where water whitening and chalking control matter. Cost rises with VeoVa level; balance VeoVa to maintain cohesion and block resistance.

11.7.2 Cost–Performance and Compliance Notes

From lowest to highest resin cost, the general order is: PVAc, then VAE, then VAc–VeoVa, and mid-range acrylics at roughly similar levels, followed by high-performance acrylics. The total formulation cost also depends on coalescent demand, the need for crosslinkers, and the breadth of the additive package.

All four binder families can meet stringent eco-label requirements when formulated with APEO-free surfactants, ammonia-free neutralization, low residual monomer achieved by redox chase and stripping, and low-VOC coalescent strategies. Where odor is critical, use carbon polishing and select counter-ions carefully to minimize amine-type notes.

11.7.3 Practical Keys

1) Start with VAE when you need flexible bonds, good peel on films, and better moisture tolerance than PVAc at a competitive cost.
2) Choose acrylates for exterior/UV and low-energy plastics.
3) Use VAc–VeoVa for alkali-resistant, moisture-robust coatings on masonry and mineral substrates.
4) Keep PVAc for cost-sensitive interior applications with moderate humidity and polar substrates.

11.7.3.1 Terminology Notes

SVOC—Semivolatile Organic Compounds: Organic substances with lower vapor pressure than VOCs. They evaporate more slowly and can persist in indoor air over longer periods. Many eco-labels monitor SVOCs because they influence indoor air quality and long-term emissions. Operationally, in coatings, common contributors include certain coalescents and plasticizers; specific cut-offs (boiling point or vapor pressure) vary by scheme—always check the relevant standard.

MVTR—Moisture Vapor Transmission Rate (also called WVTR): The rate at which water vapor passes through a film, laminate, or substrate. Typical units are $g/m^2 \cdot day$. Common test methods include ASTM E96 for coatings and building materials, and ASTM F1249 for plastic films; higher MVTR means greater water vapor permeability.

11.8 Practical Considerations

VAE performance in the plant and on the line is determined as much by what happens after polymerization as by the copolymer itself. Storage pH governs hydrolysis, viscosity drift, and odor; keeping the latex mildly acidic (≈4.8–5.0) preserves stability and extends shelf life, while any near-neutral adjustment should be made only at the point of use. Compatibility with fillers, pigments, and additives defines whether you gain body and opacity or invite water uptake, foam, and poor recoat—dispersant choice, thickener hydration, and defoamer staging are pivotal in VAE systems, especially when HEC is the primary colloid. Substrate preparation remains the final gatekeeper: low-energy films such as PE, PP, and BOPP often require corona or primer to lift surface energy, whereas porous papers and cartons demand moisture control rather than chemical activation. Throughout this section, we connect these decisions to practical test outcomes—residual monomer/odor, freeze–thaw recovery, and moisture-barrier behavior (MVTR per ASTM E96 for coatings and ASTM F1249 for films)—so that formulation tweaks translate directly to line robustness and eco-label readiness without sacrificing cost–performance (Tables 11.12–11.14).

Table 11.12 Storage pH and stability guidelines for VAE latexes (adhesives/coatings).

Scenario	Recommended storage pH	Colloid/ surfactant	Biocide strategy	Notes
Baseline VAE (HEC-stabilized)	4.8–5.2	HEC + low free surfactant	Isothiazolinone blend (per limits)	Lowest hydrolysis risk; monitor pH drift
VAE with AA (internal carboxyl)	4.8–5.2	HEC + anionic/ nonionic package	As above	Keep storage acidic; neutralize only at use
VAE for near-neutral application	Store 4.8–5.0; use pH ≤ 7	As above	As above	Do not store near-neutral; boost just-in-time
Crosslinkable systems (e.g., CDI, Zr)	4.8–5.0	Low-amine auxiliaries	As above, avoid amine donors	Avoid amine-bearing surfactants/ neutralizers
Ammonia-free requirement	4.8–5.0	HEC + APEO-free surfactants	As above	Reduces odor; improves eco-label readiness
Heat-aging robustness (50 °C, 2 weeks)	n/a	As formulated	Dual-mode wet/ dry preservative as needed	Track viscosity drift; check coagulum
Freeze–thaw stability (per spec cycles)	n/a	As formulated	n/a	HEC improves F/T; verify cycles + recovery

Table 11.13 Compatibility matrix—common fillers, pigments, and additives with VAE.

Material	Suitability with VAE	Typical dosage window	Cautions/integration notes
Calcium carbonate (GCC/PCC)	Excellent (interior)	5–40% (coatings); 5–15% (adh.)	Alkalinity can raise pH; pre-wet; control dispersant to avoid viscosity creep
Titanium dioxide (rutile)	Excellent	5–25% (coatings)	Disperse at low pH; watch surfactant demand; anti-yellowing grade for adhesives
Kaolin clay	Good	5–30%	Improves block; may reduce flexibility; shear to full plate-break
Talc	Good (block/anti-stick)	3–10%	Hydrophobic; wet with nonionic; may aid MVTR control
Barium sulfate	Good (density, burnish resistance)	3–20%	Inert; may harden film; maintain pH window
Precipitated/fumed silica	Good (rheology)	0.2–2%	Pre-gel HEC or use HEUR; avoid shock; add late
Carbon black	Moderate (adhesives)	0.1–3%	High surfactant demand; defoam aggressively; filter to remove grits
Dispersant (polyacrylate)	Excellent (pigment wetting)	0.1–0.8% on pigment	Keep pH ≤ 7 in grind; excess → water sensitivity
Dispersant (phosphate ester)	Good	0.2–1.0% on pigment	Improves TiO_2 spacing; can soften water resistance if overdosed
Thickener HEC/HMHEC	Excellent (adh./coatings)	0.2–1.2% (as product)	Hydrate fully; stage addition; F/T robustness
Thickener HEUR	Good (KU build, flow/level)	0.2–1.0% (as product)	Needs surfactant balance; watch coalescent interaction
Defoamer (mineral/silicone)	Excellent (essential)	0.05–0.5%	Seed + let-down; avoid fisheyes; test recoatability
Coalescent (e.g., Texanol®, DPM)	Good (coalescence control)	0–4% on binder	Impacts VOC/SVOC and odor; reduces with ethylene-soft VAE
Plasticizer (e.g., DINP, DOTP)	Limited (adhesives only)	0–10% on binder	Migration/odor risk; prefer internal softening via ethylene

(Continued)

Table 11.13 (Continued)

Material	Suitability with VAE	Typical dosage window	Cautions/integration notes
Crosslinker—CDI	Excellent with AA-containing systems	0.3–1.5% on polymer solids	Add late; maintain pH ≤ 7 at use; monitor pot-life
Crosslinker—Zr complex	Good (wood/paper adhesives)	0.3–1.0% on polymer solids	Dose slowly; avoid phosphate surfactants; storage pH 4.8–5.0
Crosslinker—$AlCl_3$	Good (coatings)	0.05–0.3% metal on polymer	Sensitive to pH; may reduce shelf life; add as a separate shot
Wetting agent (nonionic)	Excellent	0.1–0.5% on total	Improves film/substrate wetting; low foam grades preferred
Wetting agent (anionic)	Good	0.1–0.5% on total	Can raise water uptake; balance with nonionic
Anti-block (PE/PTFE wax)	Good (coatings)	0.3–1.5%	Add during let-down; screen to avoid specks

11.8.1 Ethylene Content vs. Performance (Monomer-basis–E:VA Where E + VA = 100)

Industry convention: ethylene percentage is quoted on a total monomer feed basis—E:VA where E + VA = 100. Raising E in the feed lowers Tg/MFFT and boosts flexibility/peel, but typically reduces block/heat resistance and can shift water tolerance and dirt pick-up. The two tables below map monomer-based E% to (i) core performance and (ii) formulation/processing implications. Polymer-bound E is usually lower than the feed % and depends on pressure, temperature, and mass transfer; confirm via off-gas and mass balance (Table 11.15).

Reading note: As E in the monomer feed increases, coalescent demand typically drops; anti-block measures (wax, silica, talc) and harder extenders may be needed to control print-stick and blocking in coatings (Table 11.16).

11.8.2 Domain Guide by Monomer-based E:VA (Table 11.17)

The tables above use monomer-basis ratios (E:VA where E + VA = 100). This domain guide labels each band as Adhesives-dominant, Shared/Coatings-viable, or Not recommended for coatings, and lists the main controls (Table 11.17).

Table 11.14 Substrate surface preparation and treatments (when required).

Substrate	Cleanliness/prep	Surface energy target	Treatment options	Quick verification
BOPP/PP/PE films	Degrease; remove slip/antistats	≥38–42 dyn/cm	Corona/plasma; adhesion primer (PEI, acrylic)	Dyne pens; 180° peel vs. time; water-break test
PET (untreated)	Clean; deionized-water rinse	≥44 dyn/cm	Corona/plasma; primer	Dyne pens; tape test; MVTR impact check
PVC (rigid/flexible)	Degrease; plasticizer migration check	≥40 dyn/cm	Primer/wash primer as needed	MEK double-rub of primer; peel after aging
Aluminum foil	Clean; light abrasion if needed	High (n/a)	Chromate-free primer if demanding	Cross-hatch adhesion (ASTM D3359); foil tear mode
Paper/carton	Control moisture; avoid sizing contaminants	n/a (porous)	Corona for films prior to lamination	Cobb60; fiber tear %; blocking after stack
Glass	Degrease; IPA wipe	High (n/a)	Silane primers when needed	Cross-hatch; water-sheen continuity
Printed/inked films	Compatibility screen with ink system	As per base film	Adhesion promoter compatible with ink	Ink pick test; solvent rub; aging under humidity

11.9 VAE Enhancement with Acrylic Acid (AA) Under Ethylene Pressure

This section clarifies the chemistry and practice of tri-monomer VAc/E/AA emulsion polymerization. Low levels of acrylic acid (AA) are copolymerized with vinyl acetate (VAc) while ethylene (E) incorporation remains pressure-controlled. AA supplies sparse carboxyl sites for postcrosslinking without excluding ethylene from the growing chains.

11.9.1 Mechanism at a Glance (VAc/E/AA)

Loci and availability: VAc is the oil phase (pre-emulsified droplets and polymer particles). Ethylene is a compressed gas that dissolves and partitions into the

Table 11.15 VAE composition map—core performance vs. ethylene content in total monomer feed (E:VA, E + VA = 100).

E:VA in feed (%)	Approx. Tg/MFFT (no coalescent)	Flexibility and peel	Block/heat	Typical applications
0:100 (PVAc)	Tg ≈ +28…+33 °C/ MFFT ≥ 15–20 °C	Low; brittle in cold	Fair–poor	Interior wood glue; paper/carton converting
5–10:95–90	Tg ≈ +15…+20 °C/ MFFT ≈ 10–15 °C	↑ vs. PVAc	Fair	Carton laminations; interior coatings
11–14:89–86	Tg ≈ +8…+15 °C/ MFFT ≈ 5–10 °C	Good; peel on films	Fair	Window patching; film-to-paper laminations
15–18:85–82	Tg ≈ 0…+8 °C/ MFFT ≈ 0–5 °C	High; excellent peel	Fair to lower	Flexible packaging; labels; low-T bonding
19–24:81–76	Tg ≈ −5…0 °C/ MFFT ≤ 0 °C	Very high; soft-tacky	Lower heat/ creep	Soft adhesives; PSA-like uses (case-by-case)
25–30:75–70	Tg ≪ 0 °C/ MFFT ≤ 0 °C	Extremely soft	Low heat resistance	Niche VAEs; soft-touch/coating aids

Table 11.16 VAE composition map—formulation and processing implications vs. ethylene content in total monomer feed (E:VA, E + VA = 100).

E:VA in feed (%)	Water resistance	MVTR/barrier	Coalescent need	Key risks and controls
0:100 (PVAc)	Moderate; sensitive to alkali	Moderate–high	Higher (to form film in cool shop)	Humidity creep → add AA/crosslinker; strip VAM
5–10:95–90	Slight ↑ vs. PVAc	Moderate	Lower than PVAc	Watch blocking at high coat weight
11–14:89–86	Moderate–good	Good for paper/film	Low–moderate	Balance anti-block/ wax; check recoat
15–18:85–82	Moderate	Good–very good	Low	Dirt pick-up/print stick → wax/PE/ PTFE; heat-age
19–24:81–76	Moderate (can drop)	Very good	Very low	Cold-flow/creep → raise solids/HEC; higher P, mass-transfer in reactor
25–30:75–70	Often reduced	Very good	Very low	Gel/blocking risk; consider alternative chemistries for exterior/UV

Table 11.17 Domain guide—application fit by E:VA in total monomer feed (E + VA = 100).

E:VA in feed (%)	Adhesives domain	Coatings domain	Key controls/notes
0:100 (PVAc)	Interior assembly; paper/carton converting	Interior primers/ paints (economy); not exterior	Coalescent need ↑; add AA/crosslinker for humidity; strip VAM
5–10:95–90	Carton/lamination; wood glues (improved flex)	Coatings-viable (interior); good block balance	Low coalescent; add wax/anti-block; keep storage pH 4.8–5.0
11–14:89–86	Film-to-paper; window-patch; laminates	Coatings-viable (interior) with anti-block	Balance PVC; PE/ PTFE wax for print-stick; check recoat
15–18:85–82	Flexible packaging; labels; cold-flex bonding	Limited (risk of blocking/print-stick)	Hard extenders/ talc; wax; heat-age viscosity; solids control
19–24:81–76	Soft/PSA-like; pressure-sensitive concepts	Generally not recommended	Creep risk; increase solids and HEC; check reactor mass transfer.
25–30:75–70	Niche soft-touch; specialty bonds	Not recommended	Gel/blocking risk; consider acrylates or VAc–VeoVa for coatings

Key: *Coatings-viable (interior)* means film formation is manageable with modest anti-block; *Limited* means feasible only with strong anti-block and hardness aids; *Not recommended* means blocking, creep, or UV limitations dominate (Table 11.18).

Table 11.18 Representative commercial VAE binder families and example grades.

Brand/family	Company	Typical focus (examples)	Example grades
VINNAPAS®, ETONIS®	WACKER	Wide VAE portfolio for adhesives, architectural and construction; ETONIS used in road/bitumen modifiers	VINNAPAS EP 707K (VAE for adhesives)
EcoVAE™	Celanese	Low-odor, low-VOC VAE for architectural paints; also used in adhesives	EcoVAE 401, EcoVAE 405
Vinamul®	Celanese	Legacy vinyl-based emulsions line including VAE grades across coatings/adhesives	Multiple regional VINAMUL VAE grades
DA-series (e.g., DA-102H, DA-265)	Dairen Chemical (DCC)	Broad VAE emulsions range for adhesives, construction, and textiles	DA-102H; DA-265
VAE CW40-series	Sinopec (regional)	Commodity/industrial VAE emulsions for coatings, adhesives, and construction	CW40-705; CW40-716

reaction medium, preferentially into VAc-rich droplets and polymer particles, with only limited dissolution in the aqueous phase. AA resides chiefly in the aqueous phase and at particle interfaces, especially if lightly neutralized.

Radical path: Radicals generated in water enter micelles/particles and propagate with the monomer they encounter. There is no "AA finishes first"—composition follows relative propagation probabilities and local availability. At low AA, chains incorporate occasional AA units while VAc remains dominant and ethylene addition is steered by pressure.

Composition levers: Ethylene content is governed by reactor pressure (30–35 bar) and temperature (≈65 °C). VAc feed rate/starvation controls heat and particle growth. The AA level is kept low to avoid excessive water-phase polymer or instability.

Interval view: E uptake is strongest in Intervals I–II; as the droplet reservoir vanishes (Interval III), composition becomes more sensitive—keep pressure and temperature steady.

Performance extension via AA (with ethylene under pressure): A low level of acrylic acid (≈0.5–1.5% on total monomer) can be copolymerized with VAc/ethylene in emulsion. AA units are incorporated sparsely (interface-favored) while ethylene uptake remains pressure-controlled; there is no exclusion of E by AA. The AA provides carboxyl sites for zirconium/aluminum/carbodiimide crosslinking, lifting water and alkali resistance beyond neat VAE.

11.9.2 Formulation and Feed Strategy (Practical)

1) Add AA to the VAc preemulsion (preferred). Do not inject neat AA into the kettle. Light neutralization (≤30% mol) is acceptable if needed for stability.
2) Keep storage pH at 4.8–5.0 during polymerization; adjust to the crosslinker's use-pH only at application.
3) Maintain ethylene at 30–35 bar; introduce ≈30 kg at initiation (5-ton case), then hold pressure during the feeds.
4) Run a starved preemulsion feed over 3.5–5 h at 65 ± 2 °C; manage heat with jacket duty and feed modulation.

11.9.3 Design Rules for VAc/E/AA (Quick Reference)

1) *AA level*: 0.5–1.0% for adhesives; up to ~1.5–2.0% for coatings with crosslinker.
2) *Ethylene pressure*: 30–35 bar; control composition by pressure, not by fixed mass (beyond the initiation amount).
3) *pH*: 4.8–5.0 during polymerization; minimize early over-neutralization to keep AA at the interface and maintain stability.

4) *Surfactants*: keep the planned package; poor early coverage raises grit/coagulum risk.
5) *Feed*: use a single, offline VAc preemulsion; seed with a small portion of the same preemulsion; no neat VAc feed.
6) *Intervals*: seed 0.5–1.0 h to ≈10–15% solids; begin the starved preemulsion ramp once jacket duty and E make-up stabilize.

11.9.4 1K Latent Crosslinking—Factory Postadd (No Two-pack Mixing)

This section describes single-component VAE(+AA) adhesives in which the crosslinking agent is incorporated at the factory and remains latent during storage. Unlike two-pack systems, no point-of-use mixing is required and the concept of pot-life does not apply; instead, in-can stability is governed by formulation pH, ionic strength, and surfactant profile. Latency is maintained at pH 4.5–5.5, and crosslinking is activated primarily during drying and film formation as water is removed and local microenvironments evolve (including modest pH drift and increased segmental mobility). The approach is compatible with PVOH-dominated stabilization and low free-surfactant levels; phosphate surfactants and amine-rich additives should be avoided when using metal-based crosslinkers.

Four representative 1K options are treated here: zirconium and aluminum complexes (coordination with carboxylate-rich domains and hydroxyl-bearing phases), polycarbodiimides (covalent conversion of carboxyl groups to N-acyl urea linkages), and zinc or other multivalent salts (ionic bridging for wet-grab enhancement). Dose, storage pH, and activation behavior are summarized in the tables that follow. Formulations should be qualified for latency by accelerated storage (for example, 40 °C) and for performance by humidity/alkali aging on the intended substrates (Tables 11.19 and 11.20).

Process note (1K): keep finished-goods pH 4.5–5.5, add crosslinker at the end under moderate shear, and minimize free surfactant (especially phosphate types with Zr/Al).

11.9.5 Why Add Acrylic Acid to VAE and What Applications Benefit

Acrylic acid (AA) is introduced at low levels into vinyl acetate–ethylene (VAE) copolymers to create pendant carboxyl functionality. These carboxyl groups increase polymer–substrate interactions on polar and treated surfaces, improve pigment and filler binding in adhesive films, and provide defined anchoring sites for latent crosslinkers used in single-component systems. In 1K VAE(+AA) adhesives, metal complexes (for example, zirconium or aluminum) coordinate with

Table 11.19 Crosslinking agents for 1K VAE(+AA)—quick selection matrix.

Crosslinker	Wet-grab/ early set	Water/alkali durability	Long-term permanence	Notes
Zirconium complexes (Zr)	Strong	Moderate	Moderate	Fast set; coordination strengthens on drying
Aluminum salts/complexes (Al)	Strong	Moderate	Limited	Economical; risk of embrittlement at high dose
Polycarbodiimide (CDI)	Moderate	Strong	Strong	Covalent links (N-acyl urea) after film formation
Zinc or other multivalent salts	Strong	Limited	Limited	Ionic bridging; mainly wet-grab enhancement

Table 11.20 Dosing and handling for 1K systems (factory postadd).

Crosslinker	Typical dose (on polymer solids)	Storage pH (1K)	Activation on substrate	Watch-outs
Zirconium complexes	0.3–1.0% as metal	4.5–6.0	Strengthens during drying	Avoid phosphate surfactants; add last under moderate shear
Aluminum salts/complexes	0.05–0.3% as Al	4.5–5.5	Coordination during drying	Excess lowers pH; verify flexibility at upper dose
Polycarbodiimide (CDI)	0.3–1.5%	4.5–5.5 (prefer 4.5–5.0 for latency)	Reacts during/after film formation	Sensitive to amines; minimize free surfactant; check 40 °C stability
Zinc or multivalent salts	0.1–0.5% as metal	5.0–6.5	Immediate ionic reinforcement	Limited permanence; verify humidity/alkali aging

carboxylate-rich regions during drying, delivering fast wet-grab and early strength, while polycarbodiimides form covalent N-acyl urea linkages with carboxyl groups after film formation to achieve durable water and alkali resistance. Ionic salts such as zinc provide additional wet-grab by carboxylate bridging. Typical AA levels are in the range 0.5–2.0% on total monomer; lower additions primarily aid wetting and anchorage, whereas higher levels increase crosslinker responsiveness

Table 11.21 Application map for (VAE + AA) 1K adhesives with latent crosslinkers.

Application	Typical substrates	Operating conditions	Primary need	Recommended system
Carton-to-film lamination (window patching, display)	Printed or lacquered carton to BOPP/PET (treated ≥38 dyn/cm)	High line speed; room temperature; moderate pressure	Fast wet-grab and clean fiber-tear without blocking	VAE(+AA 0.5–1.0%) + zirconium complex; minimize phosphate surfactant
Paper/board to aluminum foil lamination	Kraft or solid bleached sulfate (SBS) board	Room temperature; drying with warm air	Early set and humidity resistance	VAE(+AA 0.7–1.2%) + aluminum or zirconium complex
Labeling (paper to glass/metal)	Coated paper to glass or metal cans	Cold/wet fill; condensation exposure	Wet adhesion and alkali resistance	VAE(+AA 1.0–1.5%) + polycarbodiimide; low free surfactant
Paper-to-film overlamination (print protection)	Printed paper to BOPP/PET overlam	Moderate heat from dryers; nip pressure	Blocking control, scuff resistance	VAE(+AA 0.5–1.0%) + light zirconium dose; verify blocking
Wood assembly—interior humidity class D3	Hardwood/softwood joints	Press at ambient; cyclic humidity test D3	Water resistance without two-pack mixing	VAE(+AA 1.0–1.5%) + polycarbodiimide; PVOH-stabilized
Wood assembly—enhanced D4 performance (1K route)	Hardwood/softwood; finger joints	D4: boil–dry–soak cycles (EN 204/205).	Maximum water/alkali durability for 1K	VAE(+AA 1.2–2.0%) + polycarbodiimide; confirm standard-specific criteria
Paper converting (ream wrap, bag seams)	Uncoated/coated paper	Room temperature; high throughput	Quick tack and fiber-tear	VAE(+AA 0.5–1.0%) + zinc or aluminum for wet-grab bump

but can raise water uptake if not adequately crosslinked. Formulation pH is maintained at 4.5–5.5 to preserve one-component latency in-can and to activate crosslinking during drying (Tables 11.21 and 11.22).

11.9.6 Role of Acrylic Acid and Crosslinkers in Adhesives vs. Coatings

The addition of acrylic acid (AA) and latent crosslinkers plays different roles in VAE-based adhesives compared with coatings. In adhesives, AA is introduced to provide reactive carboxyl groups that anchor strongly to polar substrates and act as binding sites for latent crosslinkers such as zirconium, aluminum, carbodiimide, or zinc salts. These interactions upgrade simple VAE emulsions to

Table 11.22 Formulation guidance for (VAE + AA) 1K systems.

Element	Guideline
AA level (on total monomer)	0.5–2.0%; target 0.7–1.2% for lamination; 1.0–1.5% for labeling/wood where covalent crosslinking is used
Crosslinker choice	Zirconium/aluminum for fast wet-grab and early set; polycarbodiimide for long-term water/alkali durability; zinc salts for ionic wet-grab boost
Crosslinker dose (on polymer solids)	Zr 0.3–1.0% as metal; Al 0.05–0.3% as Al; CDI 0.3–1.5%; Zn 0.1–0.5% as metal
pH and latency	Maintain 4.5–5.5 in finished goods; add crosslinker at the end under moderate shear; avoid local over-concentration
Surfactant considerations	Minimize free surfactant; avoid phosphate surfactants with Zr/Al; prefer nonionic/low-foaming systems
Stabilization	PVOH-dominated stabilization supports 1K latency and cohesive strength
Validation tests	Accelerated storage at 40 °C for latency; humidity and alkali aging on target substrates; peel and shear per relevant standards

one-component systems with fast wet-grab, enhanced water and alkali resistance, and the durability required for D3 and D4 wood adhesives and for high-speed laminating.

In coatings, AA functions mainly to improve dispersibility, pigment binding, and adhesion to mineral or metallic substrates. At low levels, it increases colloidal stability and reduces dispersant demand. Crosslinkers can be used, but they are generally optional: zirconium or aluminum salts improve scrub resistance and block resistance; polycarbodiimides can add long-term water resistance but are rarely applied because of cost. In coatings, durability is more often secured through binder design, pigment volume concentration, and formulation balance rather than through extensive crosslinking. Excessive AA can increase water uptake and embrittlement, so usage is usually restricted to 0.5–1.5% on monomer (Table 11.23).

Table 11.23 Comparative role of AA and crosslinkers in VAE adhesives and coatings.

Aspect	Adhesives (VAE + AA)	Coatings (VAE + AA)
Primary purpose of AA	Anchorage to polar substrates; reactive sites for crosslinking	Dispersibility, pigment binding, adhesion to mineral/metal substrates
Typical AA level	0.7–2.0% (higher for D3/D4 wood adhesives)	0.5–1.5% (limited to avoid water uptake)
Crosslinkers used	Zr, Al, CDI, Zn (latent 1K postadd)	Zr, Al, Zn optional; CDI rarely used
Crosslinking benefit	Fast wet-grab, water/alkali durability, heat, and humidity resistance	Improved scrub resistance, block resistance, and modest water resistance
Formulation focus	Latency at pH 4.5–5.5; PVOH stabilization; low free surfactant	Binder design, PVC control, coalescents, balanced dispersant package
Limitations	Over-crosslinking reduces shelf stability; phosphate surfactants must be avoided with Zr/Al	Excess AA raises water uptake and brittleness; crosslinkers add cost without major durability gain

11.9.7 Representative 1K Wood Adhesives—(VAE + AA) with Polycarbodiimide (CDI)

These single-component formulations use a VAE latex containing low-level acrylic acid (AA) and a latent polycarbodiimide (CDI) postadd to achieve D3 and D4 performance without two-pack mixing. Latency is preserved at pH 4.5–5.5. Avoid amine-bearing additives and phosphate surfactants. Doses shown are by weight on total formulation; corresponding CDI levels on polymer solids are noted in the footnotes (Table 11.24).

Formulation note: CDI at 0.35% on total corresponds to approximately 0.9% on polymer solids for this formulation (Table 11.25).

Formulation note: CDI at 0.55% on total corresponds to approximately 1.3–1.4% on polymer solids for this formulation.

Procedure: Charge water and begin agitation. Add HEC solution and allow full hydration. Introduce PVOH solution. Add VAE(+AA) latex under gentle mixing. Add wetting aid, defoamer, and preservative. Adjust pH to 4.5–5.0 if needed with dilute acetic acid. Add CDI last under moderate shear; mix 20–30 min to homogeneity. Deaerate as required.

Table 11.24 Representative 1K D3 wood adhesive–VAE(+AA)/CDI (1,000 kg batch).

Ingredient	% by weight	Notes
Deionized water	21.05	Adjust for target viscosity
HEC solution (10% active)	3.00	Rheology; hydrate fully
PVOH solution (10% active)	5.00	Protective colloid; cohesion
VAE(+AA) latex (≈55% solids)	70.00	Low free surfactant
Nonionic wetting aid	0.20	Avoid phosphate types
Defoamer	0.20	Per supplier
Biocide/preservative	0.20	Per supplier
Polycarbodiimide (CDI, 100% active)	0.35	Add last, moderate shear

Table 11.25 Representative 1K D4 wood adhesive–VAE(+AA)/CDI (1,000 kg batch).

Ingredient	% by weight	Notes
Deionized water	19.90	Adjust for target viscosity
HEC solution (10% active)	3.50	Higher rheology for creep control
PVOH solution (10% active)	6.00	Strength; cohesion
VAE(+AA) latex (≈55% solids)	69.50	Low free surfactant
Nonionic wetting aid	0.15	Avoid phosphate types
Defoamer	0.20	Per supplier
Biocide/preservative	0.20	Per supplier
Polycarbodiimide (CDI, 100% active)	0.55	Add last, moderate shear

QC checks: Brookfield viscosity at 25 °C (target 3,000–6,000 mPa s, spindle per plant standard); pH 4.5–5.5; density; appearance; 24 h stability at 40 °C (no gelation or heavy viscosity drift).

D3 verification: Assemble wood joints per EN 204/205. Condition as specified and run D3 soak and humidity exposure. Assess wood failure, shear strength, and creep.

D4 verification: Perform cyclic boil–dry–soak per EN 204/205. Confirm strength retention, minimal creep, and acceptable wood failure across cycles.

Application guidance: Apply at typical spread rates; press at ambient with appropriate pressure. Allow adequate curing time for CDI reaction post-film formation. Avoid amine-containing additives and phosphate surfactants.

11.10 Conclusion

Summary: Across this chapter, VAE has emerged as a pragmatic middle-path binder: softer and more flexible than PVAc at comparable cost, yet simpler and usually lower-VOC than many acrylate options. By tuning the E:VA ratio (total monomer basis), managing protective colloids, and keeping storage and application practice disciplined, VAE delivers robust performance for adhesives and interior coatings, with credible routes to eco-label readiness. When formulated with acrylic acid (AA) at low levels, VAE becomes responsive to latent one-component crosslinkers, enabling upgrades in water resistance, alkali resistance, and durability. This dual pathway—straight VAE for interior coatings, and VAE with AA and crosslinkers for advanced adhesives—underpins its versatility.

Where VAE Excels

Adhesives dominance: E:VA bands 11–18% E (Tables 11.15–11.17) give high peel on films, cold-flex bonding, and low coalescent demand, making them suitable for film-to-paper laminations, labels, window patching, and flexible packaging. Incorporation of 0.5–2.0% AA and latent crosslinkers (zirconium, aluminum, polycarbodiimide, zinc) extends performance to wet-strength and humidity-resistant laminations and to D3/D4 wood adhesives.

Interior coatings with control. Around 9–12% E (Tables 11.12, 11.13, 11.15) supports film formation at modest coalescent demand; with anti-block agents and balanced PVC, interior wall paints are feasible. AA at 0.5–1.5% improves pigment binding and adhesion, while zirconium or aluminum additions can enhance scrub and block resistance in selected formulations.

Cost–performance balance: Resin cost typically sits between PVAc and acrylates (Table 11.11), with savings reinforced by reduced coalescent demand and simple, ammonia-free packages.

Eco-readiness: Low residual monomer via redox or stripping, APEO-free surfactants, ammonia-free neutralization, and careful preservative stewardship align VAE with major indoor-air schemes.

When to consider alternatives: Exterior or UV durability, or harsh alkali. Pure acrylates or VAc–VeoVa (Section 11.7, Table 11.11) are safer for long-term exterior exposure, high-alkalinity mineral substrates, or low-energy plastics without primers.

High heat and block resistance: PVAc (0:100) or lower-E VAEs (5–10% E) offer better blocking and print-stick control; very soft, high-E grades trade away heat resistance.

Practical Operating Guidance: pH and storage: Keep storage pH 4.8–5.0; neutralize only at the point of use (Table 11.12). For one-component crosslinkable systems, maintain 4.5–5.5 to preserve latency.

Compatibility: Use the compatibility matrix (Table 11.13) to select fillers and additives; stage the process as follows: add dispersant, build the grind, introduce the binder, adjust rheology, then add coalescent, wetting agent, and defoamer; add the crosslinker last.

Surface energy: For PE, PP, BOPP, and many printed films, raise surface energy (Table 11.14) by corona or plasma treatment or use a primer; verify with dyne pens and peel tests.

Composition choice: Use Section 11.8.1 maps (Tables 11.15–11.17) to set E:VAc by end-use domain (adhesives dominant or coatings viable), then confirm Tg, MFFT, and blocking in pilot batches.

Cost–eco perspective: Lower coalescent demand at mid-E reduces VOC and SVOC burden and odor while keeping application rheology friendly. Semibatch VAE at 65 °C with 30–35 bar ethylene (11.2A) demands good mass-transfer and relief design, but reagent lists remain simple and scalable. Addition of AA and factory post-addition of crosslinkers do not alter the core eco-profile if residual monomer is controlled and surfactants are carefully selected.

Implementation Checklist

1) *Choose the E*: VA band with Table 11.17 (domain guide).
2) Fix storage pH at 4.8–5.0 (adhesives without crosslinker) or 4.5–5.5 (with AA and latent crosslinker).
3) Plan biocide (wet and dry-film as needed).
4) Build the grind at low pH with the right dispersant; hydrate HEC fully.
5) Tune anti-block (wax, silica, talc) and PVC for coatings; specify peel and creep tests for adhesives.
6) Add crosslinker at the end of the batch under moderate shear (avoid phosphate surfactants with zirconium or aluminum).
7) *Prepare surfaces*: corona or primer for films, and control moisture for paper and carton.
8) Finish and strip to residual-monomer specification; document odor checks.
9) *Pilot and age*: heat-age viscosity, freeze–thaw cycles, peel vs. dwell, and blocking and print-stick.

Industry relevance: VAE continues to gain share where flexibility, cost control, and indoor-air compliance matter: packaging laminations, converting adhesives, and interior architectural systems. With careful formulation—especially judicious AA addition and latent crosslinker choice in adhesives, or balanced E:VA design in coatings, VAE provides a durable, eco-forward platform that complements PVAc's economy and acrylates' exterior pedigree.

12

Special Binders and Emulsion Paints for Extremely Cold Climates

12.1 Introduction

In some of the world's most unforgiving environments—regions such as Siberia, Alaska, Kazakhstan, northern Canada, and Finland—emulsion paints must perform under extreme conditions that are fundamentally different from temperate climates. These subzero regions experience prolonged winters, frequent freeze–thaw cycles, and extremely short painting seasons. Paint formulations designed for these zones must exhibit exceptionally low-temperature film formation, freeze–thaw stability, long-term flexibility, and resistance to cracking or delamination.

Conventional binders, particularly PVAc homopolymers with higher minimum film-forming temperatures (MFFT), often fail to form continuous films or become brittle when exposed to freezing conditions. This results in poor adhesion, cracking, or early coating failure. To address these limitations, specialized binder systems are required—polymers engineered to maintain flexibility and integrity at subzero temperatures while ensuring good coalescence and surface protection.

Among these technologies, vinyl acetate–ethylene (VAE) copolymers play a pivotal role as a special binder for cold climates. Incorporating ethylene segments into the vinyl acetate backbone significantly lowers the glass transition temperature (Tg) and MFFT compared to PVAc homopolymers, resulting in superior flexibility and crack resistance under severe cold. VAE binders also demonstrate excellent freeze–thaw stability and are well-suited for interior and exterior coatings where elastic recovery and durability are critical. When combined with appropriate coalescents and antifreeze agents, VAE systems deliver reliable performance in Arctic or sub-Arctic conditions without sacrificing workability or film integrity.

This chapter presents the key binder technologies, practical formulation strategies, and real-world applications of emulsion paints tailored for cold climates.

The discussion emphasizes low-MFFT polymers such as VAE copolymers, modified styrene-acrylics, and vinyl-acrylic systems augmented with coalescents and freeze–thaw stabilizers. Special attention is given to binder elasticity at low temperatures, prevention of micro-cracking, and the long-term protection of exterior surfaces exposed to severe weather.

Whether the requirement is for painting residential interiors during winter or protecting exterior masonry under Arctic conditions, this chapter provides a practical guide to producing robust and reliable coatings based on waterborne binders optimized for cold regions (Table 12.1).

Cold-climate suitability depends on achieving film formation and crack-free flexibility at or below the lowest service temperature. Reported MFFT values are indicative for coalescent-free latexes (ASTM D2354) and for practical formulations containing ~2–3% coalescent (e.g., Texanol®, DPnB). Actual values vary with molecular weight, comonomer ratio, surfactant/protective colloid package, and plasticizer level. For subzero applications without heated curing, soft pure acrylics and properly designed VA-acrylic or VAE systems are typically preferred; VAM–VeoVa grades offer additional hydrophobicity and dirt-pickup resistance

Table 12.1 Comparison of common binders for cold-climate coatings.

Binder type	Typical Tg (°C)	MFFT (°C)– 0% coalescent	MFFT (°C)– with 2–3% coalescent	Flexibility at subzero	Freeze–thaw stability
PVAc homopolymer	+28 to +32	18–25	10–14	Poor (brittle below 0 °C)	Low
VAE copolymer (15–28% ethylene)	−5 to +8	2–8	≤0–3	Good (elastic below freezing)	High
Pure acrylic (soft grade)	−25 to +5	−2 to +5	−10 to 0	Excellent	Very high
Vinyl acrylate copolymer (VAc/EA or VAc/BA)	−10 to +10	2–10	−2 to +4	Very good	High
VAM–VeoVa copolymer	0 to +15	8–15	2–8	Good (flexible, hydrophobic)	High
Styrene–acrylic (low-Tg grade)	−10 to +20	5–15	0–8	Moderate to good (composition-dependent)	High to very high

but often require modest coalescent to form film near 0 °C. PVAc homopolymers are generally unsuitable outdoors in persistent freezing conditions due to high intrinsic MFFT and low freeze–thaw stability.

12.2 Key Climatic Challenges

Paint systems used in extremely cold climates face a unique set of environmental challenges that influence both application and long-term performance. Understanding these conditions is essential for selecting appropriate binders and additives.

12.2.1 Persistent Subzero Temperatures

In regions such as Siberia and Alaska, ambient temperatures may remain below 0 °C for several consecutive months. This drastically limits the window for outdoor painting and complicates drying and film formation.

12.2.2 Freeze–Thaw Cycles

Surfaces, especially exterior walls, undergo daily and seasonal freeze–thaw cycles. Water within the paint film or substrate can freeze and expand, potentially causing cracking, delamination, or adhesion loss if the film is not sufficiently flexible.

12.2.3 Limited Painting Season

The painting season is often restricted to a few summer weeks where temperatures are suitable for application. Formulations must be engineered to dry and form a continuous film quickly during this narrow time frame.

12.2.4 High Indoor Humidity and Condensation

In colder climates, homes and buildings often suffer from indoor humidity accumulation, especially in poorly ventilated structures. This can lead to moisture build-up on painted walls and ceilings, necessitating excellent water resistance and antifungal properties.

12.2.5 Substrate Temperature Variability

In early spring or late fall, substrates like concrete or wood may remain significantly colder than the air, affecting the binder's coalescence. Binders must form films at the actual surface temperature, not just ambient air temperature.

These challenges dictate the need for specialized binders with excellent low-temperature performance, rapid film formation, and enhanced flexibility to withstand thermal stress.

12.3 Requirements for Emulsion Binders in Cold Regions

The selection of an appropriate emulsion binder for paints used in cold regions must be based on several key technical properties. These properties ensure reliable film formation, durability, and performance under subzero conditions.

12.3.1 Low Minimum Film-Forming Temperature (MFFT)

Binders should have an MFFT below 5 °C, and preferably as low as 0–2 °C. This ensures that the latex particles coalesce and form a continuous film even in cool environments.

12.3.2 High Freeze–Thaw Stability

Binders must withstand multiple freeze–thaw cycles during storage and application without coagulation or phase separation. Stabilizers such as ethylene glycol or proprietary agents like FT-100 may be incorporated.

12.3.3 Flexibility and Crack Resistance

The paint film must remain flexible at low temperatures to resist cracking caused by substrate expansion and contraction. This is especially crucial for exterior surfaces and elastomeric coatings.

12.3.4 Adhesion on Damp or Cold Surfaces

Binders must exhibit strong adhesion to substrates that may be slightly damp or cold during application. This is vital for both new construction and maintenance coatings in transitional weather.

12.3.5 Coalescent Efficiency

The binder should be compatible with efficient coalescents (e.g., Texanol, DPnB) that promote film formation without increasing VOC content excessively. Coalescents must remain effective at low ambient temperatures.

12.3.6 Water and Alkali Resistance

Exterior paints must resist water absorption and alkali attack from cement-based substrates. Binders should form dense, closed films to reduce permeability.

12.3.7 Compatibility with Freeze–Thaw Additives

The binder should tolerate the inclusion of antifreeze agents or stabilizers without destabilization or gelling.

Meeting these requirements ensures that the emulsion binder enables reliable application and long-term protection, even in some of the most climatically challenging environments on Earth.

12.4 Special Binders: Composition and Design

Designing binders for extreme cold requires careful selection and balance of monomers, plasticizing components, and stabilizing additives. Below are the key binder technologies suited for low-temperature paints:

12.4.1 Vinyl Acetate–Ethylene (VAE) Copolymers

These copolymers offer low MFFT due to the ethylene component, enhancing flexibility and coalescence at near-zero temperatures. They are particularly effective in interior paints and flexible exterior coatings.

12.4.2 Modified Styrene–Acrylic Copolymers

By adjusting the ratio of hard (styrene) and soft (butyl acrylate or 2-ethylhexyl acrylate) monomers, low-MFFT styrene–acrylics can be achieved. These systems also offer good water resistance and color retention.

12.4.3 Vinyl Acrylics with Coalescent Aids

Vinyl acrylic binders often have a relatively high MFFT, but when paired with efficient coalescents such as Texanol or DPnB, they become suitable for moderate cold conditions. These systems are cost-effective for indoor paints.

12.4.4 Internal Plasticization

Using soft monomers (e.g., ethylhexyl acrylate) or internal plasticizers within the polymer matrix provides permanent flexibility. This reduces dependence on volatile coalescents and improves low-temperature performance.

12.4.5 Surfactant System Optimization

Using polymerizable or nonionic surfactants improves freeze–thaw stability and prevents destabilization. Narrow-range ethoxylates or phosphate esters may also enhance mechanical stability.

12.4.6 Reinforcement via Crosslinking

Incorporating low levels of crosslinkers can enhance the durability of cold-climate binders without inducing brittleness. Carbodiimides (CDI) act as external crosslinkers by reacting with carboxyl groups, most effectively in copolymers that contain acrylic acid (AA). Their addition (typically 1–3% on polymer solids) improves water resistance, scrub durability, and dirt-pickup resistance while maintaining flexibility. In contrast, glycidyl methacrylate (GMA) functions as an internal self-crosslinking monomer, introduced during polymerization at low levels (1–5%). Its epoxy groups form covalent bonds with carboxyl or hydroxyl functionalities, thereby enhancing weathering and chemical resistance. Unlike *N*-methylolacrylamide (NMA), which is avoided due to formaldehyde release, CDI and GMA offer safer and more practical reinforcement strategies. The correct design of these binders allows for the production of coatings that maintain film integrity, color stability, and elasticity even when applied or cured in challenging cold environments.

12.5 Representative Paint Formulations

This section provides example formulations for waterborne emulsion paints designed for cold climates. Each formulation includes suggested materials, mixing steps, and performance characteristics. The formulations presented here were developed to address severe weathering conditions typical of cold climates, including freeze–thaw cycling, high humidity, and prolonged exposure to snow or ice. These conditions demand systems with relatively high binder and TiO_2 contents to ensure:

1) Excellent film integrity and crack resistance at low temperatures
2) High opacity and color retention in diffuse winter light
3) Improved water resistance and surface adhesion under freeze–thaw stress

These formulations are intended as performance benchmarks or starting-point recipes suitable for regions such as Central Asia, Northern Europe, northern Canada, and mountainous areas. They do not represent minimum-cost options but rather robust technical solutions that can be scaled or optimized. Adjustments in binder level, pigment volume concentration (PVC), and extender content may

Table 12.2 Interior wall paint for subzero climates.

Ingredient	Amount (by weight %)
Water	25.0
Dispersant (e.g., Tamol 731)	0.5
Wetting agent (e.g., Dynol 604)	0.3
Defoamer (e.g., BYK-024)	0.2
Titanium dioxide (rutile)	18.0
Calcium carbonate (fine)	10.0
Talc or mica	7.0
Binder (VAE latex, MFFT <2 °C)	32.0
Coalescent (e.g., Texanol)	1.0
Thickener (HEUR, KU 95–100)	1.5
Biocide and preservative	0.5
Freeze–thaw additive (FT-100)	1.0
pH adjuster (ammonia or AMP)	To pH 8–8.5

be made by R&D or technical service teams to balance local cost-performance requirements while maintaining durability and aesthetics.

A representative formulation for an interior wall paint designed for subzero climates is shown in Table 12.2.

12.5.1 Formulation—Interior Wall Paint for Subzero Climates and Mixing Procedure

Mixing Procedure

1) Disperse pigments and fillers in the water phase with dispersant, wetting agent, and defoamer.
2) Slowly add the binder under moderate agitation.
3) Adjust viscosity and pH.
4) Add freeze–thaw stabilizer and coalescent.
5) Homogenize and package.

Key Properties

1) Good freeze–thaw stability (5+ cycles)
2) Film formation at 2–5 °C
3) Matte finish, excellent adhesion on plaster and gypsum

12.6 Special Binder Formulations for Cold-Climate Paints

This section presents emulsion binder recipes engineered for low-temperature film formation, robust freeze–thaw performance, and long-term elasticity. Each entry summarizes target MFFT, polymer type, process route, formulation, polymerization procedure, and key features.

12.6.1 Formulation—Vinyl-Ethylene Copolymer (VAE) Binder-Low MFFT, Flexible (Interior/Exterior Applications)

Target MFFT: 0–2 °C (Table 12.3)

Polymerization Procedure

1) *Charge and deoxygenate*: Charge deionized water, 10% PVOH (protective colloid), surfactant, and sodium bicarbonate. Deoxygenate by nitrogen sparge/blanket for ~30 min.
2) *Seed at 65–70 °C*: Heat to 65–70 °C. Seed with ~5% vinyl acetate and a small portion of the APS/SFS initiator to form seed particles.
3) *Feeds (3–4 h)*: Meter the remaining vinyl acetate and initiator over 3–4 h while maintaining ethylene at 5–10 bar and adequate agitation. Maintain nitrogen blanket.
4) *Conversion and hold*: Maintain 70–75 °C until feeds complete, then hold 30–60 min to minimize residual monomer.
5) *Cool and finish*: Cool to ≤30 °C, adjust pH as required, and filter to remove any coagulum before packaging.

Table 12.3 VAE binder using semicontinuous emulsion polymerization process (pre-emulsified feeds).

Component	% by weight
Vinyl acetate (VAc)	45.0
Ethylene (gas, absorbed)	~8.0
Deionized water	35.0
Nonionic surfactant (NR EO)	1.2
Protective colloid (PVOH 10%)	5.0
Initiator (APS + SFS)	0.4
pH regulator ($NaHCO_3$)	0.2
Chain transfer agent (IPA)	0.1
Defoamer	0.1
Residuals/adjustment	5.0

Key Features

1) Film formation near 0–2 °C (in paint with standard coalescent levels).
2) High flexibility for crack resistance on exterior substrates.
3) Freeze–thaw stability ≥5 cycles with appropriate stabilizer package.
4) Suitable as a base binder for both interior and exterior systems.

12.6.2 Formulation—Styrene-Acrylic Binder (Modified for Exterior Use)

Target MFFT: 3–5 °C (Table 12.4)

Post-additives

1) Coalescent (e.g., Texanol or DPnB): 2.0%
2) Freeze–thaw stabilizer (e.g., FT-100): 1.0%
3) pH adjusted to 8.0 using AMP-95

Polymerization Procedure

1) *Initial charge*: Charge deionized water, anionic and nonionic surfactants, and buffer. Purge with nitrogen.
2) *Pre-emulsion*: Prepare a pre-emulsion of styrene, butyl acrylate, methyl methacrylate, and acrylic acid with remaining surfactants and water.
3) *Seeding*: Initiate at 75 °C by feeding a small amount of pre-emulsion with a portion of APS to nucleate the seed.

Table 12.4 Modified styrene-acrylic binder formulation using semicontinuous emulsion polymerization.

Component	% by weight
Styrene	22.0
Butyl acrylate	38.0
Methyl methacrylate	4.0
Acrylic acid	1.5
Deionized water	28.0
Nonionic surfactant (NR EO)	1.0
Anionic surfactant (e.g., SDS)	0.8
Initiator (APS, added gradually)	0.4
Buffer (e.g., $NaHCO_3$)	0.2
Chain transfer agent (optional)	0.2
Defoamer	0.1
Residuals/adjustment	3.8

4) *Feeds (≈3 h)*: Co-feed the pre-emulsion and initiator solution over ~3 h at 75–78 °C under steady agitation and nitrogen blanket.
5) *Postreaction hold*: Maintain temperature for 30–60 min to complete conversion and reduce residual monomer.
6) *Cooling and finishing*: Cool to ≤30 °C. Post-add coalescent and freeze–thaw stabilizer; adjust pH as needed. Filter if required.

Key Features

1) Good low-temperature flexibility and strong adhesion to masonry and concrete.
2) Moderate MFFT with improved weathering and UV resistance from the styrene/MMA balance.
3) Suitable for elastomeric exterior paints exposed to freeze–thaw cycling.

12.6.3 Formulation–Vinyl-Acrylic Binder with Internal Plasticizer

A representative formulation for a vinyl-acrylic binder incorporating an internal plasticizer is shown in Table 12.5.

Post-add: Texanol: 1.5%, and FT-100: 1.0%, and pH adjusted to ~8.0.

Polymerization Procedure

1) *Charge and deoxygenate*: Charge water, PVOH solution, nonionic surfactant, sodium bicarbonate, and defoamer. Nitrogen sparge/blanket for ~30 min.
2) *Pre-emulsion*: Prepare a pre-emulsion of vinyl acetate and 2-ethylhexyl acrylate with a portion of water and surfactant.

Table 12.5 Vinyl–acrylic binder formulation (internal plasticizer).

Component	% by weight
Vinyl acetate	40.0
2-Ethylhexyl acrylate	25.0
Methacrylic acid	1.2
Water	28.5
PVOH (10% solution)	2.5
Nonionic surfactant (NR EO)	0.8
APS + Fe^{2+} initiator	0.4
Sodium bicarbonate	0.2
Chain transfer agent	0.1
Defoamer	0.1
Residuals/adjustment	1.2

3) *Seeding (65–70 °C)*: Heat reactor to 65–70 °C. Add a small portion of pre-emulsion and a portion of the APS solution to nucleate seed particles; maintain gentle agitation.
4) *Feeds (3–4 h)*: Co-feed the remaining pre-emulsion and initiator solution over 3–4 h, maintaining 70–75 °C and nitrogen blanket. Meter Fe^{2+} activator to support redox initiation as specified.
5) *Hold and conversion*: After feeds, hold 30–60 min at the temperature to reduce residual monomer. Maintain pH with the bicarbonate buffer (≈4.8–5.2 during polymerization).
6) *Cool and finish*: Cool to ≤30 °C, neutralize to ~8.0, and post-add Texanol and FT-100. Filter if needed prior to packaging.

Key Features

1) Good elasticity and processability driven by internal plasticization from 2-ethylhexyl acrylate.
2) Cost-effective for interior paints and cold-weather primers.
3) Suitable for both manual and automated production systems.

12.6.4 Formulation—VAc–BA–VeoVa 10 Binder for Cold-Resistant Masonry Applications

Target MFFT: 2–4 °C (in paint with ~1.5–2.0% coalescent; Table 12.6)

Table 12.6 VAc–BA–VeoVa 10 binder formulation.

Component	% by weight
Vinyl acetate (VAc)	48.0
Butyl acrylate (BA)	20.0
VeoVa 10	8.0
Deionized water	18.5
Nonionic surfactant (NR EO)	1.0
Anionic surfactant	0.6
Initiator (APS/SFS)	0.4
Sodium bicarbonate	0.2
Defoamer	0.1
Residuals/adjustment	3.2

Procedure

1) *Initial charge*: Charge deionized water, nonionic/anionic surfactants, and sodium bicarbonate. Purge with nitrogen for ~30 min to remove oxygen.
2) *Pre-emulsion*: Prepare a pre-emulsion of VAc, BA, and VeoVa 10 with the remaining water and surfactants.
3) *Seeding (≈70 °C)*: Heat reactor to ~70 °C. Feed ~5% of the pre-emulsion with a small portion of the APS/SFS initiator to nucleate seed particles.
4) *Feeds (3–4 h)*: Co-feed the remaining pre-emulsion and initiator solution over 3–4 h at 70–75 °C under nitrogen blanket and steady agitation.
5) *Posthold*: Maintain temperature for 30–45 min to complete conversion and reduce residual monomer.
6) *Cool and finish*: Cool to ≤30 °C, adjust pH as needed, and filter if required prior to packaging.

Post-additives: Coalescent (e.g., Texanol): 1.5%, and freeze–thaw stabilizer (e.g., FT-100), 1.0% and pH adjusted to ~8.0 using AMP-95 or ammonia

Key Features

1) Excellent flexibility and freeze–thaw resistance for masonry/concrete.
2) High water and alkali resistance; suitable for thermal cycling and high humidity.
3) Field-validated performance on exterior walls in cold regions (e.g., Kazakhstan).

A representative formulation for an acrylic self-crosslinking copolymer based on EA/BA/MMA/GMA/AA is shown in Table 12.7.

Table 12.7 Acrylic (EA/BA/MMA-glycidel methacrylate (GMA)–AA) self-crosslinking copolymer.

Component	% by weight
Ethyl acrylate (EA)	36.0
Butyl acrylate (BA)	18.0
Methyl methacrylate (MMA)	18.0
Glycidyl methacrylate (GMA)	2.0
Acrylic acid (AA)	1.0
Deionized water	20.3
Nonionic surfactant (NR EO)	1.2
Anionic surfactant	0.6
Initiator (APS/SFS)	0.5
Sodium bicarbonate	0.2
Defoamer	0.1
Residuals/adjustment	2.1

Post-additives: Coalescent (e.g., Texanol): 1.5%, freeze–thaw stabilizer (e.g., FT-100): 1.0%, pH adjusted to 7.8–8.4 using AMP-95 or ammonia, and use low-amine auxiliaries to minimize premature epoxy side reactions

Key Features

1) Built-in crosslinking via epoxy–carboxyl/hydroxyl reactions after film formation
2) Low-temperature film formation with modest coalescent; high scrub/weather resistance
3) No formaldehyde-releasing chemistry; good storage stability when amines are minimized

Suitable Applications: Table 12.8 represents the suitable applications of acrylic GMA internal crosslinker with dual function due to the methacrylic and epoxy groups.

Table 12.8 Suitable applications and rationale—acrylic (GMA internal) binder.

Application	Why this binder	Key notes
Exterior wall topcoats (masonry, concrete, stucco, fiber-cement)	Low-Tg acrylate backbone forms film near 0–2 °C with modest coalescent; internal GMA crosslinking improves weathering, scrub, and dirt-pickup resistance.	Target matte/eggshell; ensure coalescent balance for cold starts; add silane/siloxane for enhanced water beading if needed.
Anticarbonation façade coatings (moderate elongation class)	Flexible EA/BA phase retains elasticity at low temp; GMA crosslinks boost CO_2 and water resistance without embrittlement.	Not for high-elongation (>200%) crack-bridging classes—use elastomeric systems instead.
Cold-weather exterior primers on mineral substrates	Good adhesion and early film integrity at low temperatures; GMA/AA chemistry enhances alkali and scrub resistance.	Consider adhesion promoters for dense substrates; verify with pull-off (ASTM D4541) and alkali resistance tests.
High-scrub interior topcoats in cold-applied jobs	Internal crosslinking delivers block/scrub resistance and stain holdout while maintaining application at low ambient temps.	Indoor temperatures are usually higher than 0 °C, but this binder helps when heating is limited during winter worksites.
Traffic-exposed verticals (stairwells, corridors)	Enhanced mar/scrub resistance from GMA crosslinking; good dirt-pickup resistance in soiled environments.	Avoid very glossy finishes if dirt-pickup is critical; optimize PVC and coalescent for block resistance.

Limitations: Not recommended for high-elongation elastomeric roof or bridge-deck membranes requiring crack-bridging beyond ~100–150%. For direct-to-metal or plastic substrates, dedicated adhesion promoters and corrosion/UV packages are required; consider alternative binders.

12.6.5 Formulation–VAc-BA-AA Copolymer Designed for External CDI Crosslinking

A representative formulation for a VAc-BA-AA copolymer designed for external CDI crosslinking is shown in Table 12.9.

Post-additives/Usage Notes

1) Coalescent (e.g., Texanol): 1.5%.
2) Freeze–thaw stabilizer (e.g., FT-100): 1.0%.
3) pH adjusted to ~8.0 using AMP-95 or ammonia.
4) Recommended external crosslinker at paint stage: Polycarbodiimide dispersion 1.0–2.0% on polymer solids, added last under gentle agitation; avoid high-amine additives.

Polymerization Procedure

1) *Charge and deoxygenate*: Charge deionized water, nonionic/anionic surfactants, sodium bicarbonate, and defoamer. Nitrogen sparge/blanket for ~30 min.
2) *Pre-emulsion*: Prepare a pre-emulsion of VAc, BA, and AA with a portion of water and surfactant.
3) *Seeding (65–70 °C)*: Heat reactor to 65–70 °C. Feed ~5% of the pre-emulsion with a portion of APS/SFS initiator to form seed particles.

Table 12.9 VAc-BA-AA copolymer (for CDI external crosslinking).

Component	% by weight
Vinyl acetate (VAc)	52.0
Butyl acrylate (BA)	22.0
Acrylic acid (AA)	1.5
Deionized water	18.5
Nonionic surfactant (NR EO)	1.0
Anionic surfactant	0.6
Initiator (APS/SFS)	0.4
Sodium bicarbonate	0.2
Defoamer	0.1
Residuals/adjustment	3.7

4) *Feeds (3–4 h at 70–75 °C)*: Co-feed the remaining pre-emulsion and initiator solution over 3–4 h under nitrogen. Maintain pH in the mildly acidic range (≈4.8–5.2) during polymerization.
5) *Hold and conversion*: After feeds, hold 30–60 min to complete conversion and minimize residual monomer.
6) *Cool and finish*: Cool to ≤30 °C and neutralize to pH ~8.0. Filter if required. (CDI is not added to the binder kettle; add only at the paint stage.)

Suitable applications for VAc–BA–AA (for CDI) are described in Table 12.10.

Typical applications of VAc-BA-AA binders crosslinked with carbodiimide (CDI), along with the underlying performance rationale and key formulation considerations, are summarized in Table 12.10.

Table 12.10 Suitable applications and rationale—VAc–BA–AA (CDI external).

Application	Why this binder (AA + CDI)	Key notes
Exterior masonry/concrete primers (cold-weather)	AA provides carboxyl sites for CDI, improving water resistance and adhesion to alkaline substrates while maintaining low-temperature flexibility.	Verify alkali resistance; pull-off per ASTM D4541. Allow 3–7 days for full CDI cure in cold, humid conditions.
Exterior wall topcoats (matte/eggshell)	CDI crosslinking reduces dirt pickup and enhances scrub while BA provides elasticity at low temperatures.	Balance coalescent for block resistance. Consider silane/siloxane additive for water beading if needed.
Cold-climate elastomeric-lean façade coatings (moderate elongation)	AA/BA balance gives flexibility; CDI adds durability without embrittlement typical of hard crosslinkers.	Not for very high crack-bridging classes (>150–200%); use dedicated elastomeric systems there.
Cold-weather primers for fiber-cement and plaster	Improved adhesion and early block resistance from CDI chemistry; good film formation near 0–2 °C with modest coalescent.	Surface must be clean and dry; check moisture content of substrate in winter conditions.

Key Features VAc–BA–AA (for CDI)

1) AA provides carboxyl functionality for CDI reaction, enhancing water/scrub resistance and reducing dirt pickup.
2) Retains flexibility at low temperatures; suitable for cold-climate exterior and masonry systems.
3) CDI is not part of the binder; it is added at the paint stage to preserve pot-life and storage stability.

12.7 Representative Paint Formulations

This section presents practical paint formulations based on the cold-weather binders described in Section 12.6. These examples demonstrate how binder selection influences the paint's flexibility, freeze–thaw stability, water resistance, and low-temperature application performance.

12.7.1 Formulation—Interior Wall Paint for Subzero Climates

A representative formulation of an interior wall paint designed for subzero climate application is presented in Table 12.11.

Key Properties

1) Freeze–thaw stability: ≥ 5 cycles
2) Film formation at 2–5 °C

Table 12.11 Interior wall paint for subzero climates.

Ingredient	% by weight
Water	25.0
Dispersant (e.g., Tamol 731)	0.5
Wetting agent (e.g., Dynol 604)	0.3
Defoamer (e.g., BYK-024)	0.2
Titanium dioxide (rutile)	18.0
Calcium carbonate (fine)	10.0
Talc or mica	7.0
VAE latex (MFFT <2 °C)	32.0
Coalescent (Texanol)	1.0
Thickener (HEUR, KU 95–100)	1.5
Biocide/preservative	0.5
Freeze–thaw stabilizer (FT-100)	1.0
pH adjuster (AMP-95 or ammonia)	To pH 8.0–8.5

Table 12.12 Exterior masonry paint (styrene–acrylic system).

Ingredient	% by weight
Water	22.0
Dispersant (e.g., Orotan 731A)	0.6
Wetting agent (e.g., Surfynol 104)	0.3
Defoamer	0.2
Titanium dioxide	15.0
Calcium carbonate (coarse/fine blend)	20.0
Talc	5.0
Soft, low-Tg styrene-acrylic binder (exterior-grade, elastomeric type)	30.0
Coalescent (Texanol/DPnB)	2.0
Thickener (HEUR + ASE)	1.8
Silicone additive	0.3
Biocide/fungicide	0.6
Freeze–thaw stabilizer (FT-100)	1.2
pH adjuster	To pH 8.5

3) Smooth matte finish
4) Excellent adhesion on gypsum and plaster

12.7.2 Formulation—Exterior Masonry Paint (High Flexibility)

Key Properties

1) Excellent crack-bridging on cold concrete
2) Weather-resistant with low water uptake
3) Resistant to efflorescence and alkali degradation

A representative formulation of an exterior masonry paint based on a styrene-acrylic binder system is presented in Table 12.12.

12.7.3 Formulation—Exterior Masonry Paint (VAM/BA/VeoVa Binder)

A representative formulation of an exterior masonry paint based on a VAM/BA/VeoVa binder system is presented in Table 12.13.

Key Properties

1) Exceptional adhesion on mineral substrates
2) High alkali and water resistance
3) Fast drying under cold, humid conditions

Table 12.13 Exterior masonry paint (VAM/BA/VeoVa system).

Ingredient	% by weight
Water	20.0
Dispersant	0.4
Defoamer	0.2
Filler (e.g., $CaCO_3$)	12.0
VAM/BA/VeoVa binder	40.0
Coalescent (Texanol)	1.0
Freeze–thaw stabilizer (FT-100)	1.0
Thickener (low viscosity HEC)	1.2
Antifoam and preservatives	0.5
pH adjuster	To pH 8.0

12.7.4 Formulation—Budget Indoor Paint (Cold Application)

A representative formulation of a budget indoor paint suitable for cold-climate application is presented in Table 12.14.

Key Properties

1) Cost-effective and stable at low temperatures
2) Easy to apply by brush or roller
3) Recommended for indoor renovation in unheated buildings

Table 12.14 Budget indoor paint (cold climates).

Ingredient	% by weight
Water	26.0
Dispersant	0.5
Calcium carbonate	25.0
Binder (vinyl–acrylic)	32.0
Coalescent	1.2
Thickener (HEUR)	1.5
Biocide	0.5
Defoamer	0.2
Freeze–thaw stabilizer (FT-100)	1.0
pH adjuster	To pH 8.0

Remark: This formulation is suitable for application in unheated buildings because it combines a cold-tolerant vinyl-acrylic binder, coalescent-assisted film formation, and freeze-thaw stabilization, allowing acceptable application and film development at low ambient temperatures.

12.7.5 Formulation—Economical Wall Primer for Cold Climates

A representative formulation of an economical wall primer designed for cold-climate application is presented in Table 12.15.

Key Properties

1) Low-cost and cold-weather compatible
2) Good adhesion and coverage on untreated walls
3) Film formation at ~5 °C with added coalescent
4) Stable for winter storage and transport

Remark: This formulation is suitable for winter transport and storage because the vinyl-acrylic binder system is stabilized with a freeze-thaw additive, which helps maintain dispersion stability and viscosity during low-temperature storage and transport conditions.

Table 12.15 Economical wall primer for cold climates.

Ingredient	% by weight
Deionized water	26.0
Dispersant (e.g., Orotan 731A)	0.4
Wetting agent (e.g., Dynol 604)	0.3
Defoamer (e.g., BYK-024)	0.2
Calcium carbonate (fine)	10.0
Vinyl–Acrylic Binder (see Table 12.5)	48.0
Coalescent (Texanol)	1.5
Freeze–thaw stabilizer (FT-100)	1.0
Thickener (low viscosity HEC)	1.3
Biocide and preservative	0.5
pH adjuster (AMP-95)	To pH 8.0
Residual water	Balance

12.7.6 Formulation—Economical Interior Emulsion Paint for Cold Climates

A representative formulation of an economical interior emulsion paint suitable for cold-climate application is presented in Table 12.16.

Key Properties

1) Matte white finish for interior walls and ceilings
2) Low-cost formulation with adequate scrub and hiding power
3) Forms film at 5 °C with good workability
4) Can be applied by brush or roller without sagging

12.7.7 Formulation—Eggshell Interior Emulsion Paint for Cold Climates

An eggshell-finish interior emulsion paint formulation suitable for cold-climate application is presented in Table 12.17.

Key Properties

1) Soft sheen eggshell finish
2) Scrub-resistant and easy to clean

Table 12.16 Economical interior emulsion paint for cold climates.

Ingredient	% by weight
Deionized water	24.0
Dispersant (e.g., Tamol 850)	0.5
Wetting agent (e.g., Dynol 604)	0.3
Defoamer	0.2
Titanium dioxide	10.0
Calcium carbonate (fine)	25.0
Talc or mica	5.0
Vinyl–Acrylic Binder (see Table 12.5)	28.0
Coalescent (Texanol)	1.5
Thickener (HEUR or HEC)	1.6
Biocide and preservative	0.5
Freeze–thaw stabilizer (FT-100)	1.0
pH adjuster (AMP-95 or ammonia)	To pH 8.0

Table 12.17 Eggshell interior emulsion paint for cold climates.

Ingredient	% by weight
Deionized water	23.0
Dispersant (e.g., Tamol 1124)	0.5
Wetting agent (e.g., Surfynol 104)	0.3
Defoamer	0.2
Titanium dioxide (rutile)	16.0
Calcined clay or fine talc	6.0
Calcium carbonate (fine)	10.0
Styrene–acrylic binder (see Table 12.4)	30.0
Coalescent (Texanol)	2.0
Thickener (HEUR for eggshell gloss)	1.7
Silicone emulsion (anti-mar additive)	0.3
Biocide and preservative	0.5
Freeze–thaw stabilizer (FT-100)	1.0
pH adjuster (AMP-95 or ammonia)	To pH 8.0

3) Good stain resistance for kitchens and corridors
4) Forms film at 5 °C, stable for winter storage

12.7.8 Formulation—Vinyl Satin Finish Interior Paint for Cold Climates

A vinyl satin-finish interior paint formulation suitable for cold-climate application is presented in Table 12.18.

Key Properties

1) Smooth satin finish with subtle gloss
2) Cold-weather applicable down to 5 °C
3) Good flow, leveling, and surface washability
4) Suitable for high-use areas: corridors, living rooms, stairwells

12.7.9 Formulation—Textured Emulsion Paint for Cold Climates

A representative formulation of a textured emulsion paint suitable for cold-climate application is presented in Table 12.19.

Table 12.18 Vinyl satin finish interior paint for cold climates.

Ingredient	% by weight
Deionized water	22.0
Dispersant (e.g., Tamol 1124)	0.5
Wetting agent (e.g., Surfynol 104)	0.3
Defoamer	0.2
Titanium dioxide (rutile)	14.0
Calcined clay	6.0
Calcium carbonate (fine)	8.0
Vinyl–Acrylic Binder (see Table 12.5)	33.0
Coalescent (Texanol)	2.0
Thickener (HEUR with satin control)	1.8
Silicone wax emulsion (slip/gloss)	0.3
Biocide and preservative	0.5
Freeze–thaw stabilizer (FT-100)	1.0
pH adjuster (AMP-95 or ammonia)	To pH 8.0

Table 12.19 Textured emulsion paint for cold climates.

Ingredient	% by weight
Deionized water	20.0
Dispersant (e.g., Tamol 1124)	0.5
Wetting agent (e.g., Surfynol 104)	0.3
Defoamer	0.3
Titanium dioxide (rutile)	8.0
Coarse calcium carbonate (D50 ~ 30 μm)	18.0
Mica or talc	6.0
Marble powder or silica sand (fine)	10.0
Vinyl–Acrylic Binder (see Table 12.5)	30.0
Coalescent (Texanol)	2.0
Thickener (HEUR, high viscosity)	2.0
Freeze–thaw stabilizer (FT-100)	1.0
Biocide and preservative	0.5
pH adjuster (AMP-95 or ammonia)	To pH 8.0

Key Properties

1) Rich texture with durable structure
2) Designed for trowel, roller, or spray application
3) Film formation at ~5 °C
4) Freeze–thaw stable and high mechanical strength
5) Excellent adhesion to masonry and plaster primed surfaces

12.7.10 Formulation—Wall Putty for Cold Climates

A representative formulation for a wall putty designed for cold-climate applications is shown in Table 12.20.

Key Properties

1) Smooth spreadability and fast drying
2) Excellent filling and sanding performance
3) Freeze–thaw stable and strong adhesion to plaster or concrete
4) Suitable for internal surfaces before painting

12.7.11 Formulation—Economical Wall Paint for Cold Climates

A representative formulation for an economical wall paint designed for cold-climate applications is shown in Table 12.21.

Table 12.20 Wall putty for cold climates.

Ingredient	% by weight
Deionized water	18.0
Dispersant (e.g., Tamol 850)	0.4
Wetting agent	0.3
Defoamer	0.2
Fine calcium carbonate (D50 < 5 μm)	40.0
White cement or dolomite (optional)	8.0
Mica or talc (for smoothness)	4.0
VAE binder (see Table 12.3)	26.0
Coalescent (Texanol)	1.2
Freeze–thaw stabilizer (FT-100)	1.0
Thickener (HEC or HASE for body)	0.8
Biocide and preservative	0.5
pH adjuster (AMP-95 or ammonia)	To pH 8.0

Table 12.21 Economical wall paint for cold climates.

Ingredient	% by weight
Deionized water	25.0
Dispersant (e.g., Tamol 850)	0.5
Wetting agent	0.3
Defoamer	0.2
Calcium carbonate (fine)	28.0
Talc	5.0
Vinyl–Acrylic Binder (see Table 12.5)	30.0
Coalescent (Texanol)	1.2
Thickener (HEUR or HEC)	1.5
Biocide and preservative	0.5
Freeze–thaw stabilizer (FT-100)	1.0
pH adjuster (AMP-95 or ammonia)	To pH 8.0

12.7.12 Formulation—Acrylic Elastic Putty for Filling Large Holes in Cold Climates

A representative formulation for an acrylic elastic wall putty designed for cold-climate applications is shown in Table 12.22.

Key Properties

1) Elastic and highly flexible for filling large voids
2) Maintains adhesion in subzero climates
3) Freeze–thaw stable and easy to sand after drying
4) Excellent crack-bridging performance

12.8 Wall Putty Expanded—Market, Formulations, and Competitive Strategies

12.8.1 Introduction to Wall Putty

Wall putty is a fine, white cement-based or polymer-enhanced finishing compound used to prepare internal and external walls before painting. It fills minor imperfections, evens out surfaces, and enhances paint adhesion and finish. In many markets—especially in developing regions—wall putty is often treated as a commodity product, priced aggressively and manufactured using the most economical raw materials available.

The formulation of wall putty typically revolves around mineral fillers such as calcium carbonate, talc, or dolomite, small quantities of functional additives

Table 12.22 Acrylic elastic putty for cold climates.

Ingredient	% by weight
Deionized water	17.0
Dispersant (e.g., Tamol 850)	0.4
Wetting agent	0.3
Defoamer	0.2
Fine calcium carbonate	20.0
Coarse talc or marble dust	15.0
Glass microspheres or perlite (optional)	4.0
Acrylic elastomeric binder	36.0
Coalescent (Texanol)	1.5
Freeze–thaw stabilizer (FT-100)	1.0
Thickener (HEUR for sag resistance)	2.0
Biocide and preservative	0.5
pH adjuster (AMP-95 or ammonia)	To pH 8.0
Residual water	Balance

like dispersants, defoamers, and thickeners, and a very low percentage of binder—often in the range of 5–10%, particularly in highly price-sensitive markets such as India, the Middle East, or North Africa. In some cases, manufacturers reduce binder levels even further, relying on high cement content or surface suction to compensate for the lack of polymer.

12.8.2 The Reality of Market Competition

Due to its low price point and the intense competition between local manufacturers, wall putty has become a battlefield for cost-cutting. This has led to a significant reduction in binder content, compromising flexibility and adhesion. Low-cost fillers often replace quality extenders, and coalescents, freeze–thaw stabilizers, or advanced thickeners are eliminated from the formula. Consequently, product consistency and storage stability are negatively affected.

While this approach offers short-term price advantages, it results in common failures such as cracking during drying, poor adhesion to concrete or plaster, chalking during sanding, and reduced compatibility with topcoat systems. These quality deficiencies ultimately increase maintenance costs and diminish consumer confidence, highlighting the need for balanced cost-performance strategies in modern formulations.

Table 12.23 Binder role in performance-enhanced wall putty.

Purpose	Key binder role
Cold climates	Maintain flexibility during freeze–thaw cycles
Filling large holes	Prevent shrinkage and cracking
Elastic repair putties	Bridge dynamic cracks and resist micro-movements

12.8.3 Justifying Performance-enhanced Formulations

The wall putty formulations presented in the previous chapter, particularly those tailored for cold climates, contain higher binder loads—ranging from 26% to 36% in some examples. While these levels may initially appear excessive, they are justified by their intended function, environmental durability, and long-term performance. In premium applications or extreme environments, higher binder content is not a luxury but a necessity. However, for commodity-grade wall putty, cost-optimized formulations must balance performance and affordability (Table 12.23).

12.8.4 Optimized Wall Putty Formulations for Competitive Markets

12.8.4.1 Cement-based Putty with Minimal Binder (5%)

A representative formulation of a cement-based wall putty with minimal binder content (5%) is presented in Table 12.24.

Formulation note: Best for dry powder packaging and on-site water addition. VAE redispersible powder provides minimal but essential film-forming ability for crack resistance and adhesion.

12.8.4.2 PVAc-based Interior Wall Putty (8%)

A representative formulation for a PVAc-based interior wall putty is shown in Table 12.25.

Formulation note: Economical wet mix system for interior walls. PVAc helps improve cohesion and sanding without overburdening the cost structure.

12.8.4.3 VAE Latex-based Smooth Putty (12%)

Representative ingredients for a VAE latex-based smooth wall putty formulation are summarized in Table 12.26.

Table 12.24 Cement-based putty with minimal binder (5%).

Ingredient	% by weight
White cement	35.0
Fine calcium carbonate	40.0
Talc	10.0
Redispersible VAE powder	5.0
Cellulose ether (HEC)	0.5
Dispersant (e.g., sodium polyacrylate)	0.3
Defoamer	0.2
Biocide/preservative	0.2
Residual water (for wet mix only)	Balance

Table 12.25 PVAc-based interior wall putty (total PVAc emulsion 55% represents 8% w/w).

Ingredient	% by weight
Fine calcium carbonate	50.0
Dolomite	20.0
PVAc emulsion (55%)	8.0
Water	18.0
HEC thickener	1.0
Defoamer	0.3
Dispersant (e.g., Tamol 850)	0.5
Biocide	0.2

Table 12.26 Ingredients for VAE latex-based smooth putty (VAE emulsion 12%).

Ingredient	% by weight
Fine calcium carbonate	45.0
Talc or mica	15.0
VAE emulsion (55%)	12.0
Water	24.0
HEC or HASE thickener	1.0
Dispersant	0.5
Defoamer	0.3
Biocide/preservative	0.2
Total	**100.0**

Formulation note: Balanced formula for premium-feel interior applications, achieving smoothness and good adhesion at only 12% binder.

12.9 PVAc–AA Wall Primer with ZnO and Carbodiimide (1K)

This formulation demonstrates the benefits of using PVAc copolymerized with acrylic acid combined with ZnO and carbodiimide, providing enhanced water resistance, adhesion, and durability.

Representative ingredients for a PVAc-AA wall primer containing ZnO and carbodiimide (1K) are summarized in Table 12.27.

12.9.1 Mechanism–Role of Zinc Oxide and Carbodiimide in Enhancing Water Resistance in PVAc-based Primers

The incorporation of zinc oxide (ZnO) and carbodiimide crosslinkers in PVAc–AA-based primers significantly improves performance properties, particularly water resistance, durability, and adhesion to alkaline substrates.

Table 12.27 Ingredients for PVAc–AA wall primer with ZnO and carbodiimide (1K).

Ingredient	% by weight
Deionized water	q.s. to 100.0
Dispersant (e.g., Tamol 850)	0.4
Wetting agent (e.g., Surfynol 104)	0.3
Defoamer (e.g., BYK-024)	0.2
Zinc oxide (fine grade)	3.0
Calcium carbonate (fine)	10.0
PVAc–AA latex (55% solids)	50.0
Coalescent (e.g., Texanol)	1.5
Thickener (low-viscosity HEC)	1.2
Carbodiimide crosslinker (30% solids)	0.6
In-can preservative	0.4
pH adjuster	Set to 4.8–5.2
Total	**100.0**

12.9.1.1 Role of Zinc Oxide (ZnO)

ZnO acts as a multifunctional additive:

1) *Alkalinity tolerance and pigment functionality*: Cementitious substrates exhibit high alkalinity (pH 11–13), which can hydrolyze PVAc, leading to chalking and poor adhesion. ZnO acts as a mild amphoteric oxide, reducing local alkalinity by buffering the surface pH, thereby protecting PVAc chains from base-catalyzed hydrolysis.
2) *Hydrophobic barrier*: ZnO particles impart hydrophobic character by reducing the capillary transport of water through the film. Their fine dispersion within the matrix blocks microvoids, lowering water uptake and improving the Cobb60 value: Cobb60 (g/m^2) = The amount of water (in grams) absorbed per square meter of the material in 60 s under controlled conditions.
3) *Antimicrobial protection*: ZnO provides biocidal properties, preventing fungal growth on the primer surface, which is crucial for exterior durability.

12.9.1.2 Role of Carbodiimide Crosslinker

Carbodiimide crosslinkers react with carboxylic acid groups introduced by acrylic acid (AA) units in the copolymer:

1) *Crosslink formation*: Carbodiimide reacts with –COOH groups forming urea linkages, creating a three-dimensional network within the polymer matrix, reducing polymer chain mobility.
2) *Water resistance*: Crosslinking reduces the hydrophilic character and limits water penetration, significantly improving wet adhesion and reducing film swelling.
3) *Chemical resistance*: Carbodiimide-modified systems resist alkaline hydrolysis and chemical attack, ideal for primers applied on cement-based surfaces.

12.9.1.3 Synergistic Effect

When combined, ZnO and carbodiimide offer synergistic benefits:

1) ZnO provides inorganic reinforcement and surface-level barrier properties, while carbodiimide creates a covalently crosslinked network within the polymer film.
2) This dual approach addresses both diffusion (ZnO) and polymer matrix integrity (carbodiimide), resulting in a primer with superior water resistance, long-term adhesion, and enhanced durability under severe weathering.

12.10 Nonstick, Dirt-resistant Wall Coating for Cold Climates

A representative formulation of a self-cleaning, low-soil pickup exterior coating is presented in Table 12.28.

Formulation note: This formula uses a fluoroacrylic or silicone-acrylic emulsion, providing excellent nonstick and low surface energy characteristics, critical for repelling dirt and urban pollution. It also incorporates silane-treated fillers and hollow spheres, which reduce contact area for dust particles and improve scrub resistance. The system remains flexible and freeze–thaw stable, thanks to polymer selection and stabilizer addition, making it ideal for cold climates.

For enhanced performance, optional inclusion of a surface-crosslinking additive (like a carbodiimide 0.3%) may further improve water resistance without compromising workability.

12.11 High-performance Exterior Emulsion Paint for Eastern Europe with 60-min Rain Resistance

This formulation is designed for demanding exterior applications in regions such as Kazakhstan, Russia, and Ukraine, where construction seasons are short and

Table 12.28 Formulation—self-cleaning low-soil pickup exterior coating.

Ingredient	% by weight
Deionized water	20.0
Dispersant (e.g., Tamol 1124 or OROTAN 731A)	0.6
Wetting agent (e.g., Dynol™ 604 or Surfynol 104)	0.3
Defoamer (e.g., BYK-024 or TEGO Foamex 805)	0.3
Fine titanium dioxide (rutile, treated)	12.0
Silane-treated talc or micronized silica	10.0
Hollow polymer spheres (e.g., Expancel®)	2.0
Acrylic emulsion with fluorinated segment (e.g., fluoroacrylic, 50%)	30.0
Freeze–thaw stabilizer (e.g., FT-100)	1.0
Silicone additive (e.g., Silres BS 94 or Wacker 94A)	1.0
Coalescent (e.g., Texanol or DPnB)	1.2
Thickener (HEUR or HASE type)	1.5
Biocide and preservative	0.5
pH adjuster (AMP-95 or ammonia)	To pH 8.0
Residual water	To 100%

sudden rainfall is common. The paint is optimized to resist rain within 60 min of application by incorporating VeoVa-modified acrylic technology, self-crosslinking chemistry, low MFFT design, and fast film formation. This system balances early water resistance, weatherability, and workability.

12.11.1 Design Strategy and Justification

To achieve early rain resistance, the paint system must meet the following critical criteria:

Use of a VeoVa-modified, self-crosslinking acrylic binder for rapid film hardening. Self-crosslinking functionality may be introduced via GMA epoxy groups or acetoacetoxyethyl methacrylate (AAEM) groups, the latter activated by adipic dihydrazide (ADH) as an external reactive additive. No acrylamide-based monomers are included due to health regulatory reasons.

1) Low MFFT (below 5 °C) to enable coalescence in cool spring conditions.
2) Incorporation of hydrophobic modification (siloxane and VeoVa segments) to repel water during early drying.
3) Rheology modifiers that prevent sag and wash-off while maintaining open time for workability.
4) Use of a fast-drying, low-odor coalescent to balance drying speed and eco-profile (Table 12.29).

Table 12.29 Paint formulation—high-performance exterior emulsion paint (Eastern Europe, rain resistant).

Component	Weight (%)	Function
VeoVa-modified self-crosslinking acrylic latex (50% solids, MFFT <5 °C, with GMA or AAEM functionality)	34.0	Main binder, early water resistance, weather protection
Silres BS 94 (siloxane emulsion, 50%)	1.5	Hydrophobic modifier, water repellency
TiO_2 (rutile, exterior grade)	18.0	Opacifier, UV resistance
Calcium carbonate (fine, coated)	19.0	Extender pigment
Talc (micro-fine)	5.0	Antisettling, sheen control
Nonionic dispersant	0.4	Pigment dispersion aid
Defoamer (low VOC)	0.2	Foam control
Coalescent (low odor, e.g., DPnB)	1.5	Film formation aid
HEUR thickener (KU ~95–100)	0.5	Mid-shear viscosity
HASE thickener	0.3	High-shear ICI, application feel
Biocide (in-can preservation)	0.2	Microbial protection
pH adjuster (ammonia-free)	0.1	Stability control
Water	19.3	Balance to 100%

12.11.2 Crosslinking Mechanism During Film Formation

Crosslinking in this system proceeds progressively as water evaporates and latex particles coalesce. The sequence is as follows:

1) Initial evaporation reduces water content, forcing particles into close contact.
2) Polymer particles deform and coalesce, aided by low MFFT and coalescent, enabling polymer chain interdiffusion.
3) As the film densifies, functional groups encounter each other at a higher probability:
   a) GMA epoxy groups undergo ring-opening reactions with carboxyl or hydroxyl groups, forming β-hydroxy ester linkages.
   b) AAEM (acetoacetoxyethyl methacrylate) carbonyl groups react with adipic dihydrazide (ADH), forming hydrazone linkages.
   c) These covalent bonds form during the drying stage, providing a crosslinked network that enhances early rain resistance, block resistance, and long-term durability.
   d) Unlike thermally cured systems, these reactions proceed at ambient temperature, triggered by concentration increase and loss of water plasticization.

12.11.3 Application and Performance Features

1) *Touch-dry*: ~20 min at 25 °C, 50% RH.
2) *Rain-resistant*: ~60 min postapplication under typical Eastern European spring/summer conditions.
3) *Open time*: 5–8 min (sufficient for wall work without roller marks).
4) *Wash-off resistance at 1 h*: no whitening or surface damage.
5) *Weatherability*: >3 years in accelerated QUV exposure with minimal color loss, no chalking.
6) *VOC content*: <40 g/L (EU compliant).

12.11.4 Regional Suitability

1) Ideal for spring and summer construction in Russia, Ukraine, and Kazakhstan.
2) Withstands sudden rain or dew exposure soon after painting.
3) Reduces contractor risk in variable weather patterns.
4) Supports high productivity with minimal wait time between coats.

12.11.5 Conclusion

This advanced exterior emulsion paint is tailored to meet the fast-drying and early rain resistance demands of Eastern European climates. The combination of self-crosslinking binder, hydrophobic modification, and rheology balance allows excellent field performance with industrial-level durability and application flexibility.

12.12 Cold-Climate VeoVa-based Exterior Emulsion Paint–Formulation and Technical Rationale

VeoVa (vinyl versatate) monomers are widely appreciated for their hydrophobicity, alkali resistance, and dirt pickup resistance. However, they are often overlooked in formulations designed for cold climates due to their effect on minimum film formation temperature (MFFT) and flexibility. This document addresses these limitations and shows how to formulate a VeoVa-based emulsion suitable for cold and wet regions such as Russia, Ukraine, and Kazakhstan.

12.12.1 Why VeoVa Is Not Common in Cold Climate Paints

1) VeoVa increases the MFFT of the polymer, making it harder for films to coalesce below 10 °C.
2) Films containing VeoVa tend to be rigid and less flexible, which may lead to cracking during freeze–thaw cycles.
3) Hydrophobic resins reduce water evaporation in cold, humid environments, slowing dry time.
4) These properties make VeoVa-based paints risky in early spring or late fall applications unless modified.

12.12.2 How to Make VeoVa-based Paint Suitable for Cold Climates

To use VeoVa effectively in cold-climate exterior emulsions, several formulation strategies must be employed:

1) Combine VeoVa with soft monomers such as butyl acrylate (BA) or 2-ethylhexyl acrylate (2-EHA) to reduce the MFFT.
2) Incorporate external coalescents like DPnB or Texanol to enable film formation at temperatures as low as 5 °C.
3) Use associative or cellulosic thickeners to maintain application viscosity and sag resistance.
4) Add fast-setting additives or carbodiimide crosslinkers to speed up early rain resistance, provided the copolymer contains AA/MAA to crosslink with carbodiimide.
5) Control the pigment volume concentration (PVC) to prevent brittleness while ensuring hiding power (Table 12.30).

Table 12.30 Recommended formulation—VeoVa-based exterior emulsion for cold regions.

Component	Weight (%)	Function
VAc–BA–VeoVa 10–AA latex (carboxyl-functional; 50% solids; MFFT ~3–5 °C) AA: 2% TM	35.0	Main binder
TiO_2 (rutile, exterior grade)	18.0	Opacity and weather resistance
Calcium carbonate (coated, fine)	20.0	Extender pigment
Talc (fine)	4.0	Antisettling, flatting agent
Siloxane emulsion	1.0	Water repellency
DPnB (coalescent)	1.5	Film formation aid
Nonionic dispersant	0.4	Pigment dispersion
Defoamer	0.2	Foam control
HEUR thickener	0.6	Mid-shear viscosity control
HASE thickener	0.3	ICI control
Carbodiimide crosslinker dispersion (20%)	1.0	Early rain resistance
Biocide	0.2	Preservation
Water	17.8	Balance to 100%

Performance Features

1) *MFFT*: <5 °C (with coalescent)
2) *Rain-resistant*: 60 min after application at 10 °C and 50% RH
3) *Freeze–thaw resistance*: Passes five cycles
4) *UV and alkali resistant*: Suitable for concrete and mineral surfaces
5) *VOC content*: <40 g/L (EU compliant)

12.12.2.1 Conclusion

While VeoVa-based emulsions are not traditionally used in cold climates, they can be adapted effectively by using soft monomers, plasticizing coalescents, and film-forming aids. This formulation demonstrates that with proper modification, VeoVa can deliver its hydrophobic and weather-resistant advantages even under the temperature and humidity constraints of Eastern Europe.

12.13 Antichalking Exterior Paint for Cold Climates

As exterior paint ages, exposure to UV light, moisture, and temperature cycling can cause binder degradation. This leads to pigment particles, particularly titanium dioxide or extenders like calcium carbonate, becoming unbound at the surface—a phenomenon known as chalking. It results in a powdery residue that affects appearance, transfers easily to hands or clothes, and compromises recoat adhesion.

The following formulation is designed for cold climates, where freeze–thaw cycles and reduced curing temperatures can exacerbate chalking and film failure. It uses a UV-stable acrylic binder, optimized PVC, HALS/UV absorber system, and hydrophobic modifiers for long-term durability (Table 12.31).

Table 12.31 Formulation—cold-weather acrylic exterior paint with antichalking properties.

Ingredient	% by weight
Deionized water	18.0
Dispersant (e.g., Tamol 1124)	0.6
Wetting agent (e.g., Dynol 604)	0.3
Defoamer (e.g., BYK-024)	0.3
Titanium dioxide (Rutile, surface treated)	16.0
Calcium carbonate (fine, coated)	10.0
Acrylic emulsion (UV-stable, 50%)	45.0
Coalescent (Texanol or DPnB)	1.2
Freeze–thaw stabilizer (FT-100)	1.0
HALS (e.g., Tinuvin 292)	0.3
UV absorber (e.g., Tinuvin 1130)	0.3
Silicone additive (Silres BS 94)	0.8
HEUR thickener	1.5
Biocide and preservative	0.5
pH adjuster (AMP-95 or ammonia)	To pH 8.0
Residual water	To 100

Formulation note: This cold-climate exterior paint balances excellent weatherability with resistance to chalking. The pure acrylic binder and optimized pigment volume concentration ensure that all pigments are fully encapsulated, minimizing their exposure to UV and moisture. The HALS/UV absorber combination neutralizes free radicals and absorbs harmful UV, extending film life. Silicone additive adds surface hydrophobicity, repelling water and airborne particles that lead to dirt pickup or pigment wash-off. The freeze–thaw stabilizer maintains stability in low-temperature storage and transport.

12.14 Cold Climate Painting–Practical and Technical Strategies

This section addresses the challenges and requirements of painting in cold climates such as Russia, Ukraine, and Kazakhstan. It covers practical jobsite tips, advanced formulation considerations, and real-world scenarios to help professionals achieve durable, weather-resistant coatings even in low-temperature, high-humidity environments.

12.14.1 Top 10 Tips for Painting in Cold Climates

1) Choose low-MFFT paints
   Use emulsion systems with MFFT $\leq 5\,°C$. Select soft monomer binders (e.g., BA, 2-EHA) and coalescents like DPnB.
2) Pay attention to the time allowed for application
   Apply paint during the late morning or early afternoon, after surface dew has fully evaporated and temperatures have stabilized. Avoid painting too early (risk of dew contamination) or too late in the day (risk of insufficient drying before evening temperature drops). Always allow at least 3–4 h of drying time before sunset.
3) Avoid moisture traps
   Do not paint on damp, frozen, or recently rained-on substrates.
4) Respect minimum application temperature
   Air and surface must be $\geq 5\,°C$. Cold substrates can delay film formation.
5) Increase ventilation for drying
   Enhance airflow. Avoid combustion heaters that introduce moisture and CO_2.
6) Use fast-set or crosslinking technology
   Incorporate self-crosslinking binders or carbodiimides in your formulation with PVAc/VAE/vinyl acrylate copolymered with acrylic acid for early rain resistance.
7) Surface prep is more critical
   Ensure $pH < 10$ and substrate free from salts or efflorescence.

8) Allow extended recoat time
   Increase recoat intervals by 25–50% versus normal warm-weather instructions.
9) Use additives strategically
   Apply coalescents, siloxanes, and thickener blends to support cold-weather performance.
10) Field test before mass application
   Apply test patches to assess drying, adhesion, and color uniformity.

12.14.2 Advanced Technical Considerations

1) Thermal shock resistance
   Prevents cracking from day/night temperature shifts. Use flexible binders like VeoVa hybrids.
2) Moisture vapor transmission (MVTR)
   Avoid trapping moisture. Use breathable binders with balanced hydrophobicity.
3) Freeze–thaw stability
   Ensure product survives −18 °C storage cycles. Add stabilizers and test three to five freeze–thaw cycles.
4) Alkali resistance
   Use VeoVa or siloxane-modified binders or a primer with PVAc-Acrylic Acid with post-added carbodiimide for fresh concrete or cementitious walls.
5) Inspection and Testing
   Use moisture meters, dew point checks, and plastic film tests before painting.

12.14.3 Cold Weather Jobsite Checklist

1) Surface temperature ≥ 5 °C (use IR thermometer).
2) No frost, dew, or visible moisture present.
3) Substrate moisture <12% (use moisture meter).
4) Dew point margin ≥3 °C.
5) Product MFFT ≤5 °C (check spec sheet).
6) Coalescent present in formulation.
7) Wind speed <25 km/h.
8) Weather forecast: no rain or frost within 12 h.

12.14.4 Warm vs. Cold Climate Formulation Comparison

The formulation of waterborne coatings for different climates requires careful selection of raw materials and performance modifiers. In warm climates, longer open time and moderate coalescent levels are desirable to control drying rate and

Table 12.32 Key differences between warm and cold climate formulation strategies.

Property	Warm climate focus	Cold climate focus
MFFT	Higher acceptable (8–12 °C)	Must be ≤5 °C
Drying time	Controlled, slower preferred	Accelerated, fast-set needed
Hydrophobicity	Medium to high	Balanced with breathability
Coalescent need	Often minimal	Essential
Freeze–thaw stability	Not critical	Mandatory
Application season	Long (8+ months)	Short (4–6 months)

minimize defects such as lap marks. In contrast, cold climates demand polymers with low minimum film formation temperature (MFFT), rapid drying capability, and robust freeze–thaw stability to ensure application reliability within shorter seasonal windows (Table 12.32).

12.14.5 Humid vs. Dry Climate Formulation Comparison

Formulation requirements differ significantly between humid and dry climates. Humid regions demand coatings with superior water resistance, antifungal protection, and lower VOC volatility, while dry climates prioritize slower drying rates to maintain workability and avoid lap marks (Table 12.33).

Table 12.33 Key differences between humid and dry climate formulation strategies.

Property	Humid climate focus	Dry climate focus
Water resistance	High, hydrophobic additives mandatory	Moderate
Fungal resistance	Essential (use biocides, ZnO)	Optional
Open time	Shorter to avoid wash-off	Longer to prevent lap marks
Coalescent volatility	Low-volatility coalescents preferred	Standard coalescents acceptable
VOC regulation impact	Critical	Moderate

Table 12.34 Key differences between interior and exterior coating strategies.

Property	Interior application	Exterior application
UV stability	Not required	Essential (use UV absorbers, light stabilizers)
Scrub resistance	High priority	Moderate
Dirt pickup resistance	Low priority	High priority
VOC/low odor	Critical	Important but less stringent
Moisture protection	Moderate	High (hydrophobic modifiers)

12.14.6 Interior vs. Exterior Application Formulation Comparison

Interior and exterior coatings face vastly different performance requirements. Exterior paints must withstand UV radiation, rain, and dirt pickup, while interior paints focus on scrub resistance, stain resistance, and low odor/VOC for indoor air quality (Table 12.34).

12.14.7 Case Study–Cold Weather Exterior Coating Application

Background: A construction project in Northern Europe required the application of an exterior acrylic-latex coating in early spring, when daytime temperatures ranged from 4 to 8 °C and nighttime temperatures dropped close to freezing. Conventional formulations failed to form a coherent film, resulting in poor adhesion and early blistering.

Challenges Identified

1) Low ambient temperature: Below the MFFT of standard latex binders (8–12 °C).
2) Short daylight window: Limited drying time before sunset.
3) High relative humidity (75–85%): Increased risk of condensation and slow water evaporation.
4) Freeze–thaw exposure during storage and transit: Several drums experienced partial freezing.

Formulation Strategy

To address these challenges, the following adjustments were implemented:

1) *Low-MFFT binder selection*: A styrene–acrylic copolymer with an MFFT of 2 °C was used.
2) *Coalescent optimization*: 4% of a low-volatility coalescent (DPnB) was added to aid film formation at near-freezing conditions.

3) *Freeze–thaw stability package*: Added 1.5% ethylene glycol and specialized antifreeze additives to prevent coagulation during storage.
4) *Thickener system*: A hydrophobically modified alkali-swellable emulsion (HASE) thickener was selected to maintain rheology at low temperatures.
5) *Accelerated drying*: Formulation included a quick-evaporating solvent (propyl glycol ether) in compliance with VOC limits.

Application Protocol

1) Surfaces were preheated to 5 °C using portable heaters.
2) Painting was scheduled between 10:00 AM and 2:00 PM, allowing 4 h of drying before sunset.
3) Substrate moisture was monitored using a moisture meter (<12%).
4) Dew point margin maintained at ≥3 °C.

Results

1) *Film integrity*: Uniform film formation observed without cracking or blistering.
2) *Adhesion*: Pull-off strength exceeded 1.5 MPa after 7 days.
3) *Water resistance*: Cobb60 reduced to <25 g/m^2.
4) *Durability*: Coating passed 2,000 h of accelerated weathering with no chalking.

Key Points

1) Binder MFFT and coalescent balance are critical for low-temperature applications.
2) Freeze–thaw stability must be ensured during storage and transport.
3) Proper jobsite preparation and monitoring of dew point significantly reduces failures (Table 12.35).

Table 12.35 Key data summary for cold-weather coating application.

Parameter	Value
Ambient temperature	4–8 °C
Binder MFFT	2 °C
Coalescent level	4% (DPnB)
Freeze–thaw additive	1.5% ethylene glycol
Cobb60	<25 g/m^2
Adhesion (pull-off)	>1.5 MPa after 7 days
Weathering resistance	2,000 h without chalking

12.14.8 Cold Climate Case Study: April Repaint in Karaganda

A public school exterior in Karaganda, Kazakhstan was repainted in mid-April. Temperatures ranged from −2 °C at night to 11 °C by day. The contractor used a VeoVa-modified acrylic emulsion with coalescent and siloxane. Painting began at 10:30 AM each day, and surfaces were prechecked with IR thermometer and moisture meter.

A test patch confirmed drying within 90 min and early rain resistance by 3:00 PM. Film formation was intact with no whitening. The team avoided application on days with wind >25 km/h and maintained recoat intervals of 6 h.

The coating performed well through the rainy spring and passed adhesion and blistering tests six months later.

12.14.9 Case Study—Interior Wall Coating in Heated Buildings During Winter

Background: An office renovation in Northern Canada involved applying an interior latex coating during January, when outdoor temperatures were -15 °C. The building was heated to 18 °C indoors, but wall surfaces, especially near exterior walls and windows, were significantly colder (8–10 °C). Initial attempts with standard formulations resulted in uneven gloss and poor adhesion at cold spots.

Challenges Identified

1) Substrate Temperature Gradient: Walls near windows remained 8–10 °C, below the coating's MFFT.
2) Condensation Risk: Warm indoor air and cold wall surfaces caused localized condensation, leading to surfactant leaching.
3) Ventilation Constraints: Limited air exchange due to sealed building, slowing water evaporation.
4) Drying Time: Required fast turnaround for reoccupancy without compromising film quality.

Formulation Strategy

1) *Low-MFFT polymer*: Interior-grade VAE copolymer with MFFT $\leq$3 °C to ensure proper film formation on cold surfaces.
2) *Coalescent adjustment*: Added 2.5% DPM (dipropylene glycol monomethyl ether) for enhanced coalescence under low temperatures.
3) *Antileaching additive*: Included associative thickener with hydrophobic modification to reduce surfactant migration during slow drying.

4) *pH stabilization*: Maintained pH at 8.2 for latex stability under variable temperature conditions.
5) *Freeze–thaw stable package*: Used glycol-based stabilizers to protect paint during transport and storage at subzero conditions.

Application Protocol

1) Preheated surfaces to at least 10 °C using infrared heaters before application.
2) Ensured indoor humidity <60% to minimize condensation risk.
3) Applied thin coats with a recoating interval of 4 h to accelerate curing.

12.15 Results

1) *Film appearance*: Uniform gloss achieved even near cold exterior walls.
2) *Adhesion*: Cross-cut adhesion rated 5B after 7 days.
3) *Surfactant leaching*: Minimal staining, acceptable after 24-h water wipe.
4) *Odor/VOC*: Within regulatory limits, maintaining good indoor air quality.

Key Points

1) Surface temperature control is as important as ambient temperature for successful application.
2) Low-MFFT polymers and coalescent selection are critical for cold-interior applications.
3) Moisture management reduces surfactant leaching and ensures long-term durability (Table 12.36).

Table 12.36 Key data summary for interior wall coating in cold conditions.

Parameter	Value
Indoor ambient temperature	18 °C
Wall surface temperature	8–10 °C
Binder MFFT	≤ 3 °C
Coalescent level	2.5% (DPM)
Humidity control	<60% RH
Recoating interval	4 h
Adhesion	Cross-cut 5B after 7 days

Table 12.37 Key tips for cold-climate painting applications.

Tip	Description
Choose low-MFFT paints	Use emulsion systems with MFFT ≤5 °C. Select soft monomer binders and adequate coalescent levels.
Watch the allowed time for painting	Apply during the late morning or early afternoon to avoid condensation and ensure 3–4 h of drying before sunset.
Ensure proper surface temperature	Confirm substrate ≥5 °C using an IR thermometer.
Maintain dew point margin	Ensure surface temperature is at least 3 °C above dew point.
Prevent moisture issues	Check substrate moisture <12% before painting.
Use coalescent and freeze–thaw additives	Essential for film formation and storage stability in cold conditions.
Monitor weather forecast	Avoid application if rain or frost is expected within 12 h.
Improve ventilation in interiors	Reduces drying time and minimizes condensation risk.
Use infrared heaters for critical surfaces	Preheat cold areas (e.g., near windows or exterior walls) to ensure proper film formation.
Schedule work in warmest part of day	Plan painting during late morning to early afternoon for best results.

Table 12.38 Recommended trade names for key additives in cold climate paints.

Additive type	Recommended trade names (manufacturer)
Defoamers	BYK-024 (BYK-Chemie); TEGO Foamex 805 (Evonik); Foamaster NXZ (BASF); Agitan 282 (Münzing)
Dispersing agents	Tamol 731A/Tamol 850 (Dow); Dispex AA 4040 (BASF); Orotan 1124 (Dow); Efka 4530 (BASF)
Wetting agents	Surfynol 104/Surfynol 420 (Evonik); Dynol 604/Dynol 607 (Evonik); Byk-3456 (BYK-Chemie); TEGO Wet 280 (Evonik)
Thickeners	Acrysol RM-2020 NPR (Dow); Rheovis AS 1188 (BASF); Bermocoll E 230 FQ (AkzoNobel); Natrosol Plus 330 (Ashland)
Coalescents	Texanol (Eastman); Butyl DiGlycol (BASF); DPnB (Dow); Optifilm 400 (Eastman)
Freeze–thaw stabilizers	Propylene Glycol; Ethylene Glycol or Glycerin blends; Luviskol K30 (BASF); Colloid 643 (Dow); Urea solutions, FT 100
Biocides/preservatives	Acticide MBS (Thor); Preventol A 14-D (Lanxess); Proxel GXL (Arch Chemicals)

The following table summarizes essential best practices for successful painting and coating applications in cold climates. These tips address formulation requirements, environmental conditions, and jobsite protocols to ensure film integrity and long-term durability under low-temperature conditions (Tables 12.37 and 12.38).

12.16 Conclusion

Painting and coating in cold climates is not merely a matter of lowering the minimum film formation temperature; it is an integrated process that combines science, planning, and field discipline. Success hinges on three pillars:

1) *Formulation engineering*: Selecting low-MFFT binders, optimizing coalescent packages, and ensuring freeze–thaw stability transform theoretical performance into practical reliability.
2) *Environmental control*: Monitoring surface temperature, dew point, and humidity is as critical as the chemistry itself. A well-engineered coating can still fail if applied on a cold, damp substrate.
3) *Application strategy*: Timing, surface preparation, and adherence to best practices—such as maintaining a safe dew point margin—are indispensable for durability.

Cold-climate coatings face unique challenges: limited time for application, condensation risk, and storage hazards. Yet, these challenges present opportunities for innovation—advanced latex technology, hybrid systems, and intelligent additives now make year-round painting feasible in regions where it once seemed impossible.

Ultimately, the goal is durability without compromise: films that resist cracking in subzero weather, surfaces that remain defect-free despite environmental stress, and coatings that maintain aesthetic and functional performance season after season. By combining robust formulation design with disciplined application protocols, coatings professionals can achieve this standard—even when the mercury drops.

13

Testing Methods, Durability, and Aging of Emulsion Polymers, PVAc-based Adhesives, and Paints

13.1 Introduction

In the world of emulsion polymers and water-based coatings, a formulation is only as valuable as its proven performance. Laboratory and field testing serve as the crucial bridge between theory and application—ensuring that what looks promising on paper holds up under real-world conditions.

This chapter presents the essential testing methods used to evaluate the physical, chemical, and mechanical properties of emulsion polymers and the paints derived from them. From basic parameters like pH and solids content, to more specialized assessments such as wet scrub resistance, adhesion, freeze–thaw stability, and minimum film formation temperature (MFFT), each test plays a role in predicting product success or failure.

For formulators, factory managers, and quality control personnel, mastering these testing protocols is not merely about compliance or routine—it is about decision-making, product confidence, and continuous improvement.

This chapter provides:

1) A clear description of each testing method
2) Practical procedures and equipment where applicable
3) Notes on interpretation and relevance to product performance
4) Common pitfalls and best practices

By the end of this chapter, the reader will have a practical toolkit to evaluate, troubleshoot, and optimize emulsion-based systems—whether for adhesives, paints, primers, or specialty coatings.

13.2 Typical Testing Methods

13.2.1 Solid Content

Solid content refers to the percentage of nonvolatile material (polymer + additives) present in the emulsion. It directly affects film build, viscosity, drying time, and performance properties.

Purpose: Determines the amount of the active ingredients in the formulation of the emulsion polymer and the total solid content in a paint formulation. It is crucial for consistency and cost control.
Test method: Heat a known sample weight in an oven at 105 °C for 1.5–2 h until constant weight.
Standard: ASTM D2369 (for coatings), ISO 3251.

Notes

1) Higher solid content of an emulsion polymer often improves film properties and reduces drying time.
2) Excessively high solids can increase viscosity and create application issues.
3) Ensure no foaming or splattering during heating.

13.2.2 pH Measurement

The pH of an emulsion polymer impacts stability, reaction kinetics, compatibility with additives, and application behavior.

Purpose: Monitors the emulsion's chemical balance and ensures compatibility with defoamers, thickeners, and preservatives.
Test method: Use a calibrated pH meter with a glass electrode, preferably at 25 °C.
Standard: ASTM E70, ISO 1148.

Notes

1) Most polymer emulsions are stable between pH 4–9.
2) Buffer solutions must be fresh to ensure accuracy.
3) Drifts in pH may indicate hydrolysis or microbial contamination.

13.2.3 Viscosity

Viscosity reflects the internal resistance to flow and impacts processing, application, and storage behavior.

Purpose: Ensures the product is within the desired flow and handling parameters for its intended application (e.g., brushing, spraying).
Test methods: Brookfield Viscometer (low-to-medium viscosity) and Stormer (KU) Viscometer (for paints).
Standard: ASTM D2196, ASTM D562.

Notes

1) Viscosity can be altered with associative thickeners, HEC, or polyurethane-based modifiers.
2) Shear-thinning behavior is preferred in most coatings for ease of application and sag resistance.

13.2.4 Minimum Film Formation Temperature (MFFT)

MFFT is the lowest temperature at which an emulsion forms a continuous, defect-free film upon drying.

Purpose: Determines the lowest ambient temperature at which the product can be applied without cracking or powdering.
Test method: Use an MFFT bar or a gradient plate with controlled temperature zones.
Standard: ASTM D2354, ISO 2115.

Notes

1) MFFT depends on polymer Tg and coalescent content.
2) Additives like Texanol, DPNB, or hydrophobic plasticizers can lower MFFT.
3) For cold climate applications, selecting low-Tg polymers is critical.

13.2.5 Glass Transition Temperature (Tg)

Tg is the temperature at which a polymer transitions from a brittle glassy state to a flexible rubbery state.

Purpose: Influences film hardness, flexibility, and blocking resistance.
Test method: Differential scanning calorimetry (DSC).

Notes

1) High Tg (e.g., $>30\,^{\circ}C$) → Hard films, good blocking but poor flexibility.
2) Low Tg ($<5\,^{\circ}C$) → Soft films, good flexibility but may need crosslinking.
3) Adjusted through monomer selection (e.g., BA lowers Tg; MMA raises it).

13.2.6 Wet Scrub Resistance

Wet scrub resistance measures a coating's ability to withstand repeated cleaning or abrasion while wet, without film degradation.

Purpose: Critical for interior wall paints, kitchens, bathrooms, and high-traffic areas.
Test method: Scrub tester (e.g., Gardner type) with standard brush and abrasive medium applied over painted panel.
Standard: ASTM D2486, ISO 11998.

Notes

1) Improved by using crosslinkers (carbodiimides, aziridines) and fine particle size latex.
2) Excess thickener or poor coalescence lowers resistance.

Polyvinyl alcohol (PVOH) or cellulose thickeners should be carefully optimized.

13.2.7 Adhesion Test

Adhesion measures how well a paint or adhesive film sticks to the underlying substrate.

Purpose: Essential for primers, coatings on plastics, metals, and difficult substrates.
Test methods
1) Cross-cut tape test (for adhesion to rigid substrates)
2) Pull-off strength (for structural applications)

Standard: ASTM D3359 (tape), ASTM D4541 (pull-off), ISO 2409.

Notes

1) *Surface preparation is key:* cleaning, degreasing, and abrasion improve adhesion.
2) Vinyl acrylics with functional monomers (e.g., acrylic acid) increase adhesion to polar surfaces.
3) Silanes or adhesion promoters improve bonding to glass, metal, or plastic.

13.2.8 Water Resistance

Water resistance assesses a coating's ability to resist softening, whitening, or peeling upon water exposure.

Purpose: Key for exterior paints, bathroom/kitchen applications, and adhesives exposed to moisture.

Test method: Spot testing, water immersion, or constant humidity exposure.
Standards: ASTM D870, ISO 2812-2.

Notes

1) Improved by hydrophobic monomers (e.g., VeoVa, styrene), crosslinkers, or low surfactant migration.
2) Excess surfactants or soft polymers lead to poor water resistance.
3) Use zinc oxide, silane, or carbodiimide systems for superior water durability.

13.2.9 Blocking Resistance

Blocking resistance refers to the ability of a dried film to resist sticking to itself when two coated surfaces are pressed together.

Purpose: Important for stackable coated materials, packaging, door/window frames, and roll goods.
Test method: ASTM D4946, weights placed on coated films in contact under specific conditions.

Notes

1) Affected by film hardness, coalescence, and Tg.
2) High Tg polymers (>25 °C) and low coalescent levels improve blocking resistance.
3) Silicone-based additives may help reduce blocking.

13.2.10 Gloss Measurement

Gloss is a key aesthetic and functional property of paint and coating films, reflecting the ability of a surface to reflect light in a mirror-like manner.

Purpose: Indicates surface smoothness, cleanliness, and uniformity. Gloss level affects visual perception and application quality.
Test method: Measured using a gloss meter at standard angles (20°, 60°, or 85°), depending on the gloss level of the film.

- 20°: High-gloss coatings
- 60°: Semi-gloss or universal
- 85°: Low-gloss or matte finishes

Standard methods: ASTM D523, ISO 2813.
Typical equipment: Portable or benchtop gloss meters with calibration standards.

Notes

1) Smooth, well-leveled films with low surface roughness show higher gloss.

2) Pigment particle size, dispersion quality, and binder content influence gloss.
3) Gloss improves with optimized film coalescence and slow-drying conditions.

Impact of polymerizable surfactants on gloss: Conventional surfactants may migrate to the film surface during drying, increasing surface tension gradients and lowering gloss by disrupting smoothness. Polymerizable surfactants (also called reactive or graftable surfactants) become covalently bound to the polymer backbone during polymerization. This prevents surfactant migration, resulting in a more uniform, smoother film surface and therefore higher and more durable gloss.

Formulation Tip

1) Use polymerizable surfactants to improve long-term gloss retention, especially in exterior coatings.
2) Combine with fine particle size and low PVC (pigment volume concentration) systems for best results.

13.2.11 Film Flexibility

Definition: Film's ability to bend without cracking or breaking.
Purpose: Important in applications like tapes, sealants, or elastic coatings.
Test method: Bent over a mandrel or flexed manually and inspected for cracks.
Standard methods: ASTM D522.
Typical equipment: Mandrel bend tester.
Notes: Lower Tg polymers improve flexibility.

13.2.12 Blocking Resistance

Definition: Resistance of dried films to sticking when placed under pressure.
Purpose: Essential for stacked materials, packaging, and hot climate performance.
Test method: Two coated surfaces are pressed together under weight; then peeled apart and examined.
Standard methods: ASTM D4946.
Typical equipment: Blocking tester or weights.
Notes: High Tg and hard segments reduce blocking.

13.2.13 Freeze–Thaw Stability

Definition: Latex's ability to remain stable after freeze–thaw cycles.

Freeze–thaw stability measures the ability of an emulsion polymer or waterborne paint to withstand repeated freezing and thawing cycles without coagulation, viscosity loss, or phase separation. Poor freeze–thaw stability can lead to irreversible damage, rendering the product unusable.

It is a key quality parameter for:

1) Emulsion binders shipped to or used in cold climates.
2) Paints stored in unheated warehouses.
3) Latex-based adhesives with seasonal packaging risks.

Purpose: Key for transportation and storage in cold climates.

Test method: Repeated cycles of freezing (0 °C or lower) and thawing (25 °C); observe coagulation or separation.

Standard methods: ASTM D2243.

Typical equipment: Freezer, temperature chamber.

Notes: Nonionic surfactants and FT additives like FT-100 from Troy/Arxada improve performance.

13.2.14 Chemical Resistance

Definition: The ability of the dried film to resist degradation when exposed to chemicals.

Purpose: Important for coatings in kitchens, bathrooms, or industrial applications.

Test method: Apply defined chemicals (e.g., NaOH, acids, solvents) on the film and cover. Evaluate after 1–24 h.

Standard methods: ASTM D1308.

Typical equipment: Watch glass, chemicals, inspection tools.

Notes: Epoxy-modified or crosslinked systems show better resistance.

13.2.15 Mechanical Stability

Definition: Resistance of latex to coagulation under high shear or mechanical stress.

Purpose: Ensures the latex can be pumped, stirred, or transported without destabilizing.

Test method: Latex is subjected to high-speed stirring or colloid mill treatment. Coagulation or viscosity shift is evaluated.

Standard methods: Internal QC method.

Typical equipment: High-speed stirrer, colloid mill, filter.

Notes: Stabilizer selection strongly affects mechanical stability.

13.2.16 Electrolyte Stability

Definition: Stability of latex when exposed to ionic species (salts).

Purpose: Critical for formulations mixed with hard water or containing fillers.

Test method: Add increasing salt concentrations (e.g., NaCl, $CaCl_2$) and observe for coagulation or flocculation.

Standard methods: Internal QC method.

Typical equipment: Salt solutions, stirrer, glassware.

Notes: Hydrophobically modified stabilizers often improve performance.

13.2.17 Coalescence Efficiency

Definition: Ability of latex to form a continuous film in the presence or absence of coalescing agents.

Purpose: Relevant for low-VOC paints and cold temperature applications.

Test method: Drawdowns with and without coalescents are compared for transparency and film defects.

Standard methods: Visual method.

Typical equipment: Drawdown applicator, drying chamber.

Notes: Latex with broad particle size distribution may coalesce better.

13.2.18 Storage Stability

Definition: Resistance of latex or paint to changes during extended storage.

Purpose: Assures product usability over shelf-life.

Test method: Store at room temp and/or 40 °C for 1–6 months. Observe separation, viscosity, and microbial growth.

Standard methods: ASTM D1849.

Typical equipment: Stability chamber, viscosity meter.

Notes: Biocide presence and pH buffering are essential.

13.2.19 Particle Size and Distribution

Definition: The size and spread of polymer particles in a latex.

Purpose: Affects gloss, opacity, mechanical properties, and stability.

Test method: Measured using dynamic light scattering (DLS), SEM, or laser diffraction.

Standard methods: ISO 13320.

Typical equipment: Zetasizer, SEM, or laser particle sizer.

Notes: Narrow distribution typically improves gloss and film uniformity.

13.2.20 Density

Definition: Mass per unit volume of latex or paint.

Purpose: Used for quality control, formulation consistency, and VOC calculation.

Test method: Measure the weight of known volume using a density cup or pycnometer.
Standard methods: ASTM D1475.
Typical equipment: Density cup or pycnometer.
Notes: Ensure temperature consistency (25 °C typical).

13.2.21 Foam Stability

Definition: Tendency of latex to retain foam during processing.
Purpose: Critical for film quality and pumping/mixing performance.
Test method: Latex is stirred vigorously; foam volume and collapse time are measured.
Standard methods: Foam column method.
Typical equipment: High-speed stirrer, foam column.
Notes: Add defoamers if needed; surfactant choice is key.

13.2.22 Tensile Strength and Elongation

Definition: Resistance to pulling force and the ability to stretch before breaking.
Purpose: Used in tapes, sealants, and flexible adhesives.
Test method: Cut dumbbell-shaped films and test with a UTM.
Standard methods: ASTM D412.
Typical equipment: Universal testing machine (UTM).
Notes: Crosslinking and filler type affect strength and flexibility.

13.2.23 Peel Strength

Definition: Force required to peel apart two bonded substrates.
Purpose: Key for packaging, tapes, and laminating adhesives.
Test method: Bonded substrates are peeled at 180° or 90° at a constant speed.
Standard methods: ASTM D903, ASTM D3330.
Typical equipment: UTM with peel fixture.
Notes: Surface preparation and drying time affect results.

13.2.24 Water Vapor Transmission Rate (WVTR)

Definition: Rate of water vapor passing through a film.
Purpose: Critical for packaging, breathable coatings, and waterproofing.
Test method: Film is sealed over a dish with desiccant or water and weighed periodically.
Standard methods: ASTM E96.

Typical equipment: Desiccant cups, analytical balance.
Notes: Crosslinking and film thickness reduce WVTR.

13.2.25 Oxygen Transmission Rate (OTR)

Definition: Amount of oxygen permeating through film per unit area and time.
Purpose: Important in food packaging and barrier coatings.
Test method: Specialized equipment measures oxygen passage through sealed film.
Standard methods: ASTM D3985.
Typical equipment: OTR analyzer (e.g., MOCON).
Notes: Acrylic and PVDC copolymers offer low OTR.

13.2.26 Viscoelastic Properties (Rheology)

Definition: Combination of viscous and elastic responses under deformation.
Purpose: Critical for flow, leveling, and sag resistance.
Test method: Tested by rotational rheometer for shear thinning, yield stress, G', and G''.
Standard methods: ASTM D4440, ISO 3219.
Typical equipment: Rotational rheometer.
Notes: Associative thickeners tune the rheological profile.

13.2.27 Hardness (Pencil or Shore)

Definition: Resistance of the film to penetration or deformation.
Purpose: Indicates durability and wear resistance.
Test method: Measured using pencils (6B to 6H) or durometer for elastomers.
Standard methods: ASTM D3363, ASTM D2240.
Typical equipment: Pencil set, Shore A/D durometer.
Notes: Higher Tg and filler content increase hardness.

13.2.28 Heat Aging (Thermal Stability)

Definition: Retention of properties after exposure to elevated temperature over time.
Purpose: Predicts shelf life and thermal endurance.
Test method: Film or latex is stored at 60–100 °C for 7–28 days and then evaluated.
Standard methods: ASTM D573.
Typical equipment: Oven, inspection tools.
Notes: Oxidation inhibitors or UV stabilizers may help reduce degradation.

13.2.29 ASTM D1183-03 (2021) Standard Test Methods for Flexibility and Adhesion of Adhesives After Environmental Exposure

ASTM D1183 covers procedures for evaluating the durability, flexibility, and adhesion of adhesive joints after exposure to various environmental conditions. It is used to predict service life under temperature, humidity, and cycling stresses. This method is applicable to PVAc-based adhesives, copolymer adhesives, and bonded assemblies that may face prolonged or cyclic environmental stress. Bonded specimens are subjected to controlled environmental conditions such as high humidity, water immersion, freeze–thaw cycling, or elevated temperature for a defined period. After exposure, the adhesive joints are tested for flexibility and adhesion retention compared with unexposed control samples.

Equipment Required

1) Environmental chamber capable of controlling temperature and relative humidity
2) Flexibility testing apparatus (bending or mandrel tester)
3) Adhesion testing equipment (tensile tester or peel tester)
4) Calibrated thermometers and hygrometers

Test Procedure

1) *Sample preparation:* Prepare bonded test specimens according to specified dimensions and adhesive application conditions.
2) *Conditioning:* Expose specimens to environmental cycles—common regimens include:
   a) 48 h at 50 °C and 95% RH.
   b) 10 freeze–thaw cycles (−20 to 25 °C).
   c) Continuous water immersion for specified durations.
3) *Evaluation:* After exposure, assess:
   a) *Flexibility:* Bend specimens over a mandrel and inspect for cracking or delamination.
   b) *Adhesion:* Measure tensile or peel strength reduction relative to controls.
4) *Reporting:* Document exposure conditions, test results, % adhesion retention, and visual observations (cracks, bubbles, delamination).

Relevance to PVAc-based Adhesives and Emulsion Polymers: This method simulates real-world degradation mechanisms such as moisture ingress, thermal expansion/contraction, and freeze–thaw stresses. It is particularly valuable in qualifying adhesives for exterior applications, cold-climate use, or high-humidity service (Table 13.1).

Table 13.1 Key parameters for ASTM D1183 testing.

Parameter	Typical value/range	Notes
Conditioning temp	25–50 °C	Depends on chosen environmental cycle
Relative humidity (RH)	50–95%	High humidity for moisture aging studies
Cycle duration	24–72 h per cycle	Freeze–thaw or thermal cycling regimes
Adhesion evaluation	% strength retained vs. control	Peel or tensile method as per adhesive type
Failure criteria	Visible cracks, delamination, or > 20% loss	Indicates unacceptable durability

13.3 Summary of Common Testing Methods in Emulsion Polymers and Related Paints

Table 13.2 provides a consolidated reference of commonly used test methods relevant to emulsion polymers, paints, and adhesives. This summary includes the application area, official standard methods, and typical equipment used in laboratory practice for quality control and performance evaluation.

13.4 Best Practices and Common Pitfalls in Testing

Accurate testing is the cornerstone of consistent quality and regulatory compliance. While standard methods provide detailed procedures, practical execution often determines whether results are reliable and reproducible. This section summarizes critical best practices and frequent errors encountered in the testing of emulsion polymers, adhesives, and paints.

Best Practices

1) *Instrument calibration:* Calibrate pH meters, viscometers, and gloss meters before use. Use certified calibration standards.
2) *Environmental control:* Maintain specified temperature and humidity during tests (e.g., 23 ± 2 °C and 50 ± 5% RH) unless the method calls for specific deviations.
3) *Representative sampling:* Stir emulsions thoroughly before sampling to avoid errors caused by phase separation or sedimentation.
4) *Documentation:* Record all observations, including visual defects (cracking, whitening, foaming), along with numerical data.

Table 13.2 Common testing methods for emulsion polymers, paints, and adhesives.

#	Test method	Application area	Standard methods/typical equipment
1	Solids content	Latex, paints, adhesives	ISO 3251, ASTM D1644/Oven, balance
2	pH measurement	All latex products	ASTM E70/pH meter
3	Viscosity	Latex and formulated products	ASTM D2196, ISO 2555/Brookfield viscometer
4	MFFT	Paints and coatings	ASTM D2354/MFFT bar
5	Glass transition temp (Tg)	Film properties	DSC (ASTM D3418)/DSC apparatus
6	Wet scrub resistance	Wall paints, coatings	ASTM D2486/Scrub tester
7	Adhesion	Adhesives and paints	ASTM D3359, ISO 2409/Cross-cut tester
8	Water resistance	Exterior paints, adhesives	ASTM D870/Humidity chamber
9	Gloss measurement	Paints, varnishes	ASTM D523/Gloss meter
10	Residual monomer	Latex quality	GC, HPLC, ISO 13320/Gas chromatograph
11	Film flexibility	Coatings, adhesives	ASTM D882/UTM
12	Blocking resistance	Paints (blocking prevention)	ASTM D4946/Blocking test setup
13	Freeze–thaw stability	Latex packaging, paints	ASTM D2243/Freezer/cycle test setup
14	Chemical resistance	Adhesives, coatings	ASTM D1308/Watch glass, reagents
15	Mechanical stability	Latex handling/pumping	Internal (stress test)/High shear mixer
16	Electrolyte stability	Primer and hard water resistance	Internal method/Salt solution, stirrer
17	Coalescence efficiency	Low-VOC paints	Drawdown and observation/Drawdown tools
18	Storage stability	Latex and paint QC	ASTM D1849/Stability chamber
19	Particle size and distribution	Latex particle uniformity	DLS, SEM, Laser Diffraction/Zetasizer, SEM
20	Density	Product QC and shipping	ASTM D1475/Density cup, pycnometer

(Continued)

Table 13.2 (Continued)

#	Test method	Application area	Standard methods/typical equipment
21	Foam stability	Latex processing	Foam column method/Stirrer, foam column
22	Tensile strength and elongation	Flexible adhesives, sealants	ASTM D412/UTM with extensometer
23	Peel strength	Labels, packaging adhesives	ASTM D903, D3330/UTM with peel fixture
24	Water vapor transmission rate	Packaging coatings	ASTM E96/WVTR test cell
25	Oxygen transmission rate (OTR)	Barrier coatings	ASTM D3985/OTR analyzer
26	Viscoelastic properties (rheology)	Rheology modifiers	ASTM D4440, ISO 3219/Rotational rheometer
27	Hardness (pencil/shore)	Film durability	ASTM D3363, D2240/Pencil set, Shore durometer
28	Heat aging (thermal stability)	Aging durability	ASTM D573/Oven (60–100 °C)
29	Flexibility and adhesion after environmental exposure	Adhesives under stress	ASTM D1183-03/Environmental conditioning and mechanical testing

Common Pitfalls

1) *Neglecting preconditioning:* Failure to equilibrate samples under standard conditions often leads to variability in MFFT, viscosity, or adhesion.
2) *Improper test panel preparation:* Uneven film thickness or poorly cleaned substrates can distort gloss, scrub resistance, or adhesion results.
3) *Overlooking shelf stability before testing:* Testing unstable samples can give misleading data regarding true product performance.
4) *Misinterpreting accelerated weathering:* Lab exposures do not always correlate linearly with real-world service; use caution when predicting durability.

13.5 Durability and Aging of PVAc-based Adhesives and Emulsion Polymers

Durability and aging are critical performance attributes for polyvinyl acetate (PVAc)-based adhesives and emulsion polymers. These factors influence service life and long-term reliability in diverse applications. Aging may involve both physical and chemical changes in the adhesive film or polymer matrix, leading to reduced bond strength, loss of flexibility, or visual deterioration.

13.5.1 Factors Affecting Aging

The aging process in PVAc-based adhesives and emulsion polymers is influenced by environmental and operational factors:

1) *Thermal exposure:* Accelerates oxidation, chain scission, and embrittlement.
2) *Ultraviolet (UV) light:* Causes photodegradation of the polymer backbone and tackifiers.
3) *Moisture:* Promotes hydrolysis of acetate groups and migration of water-sensitive additives.
4) *Mechanical stress:* Leads to fatigue and progressive joint weakening.
5) *Chemical exposure:* Attack by solvents, oils, acids, or alkalis.

13.5.2 Methods to Predict Adhesive Lifetime

Accelerated aging protocols are widely used to predict adhesive performance under long-term exposure:

1) Heat–humidity chambers (e.g., 70 °C, 90% RH).
2) UV weathering tests (e.g., QUV tester with UVA-340 lamps).
3) Freeze–thaw cycling.

Performance evaluation typically includes lap shear strength, peel adhesion before and after aging, and chemical resistance through immersion in water, detergents, or solvents.

13.5.3 Methods to Extend Adhesive Lifetime

Additives and stabilizers play a crucial role in improving long-term durability (Table 13.3).

13.5.3.1 Case Study 1: Humid Tropical Climate Exposure

A PVAc-based flooring adhesive in Southeast Asia exhibited bond failure within 12 months under >85% RH. Accelerated tests confirmed hydrolytic breakdown. Incorporating carbodiimide-based hydrolysis inhibitor extended service life by over 3 years.

13.5.3.2 Case Study 2: Outdoor Furniture Application

A PVAc wood adhesive for garden furniture in Southern Europe degraded rapidly under UV exposure, losing 40% of lap shear strength in 6 months. Adding a

Table 13.3 Stabilizers for durability enhancement.

Stabilizer type	Function	Examples
Antioxidants	Prevent oxidative degradation	Hindered phenols, phosphites
UV absorbers	Absorb harmful UV radiation	Benzotriazoles, triazines
HALS	Scavenge free radicals from UV exposure	Hindered amine light stabilizers
Hydrolysis inhibitors	Reduce water-induced polymer breakdown	Alkylated phenols, carbodiimides
Crosslinkers	Increase cohesive strength and water resistance	Isocyanates, melamine-formaldehyde resins

benzotriazole UV absorber with HALS maintained >85% bond strength after 12 months of outdoor exposure.

13.6 Life-Cycle Perspective

Enhancing adhesive durability reduces replacement frequency, lowering material usage and environmental impact. Life-cycle assessments indicate that improved adhesive stability through stabilizers can cut carbon footprint by up to 20% in some applications.

13.6.1 Durability and Aging in Emulsion Polymers

Durability issues also affect PVAc-based emulsions both in storage and after film formation. Common storage concerns include viscosity drift, coagulation, microbial spoilage, and freeze–thaw instability. Postapplication aging manifests as yellowing, chalking, gloss loss, embrittlement, and water whitening (Table 13.4).

13.7 Relevant Testing Standards

1) ASTM D1183—Standard Test Methods for Flexibility of Adhesives.
2) ISO 9142—Adhesives—Guide to the Selection of Standard Laboratory Aging Conditions.
3) EN 14257—Adhesives—Determination of Shear Strength of Lap Joints Durability.
4) ASTM D905—Strength Properties of Adhesive Bonds in Shear by Compression Loading.

Table 13.4 Common aging issues in emulsion polymers and remedies.

Aging phenomenon	Likely cause	Prevention/control measures
Viscosity drift	Slow polymerization, pH drift	Optimize stabilizer package, maintain pH buffers
Coagulation	Electrolyte contamination, surfactant loss	Raw material control, biocide use
Yellowing	Oxidation, residual monomers	Use antioxidants, purify latex
Chalking	UV degradation	Add UV absorbers, HALS
Water whitening	High hydrophilicity, poor coalescence	Hydrophobic monomers, coalescing agents

13.8 Strategic Keys

1) Select stabilizers based on end-use environment.
2) Balance stabilizer dosage to maintain adhesion performance.
3) Validate through accelerated and real-time aging tests.
4) Incorporate chemical modifications when prolonged UV or moisture exposure is expected.
5) Use recognized standards to substantiate durability claims.

14

Design, Safety, and Instrumentation and Automation of a 5-Ton Emulsion-Polymerization Plant

14.1 Introduction

Designing and commissioning an emulsion-polymerization plant is a multidisciplinary endeavor that integrates reaction engineering, fluid mechanics, heat and mass transfer, control strategy, and safety management. Decisions on reactor internals, agitation strategy, feed introduction, and heat-removal pathways have direct implications for conversion control, particle-size distribution, latex stability, and operator safety. Inadequate mixing or insufficient heat removal can amplify exotherms and create local hot spots; poor feed dispersion can trigger secondary nucleation, gel formation, or unstable particle growth; and ineffective pressure and vent management can compromise plant integrity.

This chapter serves as a practical guide for factory owners, technical managers, and process developers who must translate laboratory know-how into robust, large-scale operations. The emphasis is on actionable principles that govern semibatch, starved-feed emulsion polymerization: how to achieve uniform initiator and monomer distribution, how to manage exothermic intervals without overshooting, how to maintain colloidal stability across viscosity ramps, and how to embed protective instrumentation and interlocks that prevent common hazards such as initiator overfeed or runaway.

A central theme is the use of computational fluid dynamics (CFD) to complement plant experience and scale-up heuristics. Properly configured CFD models help visualize and quantify mixing efficiency, identify dead zones and short-circuiting, and predict residence-time distributions for feeds introduced via lances or recirculation loops. Coupled thermal–hydrodynamic simulations allow evaluation of heat removal during the peak polymerization interval, including jacket and coil duty, local heat-transfer coefficients, and the impact of viscosity growth on convective transport. Species-transport and reactive-source terms can be used to examine initiator decomposition, radical and inhibitor concentration fields, and the dispersion of neutralizing agents or crosslinkers—thereby linking hydrodynamics to kinetic control and gel-risk mitigation. Transient CFD scenarios further support "what-if" analyses (e.g., feed-rate excursions, agitator speed changes, or temporary loss of cooling), offering quantitative insight that informs interlock set points and emergency quench strategies.

Throughout, safety is treated as integral to process design rather than an afterthought. Pressure relief, off-gas handling, emergency quench logic, and safeguards for exothermic surges are discussed alongside control philosophy and alarm rationalization. Where relevant, CFD findings are tied to hazard identification (HAZID/HAZOP) by demonstrating how improved mixing and heat-transfer uniformity reduce the likelihood and severity of hazardous scenarios.

While grounded in industrial practice, this chapter does not replace detailed engineering or regulatory compliance work. Readers should engage qualified process engineers to validate assumptions, boundary conditions, and control narratives. The goal is to equip decision-makers with the context and technical insight needed to ask the right questions, avoid common pitfalls, and design for performance, stability, and safety—leveraging CFD and disciplined process control to de-risk scale-up and routine production.

14.2 Process Flow Overview

This section outlines the typical process flow for a 5-ton emulsion polymerization plant, covering all essential stages from raw material charging to final product storage. The sequence described below applies primarily to vinyl acetate–based emulsion polymerization using semicontinuous feed techniques, with adaptations for safety and thermal management. A schematic flow diagram should be used as a visual reference to complement the textual description provided here.

14.2.1 Process Description

1) *Raw material preparation:* Vinyl acetate monomer (VAM), water, protective colloids, and surfactants are premeasured and stored in dedicated tanks. Feed systems include calibrated metering pumps for monomer and initiator solutions.
2) *Reactor charging:* The reactor is initially charged with a water phase containing protective colloids and part of the surfactant system. This forms the aqueous continuous phase for the polymerization.
3) *Seeding stage:* A seed latex is introduced to define the initial particle number and size. The seed may be an externally prepared PVAc latex or generated in

situ by a short nucleation shot (small monomer addition with initiator) at the target temperature. Typical seed charge is a small fraction of the final solids, adjusted to the desired particle size distribution (PSD). Agitation, pH, and conductivity are verified before the main feeds so the seed is uniformly dispersed and stable.

4) *Initiator and monomer feed:* A starved-feed strategy is employed to control heat generation and particle growth. Initiator is dosed continuously or semicontinuously in parallel with monomer so that monomer remains predominantly in the particles and the reaction rate stays stable.
5) *Temperature and pressure control:* Jacket and/or coil circuits circulate chilled water or brine to remove exothermic heat of polymerization. Automated interlocks prevent temperature overshoot, and an appropriately sized pressure-relief system ensures safety.
6) *Vacuum stripping:* After polymerization, residual monomers are removed via vacuum stripping under controlled temperature to reduce VOC content.
7) *Cooling and filtration:* The latex is cooled to storage temperature (typically 25–30 °C) and filtered through a fine mesh or bag filter to remove coagulum.
8) *Product storage:* The finished emulsion is transferred to agitated storage tanks equipped with pH adjustment and biocide dosing systems.

14.2.2 Key Considerations

Heat removal: The exothermic nature of vinyl acetate polymerization requires efficient heat exchange. Jacket or external heat exchanger systems must be sized to handle peak heat release rates.

Mixing uniformity: Agitator selection must ensure no dead zones in the reactor, preventing local hot spots and incomplete polymerization.

Safety interlocks: Critical steps such as initiator addition and monomer feed should be interlocked with temperature and agitation sensors.

Environmental compliance: Vacuum stripping is essential to minimize residual monomer levels and ensure VOC compliance.

14.3 Major Equipment and Design Considerations

This section outlines the primary equipment required for a 5-ton emulsion polymerization plant and the critical design considerations for each component. Design choices for reactors, agitators, filtration, heat removal systems, and postreaction handling significantly influence product quality, operational safety, and energy efficiency.

14.3.1 Reactor

The reactor is the core unit of the emulsion-polymerization process. For a 5-ton batch capacity, the typical working volume ranges from 6,000 to 7,000 L to provide freeboard and mixing space. The length-to-diameter ratio generally falls between 1.2:1 and 1.5:1 to balance circulation against power consumption.

The material of construction is stainless steel (SS 316L) for corrosion resistance in acidic and alkaline environments. The vessel is equipped with a full dimple jacket or half-pipe coil for heating and cooling, designed for 3–5 bar utility pressure to provide flexibility in temperature control.

Provide a filtered, pressure-relieved reactor vent to keep dust out and reduce mist or odor release; route noncondensables to a condenser or carbon canister where required.

Baffles and internals (recommended practice). Four vertical baffles are fitted to the shell, equally spaced at ninety degrees. Typical baffle width is 0.08–0.12 times the internal diameter, with a short stand-off from the shell for cleaning and a clearance from the bottom head for drainage. The arrangement reduces swirling and vortex formation, shortens blend time, improves temperature uniformity and heat-transfer performance, and lowers the risk of gas entrainment and foaming.

Connections and Nozzle Arrangement (Typical)

1) *Top head:* monomer feed with dip tube below the liquid surface; initiator feed; auxiliary feed for surfactants/thickeners/seed; vertical reflux sight condenser; horizontal return condenser (slightly inclined to the reactor); pressure-relief valve or rupture disk; nitrogen inlet; filtered reactor vent; spare nozzle for future use.
2) *Sidewall:* manway for inspection; clean-in-place (CIP) spray device connection; thermowell or temperature transmitter; level measurement nozzle (differential-pressure or noncontact radar window); pH and, if used, conductivity probe ports with retractable housings; sample valve at mid-height.
3) *Bottom head:* flush-bottom valve for full drainage; transfer/discharge nozzle; optional external recirculation connection if a plate heat-exchanger loop is used; dedicated drain with isolation.

Cleaning and Drainability

1) Provide a rotating spray device sized for full internal coverage and a dedicated CIP line. Use warm water preflush followed by an alkaline wash and a final rinse; apply periodic acid passivation as required by the materials program. Lines should slope toward drains and the vessel bottom should drain by gravity through a flush-bottom valve.

2) Minimize dead-legs and pockets; as a simple rule, keep the branch length to diameter ratio short so that stagnant zones are avoided. Prompt warm-water flushing after discharge is recommended to limit fouling in high-PVOH formulations.

Instrumentation and Interlocks

1) *Temperature:* bulk thermowell; jacket/coil inlet and outlet temperature transmitters to monitor heat-removal performance.
2) *Pressure and protection:* reactor pressure transmitter; pressure-relief device sized to the credible worst case; vacuum breaker for upset conditions.
3) *Level:* differential-pressure or noncontact radar; low- and high-level alarms for feed and safety logic.
4) *Chemistry:* pH probe with retractable housing; conductivity (optional) for monitoring salt additions or stripping end-points.
5) *Flow and* utilities: flow meters on monomer, initiator, and coolant circuits; coolant pressure indicators across jacket/coil; interlocks for minimum coolant flow and high temperature.
6) *Agitation status:* variable-frequency drive speed feedback; motor current or torque monitoring as a qualitative indicator of viscosity and abnormal shear conditions; minimum-speed interlock during feeds.

14.3.1.1 Agitator and Motor Drives (Placement Note)

Provide a top-mounted gearmotor with variable-frequency control and a double-mechanical seal with barrier fluid for latex service. Where a coaxial agitator is used, two independent drives are fitted—one for the anchor-scraper and one for the axial-flow impeller. Detailed configuration, dimensions, and power selection are presented in the next subsection on the coaxial agitator system; this avoids duplication and keeps all drive details together.

14.3.1.2 Baffles and Internals (Recommended Practice)

Four vertical baffles are fitted to the shell, equally spaced at ninety degrees. Typical baffle width is 0.08–0.12 times the internal diameter, with a small stand-off from the shell for CIP access and to avoid stagnant zones, and a short clearance from the bottom head to permit drainage. The baffles extend through the straight side and terminate near the normal liquid level. This arrangement reduces swirling and vortex formation, increases axial circulation, shortens blend time, improves temperature uniformity and heat-transfer performance, and lowers the risk of gas entrainment and foaming at low liquid levels. Where internal coils, draft tubes, or coaxial agitators are used, baffle width and clearances are adjusted to maintain adequate coil clearance and to avoid interference; verification by mixing trials or CFD is recommended before scale-up. Removable or segmental baffles may be specified to facilitate cleaning in formulations that are prone to fouling, such as high-PVOH systems.

Provide a filtered, pressure-relieved reactor vent to keep dust out and reduce mist or odor release (route to a condenser or carbon filter where required).

14.3.1.3 Reflux and Return Condensers (Recommended Practice)

Two small condensers are fitted to the reactor head to control monomer vapor, return condensate, and allow visual confirmation of reflux conditions. Both are connected to the plant cooling circuit. The intent is to condense overhead vapors during normal operation and startup—not to replace the larger condenser used for vacuum stripping.

Vertical Reflux Sight Condenser A short vertical condenser with a sight glass is mounted on the reactor top nozzle to provide visible reflux. It condenses trace monomer and water vapor rising from the headspace and returns the condensate directly to the reactor. The sight section allows the operator to verify the presence of reflux and adjust agitation or feed if excessive vapor is observed. Provide isolation valves and a drain for maintenance. Typical cooling medium is chilled water or tempered brine.

Horizontal Return Condenser (Slightly Inclined) A compact shell-and-tube or coil-in-shell condenser is installed horizontally with a slight slope back to the reactor so that condensed monomer returns by gravity. Place a demister pad or vane pack upstream if mist carry-over is observed. The unit should be isolated for cleaning and fitted with vents and drains.

Indicative Heat-exchange Area for a 5-Ton Reactor For normal overhead control on a 5-ton reactor (working volume around 6,000–7,000 L), the vapor loads during starved-feed emulsion polymerization are modest. Order-of-magnitude surface areas are as follows: these figures are for guidance and should be confirmed against plant utilities and operating limits.

1) *Vertical reflux sight condenser:* about 0.5 m^2 of heat-transfer surface is typically sufficient when using 10–15 °C cooling water or 5–10 °C brine. This covers trace monomer/water vapor with high condensation coefficients and provides visible reflux without flooding the sight section.
2) *Horizontal return condenser:* about 2.0 m^2 of surface is commonly adequate for normal overhead at 55–70 °C reactor temperature with 10–20 °C cooling water. Provide a 1–2% slope toward the reactor and ensure an unobstructed condensate return path.

Integration and Safety Notes Route both condensers to the same cooling circuit where feasible and provide isolation valves for maintenance. Include a filtered, pressure-relieved reactor vent and, where required, route noncondensables to a small carbon canister or condenser-receiver downstream. Maintain a slight positive nitrogen pressure if specified by the hazard assessment. Confirm slopes, seals, and drains during commissioning to ensure uninterrupted condensate return.

14.3.1.4 Agitator Selection

Efficient mixing is critical for uniform PSD and effective heat removal during polymerization. Poor mixing can lead to localized hot spots, gel formation, and uneven polymer properties.

14.3.1.5 Coaxial Agitators for Dual Viscosity Operation

The coaxial agitator is uniquely suited to reactors that must handle both high-viscosity PVAc homopolymerizations and low- to medium-viscosity copolymer batches. It consists of two impellers mounted on concentric shafts. The outer anchor, operating at low speed and high torque, provides wall sweeping and bulk circulation, which are essential once the polymer mass exceeds 20,000 mPa s and enters the laminar regime. In contrast, the inner turbine or propeller operates at a higher speed, supplying turbulent dispersion and homogenization in the early, low-viscosity stages of the reaction, as well as during copolymer production where viscosities exceed 10,000 mPa s.

The true strength of the coaxial design lies in its flexibility. At the start of the batch, the turbine dominates, dispersing hydrophobic comonomers, emulsifiers, and initiator streams. As viscosity rises, the operator gradually reduces turbine speed and allows the anchor to take over, ensuring safe heat removal and continuous circulation even at viscosities approaching 500,000 mPa s. This smooth handover makes it possible for a single reactor to produce a wide range of emulsion polymers (see Figure 14.1)

To achieve this flexibility in practice, the system requires two independent motors. The anchor must be driven by a slow, high-torque gearbox, while the turbine requires a separate high-speed drive. This independence allows the operator to balance laminar and turbulent mixing regimes as needed, making the coaxial agitator the most versatile option for dual-viscosity operation. The operational roles of the two impellers are summarized in Table 14.1.

Coaxial Agitator—Drives, Modes, and Control Summary. Two independent VFD-driven motors rotate concentric shafts: a slow, high-torque anchor impeller for wall renewal, bulk circulation, and heat removal at high viscosities, and a higher-speed turbine impeller for efficient dispersion at lower to medium viscosities. The anchor is driven via a low-speed gearbox, while the turbine is direct-coupled

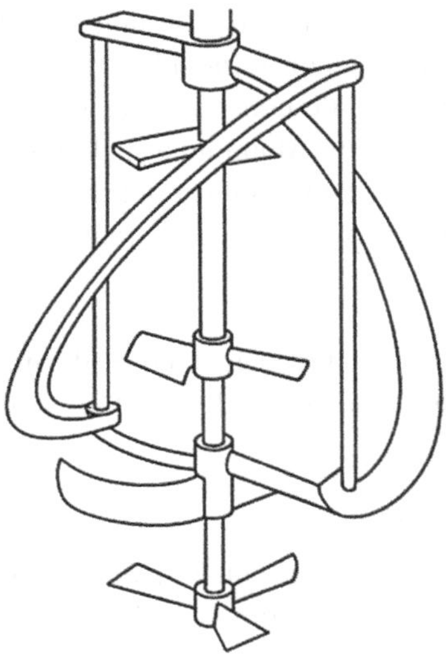

Figure 14.1 Schematic of a coaxial agitator with independent anchor and turbine impellers. The outer anchor provides slow, high-torque bulk circulation and wall-scraping action for high-viscosity PVAc systems, while the inner turbine impeller provides higher-shear mixing and efficient dispersion at lower to medium viscosities. Two separate top-mounted drives share a common mounting head: a low-speed gearbox driving the hollow outer shaft connected to the anchor, and a high-speed drive powering the concentric inner shaft for the turbine. This configuration allows effective operation across a wide viscosity range and enables smooth transition between mixing regimes. Such agitator systems are well suited for reactors used to produce both high-viscosity PVAc homopolymers and medium-viscosity VAc-based copolymers in the same vessel.

Table 14.1 Functional roles of coaxial agitator impellers.

Impeller	Speed range	Viscosity range	Primary role	Typical use
Anchor (outer)	10–60 rpm	20,000 → ≥500,000 mPa s	Bulk circulation, wall sweeping, heat removal	Mid- to late-PVAc homopolymerization
Turbine (inner)	100–600 rpm	100–10,000 mPa s	Turbulent dispersion of monomer, surfactant, initiator	Pre-emulsion, early reaction, copolymer batches
Combined	Anchor + Turbine	5,000–30,000 mPa s	Transition from turbulent to laminar mixing	Mid-feed, grade transitions

or fitted with a light reducer, allowing each impeller to operate independently or simultaneously without mechanical coupling or electronic synchronization. This configuration enables flexible operating modes across a wide viscosity range and supports the production of both high-viscosity PVAc homopolymers and medium-viscosity VAc-based copolymers in the same reactor.

1) *Drives (ratings and seals):*
   a) Anchor (outer): 11–22 kW typical (up to 30 kW for very high-viscosity grades); gearbox 25 : 1–120 : 1; 10–60 rpm with VFD; double mechanical seal with pressurized barrier fluid.

   b) Turbine (inner): 7.5–15 kW; direct-coupled or light reducer with VFD; 100–600 rpm; double mechanical seal; choose diameter to limit tip speed and shear.
2) *Utilities:* Barrier-fluid system kept 0.5–1.5 bar above reactor pressure with a small cooled reservoir; site electrical 400–480 V, 3-phase; IP55 or better enclosures; MCC with local/remote selection.
3) Operating modes (typical).
   a) *Startup/low viscosity:* both on—turbine high, anchor low.
   b) *Mid-batch/rising viscosity:* both on—raise anchor, trim turbine to maintain circulation and heat removal.
   c) *Late stage/high viscosity:* anchor dominates—turbine low or off if shear-sensitive; keep a minimum anchor speed during feeds and stripping.
   d) *Post-additions:* anchor running; short turbine pulses to wet-in thickeners, neutralizer, biocide, or coalescent.
4) *Control and interlocks:* Recipe-driven set points with operator trims ±15% within safeguards. Interlocks include minimum anchor speed for feeds, minimum coolant flow, high-high reactor temperature cutback, and motor-current/torque alarms. Manual/local places feed on hold until minimum mixing is restored.
5) *Product guidance:* PVAc homopolymer: lower turbine ceiling and rely on the anchor late in the run. Copolymers: allow higher turbine speeds and mixed operation through finish, subject to shear and foam limits.
6) *Safeguards:* Feeds hold if mixing or coolant limits are breached; turbine speed is capped above a viscosity surrogate threshold; control restores automatically when conditions return to band.

14.3.2 Blender

The blender is the postreaction unit used for final homogenization, incorporation of post-additives (e.g., ammonia, biocides, thickeners, plasticizer, coalescent), viscosity adjustment, and controlled cooling prior to filtration and packaging. It is positioned directly beneath the reactor to permit gravity transfer of the finished latex, minimizing pump shear and reducing handling time. After blending and temperature adjustment, the product is routed to filtration and then to packaging.

Capacity and percent increase relative to the reactor. For a 5-ton batch with a reactor working volume of approximately 6,000–7,000 L, specify the blender working volume at 7,500–9,000 L, corresponding to roughly a 25–35% increase over the reactor batch size. This margin accommodates post-additions, water make-up for viscosity adjustment, thermal expansion, foaming headspace, and sampling.

Specify a geometric (total) volume of 9,000–11,000 L to retain 15–20% freeboard at the target working level. Where the blender must equalize two successive reactor batches before packaging, increase the working volume proportionally (e.g., 12,000–14,000 L).

1) *Materials and pressure rating:* stainless steel SS 316L with sanitary or paint-grade finish as required; atmospheric service with allowances for slight vacuum during transfers (e.g., −0.3 bar) and low overpressure (e.g., +0.5 bar) for inerting.
2) *Geometry:* length-to-diameter ratio 1.0–1.2 to favor axial circulation at moderate power; torispherical or dished heads to aid drainage and cleaning.
3) *Agitation:* low-shear anchor or anchor-scraper as the primary impeller to manage viscosity and avoid excessive latex shear; optional secondary axial-flow impeller (e.g., pitched-blade or marine propeller) for top-to-bottom turnover; variable-speed drive to tune shear and blend time.
4) *Baffles:* two to four slim baffles (0.05–0.10 times tank diameter) when using axial-flow impellers; for anchor-scraper service, minimize or omit baffles to avoid interference, relying on the scraper for wall renewal.
5) *Heat transfer:* dimple jacket or half-pipe coil sized for controlled cooldown to 25–30 °C; optional external recirculation loop with plate heat exchanger for rapid temperature control during post-additions.
6) *Instrumentation:* Temperature measurement (bulk and outlet in the recirculation loop, if fitted), pH, conductivity (optional), level, and load cells for mass-based dosing; a sample valve at mid-height; nitrogen blanketing; and a filtered tank vent to prevent dust ingress and reduce mist or odor release. Where applicable, pressure indication and high-temperature alarms are provided for process protection.
7) *Connections:* dedicated ports for ammonia/neutralizer, thickeners, defoamer, biocide, preservatives, and wash water; spray-ball for CIP; hygienic manway for inspection.
8) *Transfer and filtration:* gravity transfer from reactor; discharge through a fine mesh or bag filter sized to operating viscosity; positive-displacement pump only if layout precludes gravity flow.
9) *Quality control:* blend-time confirmation via tracer or conductivity step test during commissioning to set minimum agitator speed and mixing time without excessive shear.

Automation note: The blender should integrate with the SCADA system for real-time monitoring of temperature, agitation speed, and optional load cell data. This integration supports accurate dosing of post-additives, ensures thermal control, and enhances batch traceability.

Typical specifications for a post-addition blender used in emulsion polymer manufacturing are summarized in Table 14.2.

Table 14.2 Post-addition blender specifications.

Parameter	Specification
Mounting	Floor-mounted beneath the reactor for gravity transfer
Capacity	7,000–7,500 L (15–30% above reactor gross volume of 6,500 L)
Material	Stainless Steel 316L, insulated
Agitation	Anchor-type agitator with full-sweep blades and PTFE scrapers
Motor and drive	Variable-speed motor (7.5–11 kW) with VFD control
Cooling system	Jacketed shell for chilled water or glycol circulation
Temperature monitoring	PT-100 probe or digital sensor integrated with SCADA
Load monitoring	Optional load cells for inline dosing or batch verification
Discharge	Bottom-discharge valve connected to filtration system
Inline filtration	100–150 μm stainless-steel mesh basket or cartridge filter
CIP access	Top manhole with CIP spray ball

14.3.3 Pre-emulsification Tank

The pre-emulsification tank is used for preparing a stable monomer-in-water emulsion before feeding it into the reactor. This step ensures uniform dispersion of hydrophobic monomers, surfactants, and protective colloids, reducing nucleation issues and improving particle size control.

Typical specifications for a pre-emulsification tank used in emulsion polymerization are summarized in Table 14.3.

Table 14.3 Pre-emulsification tank specifications.

Parameter	Specification
Type	Stainless steel 316L, unpressurized vessel
Capacity	2,600–2,800 L total (2,000–2,500 L working volume)
Mounting	On load cells for mass-based feeding accuracy
Agitation	High-shear inline emulsifier or in-tank disperser (7.5–11 kW)
Access	Top manhole with vent filter and CIP spray ball
Thermal control	Insulated shell (optional jacket for cooling/heating)
Level measurement	Sight glass or electronic transmitter
Transfer method	Metered transfer to reactor using VFD-controlled pump
Safety provisions	Vacuum breaker, rupture disc (if semisealed design is used)

14.3.4 Initiator and Catalyst Feed Tanks

Dedicated tanks are required for preparing and storing initiator and catalyst solutions to ensure accurate dosing during polymerization. APS solution tank: 300L, SS316L, fitted with a gentle agitator. Reducing-agent tank (e.g., sodium formaldehyde sulfoxylate): 300L, SS316L, sealed with a nitrogen blanket; dosing via peristaltic or diaphragm pump. Optional coinitiator tank (Fe^{2+}/ascorbic acid): 50–100L, equipped with a magnetic stirrer.

14.3.5 Common Pump Types Used in Emulsion Polymerization Plants

Selecting the right pump type ensures gentle material handling, precise metering, chemical resistance, and reliable operation.

1) *Rotary lobe pumps:* used for latex transfer and emulsion discharge; provide low shear, are reversible, and handle viscous materials well.
2) *Peristaltic pumps:* ideal for accurate dosing of initiators, acids, and surfactants; self-priming with low contamination risk because the fluid contacts only the tube.
3) *Diaphragm pumps:* air-operated or mechanical; suitable for latex, monomers, and slurries; chemically resistant and can run dry.
4) *Centrifugal pumps:* best for cooling-water circuits and other noncritical transfers; deliver high flow at low cost.
5) *Gear pumps:* provide steady flow and precise metering; well suited to coalescent and plasticizer dosing.
6) *Progressive-cavity pumps:* excellent for viscous emulsions; gentle transfer and reliable discharge.

Selection Criteria

1) *Shear sensitivity:* Use low-shear pumps (lobe or diaphragm) to protect latex stability.
2) *Viscosity:* For high-viscosity emulsions, choose progressive-cavity or gear pumps.
3) *Flow accuracy:* Use peristaltic or gear pumps for precise additive dosing.
4) *Chemical resistance:* Ensure wetted parts are corrosion-resistant (e.g., stainless steel or PTFE).
5) *Maintenance:* Prefer CIP-compatible designs to simplify cleaning and reduce downtime.

Table 14.4 Common pump types and their applications.

Pump type	Typical application	Key advantages
Rotary lobe pump	Latex transfer, emulsion discharge	Low shear, reversible, good for viscous fluids
Peristaltic pump	Initiator, acid, and surfactant dosing	Accurate dosing, no contamination, self-priming
Diaphragm pump	Latex, monomer, slurry transfer	Chemically resistant, can run dry
Centrifugal pump	Cooling water, noncritical transfers	High flow rate, low cost
Gear pump	Coalescent, plasticizer dosing	Steady flow, precise metering
Progressive cavity pump	Viscous emulsions	Gentle transfer, reliable discharge

Typical Pump Setup Example

1) *Feeding monomer emulsion to reactor:* diaphragm or lobe pump.
2) *Dosing initiator:* speed-controlled peristaltic pump.
3) *Transferring final latex:* lobe or progressive-cavity pump.
4) *Cooling-water circulation:* centrifugal pump.

Common pump types used in emulsion polymer manufacturing and their typical applications are summarized in Table 14.4.

14.3.6 Filling Line for Emulsion Polymer Production

A properly designed filling line is essential for clean, efficient, and accurate packaging of emulsions. It must minimize foaming, ensure chemical compatibility, and maintain product hygiene.

Key Considerations

1) *Filling type:* Gravity or low-pressure pump-assisted to reduce foaming.
2) *Container types:* Drums (200 L), pails (20 L), jerry-cans (5–25 L), IBCs (1,000 L).
3) *Metering system:* Load-cell-based or flowmeter-based (mass or magnetic).
4) *Material contact surfaces:* SS 304/316 or chemical-resistant polymers (PTFE, HDPE).
5) *Automation:* Semiautomatic to fully automatic depending on production volume.
6) *Defoaming precautions:* Telescopic or bottom-filling nozzles.

7) *Safety:* Grounding, overfill protection, drip-free nozzles.
8) *Cleaning:* Preferably CIP system.

14.3.6.1 Telescopic Nozzle for Defoaming in Filling Lines

A telescopic nozzle is a vertically adjustable filling nozzle used to minimize turbulence and foam formation during the packaging of foamy or sensitive liquids like emulsion polymers. It operates by descending close to the bottom of the container at the beginning of the fill cycle, then gradually rising as the product fills the container.

Advantages of Telescopic Nozzles in Emulsion Polymer Packaging

1) *Minimizes foaming:* Bottom-up filling reduces air entrapment and turbulence.
2) *Clean operation:* Prevents splashing and product overflow.
3) *Accurate filling:* Maintains a controlled, laminar flow during the entire filling process.
4) *Suitable for CIP:* Compatible with CIP systems and hygienic applications.

14.3.7 Filtration System

Filtration is a critical step in emulsion polymerization to remove coagulum, oversized particles, and foreign contaminants. Effective filtration ensures product consistency, protects downstream equipment, and improves overall reliability. In addition, product handling systems must minimize shear and prevent destabilization of the latex during transfer and storage.

14.3.7.1 Purpose of Filtration

In a 5-ton emulsion polymerization line, small amounts of coagulum, skin, gel particles, gasket shavings, and metal fines inevitably arise during polymerization, transfers, neutralization/let-down, and storage. Filtration is the final quality gate that removes these particulates before filling or application, stabilizes rheology and appearance, and protects downstream equipment and customer processes. Eventually, the following are targeted by using a filtration system:

1) Prevent blockages in coating and adhesive application equipment.
2) Enhance product uniformity and quality control.
3) Protect pumps, filling machines, and storage systems.
4) Reduce customer complaints and operational downtime.

14.3.7.2 Filtration Stages

Industrial emulsion plants generally employ a multistage filtration arrangement:

1) *Primary filtration (reactor outlet):* Removes coagulum or gel particles formed during polymerization (100–150 μm).

Table 14.5 Recommended filtration stages and micron ratings.

Filtration stage	Location	Typical micron rating	Purpose
Primary	Reactor discharge	100–150 μm	Remove coagulum, gel particles
Secondary	Before packaging	50–100 μm	Ensure smooth flow, prevent blockages
Polish (optional)	After secondary	25–40 μm	High-purity, defect-free product

2) *Secondary filtration (prepackaging or storage):* Captures finer particles before filling or bulk storage (50–100 μm).
3) *Optional polish filtration (high-purity applications):* For specialty adhesives or pressure-sensitive grades (25–40 μm).

Recommended filtration stages and typical micron ratings used in latex processing are summarized in Table 14.5.

14.3.7.3 Self-Cleaning Filtration Systems

Modern systems use automated self-cleaning filters equipped with scraping or back-flushing mechanisms, differential-pressure sensors, and stainless-steel mesh elements (40–150 μm). These systems match reactor output flow rates (typically 2–10 m^3/h) and provide continuous operation with minimal downtime.

Advantages

1) Continuous operation with minimal downtime.
2) Reduced labor for manual cleaning.
3) Improved hygiene and process safety.

Technical Note: SCADA Integration: Differential-pressure transmitters across filter housings should be connected to the supervisory control and data acquisition (SCADA) system. This enables real-time monitoring of clogging trends, triggers alarms when pressure drop exceeds set limits, and supports predictive maintenance to avoid unplanned shutdowns.

14.3.8 Cooling System and Heat Removal

Vinyl acetate monomer (VAM) polymerization is one of the most highly exothermic reactions among commonly used emulsion polymerization monomers, releasing approximately 72 kJ/mol of heat. This makes VAM-based systems the most demanding in terms of cooling capacity compared to acrylates or VeoVa derivatives.

The reactor's cooling system must be capable of removing peak heat generation during the most exothermic phase to prevent thermal runaway. For 5-ton scale plants, this typically involves a jacketed reactor combined with an external heat exchanger loop to maximize heat transfer efficiency.

Design Note

1) Cooling-system sizing must be carried out by a qualified engineer. Although chemists can estimate reaction heat from monomer feeds, only an engineer can translate that estimate into the correct capacities for the chiller, the cooling tower, and the circulation loop. Engineering design accounts for reaction upsets, fouling, and seasonal changes in utilities to maintain safe operation.
2) *Chiller capacity:* The chiller should cover the maximum expected heat load from the polymerization, with a safety margin of 20–25%. This margin absorbs unexpected spikes in reaction rate, variations in coolant temperature, and gradual loss of efficiency.
3) *Cooling tower:* For water-cooled systems, the tower is selected on total heat rejection, which is typically higher than the chiller load because it includes compressor and auxiliary heat. The tower must be specified for the local climate, with attention to hot summers, freeze protection, and water treatment for scale, corrosion, and biological control.
4) *External latex cooling loop:* After polymerization, hot latex is circulated through a heat exchanger to reach the storage temperature. The decisive factors are: (i) sufficient exchanger surface area to cool the entire batch within the allowed time; (ii) adequate latex and cooling-water flow rates to move heat effectively without damaging the latex; and (iii) the viscosity of the product, which slows cooling and may require more area or colder coolant.
5) *Why professional design matters:* If cooling is undersized, the reactor can overheat during polymerization, increasing the risk of runaway, or the latex can remain too hot during storage, increasing the risk of instability or microbial growth. Oversized equipment increases capital and operating costs. Engineering design balances these outcomes by using documented heat balances and applicable safety standards so that the plant operates safely and economically.

14.3.8.1 Cooling System Architecture and Component Functions (Chiller and Cooling Tower)

The reactor is cooled via a three-loop arrangement that transfers heat in stages without mixing fluids: a chilled-water (CHW) process loop between the reactor jacket and the chiller evaporator; the chiller's refrigerant loop, which absorbs that heat; and a condenser-water (CW) loop from the chiller condenser to the cooling tower (CT), which rejects heat to the atmosphere. This architecture decouples process dynamics from heat rejection, stabilizes reactor temperature via a buffer tank and process pump, and protects product quality by keeping the CHW loop clean and closed. VAM has one of the highest heats of polymerization among common monomers, making a conservative cooling design essential.

For a 5-ton reactor, specify a chiller of about 75 refrigeration tons (TR) and a cooling tower sized on total heat rejection, typically 1.2–1.3 × the chiller load (≈ 90–100 TR), providing margin for the peak VAM exotherm and future expansion.

14.3.8.2 Key Components and Their Functions

1) Evaporator (chiller): Absorbs heat from the chilled-water loop returning from the reactor jacket or external heat exchanger.
2) Compressor (chiller): Raises the pressure and temperature of refrigerant vapor, enabling condensation at a higher temperature.
3) Condenser (chiller): Transfers heat from refrigerant to the condenser-water loop, which then carries it to the cooling tower.
4) Cooling tower: Rejects heat from condenser water to the atmosphere via evaporative cooling.
5) Buffer tank: Provides thermal inertia, stabilizing chilled-water temperature, and ensuring minimum chiller run time.
6) Pumps: Circulate chilled water through the reactor jacket and condenser water to and from the cooling tower.

14.3.8.3 Energy Efficiency Enhancements

Variable frequency drives (VFDs) are applied to pumps, fans, and compressors to align speed with real-time load requirements. Reducing speed under part-load conditions can cut energy consumption dramatically—20% less speed can yield nearly 50% power savings due to affinity laws. Staging compressors and using VFD-driven tower fans further optimize system performance and reactor temperature control.

14.3.8.4 Cooling Media and Temperature Levels

Typical set points: chilled water (CHW) supply 6–8 °C and return 12–14 °C; condenser water (CW) supply 27–30 °C, depending on ambient conditions. For peak VAM exotherm or sub-ambient control, brine or glycol-water mixtures (−5 to 5 °C) may be required during early stages, with automatic switchover to higher-temperature utilities later in the batch.

14.3.9 Conclusion

In a water-cooled chiller, heat is managed by three separate loops:

1) *Chilled-water (CHW) loop (reactor cooling):* The reactor jacket (warm return) flows to the evaporator, then to a buffer tank, then through the CHW pump, and back to the reactor jacket as the cold supply.
2) *Refrigerant loop (inside the chiller only):* Refrigerant circulates from the evaporator to the compressor, then to the condenser, and through the expansion valve back to the evaporator.
3) *Condenser-water (CW) loop (to the cooling tower):* Water flows from the CW pump to the chiller condenser and then to the cooling tower, returning from the tower back to the pump.

14.3.9.1 External Loop Heat Exchanger (Optional): Description, Utilities, and Use Cases

Viscosity note: Because latex viscosity strongly affects loop circulation, high-viscosity grades can limit flow and raise shear. Where such products are expected, specify the external loop heat exchanger for viscous service (wide-gap plate or fouling-tolerant shell-and-tube, low-pressure drop, easy cleaning), and include a standby positive-displacement pump sized for high viscosity so the loop can be switched over when required.

Heat exchanger (for high viscosity): wide-gap plate channels or shell-and-tube with large-ID tubes; low overall pressure drop; isolation and CIP/back-flush provision; differential-pressure indication for fouling.

Pumps: primary low-shear option (rotary lobe or twin-screw) with an N + 1 standby of the same type; retain a small centrifugal only for low-viscosity campaigns. Use VFDs and generous suction piping to prevent cavitation.

Piping and layout: large-bore lines, short runs, long-radius bends; strainer on the take-off; take-off and return near the impeller zone; avoid dead legs.

Purpose: Provide a maintainable, additional heat-transfer area when jacket or coil capacity together with the overhead condenser is insufficient to remove peak exotherms with a margin. This subsection precedes the hardware comparison to clarify terminology used in Table 14.6.

When it is not needed: A correctly sized jacket or coil and condenser, supplied by a properly sized chiller and cooling tower, holds set point during feeds with a safety margin across seasons and without sustained reflux.

When it is warranted: Sustained reflux is required during normal feeds to maintain temperature; jacket or coil control valves saturate open in warm weather; recovery after feed ramps is slow; allowable jacket temperature-difference limits are reached; or maintainable surface is needed without vessel modification.

Loop description and connections: A side stream is withdrawn from the reactor through a take-off nozzle, pumped through an external heat exchanger, and returned to the reactor above the impeller zone. The simple path is, in sequence, reactor take-off; circulation pump; external heat exchanger; reactor return. Include isolation valves, a strainer, drain and vent points, and differential-pressure indication across the exchanger to detect fouling.

Utilities: Connect the exchanger to both utility circuits on separate branches—cooling-tower water for base duty and chilled water or brine for peak duty. Fit check valves or backflow preventers on each branch to avoid cross-flow.

Control approach: Use split-range control on reactor temperature. Open tower-water first; if the temperature rises, open the chilled-water or brine valve. Maintain a minimum-flow bypass on the chiller evaporator loop.

Seasonal operation: In winter or cool seasons, operate with jacket or coil and condenser only and isolate the external loop. In summer or at higher feed rates, enable the loop and use the chiller branch for peak removal; the tower branch can serve as an intermediate step in shoulder seasons.

Safeguards: Interlock monomer and initiator feeds on loss of loop flow, high reactor temperature, or chiller fault. If the loop is idle for extended periods, flush or briefly recirculate to prevent skin or coagulum formation.

14.3.9.2 Hardware Options

In a 5-ton semibatch reactor, the heat of polymerization and feed-related transients make heat removal the primary constraint on safe monomer dosing and product quality. Most plants use a hybrid cooling train: chilled brine or chilled water to handle peak exotherms, and cooling-tower water to carry the base load. The choice of heat-transfer hardware determines available area, cleanability, fouling behavior, pressure capability, and turndown, and therefore sets the ceiling for monomer feed rate, temperature control accuracy, and off-spec risk.

Scope of this section: Table 14.6 compares the principal hardware routes—dimple jacket, half-pipe coil jacket, internal coils, and external loop heat exchangers—followed by the control and redundancy practices that make a chiller–tower system reliable. Selection should reflect product fouling tendency, allowable shear, retrofit constraints, and whether the site needs high-flux brine service or can operate with tower water most of the year.

Control strategy: Use cascade temperature control with split-range valves for seamless transition between brine, chilled water, and cooling water. Implement interlocks to halt monomer and initiator feed upon loss of jacket flow, high reactor temperature, or chiller fault. Ensure minimum evaporator flow using bypass lines or low-flow trips to protect compressors.

Redundancy and safety: Provide N + 1 redundancy for critical pumps and stage fans with VFDs for part-load efficiency. Connect high-temperature alarms to

Table 14.6 Comparison of reactor cooling hardware options.

Option	Description	Advantages	Considerations/ when to use
Dimple jacket	Welded dimples on vessel shell for utility flow.	Robust, easy to clean; uniform external coverage.	Moderate heat transfer; limited area; best for medium loads.
Half-pipe coil jacket	External half-pipe channels welded to shell.	Higher pressure capability; good turbulence.	Complex fabrication; ideal for higher heat flux and brine.
Internal coil	Coils inside reactor immersed in latex.	High area density; rapid heat removal.	Difficult cleaning; fouling risk; avoid if coagulum likely.
External loop HX	Pump-around loop to plate or shell-and-tube exchanger.	Large heat-transfer area; easy maintenance.	Requires extra pumps and controls; shear risk; best for peak loads.

emergency quench and feed-cut interlocks. For cold climates, include glycol dosing and basin heaters to prevent freezing. Annex 14.A provides detailed load and component sizing calculations.

Clarification: $N + 1$ Redundancy Principle: $N + 1$ redundancy is a design principle used to ensure system reliability by providing one additional standby unit beyond the minimum required for full load operation. Here, N represents the number of units needed to meet the design load, and +1 indicates an extra unit for backup. For example, if two process pumps are required to maintain reactor cooling, an $N + 1$ configuration installs three pumps. Similarly, if three cooling tower cells are needed for heat rejection, an additional cell is included. This configuration ensures continuous operation during maintenance or unexpected failure of a unit, reducing the risk of downtime and safeguarding process safety in critical applications such as VAM polymerization.

14.3.10 Vacuum Stripping System for Reducing Free Monomer in Latex (5,000 L Reactor)

Importance: Reducing free monomer, especially vinyl acetate and butyl acrylate, directly improves product performance and acceptability. Residual monomer can plasticize the film (raising blocking and creep risk), suppress crosslinking (variable adhesion and durability), and cause tack drift. It also drives odor and VOC emissions, raising health concerns and environmental burden. Stripping to specification stabilizes storage properties, delivers predictable film formation and cure, and supports compliance with indoor-air and environmental requirements.

14.3.10.1 System Configuration

The vacuum stripping system consists of the following equipment:

1) *Stripping vessel:* Stainless steel tank (capacity 6,000–7,000 L) equipped with low-speed anchor agitator.
2) *Water ring vacuum pump:* Industrial-grade unit capable of maintaining a vacuum at 150–250 mbar. Includes water reservoir and recirculation loop.
3) *Vacuum condenser:* Positioned before the pump; condenses monomer vapors into liquid to prevent pump contamination.
4) *Vacuum receiver tank:* Collects stripped monomers and condensate.
5) *Defoamer dosing unit:* Optional system to control foam during stripping.
6) *Heating system:* Optional jacketed heating to maintain 40–60 °C latex temperature during operation.
7) *Control panel:* Includes pressure gauges, temperature indicators, safety valves, and vacuum controls.

14.3.10.2 Operating Procedure

1) Transfer the 5,000 L batch of latex to the stripping vessel.
2) Start gentle agitation (20–40 rpm anchor type).
3) Apply vacuum gradually to 150–200 mbar using the water-ring pump.
4) Optionally inject steam or nitrogen through a sparger to enhance monomer removal.
5) Maintain operation for 1–2 h depending on initial monomer content.
6) Monitor vacuum receiver for condensate volume and odor.
7) Stop vacuum once residual monomer reaches acceptable limits (e.g., <500 ppm).
8) Transfer latex to filling station or storage tank.

14.3.10.3 Technical Note: Optional Use of Steam or Nitrogen Sparging

While the vacuum stripping system is primarily driven by a water-ring vacuum pump, injecting steam or nitrogen gas during the process can enhance monomer removal efficiency in certain cases. These gases are introduced at the bottom of the stripping vessel through a sparger to displace volatile monomer vapors and improve mass transfer.

14.3.10.4 What Are Spargers

In essence, spargers are like tiny showerheads for gas. They introduce gas, typically air or oxygen, or nitrogen, or steam into liquids in a controlled manner. The key lies in their ability to disperse the gas into small bubbles, maximizing the surface area for efficient contact with the liquid.

14.3.10.5 Steam Injection Benefits

1) Helps strip residual monomers more rapidly through vapor-phase exchange.
2) Slightly increases temperature, accelerating volatilization.
3) Used when high residual monomer content (>1%) is present and rapid stripping is needed.
4) Not suitable for heat-sensitive latexes or when water content must be tightly controlled.

14.3.10.6 Nitrogen Sparging Benefits

1) Provides an inert atmosphere to sweep out monomer vapors without adding moisture.
2) Ideal for latex systems that are moisture-sensitive or heat-sensitive.
3) Safe, nonreactive, and compatible with most formulations.

Important Note: Regardless of whether steam or nitrogen is used, the core system relies on the water-ring vacuum pump for maintaining low pressure and drawing off volatile monomers. Gas injection is a performance enhancer and not a replacement.

14.3.10.7 Best Practices

1) Use a closed water circuit in the vacuum pump with integrated chiller for cooling.
2) Condenser must be efficient to protect vacuum pump and meet environmental emission limits.
3) Ensure accurate monitoring with offline testing (e.g., gas chromatography).
4) Design vessel volume 15–30% larger than batch size for headspace and safe operation.

14.3.10.8 Water Make-up Tank for Water-ring Vacuum System

A water-ring vacuum pump system inevitably loses water through evaporation, monomer vapor entrainment, and periodic purging. To ensure stable operation, maintain vacuum efficiency, and protect pump components, a dedicated water make-up tank should be integrated into the vacuum system.

Purpose and Function of Water Make-up Tank
1) Maintains consistent seal water volume in the closed-loop system.
2) Replenishes losses automatically using float or level-controlled valve.
3) Supports integration with a chiller or cooling loop.
4) Provides means to purge and refresh seal water to remove contamination.

Recommended Configuration of the Water Make-up Tank Tank material: SS 304/316 or HDPE for chemical resistance. Capacity: 200–300 L for a 5,000 L reactor system. Level control: float valve or programmable logic controller (PLC)-controlled solenoid. Monitoring: low-level indicator or alarm. Cooling: integrated coil or chilled-water loop. Overflow: bleed line to allow manual or automatic purging.

Placement The water make-up tank is connected to the closed water circuit feeding the vacuum pump. It should be located near the vacuum system skid and plumbed to allow easy access and visibility.

14.3.10.9 Operating Parameters for Vacuum Stripping of Emulsion Latex
Maintain a vacuum of about 150–250 mbar (absolute) with the latex at 40–60 °C; typical stripping time is 1–2 h, depending on residual monomer level, condenser capacity, and overall system efficiency.

Technical Justification At 150–250 mbar, residual vinyl acetate and butyl acrylate can be removed effectively without resorting to high temperatures; operating at 40–60 °C aids monomer vaporization while preserving latex stability. Temperatures above ~65 °C increase destabilization risk and should be avoided. A deeper vacuum (below ~150 mbar) is more efficient but requires higher-performance condensers and pumps to handle the vapor load.

14.3.10.10 Technical Datasheet Water Ring Vacuum Pump for Emulsion Latex Stripping (5,000 L Reactor)

1) Type: single-stage or two-stage liquid-ring vacuum pump.
2) Suction capacity: 150–250 m^3/h.
3) Ultimate vacuum: 150–200 mbar (absolute).
4) Motor power: 5.5–7.5 kW.
5) Seal-water flow: 30–50 L/min (closed loop).
6) Material of construction: SS 316 or epoxy-lined cast iron.
7) Inlet/outlet connection: DN 50–80 (2″–3″).
8) Water make-up port: provided (float valve or level-controlled inlet).
9) Vacuum receiver volume: 100–200 L.
10) Cooling requirement: seal water cooled to <35 °C.

Technical Notes

1) Vapor load: pump should handle approximately 25–40 kg/h (combined monomer and water vapor).
2) Operating window: 150–200 mbar is suitable to volatilize residual monomers at 50–60 °C.
3) Upstream condensation: proper use of condensers and receiver tanks reduces monomer carry-over and improves vacuum efficiency.
4) Seal-water temperature control: a chiller is recommended to keep seal water stable and preserve vacuum performance.

Recommended System Accessories

1) Monomer condenser: precondenses vapors before the pump.
2) Vacuum receiver tank: traps entrained liquid or foam ahead of the pump.
3) Chiller unit: maintains seal water below 35 °C.
4) Make-up water tank: restores evaporated/purged seal-water volume.

14.3.10.11 Technical Datasheet Condenser for Vacuum Stripping of Emulsion Latex (5,000 L Reactor)

1) *Type:* shell-and-tube or coil-in-tank heat exchanger.
2) *Heat-exchange surface area:* 2.5–4.0 m^2.
3) *Cooling medium:* chilled water (10–15 °C).
4) *Material of construction:* SS 316 or glass-lined steel.
5) *Design pressure:* 1–2 bar (gauge).
6) *Vacuum-side connection:* DN 50–80.
7) *Condensate outlet:* with trap to the vacuum receiver tank.
8) *Mounting orientation:* vertical or sloped horizontal.
9) *Thermal duty:* ≥10 kW (depending on vapor load).

1) 40 kg/h of vinyl acetate or butyl acrylate vapor under vacuum from a 5,000 kg latex batch.
2) *Cooling medium:* chilled water at 10–15 °C efficiently condenses vapors from ~60 to ~15 °C.
3) *Materials:* all wetted parts must resist monomers, acids, and frequent cleaning.
4) *Downstream equipment:* a vacuum receiver tank is recommended to handle foam or liquid carry-over.
5) *Installation:* mount for proper drainage and maximum vapor–surface contact.

14.3.10.12 Effect of Vacuum Stripping on Solids Content in Latex

Vacuum stripping removes both residual monomer and water vapor, so solids content rises and can affect downstream handling and stability.

Key Observations

1) Water in the latex begins to vaporize at reduced pressure, not just monomer.
2) Solids content often increases by ~1–3% depending on stripping conditions.
3) Higher solids can raise viscosity and affect pumpability.
4) In borderline-stable systems, modest coagulation or skin formation may occur.

Implications

1) Measure solids before and after stripping to detect shifts.
2) If solids are high, adjust with deionized water before packaging.
3) Expect viscosity changes that can influence defoaming and flow.

Best Practices

1) Strip under controlled vacuum (about 100–200 mbar) and temperature (~50 °C) to limit water loss.
2) Use antifoam and maintain continuous agitation.
3) Add deionized water poststripping as needed to meet product specification.

14.3.10.13 Conclusion

Vacuum stripping is effective for monomer removal but inevitably leads to water loss. Understanding and correcting for solids content increase is essential for maintaining latex quality and stability (Figure 14.2).

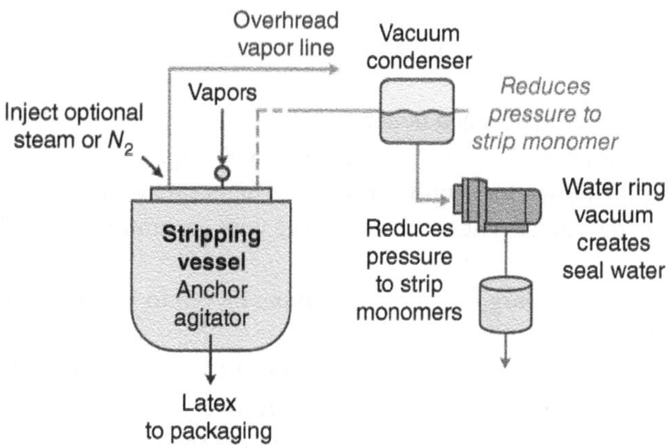

Figure 14.2 Vacuum stripping system for reducing residual monomer in latex. Latex from the blend tank is transferred to a dedicated stripping vessel equipped with a slow-speed anchor agitator to maintain bulk circulation with minimal shear. A small amount of defoamer may be added to control foam formation. Residual monomers (e.g., vinyl acetate and butyl acrylate) are removed by operating under reduced pressure; where required, a gentle stripping medium such as low-rate steam or inert gas (nitrogen) may be introduced to enhance mass transfer and monomer removal.

How the loop removes monomer: Vapors leave the vessel through the overhead line and pass first to a vacuum condenser, where most monomer–water vapor is condensed and collected. Downstream, a water-ring vacuum pump maintains the reduced pressure. The pump's seal-water tank supplies make-up water so the machine tolerates condensables and runs safely with solvent-laden vapors.

Why it works: Lowering pressure and sweeping with a light gas increases the driving force for mass transfer from the latex to the vapor phase. The condenser upstream of the pump limits solvent load to the vacuum unit and keeps off-gas handling controlled.

What leaves where: Stripped latex goes forward to packaging. Condensate from the vacuum condenser is collected for treatment or recovery. Noncondensables continue past the pump to the site's vent or abatement system.

Operating cues: Keep temperature moderate to enhance volatility without harming the latex, ramp vacuum and stripping flow to avoid foam, and interlock on foam, high condenser outlet temperature, or vacuum loss. This arrangement provides a gentle, closed-system method to reach the specified residual-monomer level while protecting equipment.

14.3.11 Steam Boiler Specification (5-Ton Reactor)

Rationale: For a 5-ton ($\approx$5 m^3) semibatch emulsion reactor, the coincident steam demand for jacket warm-up/ramps, low-pressure stripping, and minor users is modest, but a generous boiler gives resilience across seasons and startups. A 1.5–2.0 t/h unit at 6–8 barg provides ample headroom and stable control while allowing future tie-ins.

Type: horizontal, diesel/gas-fired packaged boiler.

Capacity: 1.5–2.0 tons per hour (t/h) of saturated steam at 6–8 barg.

Turndown: $\geq$8 : 1 burner turndown for semibatch swings (higher if available).

Outlet conditions: saturated steam; dryness fraction $\geq$0.95 with separator ahead of PRV stations.

Pressure reduction: PRV-A to 4–6 barg for jacket/half-pipe; PRV-B to 1–2 barg for stripping sparger and ancillary loads.

Feedwater and treatment: deaerator ($\leq$7 ppb O_2), softening/RO as required; feedwater $\approx$105 °C; continuous blowdown 3–5% with heat recovery.

Condensate: return $\geq$80–90% to the deaerator; segregate contaminated condensate from stripping service to treatment.

Efficiency and compliance: economizer, O_2 trim; stack emissions per local code; safety relief, low-water cut-off, flame supervision per standard.

Ancillaries: boiler feed pumps ($N+1$), chemical dosing, surface and bottom blowdown, sample cooler, and instrumentation for pressure, level, and conductivity.

Sizing note: The boiler is sized on connected load and desired warm-up time with a 1.2–1.3 service factor; verify against site utilities and any simultaneous users.

14.3.12 Instrumentation and Automation

Instrumentation and automation make a 5-ton emulsion-polymerization line safer, steadier, and easier to troubleshoot. In simple terms, sensors report conditions, controllers decide actions, and final elements (valves, pumps, and drives) carry them out. The automation layer coordinates dosing, temperature control, vacuum stripping, and reporting; the protection layer intervenes automatically to prevent over-temperature, over-pressure, or feed faults.

14.3.12.1 Control Architecture

A PLC executes the batch recipe, runs jacket and coil heating/cooling logic, logs key values, and manages alarms and interlocks. Above it, a SCADA system provides the human–machine interface (HMI) for operators, remote monitoring, trends and historian, electronic batch records, and recipe/version management.

For risk reduction, keep the normal control layer—the basic process control system (BPCS)—separate from the safety layer. Independent safety functions such as cutting initiator/monomer feeds, triggering emergency cooling, or stopping agitation on severe fault should be hard-wired or implemented in a safety-rated logic solver—the safety instrumented system (SIS)—so the reactor can reach a safe state even if the BPCS fails.

14.3.12.2 Critical Measurements and Placement

This subsection defines the must-have variables for the 5-ton line and, just as importantly, where to put the sensors so they read the true process rather than piping artifacts. The focus is on representative placement (well-mixed zones, correct immersion, adequate straight runs), low lag and low fouling, and clean tie-in to controls and interlocks (temperature, jacket ΔT, feed flows, vacuum/condenser duty, steam and cooling utilities, level, and selected chemistry probes). Redundancy is applied only where loss of a single measurement risks safety or product quality, and service details are specified for latex duty (materials, protection, calibration).

Temperature: Resistance temperature detectors (RTDs) or thermocouples at reactor contents, jacket inlet and outlet, emulsification tank, and (if used) monomer preheat loop. Pressure: digital transmitters on reactor vapor space and jacket side.

Level (latex in reactor and feed tanks): Measure the liquid level with guided wave radar (GWR), which uses time-domain reflectometry (TDR): a top-mounted probe sends a small microwave pulse down the probe, the liquid surface reflects it, and the instrument converts the echo time into level. GWR reads accurately in pressurized vessels, is not affected by changes in liquid density, and copes well with agitation and moderate foam; mount the probe vertically from the top nozzle (a simple stilling tube can be used if splashing is severe). As an optional backup, you may add a differential pressure (DP) level, which infers a level from the hydrostatic pressure difference between a bottom and a top connection; DP is economical for feed tanks but depends on liquid density and needs clean, short impulse lines or remote seals to avoid fouling. In practice, use GWR as the primary measurement for both the reactor and feed tanks, with DP only if continuous redundancy is required; where safety policy calls for a hard shutdown, add an independent high-level point switch to stop feeds on overfill.

Agitator: VFD feedback for speed and torque monitoring.

Technical Note: RTDs: RTDs measure temperature by correlating the resistance of a metal (typically platinum) with temperature. PT100 and PT1000 are common standards, offering high accuracy (±0.1 °C), stability, and repeatability, making them ideal for reactor contents and jacket monitoring. Install RTDs in thermowells for easy maintenance without draining the system.

14.3.12.3 Flow Measurement and Dosing

Select flow meters based on stream properties and dosing precision. Mass flow (Coriolis) is preferred for monomer and low-rate initiator dosing; positive displacement meters are an alternative for small steady flows; electromagnetic meters suit aqueous surfactant solutions. All meters should be integrated with the PLC for closed-loop control and interlocked dosing.

Technical Note about Coriolis Meters: Coriolis meters directly measure mass flow and provide intrinsic density and temperature, with typical accuracy near ±0.1% of rate. They have no moving parts and perform well with reactive chemicals like vinyl acetate and peroxide initiators. Use straight runs and vibration isolation; specify enhanced-sensitivity models for very low initiator rates.

14.3.12.4 Weighing and Mass Balance

Install high-accuracy platform load cells beneath the main reactor and the emulsification/pre-emulsion tank to enable gravimetric dosing, cross-validation of flows, and real-time mass balance. This improves batch traceability and supports reconciliation during audits.

14.3.12.5 Alarms, Interlocks, and Shutdown

Alarms, interlocks, and shutdowns turn sensor readings into timely actions. The HMI—the operator's control screen within the SCADA system—shows alarms as they occur, guides responses, and enforces safe manual overrides within defined limits. When an event needs attention away from the control room, the same priorities are mirrored by short message service (SMS) text and email alerts so field personnel see what the operator sees.

Alarm Scope and Interlocks In a 5-ton reactor, the alarm set typically covers over-pressure or vacuum protection, high reactor and jacket temperatures, loss of jacket or coil flow, agitator and variable-frequency drive faults, and feed-flow deviation (high, low, or no-flow). Interlocks act automatically without operator delay: monomer and initiator feeds are cut on high temperature, low mixing, or feed-fault conditions; emergency cooling opens, or coolant flow is boosted on approach to a high-temperature trip; and recipe steps are held until conditions return to the safe band.

Table 14.7 Alarm-to-action map (5-ton reactor).

Signal	Trigger/threshold	Automatic action (interlock)	What the operator sees (HMI)	Remote alert (SMS/email)
Over-pressure	High-high reactor pressure	Trip feeds; stop turbine; open relief path per design; hold recipe	Red trip banner; first-out identifies pressure; trend linked	High-priority alert with tag, value, and time
High reactor temperature	High-high temperature or rate of rise	Cut feeds; open emergency cooling; bias motor speeds down	Red alarm with help page; temperature and coolant trends autodisplay	High-priority alert; escalation if unacknowledged
Loss of jacket/coil flow	Flow < minimum or ΔP abnormal	Hold feeds; force safe agitator mode; request utility check	Amber alarm; interlock status shows permit denied	Medium-priority alert to utilities and operations
Agitator/VFD fault	Drive trip or high current/torque	Stop turbine; keep minimum anchor if available; hold feeds	Alarm on drive faceplate with cause; restart permitted only after reset	Medium-priority alert with drive ID
Feed deviation	High, low, or no-flow vs. set point	Cut offending feed; hold recipe; request operator check	Alarm on feed loop; link to trend and SOP	Medium-priority alert to operations
First-out trip	Any protective trip condition	Freeze current state; log first-out cause	First-out banner identifies initiating cause	High-priority alert summarizing cause

Typical alarm conditions and their corresponding automatic interlocks and operator responses for a 5-ton reactor are summarized in Table 14.7.

HMI Behavior in Practice The HMI presents a clear alarm banner with consistent colors (e.g., red for high priority) and audible tones. Each alarm shows its state—active-unacknowledged, active-acknowledged, cleared-unacknowledged, or cleared-acknowledged—and links to a plain-language help page describing likely causes, consequences, and recommended steps. From any alarm, the operator can jump to the associated trend and control faceplate to confirm cause and effect before taking action. Acknowledgments are time-stamped with username; shelving (temporary mute) is allowed with automatic unshelving; permanent suppression requires a formal procedure.

Priorities and Nuisance Avoidance Alarms are prioritized—high, medium, and low—so the most urgent events stand out. Set points, delays, and hysteresis are rationalized to prevent chatter. During known transients such as startup and shutdown, logic or short-time dead-bands filter expected deviations without globally disabling protection.

Audit Trail and Learning An event journal records time-stamped alarms, acknowledgments, set point and mode changes, and interlock actions. Links to standard operating procedures (SOPs) are available for rapid reference, and a current contact list supports callouts. After any trip, the journal and first-out record support postincident review and continuous improvement.

14.3.12.6 Data Integrity and Cybersecurity

Implement audit trails, user access levels, and recipe/version control in SCADA. Archive to a historian for trend analysis and golden-batch comparison. Segment the control network, backup PLC/SCADA configurations, and patch HMIs under change control to reduce cyber risk.

14.3.12.7 Calibration and Maintenance

Establish calibration intervals for temperature, pressure, flow, and weighing devices. Use loop checks before startup, and verify alarm/interlock logic during FAT/SAT. Maintain spare parts for critical sensors and flow meters to minimize downtime.

Technical note about FAT/SAT: Factory acceptance test (FAT) and site acceptance test (SAT) are essential stages in validating instrumentation and control systems before and after installation.

FAT is conducted at the vendor's facility before shipment. It verifies that the hardware, software, and control logic (including alarms, interlocks, and HMI screens) meet the agreed specifications. FAT ensures that errors are corrected early, minimizing costly field modifications.

SAT is performed after installation at the plant site. It confirms that the system integrates correctly with actual equipment, utilities, and safety devices. SAT also checks wiring, signal ranges, calibration, and communication between the control system and field instruments.

Together, FAT and SAT ensure that the instrumentation and automation system is reliable, safe, and compliant with operational requirements before commissioning.

14.3.12.8 Operator Interface and Reporting

Provide clear HMIs with trends for temperature, pressure, agitation speed, and feed flows. Generate automatic batch reports including set-point profiles, deviations, and interlock events to support continuous improvement.

Technical Note: HMI: The HMI is the operator's primary interface with the control system, typically displayed on SCADA or local touch panels. It provides real-time visualization of reactor temperature, jacket flow, agitation speed, pressure, and dosing rates. HMIs allow operators to start/stop equipment, adjust set points, and acknowledge alarms. Best practices include clear process graphics, alarm prioritization, trend charts for critical variables, and SOP-driven operator guidance messages.

14.4 Safety and Risk Management

Safety in VAM emulsion polymerization is critical because of the strong exothermic nature of the reaction, the volatility of monomers, and the decomposition hazards of organic peroxides. A robust safety strategy reduces the likelihood of thermal runaway, pressure excursions, and hazardous releases. This section outlines key hazards, engineering safeguards, control measures, and emergency protocols.

14.4.1 Key Hazards

Several hazards must be carefully managed in vinyl acetate polymerization. Exothermic runaway is a primary concern, as the reaction releases approximately 72 kJ/mol; loss of cooling capacity or excessive initiator feed can result in uncontrolled temperature rise. Gel formation may occur if agitation is insufficient or if dosing errors create hotspots, leading to localized polymerization and rapid viscosity increase. Pressure spikes can develop during emergency quench when steam is generated, or when monomer vaporizes under vacuum stripping faults.

Finally, initiator decomposition presents a hazard: improper storage or dosing of organic peroxides can cause rapid decomposition accompanied by sudden gas release.

14.4.2 Engineering Safeguards

A range of engineering safeguards is employed to protect against critical hazards in vinyl acetate polymerization. Pressure safety valves (PSVs) and rupture disks provide direct protection to the reactor shell by relieving overpressure. An emergency quench system, based on the direct injection of chilled water or chemical reducing agents, is used to arrest runaway reactions. To maintain heat removal capacity under fault conditions, $N+1$ cooling reliability is recommended, with redundant pumps and chiller loops ensuring continuity of service. Finally, nitrogen blanketing of the reactor vapor space reduces oxygen ingress and mitigates flammability risks.

Typical critical interlocks and corresponding operator responses implemented as engineering safeguards in vinyl acetate polymerization are summarized in Table 14.8.

14.4.3 Instrumentation and Control-based Protection

Instrumentation and automated controls provide an additional layer of safety. A high-temperature interlock immediately cuts monomer and initiator feeds if reactor temperature rises above the set point. Loss of jacket flow detection automatically trips feeds and alerts the operator, preventing thermal imbalance. In severe cases, an emergency cooling override forces the jacket circuit to switch to brine service at full valve stroke to maximize heat removal. In parallel, pressure interlocks safeguard the reactor by either activating the relief system or halting feeds upon detection of high pressure.

Table 14.8 Critical interlocks and actions.

Condition (input)	Automatic action	Operator response
High reactor temperature	Cut monomer and initiator feed; open emergency cooling	Verify circulation; prepare for quench
Loss of jacket flow	Stop feeds; alert operator	Restore flow; investigate pump/system fault
Over-pressure	Stop feeds; relieve pressure through PSV/rupture disk	Isolate sources; monitor pressure decay
Agitator trip	Hold feeds; alarm operator	Inspect impeller; clear coagulum; restart safely

14.4.4 Emergency Shutdown Procedures

In case of runaway indication:

1) Stop monomer and initiator dosing.
2) Switch jacket to maximum cooling (brine service).
3) Inject emergency quench water or reducing agent if temperature continues to rise.
4) Maintain agitation until safe to stop to prevent localized gel formation.

14.4.5 Fire and Explosion Mitigation

Fire and explosion risks are mitigated through atmospheric control, suitable equipment selection, and barriers against flame propagation. Nitrogen blanketing is maintained at a slight positive pressure to keep oxygen below flammable limits within the reactor and connected vessels. Electrical classification is applied to monomer handling zones so that explosion-proof motors, panels, and fittings are used where required. Flame arrestors are installed on vent lines to prevent flashback from external ignition sources.

14.4.6 Personnel Safety and Training

Personnel protection relies on appropriate equipment, practiced response, and procedural discipline. Operators wear chemical-resistant gloves, goggles, and antistatic overalls selected to match the hazard assessment. Emergency drills are conducted regularly to rehearse runaway response, first response actions, and evacuation routes. SOP compliance is enforced, including verification of batch sequencing and interlock status before startup.

14.5 CFD for Hazard Analysis in Emulsion-Polymerization Reactors

Computational fluid dynamics (CFD) is a computer method that predicts how liquids and heat move inside the reactor. It solves the basic conservation laws of mass, momentum, and energy on a three-dimensional model of the vessel. In simple terms, CFD is a virtual plant trial that shows mixing patterns, temperature fields, and how fast a feed spreads, before we touch the real equipment. We use it to set safe operating limits and to check what happens during startup, normal running, and credible upsets.

14.5.1 Safety and Operability Insights from CFD

1) Where hot spots and poorly mixed zones can occur during semibatch feeds.
2) How well the jacket and coils remove heat, and where coverage is weak.
3) How long does it take a feed stream to disperse and dilute to safe levels.
4) Where shear is high enough to threaten latex stability or cause coagulation.
5) Whether gas entrainment, vortexing, or foaming is likely at planned speeds and liquid levels.
6) What temperature rise to expect, and how fast, if we lose coolant, power, or a feed valve sticks open.

14.5.2 Model Inputs

1) Geometry of the reactor, baffles, impellers, coils or jacket, and feed nozzles.
2) Planned speeds for the anchor and turbine, liquid levels, and any external recirculation loop.
3) Basic physical properties versus temperature, such as density, viscosity, and heat capacity.
4) A simple time profile for heat release that represents the reaction exotherm.
5) Coolant inlet temperature and flow ranges for jacket and coils.
6) Feed schedules for monomer, initiator, and auxiliaries, including rates and where they enter.

14.5.3 Model Outputs

1) Pictures of velocity, shear rate, and temperature with clear scales and operating conditions.
2) Time plots of maximum and average reactor temperature, and coolant temperature rise, for each scenario.
3) Mixing-time and feed-dispersion numbers for each proposed nozzle location and flow rate.
4) A short table that turns findings into operating limits and interlocks.
5) A brief note showing that predictions match simple plant checks, such as tracer tests or multipoint temperature readings.

14.5.4 Safeguards Informed by CFD

1) Minimum agitator speeds that keep the temperature uniform during feeds.
2) Minimum coolant flows and set points that avoid hot spots at the peak exotherm.
3) Safe feed-nozzle positions and ramp rates to prevent local concentration spikes.

4) Emergency actions and timing for upsets, such as coolant boost, feed cut, or quench.
5) Worst-case zones for heat and gas generation to inform relief and vent analysis.

14.5.5 Common Pitfalls to Avoid

1) Ignoring the free surface when foaming or gas entrainment is possible.
2) Using unrealistically low viscosity or a single value when the batch thickens strongly.
3) Over-interpreting color plots without looking at scales, units, and operating conditions.
4) Forgetting to include coils or the correct jacket path, which changes local cooling.
5) Drawing conclusions without any plant checks or measurements.

14.6 Thermal Runaway and Safety Design in VAM Polymerization

Vinyl acetate monomer (VAM) polymerization to polyvinyl acetate (PVAc) is among the most exothermic free-radical emulsion processes, releasing approximately 72 kJ per mole of reacted monomer. If heat removal is impaired, the system is prone to a rapid, uncontrolled temperature rise known as thermal runaway. The consequences may include excessive pressure buildup, gel formation from localized micro-polymerization, and ultimately catastrophic reactor rupture.

The specific hazards of VAM polymerization are intensified by the Trommsdorff–Norrish effect, or autoacceleration, which reduces termination efficiency at elevated conversions. This leads to a sudden increase in reaction rate and heat generation when cooling demand is already high. In addition, PSD influences runaway severity: broad distributions with unstable nucleation can create hot spots, whereas well-controlled PSD moderates heat transfer and reduces localized viscosity rise.

To prevent runaway, design measures focus on reliable heat removal and rapid intervention. Key safeguards include oversized cooling jackets and $N + 1$ redundancy in chillers, precise monomer and initiator dosing under closed-loop control, and emergency quench systems capable of flooding the reactor with chilled water or reducing agents. Interlocks linked to high temperature and pressure initiate immediate feed isolation and cooling overrides.

VAM is singled out in safety design discussions because of its unique combination of reactivity, volatility, and commercial importance. Compared to acrylates or styrene, VAM exhibits higher water solubility and a narrower safe operating

Table 14.9 Causes of thermal runaway and mitigation strategies.

Cause	Consequence	Design countermeasure
Excessive initiator feed	Rapid radical generation → temperature surge	Semibatch dosing of diluted initiator under agitation
Delayed initiator decomposition	Sudden radical release overwhelms control	Controlled heating ramp; predissolution of APS
Cooling system failure	Heat accumulates, accelerating reaction (Arrhenius)	$N + 1$ cooling redundancy; interlocks to cut feeds
Poor heat transfer	Localized hotspots and gel formation	Prevent jacket scaling; ensure regular cleaning
Overfeeding monomer	Exotherm spikes beyond cooling design	Flow interlocks; mass-based dosing verification

window, making process control more critical. At the same time, its widespread industrial use—driven by the versatile applications of PVAc in adhesives, paints, and coatings—ensures that thermal safety is not an academic issue but a central design criterion. Understanding runaway mechanisms in VAM polymerization therefore provides a framework for safer operation of all emulsion polymer systems, while justifying the specific attention given to VAM in reactor design and hazard analysis.

14.6.1 Risks of Runaway Polymerization

Runaway polymerization is an uncontrolled acceleration of reaction caused by positive feedback between heat generation and reaction rate. Even small imbalances between heat production and cooling capacity can escalate into dangerous scenarios, including excessive pressure rise and possible explosion.

The principal causes of thermal runaway in VAM polymerization and their corresponding mitigation strategies are summarized in Table 14.9.

14.6.2 Trommsdorff–Norrish Effect (Gel Effect/Autoacceleration)

At high monomer conversions (>50–60%), system viscosity rises due to polymer chain growth and increased particle crowding. This reduces termination frequency while propagation continues, causing radical accumulation and autoacceleration. The effect is characterized by rapid viscosity rise, temperature spikes, and increased gel risk.

The key features of the Trommsdorff-Norrish (gel) effect during emulsion polymerization are summarized in Table 14.10.

Table 14.10 Key features of Trommsdorff–Norrish effect.

Aspect	Detail
Trigger	High conversion (>50–60%)
Mechanism	Reduced radical termination; propagation dominates
Consequences	Viscosity surge, accelerated heat release, gel formation
Control	Controlled monomer feed, chain transfer agents, early termination

14.6.3 Influence of Particle Size Distribution (PSD) on Viscosity

PSD strongly influences latex rheology and heat removal capability. Narrow PSD leads to poor packing and higher viscosity; broad PSD improves packing efficiency and reduces bulk resistance. However, PSD control must ensure proper stabilization to avoid flocculation during polymerization.

The influence of particle size distribution (PSD) on latex viscosity and its industrial implications are summarized in Table 14.11.

14.6.4 Hot Spots and Localized Overreaction

Poor agitation creates dead zones where heat removal and mass transfer are inadequate. These regions can accumulate initiator or polymer, triggering local overheating and secondary runaway. Hot spots lead to foaming, wall scaling, and localized gelation. Adequate impeller design and baffles are critical for avoiding these risks.

Note: Always combine controlled monomer feeding, semibatch initiator dosing, and emergency quench capability with robust agitation design.

Table 14.11 PSD Influence on latex viscosity.

PSD type	Effect on viscosity	Industrial relevance
Narrow PSD	Higher viscosity; difficult agitation	May increase gel risk near high conversion
Broad PSD	Lower viscosity; better heat removal	Preferred for scale-up and late-stage conversion

14.6.5 Inadequate Pressure Relief or Absence of Quenching Mechanisms

Even in waterborne systems, thermal runaway can generate severe overpressure through rapid boiling of water or volatile monomers, foam or froth that impedes the operation of safety valves, and secondary chain reactions that accelerate pressure rise. If relief devices are improperly sized, poorly calibrated, or not designed to be foam-resistant, the resulting pressure can exceed the reactor's design limits. In the absence of an effective quenching mechanism—such as rapid injection of chilled water or appropriate inhibitors—the reaction will continue to accelerate, increasing temperature and pressure until structural failure becomes likely.

14.6.6 Best Practice

Install emergency pressure-relief devices fitted with foam arrestors to maintain relieving capacity during frothing conditions. In parallel, provide dedicated quench lines capable of rapidly introducing large volumes of chilled water or a suitable radical scavenger (e.g., hydroquinone or MEHQ), sized and routed to ensure immediate dispersion into the reaction zone.

14.6.7 Early Warning Signs and Preventive Measures

Table 14.12 summarizes the principal causes of thermal runaway risk in VAM emulsion polymerization, together with their associated consequences and recommended control measures used to minimize process instability and reactor overpressure hazards.

In practice, the onset of thermal instability is rarely sudden; characteristic early warning signs usually precede runaway conditions. Recognizing these deviations at an early stage allows operators to intervene with corrective measures before

Table 14.12 Summary of causes, consequences, and control measures.

Cause	Consequence	Control measure
Excess initiator/ delayed release	Rapid radical generation → heat spike	Semibatch feed, diluted initiator, real-time temp control
Cooling failure	Heat accumulation → exothermic loop	Interlocks, backup chillers, jacket temp monitoring
Gel effect	Viscosity spike, reduced termination	End feed early, add chain transfer agents
Poor mixing	Hot spots, local overreaction	Efficient agitation, CFD-validated impeller design
No relief/quench system	Pressure buildup, rupture risk	Install PRV with foam breaker, quenching valve

Table 14.13 Early warning signs in VAM polymerization.

Parameter	Normal range	Warning/critical level
Reactor temperature	70–80 °C (stable)	>85–90 °C and rising
Temperature rise rate	<1.5 °C/min	>3 °C/min sustained
Agitator load/torque	Steady	Sudden increase (gel formation)
Residual monomer profile	Decreasing steadily	Plateau or delayed drop

hazards escalate. Table 14.13 summarizes the most common indicators observed in VAM polymerization and the preventive responses that should follow.

Typical early warning signs indicating potential instability during VAM polymerization are summarized in Table 14.13.

14.6.8 Preventive Safety Features

Preventive design focuses on smoothing heat release, maximizing heat removal, and preventing uncontrolled initiation. Pre-emulsifying the monomer promotes uniform heat distribution and reduces localized hot spots. A high-efficiency reflux condenser remains active during the feed stage to remove latent heat and limit vapor accumulation. The jacketed reactor is plumbed to both a cooling tower and a chiller, providing a wide operating envelope and redundancy for rapid heat extraction. Initiator addition is governed by metering pumps under interlock control, ensuring automatic feed cut-off on high temperature, loss of jacket flow, or agitation faults. An emergency shutdown system provides temperature-override logic with immediate isolation of feeds and steam cut-off to halt further heat input. Routine monitoring of pH and viscosity—optionally via inline viscometry—adds early visibility to process drift, enabling operators to intervene before instability develops.

14.6.9 Best Operating Practices

Safe operation relies on stable mixing, controlled addition, and assured heat-removal capacity. Continuous agitation must be maintained before, during, and after the reaction to prevent hotspots and phase separation. Monomer and initiator feeds are introduced slowly and in sync—typically over 3–5 h—to moderate the reaction rate and heat release. Cooling capacity is verified in advance by confirming that both the chiller and cooling tower are fully operational. Initiator dosing does not commence until the reactor has reached the specified polymerization temperature (e.g., 72 °C), ensuring predictable kinetics. Overpressure protection is provided through a dedicated venting system and a rupture disc sized for credible worst-case scenarios. As a practical reinforcement, semicontinuous

Table 14.14 Comparison of agitator types for emulsion polymerization.

Agitator type	Characteristics	Application
Turbine (e.g., Rushton)	High shear, strong radial flow	Low-to-medium viscosity latex systems; emulsification stages
Anchor	Sweeps reactor wall, prevents scaling	High-viscosity latex systems; late-stage polymerization

addition of pre-emulsified monomer further reduces runaway risk by distributing heat generation over time and enabling efficient thermal control via cooling and reflux systems.

14.6.10 Reactor Agitation and Mixing Efficiency

Efficient agitation is essential for the uniform distribution of monomer, initiator, and heat in emulsion polymerization. Inadequate mixing promotes hot spots, localized overreaction, and ultimately thermal runaway. Industrial practice is dominated by two agitator families—turbine and anchor—selected according to viscosity, heat-transfer demands, and scale.

Common agitator types used in emulsion polymerization reactors and their typical applications are compared in Table 14.14.

Baffles: Vertical strips along the reactor wall to break vortex formation and improve axial mixing, preventing stratification and dead zones.

Note: CFD helps identify temperature and concentration gradients before scale-up; effective agitation reduces gel formation, limits wall scaling, and supports safe semibatch operation by maintaining uniform temperature and reactant distribution.

14.6.11 Summary of Thermal Runaway Causes, Consequences, and Control Measures in VAM Polymerization

Table 14.15 summarizes the most common causes of thermal runaway in VAM-based emulsion polymerization, their potential consequences, and recommended control measures. This structured approach helps in designing preventive strategies and implementing safety systems to minimize risks (Figure 14.3).

The sequential control logic used to prevent thermal runaway during vinyl acetate emulsion polymerization is illustrated in Figure 14.3, which summarizes the monitoring of key reaction parameters, activation of safety interlocks, and initiation of emergency quench procedures.

Table 14.15 Summary of causes, consequences, and controls for runaway prevention.

Cause	Consequence	Control measure
Excess initiator/delayed release	Rapid radical generation → heat spike	Semibatch feed, diluted initiator, real-time temp control
Cooling failure	Heat accumulation → exothermic loop	Interlocks, backup chillers, jacket temp monitoring
Gel effect	Viscosity spike, reduced termination	End feed early, add chain transfer agents
Poor mixing	Hot spots, local overreaction	Efficient agitation, CFD-validated impeller design
No relief/quench system	Pressure buildup, rupture risk	Install PRV with foam breaker, quenching valve

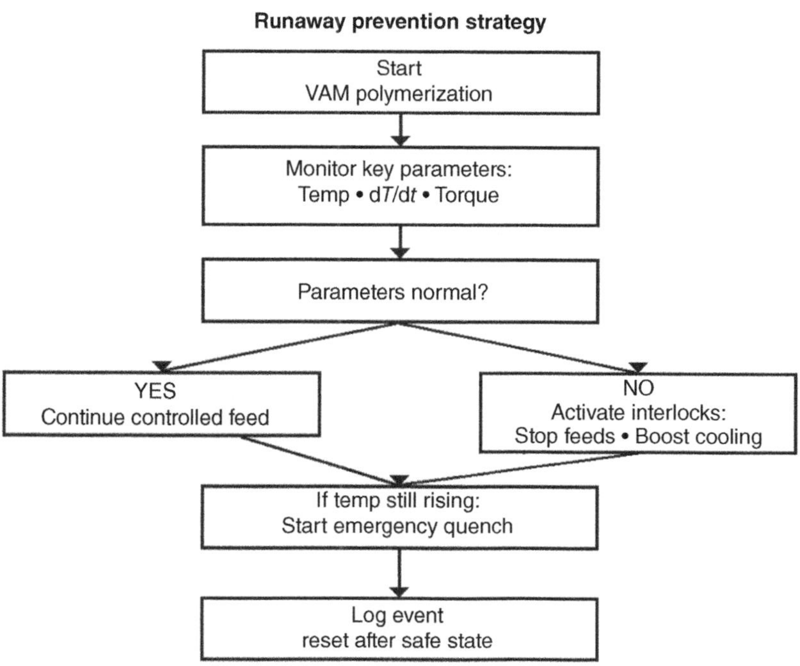

Figure 14.3 Runaway prevention strategy for vinyl acetate monomer (VAM) emulsion polymerization. The flowchart illustrates the sequential safety measures used to prevent thermal runaway during VAM emulsion polymerization, including critical process parameter monitoring, automatic interlocks, and activation of defined emergency quench actions.

15

Frequently Asked Questions in Emulsion Polymer and Adhesive Technology

15.1 Introduction

This chapter presents a collection of Frequently Asked Questions (FAQ) that address the most common problems encountered in the development, production, and application of emulsion polymers, coatings, and adhesives. The FAQ format is designed to mirror real-world practice, where formulators, process engineers, and quality control staff often frame technical challenges as direct questions.

Each FAQ provides a focused answer that combines practical experience with technical explanation. The questions cover a wide spectrum of issues, ranging from adhesion failures and film defects to foaming, block resistance, and long-term durability. The answers highlight both the underlying scientific principles and the corrective measures that can be applied in industrial practice.

By presenting problems in this accessible Question and Answer format, the chapter serves as a quick diagnostic tool for day-to-day reference, complementing the more detailed Troubleshooting Guide presented in the following chapter.

15.2 FAQ 1: Why Does My Adhesive or Coating Show Weak Adhesion to Certain Substrates?

Weak adhesion typically results from a combination of substrate characteristics and binder formulation. Substrate-related issues include low surface energy, as encountered with polyethylene (PE), polypropylene (PP), polyethylene terephthalate (PET), and untreated foils, which resist wetting. In such cases, surface treatment by corona discharge, plasma, or flame is often required. Contaminants such as grease, dust, and release agents, as well as residual moisture in porous or cementitious materials, also compromise adhesion by interfering with interfacial bonding.

Binder-related factors can be equally significant. Adhesives that lack functional groups such as carboxyl, hydroxyl, or silane exhibit limited chemical interaction with substrates. Inappropriate polymer glass transition temperature (Tg) is another common issue: polymers that are too soft may lack cohesive strength, while those that are too hard may not form continuous films. The absence of crosslinking further reduces internal strength, making interfacial failure more likely. Additionally, improper pH or ionic compatibility between binder and substrate can impair adhesion performance.

Practical Solutions: Adhesion can be improved by using silane-functional monomers, zinc oxide, or zirconium-based adhesion promoters. Binder composition should be optimized by incorporating acrylic acid, hydroxyethyl methacrylate (HEMA), or vinyl silane to provide reactive functionality. Surface cleaning is essential, and, where necessary, surface activation should be applied. Finally, careful selection of Tg and coalescent levels ensures proper wetting and cohesive integrity.

15.2.1 FAQ 2: Why Does Delamination Occur Under Humid Conditions?

Delamination refers to the separation of an adhesive or coating layer from the substrate and is often associated with moisture sensitivity, unsuitable binder design, or inadequate surface preparation. Binders that lack hydrophilic functional groups tend to exhibit poor wet adhesion because moisture weakens interfacial bonds. Improper substrate preparation, particularly the presence of dust, grease, or alkali salts, can further compromise adhesion. Low crosslink density allows moisture ingress, which softens the film and facilitates failure. In some cases, migration of additives such as surfactants or plasticizers to the interface reduces interfacial strength.

Practical solutions: To mitigate delamination, the binder should incorporate hydrophilic monomers such as acrylic acid or methacrylic acid, as well as silane coupling agents or zinc oxide to enhance wet adhesion. The application of primers specifically formulated for alkaline or damp substrates provides an additional barrier. Complete curing and uniform film formation are critical to prevent water pathways. Finally, substrate inspection and conditioning prior to application remain fundamental preventive measures.

15.3 FAQ 3: Why Is There Excessive Foaming During Polymerization or Grinding?

Foaming arises from a combination of high agitation energy, surfactant characteristics, and inadequate use of defoamers. During polymerization, excessive foam often results from the use of surfactants with high foam tendencies, particularly ethoxylated types without effective defoamer pairing. Elevated temperatures and high agitation rates exacerbate this effect, while defoamers that are either incompatible or incorrectly selected fail to suppress foam.

In the grinding phase, high-speed dispersers introduce significant air, which, when coupled with overdosing of pigment dispersants or poorly controlled addition sequences, can generate persistent foam.

Practical solutions: To prevent foaming, low-foam surfactants or defoamer-compatible stabilizers should be selected. Compatible, silicone-free defoamers should be introduced early in the process. Water addition should be gradual during grinding, and postdefoaming steps may be necessary after high-shear milling.

15.4 FAQ 4: Why Do Some Paints and Adhesives Suffer from Poor Block Resistance?

Block resistance refers to the capacity of a coating or adhesive to resist sticking to itself or to another surface under conditions of pressure and elevated temperature. Poor block resistance typically results from the use of low glass transition temperature (Tg) polymers, such as soft acrylics or VAc–2EHA copolymers, which remain tacky under heat and pressure. Overuse of coalescents or plasticizers increases film softness, further reducing block resistance. Lack of crosslinking or insufficient...

Practical solutions: Increasing Tg by copolymerizing with methyl methacrylate (MMA) or VeoVa 10 improves film hardness. Internal crosslinking or the post-addition of crosslinkers enhances structural integrity. The addition of anti–block wax dispersions provides a physical barrier, and adjusting the balance of coalescents and surfactants minimizes surface tack.

15.5 FAQ 5: What Is the "Creeping Effect" in Adhesives and Paint Films?

The creeping effect is the gradual, time-dependent deformation or movement of an adhesive or paint film under constant stress, pressure, or environmental influences such as heat and humidity. This phenomenon can manifest as slippage, sagging, distortion, or eventual failure of the bond or coating layer.

15.5.1 Why Does Creeping Occur in Adhesive Systems?

Creep in adhesives generally results from low cohesive strength within the adhesive matrix, which can be worsened by the presence of plasticizers that soften the film, excessive tackifier content leading to cold flow, or the use of soft monomers such as 2-ethylhexyl acrylate. The absence of crosslinking significantly reduces internal strength, while continuous shear or tensile stress, particularly in vertical joints, accelerates failure.

15.5.2 How Can Creeping in Adhesives Be Prevented?

Effective strategies include using harder monomers such as MMA, styrene, or VeoVa to improve rigidity; incorporating crosslinkers such as carbodiimide, aziridine, or zinc oxide to enhance structural resistance; designing formulations for plasticizer migration resistance by employing acrylics with methacrylic acid (MAA) or silane functionality; increasing molecular weight for cohesion; and ensuring clean, dry surfaces to eliminate weak boundary layers.

15.5.3 What Are the Symptoms of Creeping in Adhesive Applications?

Typical signs include slippage of labels, tiles, or decorative elements; opening of bonded joints over time; displacement under vertical load; and gradual misalignment of bonded components.

15.5.4 What Causes Creeping in Paint Films?

Creeping in coatings is commonly associated with the use of soft, low-Tg latexes, excessive film thickness, high ambient temperatures during drying, insufficient rheology control, and imbalanced pigment-to-binder ratios.

15.5.5 How to Prevent Creeping in Paint Films?

Mitigation strategies include using rheology modifiers such as HASE, HEUR, or fumed silica for sag resistance, selecting higher Tg binders for improved dimensional stability, controlling film thickness during application, optimizing coalescent selection, and incorporating thixotropic agents such as bentonite or cellulose ethers to maintain shape under gravity.

15.5.6 Is Creeping a Formulation Defect or an Application Defect?

Creep may originate from either source. Formulation-related causes include soft polymers, low crosslinking, and plasticizer uptake, whereas application issues such as excessive film thickness, high humidity, or inadequate drying can significantly worsen the problem.

15.5.7 What Types of Systems Are Most Susceptible to Creeping?

Soft acrylic or VAc–acrylate adhesives without crosslinkers, paints designed for vertical surfaces without proper rheology modification, adhesives bonding plasticized substrates such as PVC, and coatings with low solids or incorrect pigment-to-binder ratios exhibit the greatest susceptibility.

15.6 FAQ 6: What Is Plasticizer Migration in Adhesives?

Plasticizer migration refers to the movement of plasticizer molecules—commonly from a flexible substrate such as plasticized polyvinyl chloride (PVC)—into the adhesive layer or vice versa. This migration alters adhesive performance, resulting in softening, loss of strength, and, in severe cases, delamination of the bonded joint over time.

15.6.1 Why Is Plasticizer Migration a Problem?

Plasticizer migration compromises adhesive cohesion, leading to reduced bond strength, surface tackiness, bubbling, and, in some cases, staining or discoloration.

15.6.2 What Types of Adhesives Are Most Affected?

Adhesives based on polyvinyl acetate (PVAc) homopolymers and soft acrylic formulations without proper migration resistance are most susceptible. Applications involving plasticized PVC, faux leather, and vinyl flooring are particularly vulnerable.

15.6.3 How Can Plasticizer Migration Be Prevented?

Key strategies include selecting resistant polymers such as acrylics containing methacrylic acid (MAA) or silane-functional groups, incorporating barrier layers using primers or tie coats, and formulating with hard segments by adding higher-Tg monomers. Crosslinking the adhesive with carbodiimides or aziridines strengthens the network and limits plasticizer uptake. Susceptible tackifiers, such as rosin esters, should be avoided; stable alternatives should be used instead.

15.6.4 How Is Plasticizer Migration Tested?

Standard methods involve heat aging followed by peel strength retention tests, such as aging at 70 °C for seven days. Surface inspection for bubbling, softening, or staining is also conducted. Advanced techniques, such as dynamic mechanical analysis (DMA), are employed to detect reductions in modulus associated with plasticizer ingress.

15.7 FAQ 7: What Causes Discoloration of White PVAc-Based Paints During Storage?

Discoloration of white paints during storage, particularly those based on PVAc, is typically linked to the hydrolysis of residual vinyl acetate monomer (VAM). Under humid or warm conditions, unreacted VAM undergoes hydrolysis to form acetic acid and acetaldehyde. When ammonia is present—commonly used to neutralize associative thickeners (HASE, HEUR) or as a pH stabilizer—it can react with acetaldehyde to produce imine derivatives (Schiff bases) and amine–aldehyde condensation products, resulting in yellowing or brownish discoloration of the paint during storage.

15.7.1 How Can This Discoloration Be Prevented?

Effective prevention involves minimizing residual VAM through higher conversion efficiency and posttreatment steps, eliminating ammonia by substituting with alternatives such as sodium hydroxide or AMP-95, and using scavengers such as bisulfite or metabisulfite to neutralize aldehydes. Storage under cool, sealed conditions further mitigates the risk by reducing moisture ingress and hydrolysis.

15.7.2 Is This Problem Specific to PVAc-based Paints?

Yes. The issue primarily affects PVAc and its copolymers because of their hydrolytic sensitivity, residual monomer content, and frequent reliance on ammonia-based additives in waterborne formulations. By addressing both polymer chemistry and formulation practices, long-term color stability can be significantly improved.

15.8 FAQ 8: What Causes Agglomeration of Optical Brighteners in White Paints?

Agglomeration of optical brighteners (OBs), or fluorescent whitening agents, occurs when these additives lose their dispersion stability within the paint matrix.

Common causes include poor compatibility with the binder or surfactants, unfavorable pH or ionic conditions, excessive mechanical shear during processing, and interactions with polyvalent metal ions such as calcium or magnesium. Overdosing or using highly concentrated OB solutions also increases the risk of clustering.

15.8.1 What Is the Effect of Agglomerated Paints Caused by Optical Brighteners (OBs)?

Agglomerated OBs cause visible specks or streaks in the dried film, reduce overall brightness, diminish the blue undertone, and produce uneven fluorescent whitening under UV light. Such defects are highly noticeable and frequently lead to customer complaints.

15.8.2 How Can OB Agglomeration Be Prevented?

Prevention strategies include using predispersed OBs supplied in stabilized aqueous form, maintaining the pH within the range of 7.5–9.5, and avoiding high-shear environments during OB addition. The introduction of OBs is best carried out during the letdown stage rather than during high-energy grinding. The addition of chelating agents such as EDTA prevents precipitation caused by metal ions. Selecting polymers and surfactants that are tested for OB compatibility further enhances stability.

15.8.3 Is Agglomeration Reversible If It Occurs?

Not reliably. Once OBs form aggregates, redispersing them without destabilizing the entire system is difficult. In severe cases, the batch may need to be downgraded for nonwhite applications. Therefore, preventive measures are strongly recommended.

15.9 FAQ 9: What Causes Skin Formation on the Surface of Latex in Drums After Filling?

Skin formation occurs when water evaporates from the latex surface, enabling the polymer particles to coalesce into a thin film. This issue is particularly common in PVAc or VAc-based copolymer emulsions because of their relatively low minimum film-forming temperature (MFFT) and high water content.

Key contributing factors include evaporation during cooling after hot-filling, prolonged storage in partially filled or unsealed drums, exposure to warm conditions during summer, and inadequate drum agitation. The use of coalescents or surfactants that lower surface tension further accelerates skin formation.

Consequences of skin formation: Skins lead to the formation of flakes or lumps upon remixing, clogging of filters and pumps during application, inconsistent solids distribution within the drum, and an increased risk of microbial contamination beneath the surface film during long storage periods.

Prevention strategies: Skins can be avoided by filling drums completely to reduce airspace, using air-tight liners or floating disks to block air exposure, and storing products in controlled environments away from direct sunlight. The addition of small amounts of protective colloids, such as polyvinyl alcohol or cellulose derivatives, provides further surface stabilization. The use of volatile coalescents should be minimized as they promote premature film formation at the interface.

Can skin formation be reversed? Once a skin has formed, redispersion is difficult. Mechanical agitation often produces flakes, necessitating filtration and resulting in material loss. Complete restoration is rarely possible, making prevention the most reliable solution.

15.10 FAQ 10: Why Does My Latex Appear Clean Before Filtration, But Coarse Grit Forms Afterward–Sometimes Even Larger Than the Filter Mesh?

This issue commonly results from mechanical destabilization of the latex during high-shear pumping and filtration. The phenomenon is not due to preexisting contamination but is induced by turbulence and pressure gradients that cause localized coagulation.

Mechanism: High shear at pump impellers, elbows, or narrow filter screens can raise local temperatures and destabilize polymer particles. When multiple filters are installed in series, backpressure and turbulence increase significantly, intensifying the risk of grit formation. Aged latexes with low colloidal stability—due to surfactant depletion or loss of zeta potential—are particularly susceptible. The coagulated aggregates formed during this process are often larger than the original mesh size, explaining their unexpected presence after filtration.

Contributing factors include high-speed gear or centrifugal pumps, abrupt pressure drops, excessive filter stages, and the use of incompatible additives prior to filtration.

Prevention measures: The most effective strategy is to employ low-shear pumps, such as diaphragm or progressive cavity designs, for transferring sensitive latexes. Reducing the number of filtration stages and using a single, properly sized filter helps maintain laminar flow. Predilution of aged latex and avoiding harsh mixing postfiltration further reduces the risk. Filtering during letdown rather than after all additives are introduced is also recommended.

Is reversal possible? Once coagulation has occurred, grit cannot be broken down without damaging the latex structure. Refiltration may exacerbate instability, leading to further coagulation. Prevention remains the only practical approach.

15.11 FAQ 11: At What pH Do Emulsion Polymer Latexes Coagulate?

The pH coagulation point is the range where electrostatic stabilization fails, and latex particles begin to aggregate. Ionic stabilization provided by anionic surfactants or ionized functional groups is lost when these charges are neutralized by excess protons or salts. At this point, attractive van der Waals forces dominate, and the system destabilizes (Table 15.1).

Effect of acrylic acid (AA): Carboxyl groups from AA ionize at mildly acidic to neutral pH, introducing strong electrostatic repulsion. This broadens the stability window and allows latexes containing AA to remain stable even near pH 3. Conversely, systems without such functionality are highly sensitive to acidic shocks.

Practical implications: Formulators must avoid lowering pH below the stable range during thickener adjustment or latex handling. For PVAc latexes, acidic water with pH below about 4 can cause coagulation. In AA-stabilized systems, the stability window extends to lower pH, and coagulation typically occurs only when pH falls below 3 or when high electrolyte levels destabilize the latex.

Table 15.1 Typical stabilization ranges by latex type (approximate values).

Latex type	Stabilization pH range	Notes
PVAc homopolymer	4.5–5.5	Lacks internal ionic stabilization
PVAc–butyl acrylate copolymer	4.5–5.5	Limited stability beyond surfactants
Pure acrylic (noncarboxylated)	4.5–6.0	Susceptible to acidification
Acrylic or VAc copolymer with AA	3.0–4.0	Internal ionic groups enhance stability
Styrene–acrylic with AA or MAA	3.0–4.0	Carboxylic groups increase robustness
VeoVa–VAc–AA copolymer	3.5–4.5	Hydrophobicity offsets ionic effects
Polyurethane dispersions (PUDs)	4.0–5.5	Stabilized via ionic internal emulsifiers

15.12 FAQ 12: What Happens When the Monomer or Initiator Feed Rate Is Inconsistent During Semicontinuous Emulsion Polymerization?

An unstable or erratic feed rate—caused by pump pulsation, flow meter malfunction, or manual control error—can result in serious process and product issues. In semicontinuous polymerization, steady dosing is critical to controlling particle nucleation, growth, heat generation, and molecular structure (Table 15.2).

Causes of feed inconsistency include air entrainment in peristaltic or piston pumps, faulty dosing systems or flow meters, clogged tubing (especially for initiator solutions containing salts), poor temperature control affecting initiator activity, and manual dosing without automation or alarms.

Prevention strategies: Use calibrated dosing pumps and verify stroke settings regularly. Install pulsation dampeners to stabilize flow. Ensure constant agitation and temperature to avoid localized accumulation. Preemulsify monomers to prevent slug flow. For best results, integrate feed systems with automation and real-time feedback loops via PLC or SCADA.

Monitoring tools: Temperature profiles indicating exothermic spikes, online viscosity, or solids tracking to detect conversion deviations, flow and pressure alarms for early warning, and particle size analysis through frequent sampling during polymerization.

Table 15.2 Consequences of inconsistent feed rates.

Parameter affected	Impact
Particle size distribution	Broad or bimodal particle sizes due to uncontrolled or secondary nucleation.
Conversion control	Underfeeding leaves unreacted monomer; overfeeding risks runaway exotherms.
Gel formation	Localized monomer surges cause crosslinking or coagulum.
Latex stability	Sudden feed pulses destabilize micelles or surfactant layers.
Polymer structure	Copolymer composition varies across the batch, affecting Tg and properties.
Viscosity deviation	Overfeeding accelerates viscosity rise; underfeeding leads to thin latex.
Surface defects in application	Inhomogeneous particle size and MFFT distribution impair film formation.

Table 15.3 Typical odor sources in latex systems.

Source	Associated odor or note
Residual vinyl acetate	Vinegar-like, due to acetic acid from hydrolysis.
Acrylic esters (e.g., BA)	Sweet or plastic-like if unreacted.
Chain-transfer agents	Sulfurous or skunky (e.g., mercaptans).
Initiator residues	Oxidized or acidic odor from persulfate by-products.
Low molecular aldehydes	Derived from monomer or peroxide decomposition.
Surfactants/defoamers	Industrial or fatty odors depending on composition.

15.13 FAQ 13: What Causes Residual Odor in the Final Latex Product?

Residual odor in latex arises from unreacted monomers, volatile by-products, or additives remaining after polymerization (Table 15.3). These odors are common in VAc-based or acrylic latexes and may range from vinegar-like (acetic acid) to plastic-like (acrylic esters) or pungent (aldehydes or chain-transfer residues).

Consequences: Odor may lead to user complaints, regulatory concerns (VOC emissions), and perceived low quality in applications such as interior paints or food packaging adhesives.

Odor mitigation strategies: Maximize monomer conversion to >99% to minimize residual volatiles. Apply vacuum stripping or nitrogen sparging to remove volatile compounds postpolymerization. Avoid odorous chain-transfer agents (e.g., replace mercaptans with isopropanol) and use odor scavengers such as bisulfite or zinc oxide. Proper storage in sealed, cool containers limits odor development during shelf life.

15.14 FAQ 14: What Causes Fouling or Buildup on Reactor Walls During Emulsion Polymerization?

Reactor fouling occurs when polymer deposits form on internal surfaces. This is observed in two zones: (a) below the liquid level, within the reaction mixture, and (b) above the liquid level, on headspace walls where vapor condensation leads to polymerization (Table 15.4).

Table 15.4 Type 1: Fouling below the liquid level—causes and prevention.

Cause	Mechanism
Monomer impingement on wall	Polymerizes at hot metal surfaces before dispersion
Poor agitation or dead zones	Stagnant regions allow localized buildup
Localized initiator accumulation	Creates gel plugs on contact points
Uncontrolled exotherm	Accelerates polymerization near heated surfaces
Inadequate surfactant coverage	Weak colloidal stabilization near walls

Table 15.5 Type 2: Fouling above the liquid level (reactor wetting).

Cause	Mechanism
Monomer or surfactant vapor condensation	Droplets polymerize on headspace surfaces.
Poor venting or reflux control	Vapor accumulation leads to wetting and deposits.
Lack of antiwetting agents	Enables film formation above the liquid level.
No inert gas or steam sweep	Allows condensation and curing of volatile species.

Preventive strategies: Thorough preemulsification of monomers, maintaining uniform agitation, feeding monomers below the liquid surface, and using protective seed latex improve wall protection. Controlled initiator distribution prevents localized gelation (Table 15.5).

Preventive strategies: Use nitrogen or steam purges to remove vapors, maintain moisture on upper walls to prevent dry-film formation, and consider PTFE coatings or CIP spray nozzles for automated cleaning. Adding antiwetting surfactants can further reduce polymer adhesion in the headspace.

15.15 FAQ 15: Why Does High Initial Coagulum Form?

Excessive coagulum during startup generally results from poor dispersion stability, uncontrolled nucleation, or surface polymerization on reactor walls. These issues

Table 15.6 Key causes of startup coagulum.

Cause	Explanation
Insufficient surfactant or micelles	Limited stabilization for nucleated particles in aqueous phase
Poor seed latex quality or absence	No colloidal barrier during particle growth; promotes flocculation
Direct monomer addition to hot water	Monomer droplets contact reactor wall, initiating polymerization
High local initiator concentration	Causes burst nucleation and primary particle agglomeration
Rapid heating or uncontrolled dosing	Reaction begins before full emulsification
Low agitation or dead zones	Stagnant regions encourage localized reactions
VAM hydrolysis and pH drop	Formation of acetic acid destabilizes colloid early in the run

are particularly critical in vinyl acetate (VAM) homopolymerization, where the absence of internal ionic stabilization increases susceptibility to destabilization during early stages (Table 15.6).

Consequences of coagulum formation: Lower solids yield, filter and valve blockages, coarse particle development, and inconsistent batch performance. Severe cases require reactor shutdown for manual cleaning (Table 15.7).

Table 15.7 Prevention strategies.

Preventive measure	Benefit
Use seed latex or micelle buffer	Provides stabilized nucleation sites
Premix monomers as emulsions	Avoids free monomer pockets and hot spots
Add surfactants prior to initiation	Ensures micelles are available before polymer chains form
Control initiator dosing	Reduces risk of sudden burst nucleation
Maintain uniform agitation	Eliminates dead zones and ensures thermal homogeneity
Gradual heating to 70–75 °C	Enables controlled onset of polymerization

Formulation aids: Incorporate small amounts of protective colloids such as polyvinyl alcohol in the seed stage. Use staged surfactant addition for stability reinforcement. In VAM systems, rely on nonionic emulsifiers with strong wetting power and adjust pH within the optimal 4.5–5.5 range to minimize early destabilization.

15.16 FAQ 16: What Causes pH Drift During Emulsion Polymerization?

pH drift refers to unintended shifts in pH during polymerization, which compromise colloidal stability and polymer structure. For PVAc homopolymers and acrylic systems containing functional monomers, maintaining a narrow pH range is critical to ensuring stability and performance (Table 15.8).

Effects: Latex flocculation, gelation, uneven particle size distribution, poor polymer architecture, and color or odor instability from side reactions.

Control strategies: Use buffered initiator systems (e.g., persulfates with sodium acetate), dose neutralizers gradually via controlled pumps, and monitor pH continuously using compatible probes. Select surfactants tolerant to moderate pH shifts and avoid prolonged heating before initiation, as hydrolysis accelerates acidification.

Table 15.8 Common causes of pH drift and indicative ranges.

Direction	Cause	Explanation	Recommended pH Range
pH drops	Hydrolysis of VAM	Acetic acid generation at elevated temperature	≥4.5–5.0
pH drops	Acrylic acid polymerization	Carboxylic groups consume hydroxide ions	6.5–7.5 initially
pH drops	Persulfate initiators	Produce acidic species during decomposition	≥4.5
pH drops	CO_2 absorption	Carbonic acid formation in unsealed systems	≥5.0
pH rises	Ammonia-based neutralizers	Volatilization or imbalance in buffering	6.5–8.0
pH rises	Buffer breakdown	Weak phosphate buffers degrade at high temperatures	<8.5
pH rises	Overdosing on alkalis	Excess AMP-95, NaOH, or TEA	<8.5–9.0

Special considerations: PVAc systems are highly sensitive to pH drops; the ideal range is 4.5–5.5. Acrylic systems require a pH 7.5–8.5 for ionization of functional monomers such as AA or MAA.

15.17 FAQ 17: What Causes Recoagulation of Latex in the Hold Tank or Storage Drums?

Recoagulation refers to the secondary formation of coagulum during storage or transfer. It typically follows polymerization and is induced by mechanical stress, chemical changes, or contamination (Table 15.9).

Indicators: Grainy texture, lumpy aggregates, clogged filters during filling, sudden viscosity rise, and visible specks in final films.

Preventive measures: Use low-shear pumps, maintain stable pH, filter once at proper micron rating, cool and seal latex promptly, clean tanks regularly, and verify additive compatibility.

Recoagulated latex is rarely recoverable. Agitation may partially disperse lumps, but true colloidal stability is not restored.

15.18 FAQ 18: What Are the Risks of Uncontrolled Foam During Feeding?

Excessive foam during monomer or initiator feed can cause (i) apparent reactor-volume loss and poor level control; (ii) foam carryover into condensers with blockage and contamination; (iii) interfacial coagulum due to poor wetting;

Table 15.9 Common triggers and mechanisms.

Trigger	Mechanism
High-shear pumping	Breaks stabilizing layers on latex particles
Filter-induced backpressure	Multiple or tight filters create turbulence
pH drift during storage	VAM-based latex destabilizes below pH 4.5
Temperature fluctuations	Hydrolysis or evaporation accelerates instability
Contaminated or corroded tanks	Metal ions or debris induce local flocculation
Post-added incompatible ingredients	Additives disrupt colloidal equilibrium

Table 15.10 How can foam be controlled or prevented during feed?.

Strategy	Description
Select low-foam surfactants	Avoid high-foaming anionics (e.g., SDS) or highly ethoxylated surfactants
Use effective defoamers	Apply silicone- or mineral-oil-based defoamers (0.1–0.5%) compatible with the latex system
Add defoamer to preemulsion	Incorporate during preemulsification; in some cases, post-addition may be necessary
Avoid vortexing and aerated emulsions	Add monomer below the surface and prevent air entrainment during mixing
Control agitation speed during feed	Use baffles and proper impeller speed to minimize vortex formation
Defoam headspace with nitrogen/steam	Introduce inert gas or steam to collapse foam above the liquid surface

(iv) overflow with related safety and housekeeping hazards; and (v) solids-yield loss with higher risk of off-spec product rejection (Table 15.10).

Which defoamer types are most suitable?

1) *Silicone-based:* Highly efficient but risk of film defects if overdosed.
2) *Mineral-oil-based:* Economical and well-balanced performance with minimal side effects.
3) *EO/PO-based defoamers:* Self-emulsifying options suitable for acrylic systems; minimal impact on gloss.

Note: Always test for compatibility with the polymer system and evaluate potential film defects such as craters or fisheyes before full-scale use.

15.19 FAQ 19: Why Does Latex Turn Yellow During Storage?

Latex yellowing during storage signals oxidative or chemical degradation. Though it may not affect mechanical performance in darker coatings or adhesives, it is highly undesirable for white paints, paper coatings, or clear film applications (Tables 15.11 and 15.12).

Table 15.11 Common causes of latex yellowing.

Cause	Mechanism
Oxidation of residual monomers	VAM, acrylates, or styrenics oxidize, generating yellow species
Iron or metallic contamination	Fe^{3+} ions catalyze oxidation, especially under alkaline conditions
Incompatible biocides or amines	React with aldehydes forming Schiff bases or imines (yellow)
Overheating or prolonged storage	Accelerates hydrolysis and secondary discoloration reactions
Exposure to air or light	Triggers surface oxidation and photochemical degradation

Consequences: Reduced whiteness index, customer complaints, and possible brittleness if degradation affects the polymer backbone.

Table 15.12 Prevention strategies.

Preventive measure	Benefit
Add antioxidant package	Hindered phenols (e.g., BHT) delay oxidation
Store latex in sealed, opaque containers	Blocks oxygen and light exposure
Avoid contact with metals	Use stainless steel or HDPE-lined tanks
Minimize residual monomer	Optimize polymerization and apply stripping
Maintain pH in neutral range (5.0–7.5)	Avoid acid/base-catalyzed side reactions
Control storage temperature (5–25 °C)	Reduces oxidative and hydrolytic degradation

15.20 FAQ 20: Why Does My Latex Become Viscous During Storage?

Excessive viscosity during storage or handling can arise from polymer structure, colloidal instability, or external factors such as evaporation. This condition complicates pumping, mixing, and application, and often signals incipient gelation (Table 15.13).

Table 15.13 Primary causes of unexpected viscosity rise.

Cause	Mechanism
High molecular weight polymer	Over-polymerization or poor chain transfer control
Weak surfactant stabilization	Particles flocculate, forming transient networks
Neutralization of carboxylic monomers	Overdose of amines or ammonia creates electrostatic thickening
pH drift into alkaline zone	Promotes associative thickening and instability
Coagulum or gel remnants	Act as nucleation points for micro-flocculation
Additive interactions postproduction	Coalescents or thickeners may swell polymer particles
Water loss during storage	Concentrates solids and increases viscosity

Implications: Difficult pumping, poor film leveling, and higher defect risk.

15.21 FAQ 21: Why Is The Dry Film Cracking Even Though the Latex or Adhesive Is Flexible?

Cracking in flexible systems usually results from stress imbalances during drying rather than intrinsic brittleness. Even when Tg is low, nonuniform water loss or excessive film thickness can cause tension and fracture (Table 15.14).

Table 15.14 Typical causes of film cracking.

Cause	Explanation
Uneven or rapid surface drying	Skin forms before internal layers coalesce, inducing tension
Excessive film thickness	Leads to nonuniform shrinkage and internal stresses
Poor coalescence	Volatile coalescents or premature drying hinder film fusion
Incorrect plasticizer ratio	Insufficient or incompatible plasticizer causes brittleness
High pigment/filler loading	Reduces elasticity and introduces stress points
Substrate-induced stress	Dimensional changes in substrate transfer strain to the film
Temperature variations during cure	Thermal contraction creates mechanical stresses

Preventive measures include applying thin, uniform coats, using effective coalescents at correct dosage, controlling drying rates, checking filler levels, and adding nonvolatile plasticizers. Silicone resins or urethane dispersions can improve anticracking performance in demanding applications.

15.22 FAQ 22: Why Does the Gel Content Vary Significantly Between Different Batches of the Same Latex Formulation?

Gel content variation reflects inconsistencies in reaction control, monomer reactivity, or batch handling. Gel refers to insoluble fractions—crosslinked or extremely high molecular weight polymers—that resist dissolution in solvents such as tetrahydrofuran (THF) or methyl ethyl ketone (MEK). High gel levels compromise film uniformity, filterability, and application properties (Tables 15.15 and 15.16).

Gel content measurement: The Soxhlet extraction method is commonly used. A known weight of dried latex polymer film is prepared and extracted with a suitable solvent such as tetrahydrofuran (THF) or methyl ethyl ketone (MEK) for 12–24 hours to remove soluble fractions. After extraction, the insoluble residue is dried to constant weight and weighed.

Gel content is calculated as: % Gel Content = (Residue weight / Initial dry polymer weight) × 100.

Table 15.15 Primary causes of gel content fluctuation.

Cause	Mechanism
Poor initiator control	Excess radicals promote branching and crosslinking
Excessive temperature or hot spots	Accelerates chain coupling and gelation
Reactor fouling or residue	Seeds polymerization of gel in subsequent batches
Unbalanced monomer composition	Comonomers like MAA or allyl methacrylate promote network growth
CTA inconsistency	Chain transfer agent volatilization or uneven reactivity
Residual monomer after reaction	Post-crosslinking of VAM or MMA in storage
Delayed emulsifier diffusion	Secondary nucleation causes microgels

Impact on product performance: Filter clogging, gloss loss, inaccurate solids measurement, poor reactivity with crosslinkers, and shorter shelf-life.

Table 15.16 Control strategies.

Control strategy	Benefit
Maintain tight control of initiator system	Prevents excessive free radical buildup
Use consistent and calibrated CTA	Ensures predictable molecular weight control
Avoid aggressive comonomers unless needed	Reduces unwanted network structures
Apply vacuum stripping postpolymerization	Removes residual monomers to avoid secondary gelation
Standardize heating and mixing profiles	Prevents thermal or concentration gradients
Clean reactor thoroughly between batches	Eliminates contamination from previous runs

Where required, extraction time and solvent choice should be kept consistent to allow reliable comparison between samples.

The Soxhlet extraction method is commonly used:

1) Dry a known weight of latex polymer.
2) Extract with THF or MEK for 12–24 h.
3) Dry and weigh the residue.
4) Calculate: % Gel Content = (Residue weight/Initial weight) × 100

15.23 FAQ 23: How Can Foam Cause Instability in Emulsion Polymer Latexes?

Foam affects polymerization stability and final product quality by introducing gas–liquid interfaces that destabilize micelles, increase coagulation risk, and trigger secondary nucleation (Table 15.17 and 15.18).

Table 15.17 Foam-induced conditions and impacts.

Foam-induced condition	Impact on latex stability
Entrapped air bubbles during polymerization	Acts as nuclei for uncontrolled growth and coagulation
Disruption of surfactant film	Weakens particle and micelle integrity
Local monomer concentration near foam zones	Promotes runaway polymerization in microdomains
Accumulation of contaminants in foam	Leads to sticky deposits and coagulum formation
Increased apparent viscosity	Causes pumping and mixing inefficiencies

Table 15.18 Prevention measures.

Prevention strategy	Description
Add compatible defoamers early	Use silicone or EO/PO-based agents during preemulsion
Select low-foam surfactants	Avoid SDS and high-HLB emulsifiers
Degas feeds before addition	Removes entrained air from water and monomer streams
Monitor agitation speed	Prevents vortexing and excessive aeration
Apply defoamer drip to reactor headspace	Collapses foam without overdosing

15.24 FAQ 24: Why Does Grit or Coagulum Form During Prolonged Monomer Feeds?

Late-stage grit typically arises from colloidal instability as solids content rises and system dynamics shift. Surfactant depletion, compositional drift, and uneven thermal profiles are common triggers during prolonged feeds (Table 15.19).

Table 15.19 Primary causes of late-stage grit formation.

Cause	Mechanism
Surfactant depletion	Insufficient coverage for new particle surfaces
Residual monomer imbalance	Hydrophobic blobs destabilize the dispersion
Thermal gradients	Localized heating accelerates gelation
Drop in radical flux	Delays polymerization, leaving monomer pockets
Reduced shear or mixing uniformity	Dead zones promote coagulation
Coalescent accumulation	Plasticization lowers colloidal stability

Preventive strategies:

Action	Benefit
Balance surfactant and monomer feeds	Maintains micelle integrity throughout reaction
Stage emulsifier additions	Reinforces stability in high-solids phase
Adjust initiator dosing ramp	Sustains adequate radical flux to prevent monomer pooling
Install inline cooling	Avoids late-stage overheating and hot spots
Monitor mixing profile	Ensures consistent shear distribution and avoids dead zones

15.25 FAQ 25: Why Do I Observe Unexpected Secondary Nucleation or a Bimodal Particle Size Distribution During Semicontinuous Emulsion Polymerization?

Secondary nucleation or the formation of bimodal particle size distributions is usually caused by excess surfactant in the initial charge. In many semicontinuous processes, a substantial portion of the emulsifier is added before polymerization begins to stabilize the monomer preemulsion and ensure micelle-based nucleation. However, when the initial surfactant level remains significantly above the critical micelle concentration (CMC), unadsorbed emulsifier persists in the aqueous phase and initiates new micelles during the monomer feed stage.

Consequences of excess surfactant at startup:

1) Broadened particle size distribution
2) Formation of bimodal latex systems
3) Reduced predictability of polymer morphology and viscosity (Table 15.20)

Maintaining a controlled emulsifier profile across the polymerization minimizes the risk of late-stage particle nucleation and ensures a uniform particle size distribution for consistent performance.

15.26 Conclusion

The troubleshooting strategies presented in this chapter bridge the gap between laboratory success and industrial reality. By systematically addressing root causes of common failures—ranging from coagulum formation and pH drift to late-stage instability and application defects—these FAQs serve as a rapid diagnostic tool for production teams and formulators.

Table 15.20 Preventive strategies.

Action	Benefit
Balance initial surfactant dose	Matches particle surface area and total monomer charge
Target near-CMC concentration	Promotes nucleation without excess micelle formation
Monitor particle size evolution early	Detects secondary nucleation before it escalates
Stage surfactant addition during feed	Enhances stability while limiting new particle growth

Three key principles emerge repeatedly across the issues discussed. First, process control is essential: accurate dosing, pH regulation, and thermal stability form the foundation of batch consistency. Second, colloidal stability remains central, with balanced surfactant systems and controlled particle nucleation underpinning robust latex performance. Finally, preventive measures play a decisive role, since early intervention through proper formulation, equipment calibration, and raw material quality reduces the likelihood of failures.

While this chapter focuses on immediate problem-solving, real-world troubleshooting often requires deeper context. The following chapter, Industrial Case Studies in Emulsion Polymers and Adhesive Technology, complements these FAQs by presenting practical scenarios where theory, observation, and corrective actions converge to restore product quality and operational reliability.

16

Industrial Case Studies in Emulsion Polymers and Adhesive Technology

16.1 Introduction

While Chapter 15 presented a concise FAQ format for common problems, practical experience demonstrates that real-world failures often involve multiple variables that cannot be captured in brief answers. This chapter expands on the troubleshooting framework by documenting industrial case studies drawn from actual emulsion polymer and adhesive operations.

These cases illustrate a spectrum of issues, from reactor-scale process deviations and raw material inconsistencies to storage anomalies and application defects. Each case demonstrates how careful observation, systematic diagnosis, and corrective action transform a critical failure into a controlled solution.

Rather than following a rigid grouping, the case studies are presented as a collection of scenarios that reflect practical challenges encountered in diverse operational environments. This format allows the reader to follow the technical logic applied in each situation, understand decision-making processes, and draw parallels with their own production contexts.

Structure of Each Case Study

Each case follows a five-part structure:

1. Problem description
2. Observations and data collected
3. Root cause analysis
4. Corrective and preventive measures
5. Outcome and lessons learned

All cases are anonymized for confidentiality. Technical content remains accurate to ensure maximum relevance for plant engineers, formulation scientists, and troubleshooting professionals.

16.2 Case Study 1—Batch-to-Semicontinuous Transition in a PVAc Adhesive Plant

Background: A mid-sized adhesive manufacturer producing PVAc homopolymer emulsions for woodworking operated exclusively with batch polymerization reactors. Rising demand for higher solids, improved product consistency, and better energy efficiency motivated a transition to semicontinuous emulsion polymerization.

Problem statement: During initial semicontinuous trials, the plant encountered a rapid viscosity rise after approximately 70% conversion, difficulty controlling the exotherm that prompted unplanned cooling-water surges, and inconsistent particle-size distribution (PSD) that produced variable adhesive performance.

Investigation: A process review compared batch versus semicontinuous feed protocols. Real-time temperature monitoring revealed periodic spikes coinciding with monomer-feed pulses. Laser-diffraction particle sizing showed a bimodal distribution in early trials, and conversion profiles indicated mid-reaction secondary-nucleation events.

Root cause: Feed-rate fluctuations and insufficient mixing in the reactor's upper zone during semicontinuous operation allowed local monomer supersaturation, which triggered secondary nucleation and uneven heat generation.

Solution implemented: Monomer-feed control was improved by installing a mass-flow controller with ±0.5% accuracy. The agitation system was upgraded with a secondary impeller at mid-height to prevent stratification and improve circulation, and heat-removal capacity was increased by retrofitting an external loop with a plate heat exchanger.

Table 16.1 Summary of performance before and after optimization in transition from batch to semicontinuous system.

Parameter	Before optimization	After optimization
Viscosity rise after 70% conversion	+45% (runaway risk)	+5% (controlled)
Particle size distribution	Bimodal (0.15 and 0.45 μm peaks)	Unimodal (0.30 μm average)
Conversion yield	94%	99%
Cooling water usage	High (frequent surges)	Steady, reduced by 20%

Outcome: A stable viscosity profile was maintained throughout the reaction, a uniform PSD (0.25–0.35 μm) was achieved, and consistent adhesive viscosity and film strength were delivered across six consecutive production runs.

Lessons learned: Semicontinuous stability relies on precise monomer-feed control; mixing profiles typically require adjustment when scaling from batch to semicontinuous operation, and heat-removal systems must be sized for peak exothermic loads rather than averages (Table 16.1).

The key performance improvements observed during the transition from batch to semicontinuous polymerization are summarized in Table 16.1.

16.3 Case Study 2–Surfactant Ratio Effects on VAc–BA Stability

Background: A manufacturer of vinyl acetate–butyl acrylate (VAc–BA) copolymer emulsions for flexible coatings received customer complaints of product thickening after three to 4 months of storage, which reduced workability and produced visible coagulum in the pail. The emulsion had been stabilized with a fixed nonionic/anionic surfactant ratio for more than 5 years without prior issues.

Problem statement: After switching to a different nonionic-surfactant supplier, shelf life declined from approximately 12 months to less than 6 months. Microscopy showed gradual particle flocculation accompanied by settling.

Investigation: Surfactant characterization revealed that the replacement nonionic differed slightly in hydrophilic–lipophilic balance (HLB) and degree of ethoxylation from the incumbent material. Zeta-potential measurements decreased from about −38 mV (stable) to roughly −24 mV (borderline), and accelerated aging at 50 °C confirmed faster viscosity rise in the modified batches.

Root cause: The nonionic/anionic surfactant balance became suboptimal following the supplier change. The new nonionic provided reduced steric stabilization, weakening the combined electrosteric barrier and allowing gradual particle aggregation during storage.

Solution implemented: The surfactant ratio was reoptimized by increasing the anionic fraction from 30% to 45% of the total surfactant system. Screening trials were conducted with three nonionics of similar HLB but varied ethylene-oxide chain length to restore steric protection, and the formulation was adjusted until the zeta potential returned to −35 mV or lower.

Outcome: Shelf life was restored to 12 months with no coagulum observed under accelerated testing. Storage stability improved across both summer and winter conditions.

Lessons learned: Changes in surfactant suppliers can impact latex stability even when nominal HLB appears similar; zeta potential is a practical diagnostic for emerging destabilization, and maintaining electrosteric balance is essential for long-term VAc–BA stability (Table 16.2).

Table 16.2 Summary of key parameters before and after optimization of long-term stability in VAc–BA copolymer emulsions.

Parameter	Before optimization	After optimization
Zeta potential	−24 mV (borderline)	−36 mV (stable)
Shelf life	<6 months	12 months
Coagulum formation	Visible by month 4	None after 12 months
Surfactant system ratio	70% nonionic/30% anionic	55% nonionic/45% anionic

The key parameters influencing long-term stability of VAc–BA copolymer emulsions before and after optimization are summarized in Table 16.2.

16.4 Case Study 3—High-solids PVAc Emulsion Production and Viscosity Control

Background: A coatings manufacturer sought to produce PVAc emulsions at 65% solids to reduce drying time in waterborne paints and adhesives. Previous products were formulated at 55% solids; the higher-solids target aimed to improve coverage per coat and lower transport costs.

Problem statement: Initial high-solids runs produced excessive viscosity—greater than 20 000 cP—complicating pumping and transfer, along with poor flow and leveling in paint applications and occasional coagulum formation during storage.

Investigation: Rheology showed a steep viscosity rise beyond about 60% solids due to reduced inter-particle spacing. Review of the surfactant system indicated that the existing level (3.5% on monomer) provided insufficient stabilization at higher particle concentrations. Polymerization kinetics pointed to slower heat dissipation at high solids, causing localized temperature spikes that affected particle morphology.

Root cause: Increased particle crowding and incomplete stabilization led to flocculation and strong viscosity build-up, exacerbated by heat-induced chain branching during polymerization.

Solution implemented: Total surfactant was raised from 3.5% to 4.2% on monomer with a higher nonionic fraction to boost steric stabilization. The monomer feed rate was slowed during the final 20% of the reaction to limit heat build-up and particle-packing rate. Postpolymerization, a low-level dispersant (0.2% sodium polyacrylate) was introduced to improve flow and prevent settling.

Outcome: Final viscosity was controlled between 8,000 and 10,000 cP, the PSD stabilized at 0.25–0.30 μm, and no coagulum was observed after 6 months of ambient storage.

Lessons learned: At high solids, both stabilization and heat management are critical to viscosity control; surfactant demand rises disproportionately as

Table 16.3 Summary of performance before and after optimization in high-solids PVAc emulsion production and viscosity control.

Parameter	Before optimization	After optimization
Solids content	65%	65%
Final viscosity	> 20,000 cP	8,000–10,000 cP
Coagulum formation	Present after 3 months	None after 6 months
Particle size distribution	0.20–0.40 μm (broad)	0.25–0.30 μm (narrow)

solids approach close-packing, and controlling monomer feed in late stages helps prevent runaway viscosity (Table 16.3).

The performance of high-solids PVAc emulsion production before and after optimization of viscosity control is summarized in Table 16.3.

16.5 Case Study 4—Temperature Impact on PVAc Particle Size

Background: A PVAc adhesive manufacturer experienced unexpected variations in product viscosity and adhesion performance. Investigation showed polymerization temperatures fluctuated between 65 and 80 °C due to malfunctioning temperature-control equipment. This case highlights the impact of temperature on PSD and overall latex performance.

Problem statement: Temperature fluctuations during polymerization produced inconsistent particle size, a broader molecular-weight distribution, and reduced storage stability of the adhesive emulsion.

Investigation: Laboratory replication demonstrated that runs at 65 °C yielded a narrow PSD with an average of roughly 0.25 μm, whereas runs at 80 °C broadened the PSD to approximately 0.20–0.40 μm. Gel-permeation chromatography (GPC) indicated lower molecular weights at higher temperatures, attributable to increased chain transfer, and storage-stability testing showed elevated coagulum levels in the high-temperature batches.

Root cause: Elevated polymerization temperature increased radical generation rates, shortening polymer chains and altering nucleation dynamics; the result was greater variability in particle number and reduced colloidal stability.

Solution implemented: The temperature-control system was repaired and recalibrated, and the polymerization temperature was standardized at 70 °C ± 1 °C. Inline temperature monitoring with automatic alarms was installed, and the initiator-feed profile was modified to align with the optimized temperature to maintain consistent polymer growth rates.

Table 16.4 Summary of performance before and after optimization in temperature impact on particle size distribution in PVAc emulsion polymerization.

Parameter	Before optimization	After optimization
Polymerization temperature	65–80 °C (variable)	70 °C ± 1 °C
Particle size distribution	0.20–0.40 μm (broad)	0.25 μm ± 0.02 μm (narrow)
Molecular weight distribution	Broad	Narrow
Storage stability	3–4 months with coagulum	6+ months, no coagulum

Outcome: PSD stabilized at approximately 0.25 μm ± 0.02 μm, molecular-weight distribution narrowed, and adhesive performance (peel strength and setting time) returned to specification. Storage stability improved to 6 months with no coagulum formation.

Lessons learned: Precise temperature control is essential for consistent particle size and polymer properties; even short-term deviations can significantly impact performance. Inline monitoring and preventive maintenance reduce the risk of recurrence (Table 16.4).

The effect of temperature control on particle size distribution in PVAc emulsion polymerization before and after optimization is summarized in Table 16.4.

16.6 Case Study 5—Agitation and Coagulum Formation in PVAc Emulsion Polymerization

Background: An industrial PVAc emulsion production line began experiencing frequent coagulum formation in the reactor and downstream filters. Coagulum levels exceeded approximately 1.5% by weight, causing cleaning downtime, yield loss, and higher maintenance costs.

Problem statement: Despite adherence to standard recipe parameters, batches displayed unstable viscosity, and visual checks showed coagulum deposits on the reactor wall and agitator shaft.

Investigation: A review of the agitation system found worn impeller blades that reduced effective pumping capacity. Computational fluid-dynamics simulations revealed dead zones at the reactor base where monomer droplets were insufficiently dispersed. Batch-to-batch agitation-speed variation further contributed to nonuniform shear distribution.

Root cause: Inefficient mixing created local regions of elevated monomer concentration that initiated uncontrolled polymerization at the wall, leading to particle agglomeration. Excess shear in isolated zones also destabilized the latex and promoted coagulum formation.

Table 16.5 Summary of key parameters before and after optimization in agitation and coagulum formation in PVAc emulsion polymerization.

Parameter	Before optimization	After optimization
Coagulum content	>1.5% by weight	<0.2% by weight
Reactor cleaning frequency	Every 3 batches	Every 15 batches
Agitation system	Worn flat-blade	New pitched-blade turbine
Production yield	Baseline	+3% increase

Solution implemented: The worn impeller was replaced with a pitched-blade turbine to improve axial and radial flow, agitation speed was standardized to maintain uniform shear, and baffles were installed to break vortex formation and enhance top-to-bottom circulation. An in-process filtration step was introduced prior to cooling to capture nascent coagulum.

Outcome: Coagulum content fell to below 0.2% by weight, reactor-cleaning frequency decreased from every three batches to every fifteen batches, viscosity and solids stabilized within target specifications, and overall production yield improved by about 3%.

Lessons learned: Impeller type, speed, and positioning are critical determinants of latex stability; even small dead zones can become hotspots for coagulum. Regular inspection and preventive maintenance of agitation systems are essential to sustain stable operation (Table 16.5).

The influence of agitation system optimization on coagulum formation in PVAc emulsion polymerization is summarized in Table 16.5.

16.7 Case Study 6–NPE Replacement for REACH Compliance

Background: A European coatings manufacturer needed to comply with EU REACH regulations prohibiting nonylphenol ethoxylates (NPEs) due to environmental and health risks. The existing vinyl acetate–butyl acrylate (VAM/BA) copolymer latex relied on NPE-based surfactants for emulsion stability and performance, creating a need to eliminate NPEs without compromising product quality or application properties.

Problem statement: Regulatory deadlines required immediate reformulation of the surfactant system while maintaining colloidal stability, PSD, gloss, and water resistance. Customer performance expectations and documentation needs for compliance further constrained development timelines.

Investigation: The role of NPE-based surfactants in particle stabilization and film formation was assessed, and alternative chemistries—alcohol ethoxylates

Table 16.6 Key performance parameters before and after NPE replacement.

Parameter	With NPE-based surfactants	After NPE replacement
Particle size distribution	0.20–0.30 μm (narrow)	0.20–0.30 μm (narrow)
Storage stability	12 months	12 months
Gloss (60°)	≥85	≥85
Water resistance	Pass (24 h immersion)	Pass (24 h immersion)
Regulatory compliance	Noncompliant (NPE present)	Fully REACH compliant

(nonionic) and phosphate-ester anionics—were screened. Accelerated storage-stability tests and particle-size monitoring were performed alongside standardized evaluations of gloss, water resistance, and adhesion per ISO/ASTM methods to benchmark film performance.

Root cause: In the original formulation, NPE-based surfactants provided strong electrosteric stabilization. Their removal disrupted the balance of steric and electrostatic forces, elevating the risk of particle flocculation and gloss loss unless adequately compensated by a redesigned surfactant package.

Solution implemented: NPEs were replaced with a combination of alcohol-ethoxylate nonionics and phosphate-ester anionics to restore electrosteric stability. The nonionic/anionic ratio was optimized to match the original HLB and to ensure robust micelle formation and particle stabilization. The final surfactant package was verified to support consistent particle size and film performance under varied storage and application conditions.

Outcome: The latex maintained a narrow PSD (0.20–0.30 μm) with no phase separation or significant viscosity drift during 6-month accelerated aging. Coating performance met benchmarks for gloss, water resistance, and adhesion, and full REACH compliance was achieved ahead of deadline with supporting documentation.

Lessons learned: Regulatory-driven changes benefit from proactive R&D to avoid last-minute reformulation crises. Alternative surfactants must be vetted for both stability and film performance to secure customer acceptance, and maintaining electrosteric balance remains central to latex design even when replacing functional surfactants (Table 16.6).

The key performance parameters before and after replacement of NPE-based surfactants are summarized in Table 16.6.

16.8 Case Study 7—Alcohol Ethoxylate Optimization for Improved Adhesion and Cost Efficiency

Background: A packaging-adhesive manufacturer sought to improve performance and cost efficiency in a PVAc-based system. The existing formulation used a broad-range alcohol-ethoxylate (AE) surfactant for stabilization and wetting, but regulatory pressure and cost targets motivated a search for a more efficient alternative without sacrificing adhesion or water resistance.

Problem statement: The broad-range AE provided adequate stabilization but required a relatively high dosage due to its higher critical micelle concentration (CMC). This increased cost, limited water resistance, and delayed early film formation reduce green bond strength on packaging substrates.

Investigation: Comparative testing evaluated broad-range AE against a narrow-range secondary AE for CMC and micelle-formation efficiency; adhesion on paper and coated board was measured via ASTM D903 peel; and water resistance was assessed using Cobb60 and immersion trials.

Root cause: The broad-range AE exhibited a higher CMC and broad ethoxylation distribution, which reduced micelle stability and slowed film formation. Consequently, the set speed decreased and higher surfactant levels were needed versus a narrow-range AE with more uniform ethoxylation.

Solution implemented: The broad-range AE was replaced with a narrow-range secondary AE to improve micelle efficiency and lower CMC. Dosage was optimized to maintain colloidal stability while reducing total surfactant by approximately 15%, and improvements in early film formation and wet adhesion were confirmed in lab and pilot runs.

Outcome: Adhesion on coated board and flexible packaging films improved, delivering faster green bond strength. Water resistance increased due to reduced free-surfactant migration during film formation. Surfactant usage fell by about 15%, lowering cost and environmental load, while overall performance and stability met customer specifications for scale-up.

Lessons learned: Surfactant ethoxylation range strongly influences micelle formation, film integrity, and adhesive performance. Lower-CMC surfactants offer both economic and technical advantages by enabling reduced dosage without compromising stability, and targeted surfactant optimization is a cost-effective lever for improving packaging-adhesive performance (Table 16.7).

Table 16.7 Key performance indicators before and after optimization in alcohol ethoxylate for improved adhesion and cost efficiency.

Parameter	With broad-range AE	With narrow-range AE
Surfactant dosage	Baseline (100%)	Reduced by 15%
Critical micelle concentration (CMC)	Higher	Lower
Early film formation	Slower	Faster
Water resistance	Moderate	Improved
Adhesion strength	Standard	Enhanced

The key performance indicators observed after optimization of alcohol ethoxylate surfactants are summarized in Table 16.7.

16.9 Case Study 8—Polymerizable Surfactants in VAM/BA/VeoVa 10 Polymerization

Background: A mid-sized coatings manufacturer experienced recurring surfactant leaching in exterior architectural paints. During film formation, mobile surfactants migrated to the surface, producing water whitening, application foaming, and reduced weathering resistance.

Problem statement: Even with optimized doses of conventional nonionic and anionic surfactants, exterior paints showed leaching under wet conditions and progressive surface defects. Customers reported gloss loss and diminished durability in high-humidity environments.

Investigation: "Root cause" analysis identified high surfactant mobility as the driver of water whitening and dirt pick-up. A review of literature and supplier data highlighted polymerizable surfactants (surfmers) bearing (meth)acrylate groups as candidates to reduce migration. Laboratory trials compared standard formulations with variants incorporating reactive surfactants.

Root cause: Conventional surfactants remain physically associated with latex particle surfaces and migrate during coalescence. This migration weakens the film–substrate interface and creates pathways for moisture ingress, compromising gloss and durability.

Solution implemented: Polymerizable surfactants with (meth)acrylate functionality were introduced into the monomer preemulsion for VAM/BA/VeoVa-10 copolymerization. Surfactant levels were optimized to balance particle nucleation and film integrity without destabilizing the emulsion process. Covalent incorporation was verified by FTIR and solvent-extraction tests, confirming reduced free-surfactant content.

Outcome: Surfactant leaching was eliminated, with a marked reduction in water whitening in exterior exposure. Gloss retention and dirt resistance improved after 1,000 h of accelerated weathering, and application foaming decreased, reducing defoamer demand by approximately 20%. Latex stability and PSD remained comparable to conventional systems.

Lessons learned: Polymerizable surfactants enhance exterior durability by chemically bonding to the polymer backbone, limiting migration. They are especially effective where resistance to water whitening and dirt pick-up is critical, and their dosage must be tuned to avoid instability or unnecessary cost (Table 16.8).

The performance differences between conventional surfactants and polymerizable surfactants are summarized in Table 16.8.

Table 16.8 Performance comparison before and after polymerizable surfactant incorporation.

Parameter	Conventional surfactants	With polymerizable surfactants
Water whitening (24 h immersion)	Visible whitening	No whitening
Gloss retention (after 1,000 h)	75%	90%
Foaming tendency during application	High	Low
Dirt pick-up resistance	Moderate	Excellent
Latex stability	Good	Good

16.10 Case Study 9–Resolving Excessive Foaming by Switching to Low-foam Surfactants

Background: A VAM/ethyl acrylate-based adhesive batch exhibited severe foaming during both polymerization and application. Foam accumulation in the reactor extended deaeration time and produced surface defects in dried adhesive films, leading to customer complaints and higher production costs.

Problem statement: The formulation relied on sodium lauryl ether sulfate (SLES) and linear alkylbenzene sulfonate (LAS) as surfactants. Although these provided strong emulsification and particle stabilization, they generated high foam levels under agitation and thermal conditions.

Investigation: Reactor foam height reached up to ~30 cm during polymerization and mixing, and each batch required an additional 45 min of deaeration, reducing productivity. Film-quality defects such as pinholes and craters were traced to entrapped air during application. Compatibility testing identified low-foam surfactants capable of maintaining colloidal stability without excessive foam generation.

Root cause: High-foaming anionic surfactants (SLES and LAS) produced resilient foam layers that resisted collapse under shear, retaining air in the latex and causing drying defects while prolonging production cycles.

Solution implemented: SLES and LAS were replaced with low-foam secondary-alcohol ethoxylates (narrow-range) in combination with phosphate-ester anionic surfactants. Surfactant concentration was optimized to preserve PSD and stability, allowing a reduction in silicone-based defoamer usage, which improved surface aesthetics and lowered additive cost.

Table 16.9 Comparison of key parameters before and after surfactant optimization in resolving excessive foaming.

Parameter	Before optimization (SLES/LAS)	After optimization (low-foam system)
Foam height during polymerization	30 cm	<5 cm
Deaeration time	45 min	5 min
Film defects (pinholes/craters)	Frequent	Minimal
Defoamer demand	High	Reduced
Production throughput	Baseline	+12% increase

Outcome: Foam height during polymerization decreased from ~30 to <5 cm. Deaeration time was shortened by about 40 min, increasing throughput and energy efficiency. Adhesive film quality improved with fewer pinholes and surface craters, and foaming during application diminished, reducing operator intervention and rework.

Lessons learned: Surfactant selection strongly influences foam stability, processing time, and final film quality. Adopting low-foam surfactants can raise productivity and reduce costs without sacrificing emulsion stability, and addressing foam early in the process minimizes downstream defects and quality complaints (Table 16.9).

The effect of surfactant optimization on excessive foaming and production performance is summarized in Table 16.9.

16.11 Case Study 10—Styrene–Acrylic Batch Failure Due to VAM and PVAc Contamination

Background: A coatings manufacturer experienced a catastrophic batch failure during production of a styrene–acrylic latex. The reactor, previously used for PVAc-based adhesive polymerization, had not been fully decontaminated before switching chemistries.

Problem statement: Residual vinyl acetate monomer (VAM) and polyvinyl acetate (PVAc) from the prior run remained in the reactor, interfering with styrene–acrylic polymerization. The batch exhibited incomplete conversion and severe coagulation, culminating in the disposal of the latex and significant production downtime.

Investigation: Visual inspection identified PVAc gel residues on reactor walls and agitator shafts. Polymerization-kinetics analysis showed erratic temperature and conversion profiles consistent with chain-transfer effects and copolymerization anomalies. Laboratory replication confirmed that even trace levels of VAM and PVAc destabilized the styrene–acrylic system, triggering flocculation and gel formation. Cleaning records indicated insufficient alkaline cleaning and passivation prior to the chemistry changeover.

Root cause: Two contamination modes drove the failure. Residual VAM, as a reactive monomer, copolymerized with styrene and acrylates, altering composition drift, reaction kinetics, and resultant film properties. PVAc residues, as polymeric carryover, acted as physical contaminants and unintended nucleation sites that trapped surfactants and promoted localized coagulation within the reactor.

Solution implemented: A hot alkaline cleaning step was implemented using 2% NaOH at 80–90 °C to remove PVAc residues, followed by acid passivation with 1% citric-acid solution to neutralize residual alkalinity and mitigate metal-catalyzed side reactions. Strict cleaning validation was instituted via swab tests and conductivity checks prior to initiating the new chemistry. Where feasible, reactors were dedicated separately for PVAc-based adhesives and styrene–acrylic systems to eliminate cross-contamination risk.

Outcome: Subsequent styrene–acrylic batches reached full conversion without coagulation. The revised cleaning and passivation SOPs reduced turnaround time while ensuring contamination-free processing. Customer complaints regarding off-spec coatings ceased, and overall production reliability improved.

Lessons learned: Residual monomers and polymers can severely destabilize emulsion polymerizations when switching chemistries. Thorough cleaning and passivation are mandatory when multi-purpose reactors are used across different latex platforms, and dedicated equipment is preferred when possible to minimize contamination risk and optimize scheduling.

Technical note: Residual VAM interferes chemically by copolymerizing with styrene and acrylates, causing deviations in glass-transition temperature (T_g), polarity, and molecular-weight distribution. PVAc residues, though largely

Table 16.10 Key process parameters before and after corrective actions in styrene–acrylic batch failure.

Parameter	Before cleaning protocol	After cleaning and passivation
Batch stability	Coagulation and gel formation	Stable polymerization
Conversion efficiency	<70%	>98%
Cross-contamination risk	High	Eliminated
Reactor cleaning procedure	Basic rinse	Hot alkaline + acid passivation
Production reliability	Frequent failures	Consistent performance

inert under reaction conditions, destabilize the colloid by creating localized high-viscosity zones and serving as seeds for uncontrolled particle growth. Together, these contaminants drive latex instability, grit formation, and compromised film performance (Table 16.10).

The key process parameters before and after corrective cleaning and passivation actions in a styrene-acrylic batch failure are summarized in Table 16.10.

16.12 Case Study 11—Surfactant Chain Length Impact on Particle Size Control

Background: A VAM/2-ethylhexyl acrylate (2-EHA) copolymer was produced for flexible adhesive applications. The original formulation employed C12–C14 alcohol-ethoxylate nonionic surfactants. Following a supply-chain substitution, the manufacturer replaced them with C16–C18 alcohol ethoxylates, assuming minimal performance impact.

Problem statement: The substitution led to unexpected instability during polymerization and storage. Latex batches showed increased particle size and viscosity drift with ensuing coagulum and filtration issues, and customers reported variability in adhesive-film smoothness and drying behavior.

Investigation: Laser-diffraction measurements showed the volume-average particle size shifting from approximately 0.25 μm to about 0.40 μm. Zeta-potential testing indicated reduced surface charge, consistent with weakened electrosteric stabilization. Property analysis of the surfactants confirmed that the longer C16–C18 alkyl chains increased hydrophobicity and lowered micelle population, reducing nucleation sites during polymerization.

Root cause: Surfactant chain length directly affected HLB and micelle formation. Longer C16–C18 chains promoted greater hydrophobicity, decreasing micelle number and shifting the polymerization locus toward monomer droplets, which produced fewer but larger particles, a broadened PSD, and colloidal instability.

Solution implemented: The process reverted to a C12–C14 alcohol-ethoxylate blend to restore micelle characteristics and nucleation behavior. Surfactant-system efficiency was increased by adding a small fraction of a shorter-chain nonionic to fine-tune HLB, and supplier-qualification procedures were instituted to prevent unverified surfactant substitutions in critical formulations.

Outcome: Particle size returned to the target range (0.25–0.28 μm) with a narrowed distribution. Latex stability improved, eliminating coagulum formation and reducing viscosity drift during storage. Adhesive performance met the required flexibility, smoothness, and drying-time specifications, and customer complaints ceased as production consistency was restored.

Lessons learned: Surfactant chain length significantly influences nucleation behavior, PSD, and long-term latex stability. Even seemingly minor surfactant

Table 16.11 Comparison of key parameters before and after surfactant change.

Parameter	C16–C18 alcohol ethoxylates	C12–C14 alcohol ethoxylates (optimized)
Average particle size	0.40 μm	0.25–0.28 μm
Particle size distribution	Broad	Narrow
Latex stability	Poor (coagulum present)	Excellent
Viscosity drift (after 30 days)	+25%	<5%
Adhesive performance	Variable	Consistent

substitutions can disrupt polymerization dynamics and application performance, making rigorous raw-material control and supplier management essential (Table 16.11).

The effect of changing from C16–C18 to optimized C12–C14 alcohol ethoxylates on latex performance is summarized in Table 16.11.

16.13 Case Study 12–Impact of Surfactant Purity on Latex Stability

Background: A paint-resin manufacturing plant producing styrene–acrylic emulsions experienced an unexpected rise in coagulation levels during routine production runs, despite adherence to standard formulations and process parameters.

Problem statement: Latex batches displayed visible coagulum, increased filter clogging, and unstable viscosity profiles. Product failure rates rose, causing costly rework and production delays.

Investigation: Contamination was ruled out and agitator integrity was verified. pH and solids remained within specification during polymerization. Raw-material analysis revealed that the alcohol-ethoxylate (AE) surfactant contained approximately 5% unreacted fatty alcohol due to a supplier quality-control issue. Laboratory replication confirmed that batches made with the impure surfactant showed early-stage flocculation and rapid viscosity drift.

Root cause: Unreacted fatty alcohol reduced the effectiveness of the AE surfactant and introduced hydrophobic domains in the aqueous phase. This disrupted micelle formation, destabilized the latex, and increased the likelihood of coagulum formation during polymerization and storage.

Solution implemented: The suspect surfactant lot was quarantined and rejected. A new AE surfactant supplier with certified purity was qualified. Incoming quality-control checks were implemented for surfactant

Table 16.12 Effect of surfactant purity on latex stability.

Parameter	Impure surfactant (5% alcohol)	Pure surfactant (certified)
Coagulum content	High (visible in reactor)	None
Viscosity stability	Unstable (rapid drift)	Stable
Particle size distribution	Broad, bimodal	Narrow, uniform
Customer complaints	Frequent	Resolved
Supplier control	Weak	Strengthened with audits

purity using gas chromatography and hydroxyl-value determination, and the supplier-qualification protocol was updated to include routine audits and certificate-of-analysis verification.

Outcome: Coagulation incidents were eliminated in subsequent production. Latex stability improved, maintaining target particle size and viscosity over storage. Customer complaints related to instability were resolved and product quality was restored. Enhanced supplier-quality management reduced raw-material risk in future production.

Lessons learned: Surfactant purity strongly influences micelle formation, colloidal stability, and final latex performance. Unreacted raw materials in surfactants can act as destabilizers, leading to early flocculation. Strict raw-material QC and supplier validation help prevent costly downstream failures (Table 16.12).

The effect of surfactant purity on latex stability is summarized in Table 16.12.

16.14 Case Study 13—Temperature Sensitivity of Surfactant Packages

Background: A manufacturer producing VAM/butyl acrylate (BA) copolymer latex for adhesives experienced batch failures during winter operations. A reaction protocol that was stable under normal conditions showed low conversion and an unstable emulsion in colder months.

Problem statement: One batch exhibited incomplete conversion, with solids content approximately 4% below specification. Visual inspection revealed signs of phase separation and foaming despite an unchanged recipe and dosing schedule, leading to longer production time and off-spec material requiring disposal.

Investigation: Reactor logs showed initial polymerization temperatures averaging about 67 °C—roughly 3–4 °C below the normal range. Surfactant review identified a primary nonionic component with a cloud point of 65–68 °C, close to the actual reaction temperature. Laboratory replication confirmed that at 67 °C, the nonionic surfactant partially phase-separated, reducing micelle

Table 16.13 Effect of polymerization temperature on latex stability.

Parameter	Before adjustment (67 °C)	After adjustment (72 °C)
Conversion efficiency	Low (−4% solids)	Within spec
Latex stability	Phase separation observed	Stable
Foaming	Severe	Controlled
Surfactant behavior	Near cloud point	Well above cloud point
Batch rejection risk	High	Eliminated

stability and causing premature coagulation. Zeta-potential measurements at low temperature indicated diminished electrostatic protection, further destabilizing the latex.

Root cause: The temperature sensitivity of the surfactant package led to loss of micelle integrity when reactor temperature approached the nonionic surfactant's cloud-point window. This undermined colloidal stability and lowered conversion efficiency.

Solution implemented: Polymerization temperature was raised by approximately 5 °C to maintain a safe margin above the surfactant cloud point, and reactor-heating protocols were adjusted for seasonal conditions to ensure consistent temperature control. Alternative surfactants with higher cloud points were evaluated for longer-term optimization in cold climates.

Outcome: Conversion returned to specification, with solids content meeting the 55% target. No phase separation or foaming was observed in subsequent batches after adjustment, and seasonal variability was mitigated, improving winter production reliability.

Lessons learned: Surfactant cloud point is a critical parameter influencing micelle stability under varying temperature conditions. Process parameters for emulsion polymerization must account for seasonal changes, and proactive temperature monitoring helps ensure consistent quality year-round (Table 16.13).

The effect of polymerization temperature adjustment on latex stability and process performance is summarized in Table 16.13.

16.15 Case Study 14—Cloud Point Failure in Winter Production

Background: A VAM/butyl acrylate (BA) copolymer latex was produced in a facility without temperature-controlled monomer feed tanks. In winter, colder monomer feeds suppressed reactor temperature. The process used a persulfate/ascorbic-acid redox system to initiate at 30–35 °C. Under normal ambient conditions, the process was stable, but during winter, the reduced feed temperature brought the operating range dangerously close to the cloud point of the primary surfactant.

Problem statement: During a winter batch, the latex exhibited mid-reaction phase separation accompanied by foaming and uncontrolled particle nucleation. Final product testing showed a broadened PSD, reduced gloss, and compromised adhesion, yielding off-spec material unsuitable for high-performance applications.

Investigation: Reactor logs indicated polymerization temperatures averaging ~30 °C early in the batch and ~35 °C at mid-feed. The primary nonionic surfactant had a cloud point of 28 °C—too close to the operating window. Microscopy and particle-size analysis revealed secondary nucleation after ~40% monomer conversion, coincident with surfactant clouding. Surfactant screening confirmed that the selected nonionic provided poor stability near its cloud point.

Root cause: Cold monomer feeds lowered reactor temperature to 30–35 °C, a range close to the nonionic surfactant's 28 °C cloud point. Although the redox system sustained polymerization at this temperature, the nonionic lost water solubility near its cloud point, destabilizing micelles and triggering secondary nucleation that produced a bimodal PSD.

Solution implemented: The nonionic surfactant was replaced with a higher-EO grade to raise the cloud point above 40 °C. A seasonal SOP was implemented requiring monomer-feed preheating in winter to maintain a consistent process temperature. Raw-material specifications were updated to include cloud-point verification for all nonionic surfactants.

Outcome: No recurrence of phase separation or foaming occurred in subsequent winter batches. PSD stabilized at 0.25–0.30 μm, supporting consistent film formation and gloss. Adhesive and coating performance returned to specification across multiple production cycles.

Table 16.14 Effect of cloud point adjustment on latex stability.

Parameter	Before adjustment (cloud point 28 °C)	After adjustment (cloud point >40 °C)
Process stability	Phase separation at mid-feed	Stable throughout
Particle size distribution	Bimodal (0.15 and 0.45 μm)	Unimodal (0.25–0.30 μm)
Foaming	Severe	Controlled
Final product quality	Off-spec, poor gloss	Within specification
Preventive measure	None	Cloud point specification enforced

Lessons learned: In redox-initiated systems, the surfactant cloud point should exceed the operating temperature by at least 10 °C to maintain stability. Seasonal temperature variations can significantly affect surfactant performance if not proactively managed, and raw-material verification (including cloud-point checks) prevents costly production disruptions (Table 16.14).

The effect of cloud-point adjustment on latex stability and process performance is summarized in Table 16.14.

16.16 Case Study 15—Surfactant Migration Causing Gloss Reduction

Background: A waterborne styrene–acrylic paint formulated for architectural use exhibited premature gloss loss after exterior exposure, despite meeting initial specifications for viscosity, hiding power, and application properties.

Problem statement: Within 2 weeks of outdoor exposure under warm, humid conditions, gloss decreased markedly. Microscopy of dried films revealed crystalline surface deposits producing a whitish haze that degraded the appearance and customer satisfaction.

Investigation: Surface scrapings analyzed by ATR-FTIR identified the deposits as ethoxylated species originating from nonionic surfactants. Differential scanning calorimetry (DSC) confirmed that these surfactants crystallized as water evaporated during film formation. Migration studies showed excess free surfactant remaining in the aqueous phase migrated to the film surface upon drying, and screening indicated that low-molecular-weight alcohol ethoxylates (AEs) were particularly prone to migration and crystallization.

Root cause: Excess free nonionic surfactant migrated during drying and crystallized at the paint surface, reducing gloss and impairing surface uniformity and dirt-pick-up resistance. The absence of polymerizable surfactant limited the ability to anchor surfactant functionality within the polymer network.

Solution implemented: Total surfactant level was reduced by approximately 20% to minimize free surfactant. A portion of the nonionic package was replaced with a polymerizable surfactant bearing (meth)acrylate reactive groups to covalently anchor it in the polymer network. Latex stability was verified postadjustment to confirm no coagulation or unwanted particle growth, and accelerated weathering tests were run to validate long-term gloss retention.

Outcome: Gloss retention improved significantly under outdoor conditions, with no surface haze after 60 days of exposure. Application properties and mechanical stability were maintained, and customer complaints ceased following implementation.

Lessons learned: Surfactant migration can compromise film aesthetics even when colloidal stability appears satisfactory. Optimizing total surfactant content

Table 16.15 Effect of surfactant optimization on gloss retention.

Parameter	Before optimization	After optimization
Gloss retention (60 days)	Severe reduction; whitish haze	High gloss; no haze
Surface defect	Crystalline deposits visible	None detected
Surfactant system	100% nonionic AEs	80% nonionic + 20% polymerizable surfactant
Customer complaints	Frequent	None postchange

for both polymerization stability and final film performance is essential, and polymerizable surfactants offer a durable solution to migration and leaching in waterborne coatings (Table 16.15).

The effect of surfactant optimization on gloss retention and surface quality is summarized in Table 16.15.

16.17 Case Study 16—Mixed Surfactant Synergy Improves Conversion

Background: A VAM/ethyl acrylate copolymer latex was developed for flexible-packaging adhesive applications. The original process used a single nonionic surfactant system, which provided acceptable stability but limited conversion rate and control of PSD.

Problem statement: Batches employing only nonionic surfactants exhibited prolonged reaction times, with conversions lagging target schedules by more than 2 h. The final latex displayed a broad PSD (0.15–0.45 μm), leading to inconsistent film properties and poor mechanical stability during post-processing.

Investigation: Kinetic profiles showed slow initial nucleation and delayed attainment of the CMC. Zeta-potential measurements indicated weak electrostatic stabilization, consistent with nonionics providing primarily steric stabilization. A literature review and lab screening suggested that combining anionic surfactants (for electrostatic repulsion) with nonionics (for steric stabilization) would improve nucleation dynamics and growth control. A phosphate-ester surfactant (anionic) and a secondary-alcohol ethoxylate (nonionic) were selected for trial.

Root cause: Exclusive use of nonionic surfactants yielded insufficient electrostatic repulsion during nucleation, producing nonuniform micelles and a broad PSD. Slow nucleation delayed polymerization, extending reaction times and risking incomplete monomer conversion within the scheduled batch duration.

Table 16.16 Effect of mixed surfactant system on VAM/EA polymerization.

Parameter	Before optimization (nonionic only)	After optimization (mixed system)
Conversion time	Extended; +2 h beyond target	Reduced by 1.5 h
Final conversion	88–90%	98–99%
Particle size distribution	Broad (0.15–0.45 μm)	Narrow (0.25–0.32 μm)
Mechanical stability	Moderate	High (ASTM D1300 compliant)

Solution implemented: A mixed surfactant system was introduced: approximately 60% nonionic (secondary alcohol ethoxylate) and 40% phosphate-ester anionic. Total surfactant dosage was adjusted to maintain an equivalent HLB while increasing electrostatic stability. Feed strategy was optimized to allow gradual emulsifier addition during monomer dosing, ensuring consistent micelle availability. Conversion and PSD were monitored via inline sampling to validate improvements.

Outcome: Conversion improved by ~12%, shortening batch time by roughly 1.5 h. PSD narrowed to 0.25–0.32 μm, improving film uniformity. Mechanical stability increased, evidenced by improved ASTM D1300 sedimentation resistance. No adverse foaming or coagulation occurred, confirming compatibility of the mixed system.

Lessons learned: Mixed anionic–nonionic surfactant packages provide synergistic stabilization by balancing electrostatic and steric forces. Surfactant optimization significantly impacts polymerization kinetics, PSD, and downstream latex performance, and compatibility testing with controlled dosing is critical to avoid foaming or destabilization (Table 16.16).

The effect of using a mixed surfactant system on VAM/EA polymerization performance is summarized in Table 16.16.

16.18 Case Study 17–Excessive Foam Formation During Polymerization

Background: A large-scale VAM/BA emulsion polymerization process experienced persistent foam buildup during monomer and initiator feeding. Despite adherence to the standard recipe and operating parameters, the reactor headspace was filled with foam, creating operational inefficiencies and contamination risk.

Problem statement: Excessive foam overflowed into the condenser line, causing partial blockage and reducing cooling efficiency. Operators struggled to maintain reaction volume, and one batch was scrapped after foam contaminated the sampling line and filters. Foam persisted even after multiple antifoam additions, extending batch time and increasing production cost.

Investigation: A review of the surfactant package identified high-foaming anionic species (e.g., SDS) combined with ethoxylated nonionics. Foam-height measurements during initiator feed peaked at approximately 35 cm within 20 min of monomer-feed start. Evaluation of defoamer addition showed a delayed impact due to poor dispersion in the preemulsion. Laboratory trials demonstrated that pairing silicone defoamers with EO/PO-based antifoams improved foam knock-down in simulated runs.

Root cause: The surfactant system promoted high surface activity, and the absence of an effective defoamer during initial emulsification allowed foam to form and stabilize. Agitation intensity and initiator decomposition further contributed to gas release, sustaining foam during the feed stage.

Solution implemented: The primary anionic surfactant was replaced with a lower-foaming phosphate-ester derivative while maintaining overall HLB balance. A silicone-based defoamer (0.15% on monomer) was incorporated into the preemulsion rather than added at a late stage to ensure early distribution. A nitrogen blanket was introduced in the reactor headspace to reduce oxygen solubility and foam stabilization, and agitator speed was optimized to avoid vortex formation and air entrainment.

Outcome: Foam height was reduced from about 35 cm to less than 5 cm during critical feeding stages. Foam overflow into the condenser and associated line contamination were eliminated. Batch time improved by approximately 20% owing to fewer interruptions for foam control, while mechanical stability and film quality were maintained without surfactant loss.

Lessons learned: Foam control should be embedded in recipe design rather than treated solely as an in-process corrective. Silicone defoamers used in combination with EO/PO copolymer antifoams provide synergistic performance in

Table 16.17 Effect of foam control strategy on polymerization process.

Parameter	Before optimization	After optimization
Foam height during feed	35 cm	<5 cm
Foam overflow incidents	Frequent	None
Batch time	Extended (+20%)	On schedule
Defoamer strategy	Late-stage only	Pre-emulsion + in-process
Cooling efficiency	Reduced due to condenser fouling	Normal

high-foaming systems. Preemulsion incorporation of defoamers ensures better distribution and early-stage prevention, and reactor-hardware choices—such as baffles and agitation profile—strongly influence foam generation (Table 16.17).

The effect of foam-control strategy optimization on polymerization process performance is summarized in Table 16.17.

16.19 Case Study 18–Foaming During Film Formation Leads to Adhesion Loss

Background: A pressure-sensitive adhesive (PSA) based on a VAM/2-ethylhexyl acrylate (2-EHA) copolymer was designed for high-tack label and tape applications. While the formulation met internal shear and peel-strength targets in laboratory testing, large-scale production and application revealed inconsistent adhesion.

Problem statement: Adhesion loss occurred on release paper and various low-energy substrates during application. Visual inspection showed microbubbles embedded within the adhesive layer, indicating incomplete substrate contact. Despite high tack and an appropriate polymer T_g, peel strength fell below specification.

Investigation: Optical microscopy confirmed numerous microbubbles trapped within the wet adhesive film before drying. Rheological assessment found no intrinsic viscosity anomaly, ruling out polymer structural issues. Analysis of the surfactant package identified a high-foaming ethoxylated surfactant prone to air entrapment during coating. Laboratory coating simulations reproduced the adhesion failure when foaming was induced under shear during application.

Root cause: The high-foam surfactant promoted excessive air entrainment during coating and drying. Microbubbles in the wet film reduced the effective contact area with the substrate, limiting wet-out and lowering peel strength—a mechanical failure mechanism independent of polymer composition.

Solution implemented: The high-foaming ethoxylated surfactant was replaced with a low-foam secondary-alcohol ethoxylate of similar HLB. A silicone-based defoamer, optimized for PSA use, was incorporated at 0.05% to prevent crater formation while suppressing foam. Coating speed was adjusted and the adhesive was vacuum-degassed prior to application to enhance air removal. Performance improvements were validated by peel testing according to ASTM D3330.

Outcome: Adhesion failures on release liners and plastic substrates were eliminated following foam-control adjustments. Peel strength improved by approximately 18% versus prior batches, meeting specification. The coating process achieved stable application without bubble trails or surface defects, and customer complaints were resolved in subsequent production.

Table 16.18 Effect of foam control on PSA adhesion performance.

Parameter	Before foam control	After foam control
Adhesion on release liner	Frequent failure	Consistent; passed ASTM D3330
Peel strength	Below spec by 18%	Improved by 18%; within spec
Surface defects	Visible microbubbles	None detected
Surfactant system	High-foaming ethoxylate	Low-foam AE + 0.05% silicone defoamer

Lessons learned: Foam management during coating and film formation is as critical as during polymerization. Surfactant selection must consider both polymerization stability and end-use air-release behavior. Silicone-based defoamers can be used safely in PSAs when correctly dosed to avoid surface defects, and in-line degassing offers an effective preventive measure for high-viscosity, foam-prone adhesives (Table 16.18).

The effect of foam-control optimization on PSA adhesion performance is summarized in Table 16.18.

16.20 Case Study 19—Yellowing of PVAc-based Copolymer Paint Binder Due to APS Residues

Background: A decorative exterior paint formulated with a vinyl acetate–butyl acrylate (VAM/BA) copolymer binder developed yellow discoloration after 2 months of storage, compromising aesthetic quality and raising concerns about long-term stability.

Problem statement: Although initial whiteness and hiding power met specification, the paint showed progressive yellowing during storage—especially in containers exposed to light and moderate heat—prompting customer complaints and a detailed root-cause investigation.

Investigation: Accelerated-aging tests recorded a steady increase in yellow index within 8 weeks. Spectroscopic analysis suggested oxidative changes at titanium dioxide pigment surfaces. Residual analysis of the latex detected ammonium persulfate (APS) and sulfate ions from the initiator system, and the paint pH was slightly alkaline ($\approx$8.0), conditions known to promote persulfate-driven oxidation.

Root cause: Residual APS and its decomposition products acted as oxidants in the alkaline paint medium, catalyzing surface reactions on titanium-dioxide pigments that produced colored complexes and yellowish species. The effect intensified under storage with heat and intermittent light exposure.

Table 16.19 Effect of APS reduction and redox system on yellowing.

Parameter	Before optimization	After optimization
Yellowing index after 60 days	High (visible discoloration)	Reduced by 85%
Initiator system	APS only	APS + redox (low APS level)
Chelating agent	Not used	EDTA included
Paint stability	Compromised after 8 weeks	Stable for 12 weeks+

Solution implemented: APS dosage during polymerization was reduced by approximately 25% without sacrificing conversion or particle stability. A redox initiation package (APS with sodium formaldehyde sulfoxylate) was adopted to lower overall persulfate demand and enable low-temperature polymerization. Chelating agents (e.g., EDTA) were incorporated into the latex to bind trace metals that could catalyze oxidation, and postpolymerization stripping was implemented to minimize residual initiator and sulfate species.

Outcome: Whiteness index remained stable after 12 weeks in accelerated aging, and the yellowing index declined by roughly 85% after 60 days of storage compared with the prior formulation. No adverse effects were observed on polymerization kinetics, PSD, or binder performance.

Lessons learned: Initiator residues can interact with pigments, fillers, or other additives in downstream applications to drive color shifts or degradation; the initiation strategy should consider both polymerization efficiency and compatibility with the final paint. Redox initiation systems offer a practical route to reduce persulfate levels without compromising polymerization control (Table 16.19).

The effect of APS reduction and the introduction of a redox initiator system on yellowing and paint stability is summarized in Table 16.19.

16.21 Case Study 20–Slow Polymerization Onset with Hydrogen Peroxide Redox System

Background: A VAM/BA emulsion polymerization using a hydrogen peroxide–iron (II)–ascorbic acid redox system exhibited an unusually long induction period (>40 min) before the exothermic polymerization began.

Problem statement: The reaction lagged significantly during startup, delaying batch completion and increasing energy consumption required to maintain the reaction temperature.

Investigation: Oxidation–reduction potential (ORP) measurements indicated low redox activity during the induction period. Inductively coupled plasma (ICP) analysis confirmed partial oxidation of Fe^{2+} to Fe^{3+} in the stored activator

Table 16.20 Effect of fresh $FeSO_4$ solution on polymerization onset.

Parameter	Before optimization	After optimization
Induction time	>40 min	<10 min
Batch duration	Extended by 20%	On schedule
Activator storage	Ambient air	Nitrogen + ascorbic acid
Conversion profile	Irregular	Consistent

solution. Simulated runs using freshly prepared $FeSO_4$ restored normal initiation within approximately 8–10 min.

Root cause: The iron(II) sulfate activator solution oxidized to iron(III) during storage, diminishing the efficiency of the hydrogen peroxide–ascorbic acid redox initiator system.

Solution implemented: $FeSO_4$ solution was prepared freshly prior to each batch, with a trace amount of ascorbic acid added to help maintain Fe^{2+}. The activator solution was stored under a nitrogen blanket to minimize oxidative conversion to Fe^{3+}.

Outcome: Polymerization onset time decreased from more than 40 min to less than 10 min, total batch time improved by approximately 18% and a consistent conversion profile was achieved across multiple runs.

Lessons learned: Redox initiator systems are highly sensitive to the oxidation state of the metal activator. Fresh preparation and controlled storage of the activator are essential to achieve predictable initiation kinetics (Table 16.20).

The effect of using a freshly prepared $FeSO_4$ activator solution on polymerization onset and batch performance is summarized in Table 16.20.

16.22 Case Study 21—Coarse Particle Size Due to Incorrect Initiator Phase Location

Background: A styrene–acrylic emulsion polymerization employed sodium persulfate added entirely in the aqueous phase. The resulting latex showed a broad PSD and produced lower gloss in coatings.

Problem statement: Although solids and viscosity met specification, gloss tests failed, and microscopy revealed oversized particles and traces of coagulum.

Investigation: Particle-size analysis measured a PSD spanning 0.15–0.60 μm. FTIR[1] and TEM[2] observations were consistent with secondary nucleation events. A comparative trial with split initiator dosing markedly reduced PSD breadth.

1 FTIR (Fourier-transform Infrared spectroscopy) measured the latex's IR "fingerprint," where changes (e.g., residual monomer/sulfate end-group bands on fresh polymer surfaces) supported the presence of newly formed particles—evidence of secondary nucleation.
2 TEM (transmission electron microscopy) directly imaged the dried latex, revealing a bimodal particle population (new sub-200 nm particles alongside larger primaries), confirming secondary nucleation and explaining the broadened PSD and lower gloss (Table 16.21).

Table 16.21 Effect of initiator feed distribution on PSD and gloss.

Parameter	Before optimization	After optimization
Particle size distribution	Broad (0.15–0.60 μm)	Narrow (0.20–0.32 μm)
Gloss level	Low (failed QC)	Improved by 15%
Initiator addition	100% aqueous phase	80% aqueous/20% monomer
Coating quality	Variable	Consistent

Root cause: Locating all initiators in the aqueous phase limited nucleation control. Insufficient radical concentration in the monomer phase delayed initiation within droplets, promoting secondary nucleation and a broadened PSD.

Solution implemented: The initiator feed was split between phases—approximately 80% to the aqueous stream and 20% to the monomer phase—with controlled addition to maintain uniform radical generation across loci. Gloss and PSD were verified by standard quality-control testing.

Outcome: Gloss on black-glass panel tests improved by about 15%. PSD narrowed to 0.20–0.32 μm, and both filterability and mechanical stability were enhanced.

Lessons learned: Initiator location is critical for managing nucleation mechanisms in emulsion polymerization. Balancing radical presence across aqueous and monomer phases promotes uniform particle growth and improved film quality.

The effect of initiator feed distribution on particle size distribution and gloss performance is summarized in Table 16.21.

16.23 Case Study 22–Reactor Runaway Due to Overdosing Thermal Initiator

Background: In a 10-ton VAM-based copolymer batch, the ammonium persulfate (APS) initiator charge was mistakenly doubled during preparation due to a manual weighing and transfer error into the initiator feed tank.

Problem statement: Excess initiator accelerated the polymerization rate and produced a strong exothermic reaction. Heat release exceeded the available cooling capacity, leading to uncontrolled temperature rise and localized boiling.

Investigation: The temperature profile recorded a rapid spike of approximately +25 °C within 5 min of initiator-feed initiation. Pressure sensors registered venting of steam and residual-monomer vapors through the safety-relief system. The batch was scrapped following coagulation and contamination of the vent and relief lines.

Root cause: A manual dosing error during APS addition resulted in twice the required initiator quantity. The uncontrolled radical flux triggered an exothermic chain reaction that overwhelmed the heat-removal capacity.

Solution implemented: Automated gravimetric dosing with interlock-based verification was installed to prevent mischarges. Real-time temperature monitoring

Table 16.22 Impact of initiator overdose on reaction safety and mitigation.

Parameter	Before optimization	After optimization
APS dosage error	Double intended amount	Automated gravimetric dosing
Temperature spike	+25 °C uncontrolled	<±2 °C of set point
Safety risk	Severe (steam + monomer venting)	Minimal with PLC interlocks
Batch outcome	Complete loss	100% recovery postfix

with PLC alarms was added, and an emergency shutdown logic was configured to halt both monomer and initiator feeds if the temperature exceeds a preset threshold.

Outcome: No similar incidents have occurred since automation. Operator safety and batch-to-batch consistency improved, and reaction exotherms have been maintained within ±2 °C of the control set point in subsequent runs.

Lessons learned: Initiator dosage exerts a critical influence on heat generation and polymerization kinetics. Large-scale processes require strict dosing control with redundant verification, while automated safeguards minimize human error and enhance operational safety (Table 16.22).

The impact of initiator overdose and the effectiveness of automated dosing and control measures on reaction safety are summarized in Table 16.22.

16.24 Case Study 23—pH Drift Reducing Persulfate Efficiency and Causing Hydrolysis

Background: A VAM/2-ethylhexyl acrylate (2-EHA) emulsion polymerization produced incomplete monomer conversion and a latex with poor storage stability. Downstream adhesives formulated from this latex showed visual defects and noticeable odor.

Problem statement: At scheduled termination, the batch achieved only about 91% conversion and exhibited viscosity below the target specification. Latex performance deteriorated within 2 months of storage, indicating polymer degradation.

Investigation: pH logs recorded a drop from 5.0 to 3.2 during the polymerization cycle. Residual vinyl acetate content remained elevated (>1.5%) after reaction completion. ATR-FTIR[3] indicated hydrolysis of acetate groups with formation of free acetic acid, consistent with backbone destabilization.

3 ATR-FTIR (attenuated total reflection–Fourier transform infrared) confirmed PVAc hydrolysis to **acetic acid** by a **decrease in ester C=O (~1,735 cm^{-1})** and **growth of acid C=O (~1,710 cm^{-1})** plus a broad **O—H band (2,500–3,300 cm^{-1})**. The absence of strong **COO^{-}** bands (~1,550/1,410 cm^{-1}) indicated mostly **free acid**, not acetate salt (Table 16.23).

Table 16.23 Effect of pH control on conversion and polymer stability.

Parameter	Before optimization	After optimization
pH during polymerization	Dropped to 3.2	Maintained at 4.5–5.0
Final conversion	91%	>99%
Residual VAM	>1.5%	<0.2%
Shelf life	2 months with defects	6+ months stable

Root cause: An uncontrolled acidic drift reduced persulfate initiator efficiency, lowering radical generation and slowing conversion. Concurrently, the low pH accelerated hydrolysis of VAM units, weakening the polymer backbone and compromising colloidal stability.

Solution implemented: A controlled alkaline feed (sodium bicarbonate solution) was introduced during reaction to neutralize acidity, maintaining pH in the range 4.5–5.0 throughout polymerization. Inline pH monitoring with an automated dosing pump was implemented to add buffer as required.

Outcome: Monomer conversion improved to greater than 99% with no residual monomer issue. Viscosity returned to the target range, supporting proper film-forming behavior. Shelf life extended beyond 6 months without signs of degradation.

Lessons learned: Persulfate efficiency is strongly pH-dependent and requires precise control during emulsion polymerization. Acidic drift can promote the hydrolysis of vinyl-acetate units, degrading performance and storage stability. Automated pH-control systems help prevent downstream quality and shelf-life failures.

The effect of pH control during polymerization on conversion efficiency and latex stability is summarized in Table 16.23.

16.25 Case Study 24—Loss of Bond Strength in Paper-to-carton Adhesive During Rainy Season

Background: A PVAc homopolymer adhesive used for paper-to-carton lamination began to exhibit slow drying and weak initial tack during periods of high ambient humidity.

Root cause: Elevated humidity reduced the water-evaporation rate and slowed coalescent action, delaying film formation and weakening early tack.

Solution implemented: To counteract humidity effects, the coalescent level was increased by 0.5% to accelerate film formation, and the thickener dosage was reduced to improve substrate wetting under humid conditions.

Outcome: Initial tack returned to the target range, and drying time decreased by approximately 20% relative to the baseline.

Table 16.24 Seasonal adjustment for humid-weather bonding.

Parameter	Before optimization	After optimization
Drying time	Extended in humid conditions	Reduced by ~20%
Initial tack	Weak under high RH	Restored to target
Coalescent level	Baseline	+0.5% increase
Thickener dosage	Standard	Reduced for better penetration

Lessons learned: Adhesive drying rate and early tack are highly sensitive to seasonal humidity. Seasonal adjustments to coalescent loading and rheology modifiers provide an effective means to maintain performance (Table 16.24).

The effect of seasonal formulation adjustments on adhesive performance under humid conditions is summarized in Table 16.24.

16.26 Case Study 25—Foaming in High-speed Carton-to-carton Adhesive

Background: A VAM/BA copolymer adhesive was developed for high-speed carton-to-carton folding machines. During production trials, operators reported excessive foaming in the gluing section. Foam overflow caused line stoppages, poor glue laydown, and coating defects, lowering line efficiency and prompting customer complaints.

Problem statement: Despite the inclusion of a silicone defoamer, the adhesive generated persistent foam at machine speeds above 300 m/min. Conventional process adjustments—slower agitation and reduced pump pressure—provided only temporary relief.

Investigation: The silicone defoamer in use was a low-viscosity polydimethylsiloxane (PDMS) emulsion with very high surface activity. Laboratory shear simulations showed that, under high shear, defoamer droplets fractured into submicron domains that migrated rapidly to the air–liquid interface. Rather than rupturing bubbles, the high local concentration of silicone disrupted cohesion in the thin adhesive film at the application nip, creating micro-ruptures where air was entrapped. The disrupted film then trapped and stabilized new bubbles, giving rise to an apparent "foam-boosting" effect.

Root cause: The selected silicone defoamer was excessively surface-active and insufficiently compatible with the VAM/BA copolymer matrix. At high shear, it destabilized the wet adhesive film instead of suppressing foam, acting as a nucleation site rather than a bubble breaker.

Solution implemented: The silicone defoamer was replaced with a modified polysiloxane defoamer containing hydrophobic silica, designed for higher

Table 16.25 Effect of defoamer replacement on foam control.

Parameter	Before (silicone defoamer, high surface activity)	After (modified silicone/ optimized addition)
Foam level at 300 m/min	Severe, overflow in glue pan	Controlled, no overflow
Adhesive film uniformity	Microruptures, bubble entrapment	Smooth film, no entrapped bubbles
Line efficiency	Frequent stoppages	Continuous operation
Customer complaints	High (poor glue laydown)	None reported

shear resistance and better compatibility with polar copolymer emulsions. A comparative screening program of mineral-oil–based and silicone-free defoamers for carton adhesives was conducted to establish performance benchmarks. The dosing protocol was revised so that defoamer was added gradually during letdown rather than as a single batch addition, preventing localized overconcentration.

Outcome: Foam levels decreased by more than 80% in full-speed trials. Adhesive-film uniformity was restored with no visible bubbles or defects, and line efficiency improved with no further stoppages attributable to foam.

Lessons learned: Silicone defoamers can paradoxically enhance foam formation if they are too surface-active, poorly dispersed, or overdosed. Compatibility matters: VAM/BA adhesives—with relatively soft backbones and higher surfactant loads—are especially sensitive to overactive silicones. In harder, less surfactant-rich systems (e.g., PVAc homopolymers or some styrene–acrylic adhesives), the problem is less common but still possible if the defoamer is mismatched. Modified silicone defoamers (silica containing or organically modified) or carefully selected mineral-oil–based defoamers generally provide better control in high-speed folding-carton applications (Table 16.25).

The effect of defoamer replacement and optimized addition strategy on foam control and adhesive film quality is summarized in Table 16.25.

16.27 Case Study 26—Failure in Untreated PVC-to-wood Lamination (Role of PMA, AA, and CDI)

Background: A PVAc–PMA–AA terpolymer adhesive (polyvinyl acetate–poly (methyl acrylate)–acrylic acid), stabilized with PVOH and crosslinked with carbodiimide (CDI), was developed for bonding untreated PVC film to wood substrates. The design intent assigned PMA segments to improve wetting/compatibility with PVC, PVAc/AA to bond to wood, and CDI to introduce covalent crosslinks within the adhesive network for durability (targeting D3-class water resistance). Despite this design, laminated panels failed at the PVC interface under humidity and peel tests.

Problem statement: Panels failed by adhesive failure at the PVC interface during humidity cycling and 180° peel testing. Adhesion to wood was strong; failure originated at the PVC/adhesive boundary. The formulation used untreated PVC film (no surface treatment).

Clarification 1—PMA–PVC wetting and interfacial entanglement: Polarity/solubility basis: PVC has a Hildebrand solubility parameter $\delta \approx 19–19.5$ (MPa)½; PMA is close at $\delta \approx 18.5–19$ (MPa)½, whereas PVAc is slightly higher ($\delta \approx 20–20.5$). Closer δ values favor compatibility; accordingly, PMA-rich segments preferentially migrate to the PVC interface during film formation, reduce interfacial tension, and lower contact angle (improved wetting). Interfacial entanglement: PMA is a soft, nonionic acrylate; PMA-rich chains can interdiffuse a few nanometers into amorphous PVC, generating physical chain entanglement and dispersion–dipole interactions (PMA C=O with PVC C—Cl). Limitation: these interactions are physical (noncovalent) and are sensitive to plasticizers, temperature, and moisture; PMA alone cannot guarantee durable adhesion to untreated PVC, so high cohesive strength and, where feasible, increased PVC surface energy (e.g., light corona) are required.

Clarification 2—Exact bonding mechanism to wood for the PVC–PMA–PVAc–AA system: Bonding to wood proceeds via three concurrent mechanisms. First, hydrogen bonding: PVAc carbonyls and AA carboxyls interact with the abundant hydroxyls of wood (cellulose/hemicellulose); PVOH in the formulation contributes additional hydroxyl groups. Second, mechanical interlocking: the low-viscosity adhesive penetrates wood porosity and cell lumens; upon drying/curing, micro-anchoring reinforces adhesion. Third, internal covalent crosslinking near the interface: CDI reacts with AA carboxyls within the polymer network to form N-acyl urea linkages, increasing cohesive strength adjacent to the interface and resisting peel under humidity and stress. Note: under aqueous ambient conditions, CDI does not covalently bond to wood hydroxyls. If direct covalent anchoring to wood is required, alternative chemistries (e.g., multifunctional epoxies or isocyanates) or Zr/Al coordination complexes may be used; these were not part of the base formulation.

Investigation: Contact-angle measurements on untreated PVC were marginal at the initial PMA level; increasing PMA reduced contact angle and improved wetting. ATR-FTIR[4] on dried films confirmed CDI–AA crosslinking via

4 ATR-FTIR (attenuated total reflection–Fourier transform infrared), a quick, nondestructive test where you press a dried film onto a crystal and the instrument reads the top ~1–2 μm; it shows peaks for specific chemical groups, so you can see if a "marker peak" (e.g., epoxide ~910 cm^{-1} or CDI N=C=N ~2,115 cm^{-1}) stays the same (stable) or drops (reacting) (Table 16.26).

the appearance of N-acyl urea bands within the adhesive network, with no new covalent signatures at the PVC interface. Cross-section microscopy of failed panels showed clean separation at the PVC boundary, consistent with insufficient interfacial interaction on the PVC side despite good cohesion to wood.

Root cause: Untreated PVC exhibits low surface energy and limited polarity. PMA provided better wetting and limited interfacial entanglement, but this was insufficient for durable adhesion under humidity and stress. Although CDI established internal crosslinks with AA (boosting cohesion toward the wood side), the PVC/adhesive boundary remained the weakest link.

Solution implemented: PMA content in the terpolymer was increased from 15% to 22% to enhance PVC compatibility and reduce contact angle. CDI dosage was optimized to 0.6–0.8% on polymer solids to maximize cohesion without embrittlement; AA was maintained at 1.0–1.5% (on monomer) to provide sufficient carboxyl sites. For robustness, light corona treatment of PVC was recommended to raise surface energy; where surface treatment was not permitted, a small dose of low-foam wetting agent compatible with carton equipment was included.

Outcome: Adhesion to PVC improved and failure modes shifted from interfacial to cohesive. Wood adhesion remained strong via hydrogen bonding and micro-anchoring, and water resistance aligned with D3 classification after CDI cure. Panels passed humidity cycling and peel tests across repeated production trials.

Lessons learned: PMA does not chemically crosslink with PVC; it improves adhesion via polarity/solubility matching and limited interfacial entanglement. Durable PVC-to-wood bonds require a PVC-compatible phase (PMA-rich) and a wood-affinitive, crosslinkable phase (PVAc/AA with CDI) to ensure high cohesion near the interface. Where possible, modest PVC surface activation significantly widens the process window and improves reliability.

The effect of formulation optimization and optional PVC surface treatment on adhesion performance is summarized in Table 16.26.

Table 16.26 Effect of formulation optimization and optional PVC treatment.

Parameter	Initial (15% PMA, CDI 0.4%, untreated PVC)	Optimized (22% PMA, CDI 0.7%, optional corona)
Wet-out on PVC (contact angle)	Marginal (higher angle)	Improved (lower angle)
Failure mode	Adhesive at PVC interface	Cohesive in adhesive layer
Peel after humidity cycling	Fail	Pass
Water resistance (D3)	Borderline	Pass

Table 16.27 Effect of moisture-control measures on D4 adhesive performance.

Parameter	Before optimization	After optimization
EN-204 test result	Failed immersion test	Passed all cycles
Crosslinker storage	Exposed to humidity	Moisture-proof packaging
Mix preparation	Occasional delays	Fresh daily preparation

16.28 Case Study 27—D4 Wood Adhesive Showing Poor Water Resistance

Background: A D4-class PVAc copolymer/polyisocyanate adhesive failed EN-204 immersion testing.

Root cause: The polyisocyanate crosslinker had been stored in a humid environment, where it partially reacted with moisture, reducing available —NCO functionality and lowering cure reactivity at application.

Solution implemented: Moisture ingress was addressed by switching to moisture-proof packaging for the polyisocyanate and preparing fresh mixes daily immediately before application, thereby preserving reactive isocyanate content at the point of use.

Lessons learned: Polyisocyanate-based systems are highly moisture-sensitive; effective moisture control is essential to maintain crosslinker stability. Packaging and handling protocols—particularly barrier packaging and just-in-time mixing—directly determine compliance with EN-204 performance requirements (Table 16.27).

Table 16.27 summarizes the effect of moisture-control measures on D4 adhesive performance.

16.29 Case Study 28—Gel Formation in PVAc–GMA Crosslinkable Adhesive

Background: A PVAc–GMA copolymer adhesive was developed for wood-bonding applications requiring enhanced water resistance and durability. Glycidyl methacrylate (GMA) was incorporated as a dual-function reactive comonomer to introduce pendant epoxide groups. These epoxides were intended to react during curing with hydroxyl groups from PVOH, carboxyl groups from acrylic acid (AA), or with external crosslinkers such as carbodiimide (CDI) or zirconium complexes, enabling a stable one-component (1K) system with controlled crosslinking upon application and drying.

Problem statement: After 3 months of storage under elevated summer temperatures, multiple adhesive drums exhibited partial gelation. Products showed a marked viscosity increase, reduced shelf life, and in some cases full coagulation, resulting in poor brushability and defective bonding performance.

Investigation: Rheological testing revealed an exponential viscosity rise beginning after approximately 8–10 weeks, progressing to bulk gelation by 12–14 weeks. FTIR[5] analysis showed a reduction in the epoxide band even without application, indicating premature GMA reactivity. Formulations with higher residual PVOH content gelled more rapidly, implicating hydroxyl-induced epoxide ring opening. Trace iron contamination from unlined storage drums was detected and found to catalyze additional epoxide ring-opening reactions. Accelerated storage at 35–40 °C further amplified the gelation rate, confirming the thermal sensitivity of GMA-functionalized systems.

Root cause: Pendant epoxide groups introduced by GMA underwent premature side reactions during storage. Hydroxyl donors (PVOH), trace acids, and catalytic metal ions triggered uncontrolled epoxide ring opening, progressively forming a crosslinked network in the drum and driving viscosity increase and eventual gelation.

Solution implemented: A chelating agent (EDTA) was introduced at 0.05% to suppress metal-ion catalysis; residual PVOH content was reduced by approximately 20% to minimize hydroxyl-initiated reactions; formulation pH was adjusted to 6.0–6.5 to reduce acid-catalyzed epoxide reactivity; and storage conditions were controlled at ≤25 °C. In parallel, accelerated storage trials with alternative external crosslinkers (e.g., CDI) demonstrated controlled reactivity without premature gelation, preserving the 1K workflow.

Outcome: Shelf life improved from roughly 3 months to at least 12 months at ambient conditions. The adhesive retained application viscosity and film-forming characteristics throughout storage, and crosslinking was successfully activated during drying/curing. Resulting bonds met D3 classification requirements for water resistance.

Lessons learned: PVAc adhesives containing GMA are highly sensitive to storage conditions due to residual epoxide reactivity. Stability depends on minimizing free hydroxyl donors and eliminating trace metal ions, as well as moderating acidity. For 1K GMA systems, the balance between latent storage stability and effective curing during application must be carefully engineered.

Table 16.28 summarizes the effect of stabilization on the storage stability and performance of PVAc–GMA adhesives.

---

5 FTIR: Stands for Fourier-transform infrared (spectroscopy), which is a quick light-based test that shows which chemical groups are present. We looked at the "epoxide peak" near 910 cm^{-1} in dried films and saw it slowly shrink during warm storage, which means those epoxide groups were reacting too early. In the CDI trial, the CDI peak (~2,115 cm^{-1}) stayed unchanged in storage and only dropped after curing, showing the crosslinking stayed latent until application (Table 16.28).

Table 16.28 Effect of stabilization on PVAc–GMA adhesive storage stability.

Parameter	Initial formulation (unstabilized)	Optimized formulation (chelated, pH 6.5)
Shelf life at 30–35 °C	3 months (gel formation)	≥12 months (stable viscosity)
Viscosity rise (8 weeks)	>200%	<20%
Gelation occurrence	By 12–14 weeks	None after 12 months
Bond strength after cure	Poor, inconsistent	Strong, passed D3 classification

16.30 Case Study 29—D3 vs D4 Wood Adhesive—Activated PVOH + CDI, AA-functional VAM/BA/VeoVa

Background: A VAM/BA/VeoVa binder failed EN-204 classification under prolonged humidity exposure. End users requested D4 performance without moving to a two-component (2K) system or significantly increasing cost.

Root cause: The binder lacked acrylic acid (AA), leaving too few carboxyl (—COOH) groups to support covalent crosslinking. As a result, the film relied primarily on physical cohesion and failed under hydrolytic stress.

Solution implemented: The backbone was modified to include 0.8–1.0 wt% AA on total monomers, introducing carboxyl functionality for subsequent crosslinking. N-methylolacrylamide (NMA) was replaced by an external carbodiimide (CDI) crosslinker to retain a 1K system with latent cure at pH 4.8–5.2. Reactive polyvinyl alcohol (e.g., acetoacetyl-functional PVOH) was employed as the protective colloid to furnish additional reactive sites, enabling a lower CDI dose at equivalent D4 performance.

Lessons learned: Acrylic acid is required to supply —COOH groups that react with carbodiimide (CDI); without AA, achieving EN-204 D4 is unlikely. Using activated (acetoacetyl-functional) PVOH increases the density of crosslinkable sites, allowing a lower CDI dose for the same D4 margin. Maintaining pH 4.8–5.2 preserves CDI latency during storage and enables activation during film formation and wood penetration (Table 16.29).

Table 16.29 summarizes the formulation requirements and typical composition ranges identified in this case study for achieving D3 and D4 performance using an activated PVOH+CDI one-component adhesive system.

16.31 Case Study 30—Roof Coating Chalking in Tropical Climates

Background: An acrylic–VeoVa–VAM roof coating exhibited chalking and noticeable color fading after 6 months of tropical sunlight exposure.

Table 16.29 Requirements and typical formulation ranges (activated PVOH + CDI, 1K).

Parameter	Initial formulation	Optimized for D3	Optimized for D4
AA in binder (on monomers)	None	0.5–0.8%	0.8–1.2%
Activated PVOH (10–12% active; e.g., AcAc-PVOH)	None	14–16%	16–18%
VAM/BA/VeoVa/AA latex (55% solids)	60%	58–60%	56–58%
CDI dispersion (30–40% active; wet on total)	None	1.0–1.2%	1.2–1.5%
Nonionic wetting aid (low-foam)	0.2%	0.2%	0.2%
Defoamer	0.2%	0.2%	0.2%
Biocide	0.2%	0.2%	0.2%
DI water	Balance	Balance	Balance
Storage/application pH	Uncontrolled	4.8–5.2	4.8–5.2
EN-204 classification (typical)	Failed	D3 pass	D4 pass

Notes: CDI level is given as wet addition of a 30–40% dispersion; values indicate typical working ranges. Activated PVOH examples: acetoacetylated PVOH (preferred).

Root cause: Insufficient ultraviolet (UV) protection together with a low VeoVa level reduced hydrophobicity and weathering resistance under intense radiation and humidity.

Solution implemented: VeoVa 10 content was increased to 20% to enhance hydrophobicity and resist water uptake. A hindered amine light stabilizer (HALS) was added at 1% to improve resistance to UV-induced degradation and to slow pigment binder-matrix oxidation.

Lessons learned: VeoVa incorporation contributes to alkali and water resistance in roof-coating binders, supporting longer-term film integrity. In high-radiation climates, UV stability must be reinforced with HALS or equivalent light-stabilization packages to mitigate chalking and color fade (Table 16.30).

Table 16.30 summarizes the effect of VeoVa level and HALS stabilization on the weathering durability of the roof coating.

Table 16.30 Effect of VeoVa and HALS on roof coating durability.

Parameter	Before optimization	After optimization
Weathering resistance	Chalking after 6 months	No chalking after 12 months
VeoVa level	12%	20%
UV stabilizer	None	1% HALS

Table 16.31 Impact of monomer ratio on PSA tack retention.

Parameter	Before optimization	After optimization
Tack after 3 weeks	Lost	Maintained
2-EHA level	35%	40%
Tackifier	Not included	Included

16.32 Case Study 31—PSA Losing Tack on Film-to-paper Lamination

Background: A VAM/2-ethylhexyl acrylate (2-EHA) copolymer PSA exhibited loss of tack after about 3 weeks on laminated film-to-paper labels.

Root cause: An excess of crystalline VAM-rich segments increased the effective glass-transition temperature (Tg) over time, shifting the viscoelastic balance toward rigidity and reducing tack retention in service.

Solution implemented: The soft-monomer (2-EHA) content was raised from 35% to 40% to lower Tg and restore the viscoelastic window for pressure sensitivity. In parallel, a compatible, noncrystallizing tackifier was introduced to sustain long-term adhesion without phase separation.

Lessons learned: PSA performance relies on maintaining a sufficiently low T_g to preserve the appropriate viscoelastic balance. Avoiding excessive crystallinity in VAM-rich domains is essential for sustained tack over time (Table 16.31).

Table 16.31 summarizes the impact of monomer ratio optimization on PSA tack retention.

16.33 Case Study 32—High-PVC Exterior Paint Binder Showing Efflorescence

Background: An exterior wall paint formulated with a VAM/BA/VeoVa copolymer binder exhibited white, salt-like deposits following rainfall.

Root cause: Surfactant molecules leached from the latex film and interacted with alkaline salts originating from the cementitious substrate, producing visible efflorescence on the coating surface.

Solution implemented: The binder system was reformulated to employ polymerizable (reactive) surfactants to minimize post-film-formation surfactant migration. In addition, an alkali-resistant primer was specified prior to the topcoat to block upward migration of soluble salts from the substrate.

Lessons learned: Even with hydrophobic copolymer binders, exterior durability is strongly influenced by surfactant selection and migration behavior. Appropriate

Table 16.32 Strategies to mitigate efflorescence in high-PVC exterior paints.

Parameter	Before optimization	After optimization
Efflorescence after rain	Severe	None observed
Surfactant type	Conventional anionic/nonionic	Polymerizable surfactant
Primer application	Not applied	Alkali-resistant primer used

substrate preparation—including the use of alkali-resistant primers—reduces efflorescence risk and helps maintain long-term appearance (Table 16.32).

Table 16.32 summarizes practical formulation and application strategies used to mitigate efflorescence in high-PVC exterior paints.

16.34 Case Study 33—Excessive Foaming During Emulsion Polymerization

Background: A VAM/BA emulsion-polymerization plant experienced frequent batch interruptions due to foam overflow, leading to latex loss, condenser fouling, and extended cleaning downtime. Foam persisted despite routine defoamer dosing and adversely affected product stability and safety compliance.

Root cause: High-foaming surfactants were employed without prior compatibility testing. In addition, the silicone defoamer used was overdosed, producing crater defects in finished coatings while still failing to control foam during reaction. Insufficient headspace control and excessive agitation further aggravated the issue by allowing foam to accumulate and overflow into vent lines.

Solution implemented: The surfactant package was replaced with low-foam secondary-alcohol ethoxylates and phosphate esters. A silicone-free, mineral-oil–based defoamer was introduced and predispersed in the monomer preemulsion to ensure early, uniform availability. Agitation speed was adjusted to minimize vortex formation while preserving dispersion efficiency. A nitrogen blanket was installed in the reactor headspace to collapse foam during feed and peak-reaction stages.

Outcome: Foam height decreased from approximately 35 cm to less than 5 cm at peak polymerization. No batch interruptions or vent fouling occurred over 12 consecutive runs. Film quality improved, with no crater or fisheye defects attributable to prior defoamer overdosing.

Lessons learned: Effective foam control must integrate surfactant chemistry, mechanical agitation, and reactor-headspace management rather than relying solely on defoamer addition. Overdosing incompatible defoamers can introduce film defects while failing to address the root causes. Nitrogen blanketing

Table 16.33 Effect of foam control optimization on process stability.

Parameter	Before optimization	After optimization
Foam height during peak reaction	35 cm	<5 cm
Batch interruption frequency	2–3 per week	None for 12 runs
Defoamer type	Silicone based (overdosed)	Mineral-oil-based (optimized dosage)
Film defects	Crater/fisheyes observed	No defects

provides a robust mechanical measure for persistent foam in high-solids or high-shear processes (Table 16.33).

The effect of foam control optimization on process stability is summarized in Table 16.33.

16.35 Case Study 34—Cost Reduction in Emulsion Paints without Losing Performance

Background: A construction-paint manufacturer sought to reduce binder cost by lowering VeoVa content in a VAM/BA/VeoVa copolymer designed for exterior applications.

Root cause: VeoVa is more expensive than VAM and BA. Reducing VeoVa lowers hydrophobicity and can diminish dirt-pick-up resistance and gloss retention, compromising long-term exterior durability.

Solution implemented: VeoVa content was reduced from 15% to 8% while butyl acrylate was increased slightly to preserve film flexibility. To restore water repellency, a silane–siloxane emulsion was incorporated at 1.5% (e.g., Wacker Silres BS 290 or Dow Silres BS 94). Surfactant selection and HLB balance were adjusted to maintain colloidal stability of the latex after the monomer shift and additive introduction.

Lessons learned: Binder cost savings are achievable by rebalancing monomer composition and introducing functional additives; however, any loss in hydrophobicity must be compensated to maintain exterior durability (Table 16.34).

Table 16.34 Effect of VeoVa reduction and additives on paint performance.

Parameter	Before optimization	After optimization
VeoVa level	15%	8%
Water absorption	High after VeoVa cut	Restored with silicone additive
Gloss retention	Dropped by 20%	Restored to acceptable level

Table 16.35 Tackifier optimization to reduce blocking.

Parameter	Before optimization	After optimization
Blocking at 40 °C	Severe	None
Tackifier softening point	Low	High
Tackifier content	Original level	Reduced by 10%

Table 16.34 summarizes the effect of VeoVa reduction and the use of performance additives on key paint properties.

16.36 Case Study 35—Label Adhesive Blocking in Roll Storage

Background: A rosin-tackified PVOH adhesive for paper labels exhibited blocking when rolls were stored at 40 °C.

Root cause: A low-softening-point rosin ester migrated to the adhesive surface under elevated temperature, increasing surface tack and promoting blocking during storage.

Solution implemented: The tackifier package was reformulated by replacing the rosin ester with a higher-softening-point hydrogenated grade to reduce thermal migration. In addition, the total tackifier content was reduced by approximately 10% to further limit mobile species at the surface while preserving required adhesion.

Lessons learned: Tackifier selection for hot-storage conditions must account for thermal stability and migration tendency. Balancing tackifier loading against performance targets is critical to preventing blocking in rolled label stock (Table 16.35).

Table 16.35 summarizes the effect of tackifier optimization on reducing blocking in the adhesive system.

16.37 Case Study 36—Poor Wet Tack on High-speed Film Bonding

Background: A hydrocarbon-resin-tackified PVOH adhesive for film-to-carton window patching failed to achieve adequate grab at machine speeds above 200 m/min.

Root cause: The tackifier dispersion contained oversized particles that dissolved too slowly in the aqueous phase, delaying tack development at the lamination nip. Contributing factors included suboptimal wetting on a low-energy film surface and excessively high-shear viscosity at the applicator, which caused bond-line starvation at speed.

Solution implemented: The formulation and process were returned to accelerate early viscoelastic response and ensure continuous wet-out at speed. A finer-particle hydrocarbon-resin dispersion (typical D50 ≤0.2–0.3 μm) or a microemulsion grade was selected to increase the diffusion rate into the PVOH phase. Resin selection was shifted to a lower softening-point hydrocarbon resin—or a rosin-ester blend—to narrow the effective glass-transition window and sharpen early tack. Wetting was improved with 0.1–0.3% of a low-foam nonionic wetting aid, and film surface energy was verified and, where feasible, raised to ≥38 dyn/cm by corona or plasma treatment. Rheology was retuned to prevent applicator starvation by targeting a high-shear viscosity of 150–250 mPa s at approximately 10,000 s^{-1} while maintaining 1,500–3,000 mPa s Brookfield at application. Total solids were increased by 2–3%—without raising Brookfield viscosity—to reduce the water-layer thickness and shorten set time. Where regulations permitted, 0.5–1.5% of a fast-evaporating glycol-ether co-solvent was added to further intensify early tack without elevating application temperature.

Lessons learned: At high line speeds, wet-tack kinetics are governed by tackifier particle size and the solubility window. Wetting and surface-energy control are as critical as resin selection when bonding to low-energy films. Maintaining the specified high-shear viscosity range is essential to avoid applicator starvation above 200 m/min (Table 16.36).

Table 16.36 summarizes the influence of nonthermal formulation and surface modifications on maintaining wet tack at machine speeds above 200 m/min.

Table 16.36 Influence of nonthermal changes on wet tack at more than 200 m/min.

Parameter	Before optimization	After optimization
Machine-speed performance	Failed at >200 m/min	Passed at >200 m/min
Resin particle size	Large	Fine (typical D50 ≤0.3 μm)
Resin grade	Baseline hydrocarbon resin	Lower softening-point hydrocarbon resin or rosin-ester blend
Wetting aid (low-foam nonionic)	None	0.1–0.3%
Film surface energy	Approximately 34–36 dyn/cm	38 dyn/cm or higher (treated)
High-shear viscosity (≈10,000 s^{-1})	Greater than 300 mPa s	150–250 mPa s
Brookfield viscosity (application)	1,500–3,000 mPa s	1,500–3,000 mPa s (held constant)
Total solids	Baseline	Baseline plus 2–3%
Application temperature	Baseline	Baseline (unchanged)

16.38 Case Study 37—Gel Formation in Borate-modified PVOH Adhesive

Background: A borate-modified polyvinyl alcohol (PVOH) adhesive used for tissue-core bonding developed visible lumps after 2 weeks of storage.

Root cause: Gelation arose from reversible complexation between borate species and vicinal diols on PVOH chains under elevated pH conditions. Borate functions as a tetrahedral Lewis-acid center that coordinates two vicinal-diol oxygens on PVOH, forming reversible di-diol borate complexes; PVOH donates the diol ligands while borate provides the bridging center. These interactions generated transient interchain bridges rather than permanent covalent bonds. High borate dosage, storage pH above neutrality, and the use of highly hydrolyzed, higher-molecular-weight PVOH increased the density of reversible linkages and promoted partial gel formation. Localized alkaline pockets during borate addition and the presence of polyol plasticizers further aggravated lumping.

Solution implemented—Track A (controlled-borate formulation): Storage pH was adjusted to 4.8–5.4 using a mild buffer (e.g., acetate), a range in which borate–diol complexes largely dissociate during storage. The total borate content was limited to ≤0.08% as B_2O_3 equivalent on the total formulation and was dosed slowly into a well-diluted, vigorously mixed adhesive to avoid localized high-pH zones. The PVOH grade was changed to a partially hydrolyzed polymer (approximately 87–89 mol% hydrolysis) or a lower-molecular-weight grade in order to reduce the density of available vicinal diols and solution elasticity. A chelating agent (0.05% EDTA or equivalent) was included to bind calcium and magnesium ions that may influence complexation kinetics. Polyol plasticizers were removed or minimized while borate was present; where plasticization was necessary, small amounts of compatible glycol ethers were used instead. Track B—Borate-free formulation: Borate was eliminated entirely, and body was established using an associative thickener—0.2–0.4% HEUR or HASE—or 0.1–0.3% HEC, tuned to deliver the required shear profile at the applicator. Storage and application pH were maintained at 4.8–5.4 to support preservative efficacy and PVOH stability. Plasticization was provided with 2–5% dipropylene glycol (or similar), and when a faster wet-grab was required, 3–8% carbohydrate tackifier (e.g., dextrin) was incorporated, balanced to avoid foaming and excessive drying times.

Outcome: Both tracks produced lump-free storage for 8 weeks at 40 °C, viscosity drift within 10%, no screen clogging, and acceptable tissue-core bond strength at typical converting line speeds.

Lessons learned: Borate modifies PVOH through reversible diol complexation, and the risk of gel formation increases with pH, borate level, and diol-site density.

Table 16.37 Influence of borate level and pH on storage stability of PVOH adhesive.

Parameter	Problem batch	Track A: controlled borate	Track B: borate-free
PVOH grade	Fully hydrolyzed, high MW	Partially hydrolyzed or lower MW	Partially hydrolyzed or lower MW
Borate dosage (B_2O_3 eq., on total)	Approximately 0.15%	No more than 0.08%	0%
Storage pH	About 8.5–9.0	4.8–5.4 (buffered)	4.8–5.4 (buffered)
Polyol plasticizer	Present	Removed or minimized	Allowed, optimized
Chelant (EDTA)	None	0.05%	Optional
Eight-week stability at 40 °C	Lumps and partial gel	No gel; viscosity drift within 10%	No gel; viscosity drift within 10%

When borate must be used, tight pH control and capping the borate level are decisive. For long shelf life and minimal storage risk, a borate-free rheology strategy is often the most robust path (Table 16.37).

The influence of borate level and pH on storage stability of PVOH adhesive is summarized in Table 16.37.

16.39 Case Study 38—Loss of Adhesion in Metallized Film Lamination

Background: A flexible packaging converter producing PET films metallized with aluminum by physical vapor deposition (PVD) reported unbonded zones with local delamination after lamination. These unbonded zones appeared as irregular bubble-like defects between the metallized layer and the adhesive layer within 24–48 h of production. The adhesive system was based on a standard water-borne polyurethane dispersion without a crosslinker. Such defects compromised optical clarity and barrier performance, thereby reducing the shelf life of packaged goods.

Root cause: Metallized films present an extremely smooth surface and very low surface energy, which reduces adhesive wetting. The adhesive lacked reactive chemistry to establish durable bonds with the aluminum layer. In addition, the adhesive pH was high (≈8.5), a condition that could attack thin aluminum coatings, while the lamination nip temperature was insufficient to promote proper adhesive flow and activation. These combined factors weakened the interface, and stress release during curing generated local delamination zones.

Table 16.38 Performance before and after optimization in metallized film lamination.

Parameter	Before optimization	After optimization
Adhesive system	PU dispersion only	PU dispersion + blocked isocyanate
pH	8.5	6.8
Lamination nip temperature	70 °C	95 °C
Unbonded zones after 48 h	Severe	None
Barrier stability after humidity aging	Failed	Passed

Solution implemented: The adhesive was upgraded by introducing a blocked isocyanate crosslinker at 4% on adhesive solids, activated during heat lamination. To improve adhesion, 0.8% of a polyester-based adhesion promoter was added to enhance chemical interaction with the metallized surface. The adhesive pH was adjusted to 6.5–7.0 to prevent aluminum degradation. Lamination parameters were optimized by increasing nip temperature to 95 °C, ensuring blocked isocyanate activation. Coating was applied uniformly at 8 g/m^2 wet weight via gravure, followed by conditioning of laminated rolls for 72 h at 25 °C and 50% RH before slitting and conversion.

Lessons learned: Lamination of metallized films requires adhesives with reactive functionality to form strong and durable bonds. Adhesive pH and lamination nip temperature exert a critical influence on both adhesion and aluminum layer integrity. Incorporating crosslinking chemistry ensures stability against humidity and mechanical stress, thereby preventing the formation of unbonded zones (Table 16.38).

Table 16.38 summarizes the performance changes observed before and after optimization in metallized film lamination.

16.40 Case Study 39–Open Time Too Short in Tackified PVOH/PVAc Blend

Background: A PVAc/PVOH blend adhesive tackified with gum rosin showed an open time of approximately 45 s, which was unsuitable for manual carton assembly.

Root cause: The rosin tackifier accelerated water loss from the adhesive film, shortening the workable open time.

Solution implemented: A humectant strategy was adopted by adding 2% sorbitol to slow water evaporation. The modification extended the open time from 45 s to about 2 min without compromising bond strength.

Table 16.39 Effect of humectant addition on open time.

Parameter	Before optimization	After optimization
Open time	45 s	2 min
Humectant	None	2% sorbitol
Bond strength	Adequate	Adequate

Lessons learned: Open time is governed by both tackifier chemistry and water-retention additives. Properly balancing these factors is essential for adhesives intended for manual application (Table 16.39).

Table 16.39 summarizes the effect of humectant addition on extending adhesive open time while maintaining bond strength.

16.41 Case Study 40—Adhesive Failure on Untreated Polypropylene (PP) Film

Background: A PVAc–acrylic copolymer adhesive for film lamination exhibited complete delamination from untreated polypropylene (PP) within 24 h of bonding.

Root cause: Untreated PP possesses very low surface energy (≈29 mN/m) and lacks polar groups capable of hydrogen bonding, which prevents adequate wetting and chemical anchorage. In the absence of surface activation and reactive functionality within the adhesive, bond strength is severely compromised.

Solution implemented: The polymer was modified to a VAM/BA/AA terpolymer to introduce carboxylic groups and improve reactivity. An adhesion promoter, vinyltriethoxysilane, was incorporated at 2% to facilitate interfacial coupling, and a carbodiimide crosslinker was added at 0.5% to enhance cohesion and hydrolysis resistance. Process controls maintained pH in the range 4.8–5.2 to preserve carbodiimide stability. Prior to lamination, the PP film received corona treatment to increase surface energy and create reactive sites.

Table 16.40 Performance before and after optimization in untreated PP film for lamination.

Parameter	Before optimization	After optimization
Polymer type	VAM/BA copolymer	VAM/BA/AA terpolymer
Adhesion promoter	None	2% vinyltriethoxysilane
Crosslinker	None	0.5% carbodiimide
Surface treatment	None	Corona treatment
Bond performance	Failed in 24 h	Stable after humidity aging

Lessons learned: Achieving adhesion to untreated polyolefins requires both chemical functionality in the adhesive and physical surface activation. The combination of silane adhesion promoters with carbodiimide crosslinkers improves bond strength and durability, while corona or plasma treatment is essential to raise wettability and promote interfacial chemical bonding (Table 16.40).

Table 16.40 summarizes the formulation and surface-treatment optimization required to achieve reliable lamination performance on untreated PP film.

16.42 Case Study 41 – Blocking on Lacquered Carton Lamination

Background: Carton-to-lacquered-carton lamination experienced blocking during stacked storage, causing sheets to adhere and disrupting downstream converting. The lacquered surface presented a low-porosity, hydrophobic interface.

Root cause: Low porosity and hydrophobicity on the lacquered surface slowed adhesive film formation and extended open time. In parallel, the wet adhesive exhibited limited substrate wetting, leaving regions with residual surface tack even after partial drying.

Solution implemented: Film-formation kinetics were accelerated by increasing the coalescent level by 0.5% without sacrificing clarity. Wetting on lacquered substrates was improved by adding 0.2% silicone-based dynamic wetting agent (e.g., TEGO Wet 270 or BYK-346), which lowers dynamic surface tension to promote uniform wet-out while remaining nonreactive and low-foaming with good recoatability. Where blocking persisted, an optional 0.1–0.2% alkoxysilane adhesion promoter or silane-functional surfactant (e.g., Silres BS type or glycidoxy/aminosilane) was introduced; during drying, alkoxysilanes hydrolyze and condense, forming siloxane bonds to hydroxyl-bearing or mineral surfaces and anchoring within the polymer matrix to reduce migration and surface tack. Adhesive pH was maintained between 4.8 and 5.2 to balance rapid film formation, preservative efficacy, and suitable conditions for silane hydrolysis/condensation when used. When appearance targets permitted, 0.1–0.3% micronized polyethylene or PTFE wax emulsion was incorporated as an antiblock additive with ongoing gloss monitoring.

Lessons learned: On low-porosity lacquered boards, rapid wetting coupled with controlled open time is essential to avoid blocking. Dynamic wetting agents act during application by lowering surface tension and improving film uniformity; they are typically nonreactive and selected for minimal migration. Reactive alkoxysilane options can provide postdry anchoring either to the substrate or within the polymer matrix, suppressing long-term surface tack and blocking. Antiblock additives such as micronized wax emulsions can further improve stacked performance when gloss requirements allow (Table 16.41).

Table 16.41 summarizes the formulation adjustments used to resolve blocking and improve lamination performance.

Table 16.41 Performance before and after optimization in blocking on lacquered carton lamination.

Parameter	Before optimization	After optimization
Blocking issue	Severe	Resolved
Wetting agent	None	0.2% silicone dynamic wetter; optional 0.1–0.2% alkoxysilane promoter
Coalescent	Standard level	+0.5% for faster film formation
pH control	Uncontrolled	Maintained at 4.8–5.2
Anti-block additive	None	0.1–0.3% micronized PE/PTFE wax emulsion (if gloss permits)

16.43 Case Study 42—Loss of Bond on PVC-to-wood Lamination

Background: A postforming process used for PVC-to-wood lamination failed peel-strength tests after hot–cold cycling. The adhesive system was intended for furniture-grade panels exposed to humidity and thermal stress. Initial tack appeared satisfactory; however, after conditioning and temperature cycling, severe delamination was observed.

Root cause: Plasticizer migrated from the PVC into the adhesive layer, softening the bond line. The original adhesive did not provide sufficient reactive crosslinking to resist plasticizer ingress or dimensional changes during thermal cycling. In addition, pH was not maintained within the recommended range, reducing crosslinker efficiency.

Solution implemented: The adhesive was reformulated as a water-borne polyurethane dispersion (PUD) with high polarity to promote PVC anchorage. A blocked-isocyanate dispersion was incorporated at 3.5% (~20% NCO equivalent) to provide reactive crosslinking during lamination. Bonding to the wood substrate was enhanced by adding 0.5% amino-functional silane (prehydrolyzed). Process controls maintained adhesive pH between 4.5 and 5.5 to preserve crosslinker latency and avoid PVC hydrolysis. The application used a roller coaster at 10 g/m^2 wet film; the coating was dried at 70 °C, and the blocked isocyanate was activated at 95 °C in the lamination nip. Laminated panels were conditioned for 72 h at 25 °C and 50% RH to complete cure.

Lessons learned: PVC lamination reliability improves when the adhesive resists plasticizer migration and maintains dimensional stability under heat and humidity. Combining a polyurethane dispersion with a blocked isocyanate

Table 16.42 Performance before and after optimization in PVC-wood lamination.

Parameter	Before optimization	After optimization
Adhesive type	PVAc–acrylic	PU dispersion + blocked isocyanate
Crosslinker	None	3.5% blocked isocyanate
Adhesion promoter	None	0.5% silane
pH control	Uncontrolled	4.5–5.5
Bond after hot–cold cycling	Failed	Passed (>7 N/25 mm)

crosslinker and an amino-silane coupling agent provided durable bonding and resistance to thermal cycling. Maintaining the correct pH window and activation temperature is critical for full crosslinking and long-term performance (Table 16.42).

16.44 Case Study 43—Delamination of PE-to-carton Lamination Despite Surface Treatment

Background: A packaging operation producing window-patched cartons for high-end products employed a VAM/BA copolymer adhesive to laminate polyethylene (PE) film to printed carton board. Despite the use of corona treatment on the PE film, customers reported delamination within 24 h, particularly when cartons were stored under humid conditions.

Root cause: Corona treatment raised the surface energy of PE only transiently; hydrophobic recovery reduced its effectiveness with time. In parallel, the adhesive polymer lacked sufficient polar functionality to secure durable anchorage. Humidity further weakened the interface in the absence of chemical bonding.

Solution implemented: The adhesive was upgraded to a VAM/BA/AA terpolymer to introduce carboxylic functionality. A dual-reactive package was added, comprising 1.0% silane adhesion promoter (aminopropyltriethoxysilane) and 0.5% carbodiimide crosslinker, enabling covalent bond formation at the interface and increased cohesive strength via carboxyl–carbodiimide reactions. Process controls maintained adhesive pH at 4.8–5.2 to preserve carbodiimide stability and to promote silane hydrolysis and condensation. Corona treatment was verified immediately before lamination to minimize hydrophobic recovery of the PE surface. The adhesive was applied at 8–10 g/m^2 wet film, dried to tack, and laminated under standard nip conditions to ensure complete wet-out of the treated substrate.

Table 16.43 Performance before and after optimization in delamination of PE-to-carton lamination.

Parameter	Before optimization	After optimization
Polymer	VAM/BA	VAM/BA/AA
Adhesion promoter	None	1.0% silane
Crosslinker	None	0.5% carbodiimide
PE surface	Corona (timing not controlled)	Corona applied immediately prelamination
pH	Uncontrolled	4.8–5.2
Peel strength after 24 h humidity	Failed (<1.0 N/25 mm)	Passed (>2.5 N/25 mm)

Lessons learned: Reliable bonding to PE requires synergy between surface activation and adhesive chemistry. Introducing acrylic acid functionality and pairing it with silane promoters and carbodiimide crosslinkers markedly improves humidity resistance. Surface treatment alone is insufficient; the timing of corona and the adhesive reactivity must align. Maintaining the pH window at 4.8–5.2 preserves the effectiveness of both silane and carbodiimide during application and cure (Table 16.43).

The performance before and after optimization in PE-to-carton lamination is summarized in Table 16.43.

16.45 Case Study 44—Poor Adhesion on UV-cured Gloss Lacquer

Background: A carton converter using UV-cured glossy lacquer observed their standard PVAc–acrylic adhesive peeling cleanly after curing.

Cause: The UV-cured lacquer created a crosslinked network with extremely low surface energy and no available polar functional groups, preventing hydrogen bonding and strong adhesion.

Solution: Switched to a styrene–acrylic adhesive with better compatibility for nonpolar surfaces and added an adhesion promoter specifically designed for UV-cured coatings.

Lesson learned: When bonding UV-cured lacquered substrates, adhesives require nonpolar compatibility and often need specialty adhesion promoters to overcome low surface energy.

Technical insight: UV-cured coatings are highly crosslinked, eliminating reactive sites and creating surfaces with critical surface tension as low as 20–25 dynes/cm. Traditional PVAc-based adhesives depend on hydrogen

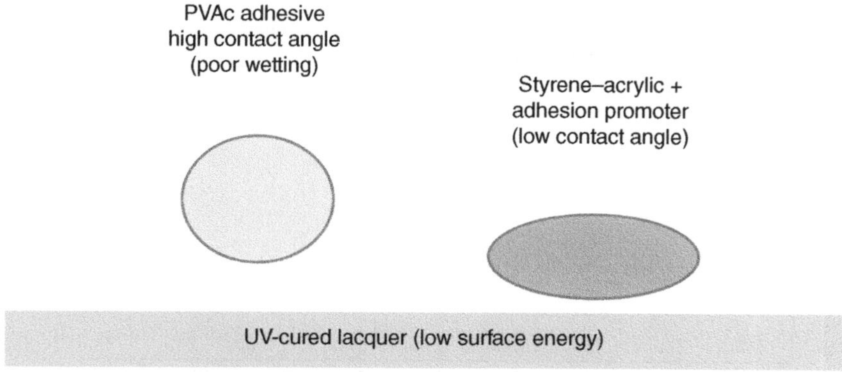

Figure 16.1 Wetting behavior on UV-cured lacquer. *Discussion:* Before modification (left), PVAc–acrylic adhesive forms beaded clusters due to poor wetting; after modification (right), styrene–acrylic with adhesion promoter wets uniformly, enabling adhesion. *Explanation:* On low-energy UV-cured lacquered surfaces, conventional PVAc–acrylic adhesives fail to spread (left panel, "Before"), producing beaded or clustered droplets that do not wet the substrate. This poor wetting prevents intimate contact and hydrogen bonding, leading to clean adhesive failure. After switching to a styrene–acrylic system with an adhesion promoter (right panel, "After"), the adhesive exhibits uniform wetting and spreading (small, evenly distributed drops), enabling stronger interaction with the nonpolar lacquered surface.

bonding; hence they fail. Styrene-acrylic systems incorporate nonpolar styrene segments that improve wetting on low-energy substrates. Adhesion promoters, often organosilanes or polymeric adhesion promoters, create a chemical bridge between the polar adhesive and the nonpolar substrate.

Example trade names (for illustration only): Styrene–acrylic dispersion: BASF Acronal® 290D; Adhesion promoter: Evonik Dynasylan® SIVO 160; Wetting aid: BYK-348 (Figure 16.1).

The effect of formulation modification on wetting behavior over UV-cured lacquered surfaces, and its direct impact on adhesion performance, is illustrated in Figure 16.1.

16.46 Case Study 45–Film Cracking in Subzero Exterior Masonry Paint

Background: A VAE-based exterior paint cracked within 2 weeks of exposure to sub-zero temperatures in a Northern European city.

Cause: The binder's glass transition temperature (T_g) was too high, reducing flexibility at low temperatures. At sub-zero conditions, the film became brittle.

Solution: Reformulated the binder by incorporating 12% ethylene to lower the T_g and increased the coalescent level to ensure film formation at 0 °C.

Lesson learned: For coatings in cold climates, achieving low MFFT and high flexibility is essential to prevent brittle failure.

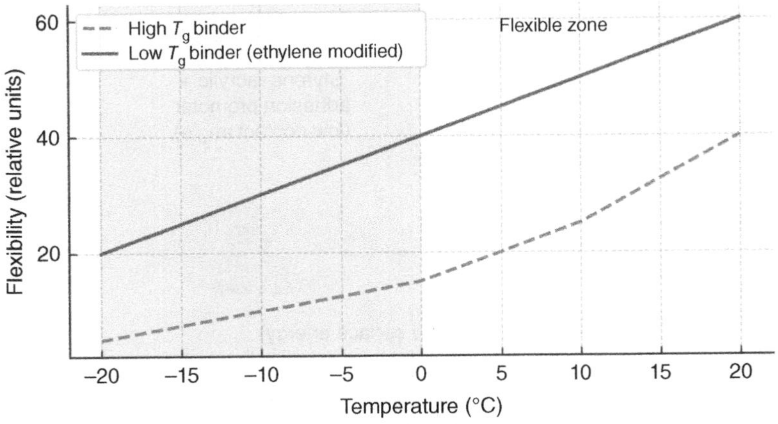

Figure 16.2 Effect of T_g on binder flexibility. The dashed curve (high-T_g) remains rigid at low temperatures, whereas the solid curve (low-Tg, ethylene-modified) maintains higher flexibility within the shaded/hatched "Flexible Zone." *Explanation:* Glass transition temperature (T_g) governs binder flexibility. The dashed curve (high-T_g binder) remains comparatively rigid at low and moderate temperatures, showing limited flexibility in cold conditions. The solid curve (low-T_g, ethylene-modified binder) retains substantially higher flexibility across the same temperature range. The shaded/hatched region labeled "Flexible Zone" indicates temperatures at which sufficiently low-T_g binders remain elastic—critical for coatings and adhesives used in sub-ambient or fluctuating climates.

Technical insight: Film cracking in VAE-based coatings results from insufficient flexibility due to high T_g. Ethylene reduces T_g and MFFT, improving low-temperature performance. Adding plasticizers or coalescents further enhances flexibility, though balance with scrub resistance is necessary. Alternatively, using VeoVa-modified binders or soft acrylates can improve flexibility.

Example trade names (for illustration only): VAE binder: Wacker VINNAPAS® EF 101; Coalescent: Dow Texanol™; Freeze-resistance additive: BASF Butyl CARBITOL® (Figure 16.2).

The influence of glass transition temperature (Tg) on binder flexibility, particularly the improved low-temperature performance of ethylene-modified systems, is illustrated in Figure 16.2.

16.47 Case Study 46—Freeze–Thaw Instability in Distribution

Background: A cold-climate wall primer coagulated after three freeze–thaw cycles during transport.

Cause: The binder lacked an adequate freeze–thaw stabilizer system, causing irreversible particle flocculation under repeated freezing and thawing.

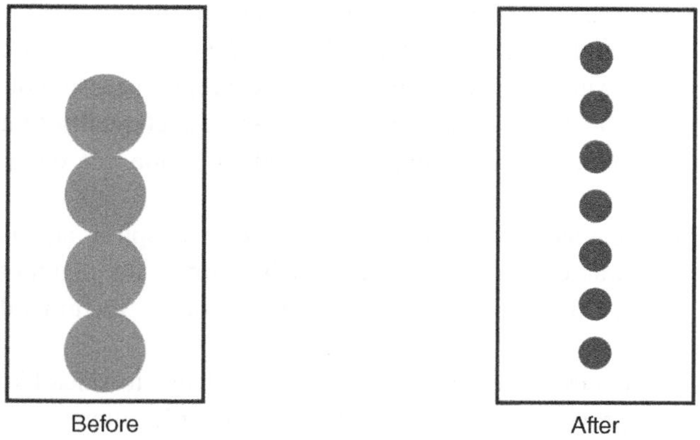

Figure 16.3 Freeze–thaw stability (before and after). *Explanation:* During freezing ("Before"), water crystallization compresses the dispersed polymer particles, forcing them into aggregated clusters. After thawing ("After"), a stable emulsion will re-disperse into its original uniform particle distribution, indicating that the system can withstand freeze–thaw cycling without irreversible coagulation or phase separation.

Solution: Added propylene glycol as a cryoprotectant and optimized surfactant balance to enhance colloidal stability at sub-zero temperatures.

Lesson learned: Freeze–thaw stability must be validated under conditions simulating real transport and storage, not only laboratory tests.

Technical insight: Freezing forces water to crystallize, concentrating salts and destabilizing the electrostatic double layer around polymer particles. Propylene glycol and glycols act as antifreeze agents, reducing ice formation and osmotic shock. Correct surfactant ratio helps maintain steric and electrostatic stability. For highly demanding climates, specialized freeze–thaw stabilizers (e.g., polyurethane thickeners) may be used.

Example trade names (for illustration only): Cryoprotectant: Dow Propylene Glycol USP; Surfactant package: BASF Disponil® series; Thickener: Rheovis® PU 1211 (Figure 16.3).

The effect of freeze-thaw cycling on emulsion stability, including particle aggregation during freezing and re-dispersion upon thawing, is illustrated in Figure 16.3.

16.48 Case Study 47—Poor Adhesion to Damp Substrates (Corrected)

Background: A vinyl–acrylic exterior paint applied to a cold, slightly damp concrete wall peeled after 1 month.

Cause: Insufficient binder polarity and lack of chemical coupling to the hydroxyl-rich, alkaline cement surface led to poor wet adhesion under damp conditions.

Solution: Introduced an organofunctional silane coupling agent (preferably amino- or epoxy-silane) and modestly increased polar functionality in the binder (e.g., acrylic acid). Amino/epoxy silanes condense with surface silanols on concrete to form Si—O—Si bonds and present organic functionality that interacts with the polymer, markedly improving wet adhesion on damp masonry.

Lesson learned: For damp cementitious substrates, use silane coupling agents (amino/epoxy) plus adequate binder polarity, which mainly aid pigment wetting/dispersing and do not create durable chemical bridges to mineral surfaces.

Technical insight: Amino-silanes (—NH_2) hydrolyze and condense to silica-like networks on concrete, providing hydrogen-bonding and ionic interactions with polar binders. Epoxy-silanes enable covalent linkage via epoxy reactions, particularly with amine, acid, or hydroxyl groups present in the polymer. Increasing binder polarity with AA/MAA improves interfacial wetting and hydrogen bonding without over-plasticizing the film. Surfactant level should be carefully controlled, and low-water-sensitivity grades selected, to avoid re-emulsification under damp service.

Example trade names (illustrative): Amino-silanes include Evonik Dynasylan® AMEO/AMMO and Momentive Silquest A-1100. Epoxy-silanes include Evonik Dynasylan® GLYMO and Momentive Silquest A-187.

16.49 Case Study 48—Efflorescence and Staining in Early Spring Applications

Background: A budget cold-climate exterior paint developed white stains after early-season rain.

Cause: Surfactant leaching from the paint film combined with salt migration from the masonry surface, forming visible efflorescence.

Solution: Replaced free surfactants with polymerizable surfactants in binder synthesis and mandated an alkali-resistant primer to block salt migration.

Lesson: Cold climates with frequent thaw cycles exacerbate surfactant leaching and salt migration, making primer systems essential.

Technical insight: Polymerizable surfactants ("surfmers") become part of the polymer backbone, minimizing leaching risk. Efflorescence control also requires surface preparation and sealing. Using hydrophobic modifiers and film-forming additives improves early rain resistance.

Example trade names (for illustration only): Binder: Wacker VINNAPAS® EP series with surfmers; Primer: BASF Acronal® 290D-based alkali-resistant primer; Additive: Evonik TEGO® Wet 510.

16.50 Case Study 49—Rain Damage Within an Hour of Application

Background: An Eastern European exterior emulsion paint failed after light rain 1 h postapplication in cold conditions.

Cause: Binder film had not fully coalesced at low temperature, leaving micropores that allowed water ingress.

Solution: Developed a quick-set cold-climate binder using a blend of low-MFFT polymer and fast-evaporating coalescent for rapid early film formation.

Lesson learned: Regions with unpredictable rain demand systems with accelerated film formation and early water resistance.

Technical insight: Low-MFFT binders and volatile coalescents speed coalescence and minimize vulnerability. Hydrophobic additives and wax dispersions can further enhance resistance. Careful balance with open time is critical to avoid lap marks.

Example trade names (for illustration only): Binder: Dow Rhoplex™ Cold-Weather Grade; Coalescent: Eastman Optifilm™ Enhancer 300; Wax additive: Michelman Michem® Emulsion 62330.

16.51 Case Study 50—Solid Content: Oven vs. Infrared Balance

Background: A PVAc–acrylic copolymer latex was tested for solid content using both a conventional 105 °C oven method (ASTM D2369) and a rapid infrared moisture analyzer. The oven method gave 54.8%, while the infrared balance read 55.2%. The small difference was within tolerance, but the IR method reduced testing time from 1 h to 7 min, enabling faster batch release.

Cause: The slight difference between methods was due to variation in heat transfer and possible incomplete drying of volatiles in rapid IR analysis compared with oven drying.

Solution: Validated the IR method as an acceptable alternative for routine batch control, while maintaining the oven method for reference or arbitration testing.

Lesson: Rapid moisture analysis can significantly improve production efficiency, but method validation is essential to maintain accuracy and compliance.

Technical insight: Infrared balances heat samples quickly using radiant energy, which can cause localized overheating or leave trace volatiles compared to the uniform heat of an oven. For polymer emulsions, differences within ±0.5% are generally acceptable if both methods are properly standardized.

Example trade names (for illustration only): Infrared analyzer: Mettler Toledo HB43-S; Reference oven method: ASTM D2369 standard.

16.52 Case Study 51—pH Drift During Storage

Background: A styrene–acrylic binder stored at 40 °C for accelerated aging showed a pH drop from 8.5 to 7.2 after 4 weeks. Measurement was performed with a bench pH meter calibrated at pH 7 and 10 buffers. The drop indicated ammonia loss and partial hydrolysis of acrylate ester groups, prompting the addition of a more stable amine neutralizer.

Cause: Ammonia, commonly used for pH adjustment, is volatile and evaporates over time, especially under heat. Hydrolysis of ester groups in the presence of water and elevated temperature can release acidic species, compounding the pH decline.

Solution: Replaced ammonia with a tertiary amine neutralizer of lower volatility and included a pH buffer system to maintain stability during storage.

Lesson: Binder pH stability under heat aging is critical to preventing viscosity drift and microbial growth. Neutralizer selection plays a key role in long-term pH control.

Technical insight: Low-volatility amines such as 2-amino-2-methyl-1-propanol (AMP-95) provide sustained pH control without significant evaporation. Maintaining pH above 7.5 reduces the risk of acid-catalyzed hydrolysis and improves dispersion stability.

Example trade names (for illustration only): Neutralizer: Angus AMP-95; Buffer additive: Dow Optibase™ 450.

16.53 Case Study 52—Viscosity: Stormer vs. Brookfield

Background: An interior wall paint gave 92 KU on a Stormer viscometer but 1,200 cP at 20 rpm using a Brookfield RV spindle #4. The discrepancy reflected shear-thinning behavior, with the Stormer value representing mid-shear conditions and Brookfield showing low-shear viscosity important for sag resistance.

Cause: Shear-thinning rheology in paints causes viscosity to vary with the applied shear rate. Stormer viscometers simulate application shear, while Brookfield instruments measure low-shear conditions relevant for storage and leveling.

Solution: Used both measurements for comprehensive rheology control: Stormer for brushability and Brookfield for sag resistance and settling stability.

Lesson learned: Interpreting viscosity requires understanding the shear conditions simulated by each instrument. Single-point data can be misleading for pseudoplastic systems.

Technical insight: KU values (Krebs units) correlate with application feel, while Brookfield data reflect structure at rest. Associative thickeners (e.g., HEUR) strongly influence low-shear viscosity but minimally affect mid-shear values.

Example trade names (for illustration only): Thickener: BASF Rheovis® PU 1214; Brookfield Viscometer: Brookfield DVNext.

16.54 Case Study 53–MFFT Optimization

Background: A vinyl–acrylic binder intended for cold-climate paints was tested with an MFFT bar. The unmodified binder had an MFFT of 14 °C. Addition of 2% Texanol™ reduced MFFT to 4 °C, allowing safe application at 5 °C ambient temperature.

Cause: High MFFT results in incomplete film formation at low temperatures, leading to defects such as powdering or poor adhesion.

Solution: Introduced a low-volatility coalescent to lower MFFT without excessive odor or VOC impact.

Lesson learned: Balancing coalescent content ensures good film formation under cold conditions while maintaining regulatory compliance.

Technical insight: Coalescents reduce the glass transition temperature (T_g) of polymer particles during drying, facilitating coalescence below the binder's intrinsic MFFT. Texanol™ is widely used for low-temperature applications due to its efficiency and permanence.

Example trade names (for illustration only): Coalescent: Eastman Texanol™; Binder: Wacker VINNAPAS® EP 3360.

16.55 Case Study 54–T_g Measurement via Differential Scanning Calorimeter (DSC)

Background: Differential scanning calorimetry (DSC) was used to determine the glass transition temperature (T_g) of a pure acrylic binder. A T_g of −8 °C indicated high flexibility at low temperatures, making the binder suitable for elastomeric coatings. Increasing the hard monomer content raised T_g to +15 °C, improving block resistance.

Cause: Binder flexibility and blocking resistance depend on polymer chain mobility, which correlates with T_g. Higher soft-monomer content lowers T_g, improving low-temperature flexibility but reducing block resistance.

Solution: Balanced monomer composition by introducing hard segments for improved blocking while maintaining adequate flexibility for the intended application.

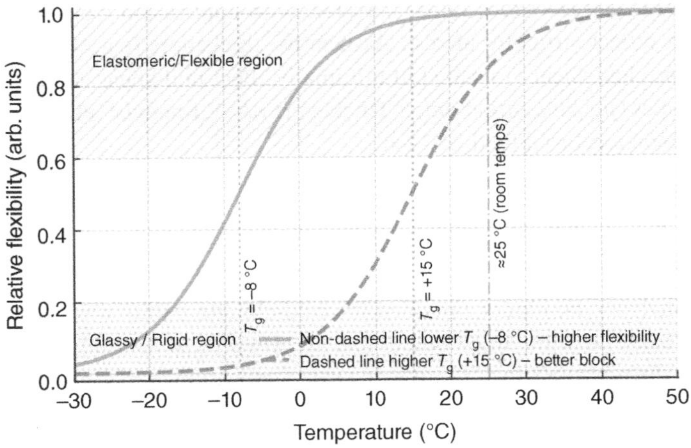

Figure 16.4 **T_g governs binder flexibility (DSC-informed).** *Explanation:* Differential scanning calorimetry (DSC) shows that T_g sets the temperature at which chain mobility increases. The nondashed curve (lower T_g, −8 °C) enters the flexible regime below 0 °C, supporting elastomeric crack resistance in cold weather. The dashed curve (higher T_g, +15 °C) remains relatively rigid at subambient temperatures but provides better block resistance near 25 °C. Balancing monomer composition shifts T_g to trade off low-temperature flexibility against blocking, consistent with Case Study 17.54.

Lesson learned: T_g optimization is critical in coating design to achieve the desired balance between flexibility, hardness, and resistance to sticking.

Technical insight: DSC provides accurate T_g data by detecting heat flow changes as polymer transitions from the glassy to the rubbery state. T_g strongly influences film performance: low T_g enhances flexibility and crack resistance, whereas high T_g improves hardness and mar resistance.

Example trade names (for illustration only): DSC instrument: TA Instruments Q2000; Pure acrylic binder: BASF Joncryl® series (Figure 16.4).

The relationship between glass transition temperature (Tg) and binder flexibility, as informed by DSC analysis and monomer composition, is illustrated in Figure 16.4.

16.56 Case Study 55—Wet Scrub Resistance: ASTM vs. ISO

Background: Two test methods were applied to the same semi-gloss wall paint: ASTM D2486 (10,000 cycles) and ISO 11998 (Class 1, <5 μm film loss after 200 cycles). Results confirmed compliance with both standards, supporting multi-market claims.

Cause: Different standards define scrub resistance in terms of either cycles to failure (ASTM) or film thickness loss under fixed cycles (ISO). Variability may arise from different brush types, applied pressure, and evaluation criteria.

Solution: Designed paint with a hard polymer matrix and optimized pigment volume concentration (PVC) to ensure film cohesion under both test conditions.

Lesson learned: Understanding differences between regional test standards is crucial for meeting international product claims.

Technical insight: High wet scrub resistance requires low PVC for good binder continuity, crosslinkable binder systems, and proper pigment dispersion to avoid weak points.

Example trade names (for illustration only): Crosslinking binder: Dow Rhoplex™ PR-216; Additive for scrub resistance: BASF Loxanol® EFC 100.

16.57 Case Study 56–Adhesion Testing: Cross-hatch vs. Pull-off

Background: A water-based primer for galvanized steel passed ISO 2409 cross-hatch adhesion with a 0 rating but showed only 3.5 MPa pull-off strength (ASTM D4541). This indicated the coating film adhered well under shear stress but failed under tensile loading due to internal cohesion limitations.

Cause: Cross-hatch evaluates adhesion under cutting and tape pull (shear and peel stresses), while pull-off measures tensile adhesion, which is sensitive to cohesive strength of the coating layer itself.

Solution: Improved cohesive strength by adjusting binder chemistry and adding functional adhesion promoters for better stress distribution.

Lesson learned: Multiple adhesion tests provide complementary information: passing one method does not guarantee overall adhesion performance under all stress types.

Technical insight—Adhesion promoters and cohesive strength for metal substrates

Adhesion promoters: On metal substrates, silane and phosphate/phosphonate functionalities enhance bonding by reacting with the native oxide layer to form durable M—O—Si or M—O—P linkages. Representative silane examples include Momentive Silquest A-187 (3-glycidyloxypropyltrimethoxysilane; epoxy-silane), Evonik Dynasylan GLYMO (3-glycidyloxypropyltrimethoxysilane; epoxy-silane), Momentive Silquest A-1100 (γ-aminopropyltriethoxysilane; amino-silane), and Evonik Dynasylan AMEO/AMMO (aminosilanes). For acrylic systems requiring direct copolymer anchoring, Momentive Silquest A-174 (3-methacryloyloxypropyltrimethoxysilane) is frequently used. Phosphate/phosphonate adhesion promoters can be introduced via phosphate-functional mono- or di-esters of acrylates; illustrative commercial examples include Arkema/Sartomer Sipomer PAM-200 or PAM-300 (phosphate-functional acrylate monomers) and Allnex Ebecryl 168 (phosphate acrylate oligomer) used at low dosage to improve adhesion and corrosion-creep resistance on aluminum and steel.

Cohesive strength (crosslinking) without undue loss of flexibility: To raise tensile adhesion while preserving elongation, use modest crosslink density and maintain a low-to-moderate T_g. Practical approaches include water-dispersible polycarbodiimide crosslinkers such as Nisshinbo Carbodilite grades (for example, Carbodilite E-02/E-03) for polyurethane-free wet adhesion boost; zirconium crosslinkers such as ammonium zirconium carbonate (e.g., AZC solutions) for improved wet adhesion and blocking; and water-dispersible blocked isocyanates (e.g., Covestro Bayhydur BL series) for heat-deblocked covalent crosslinking. Use the lowest effective dosage to avoid embrittlement and verify balance by correlating T_g (via DSC) with tensile/elongation tests.

Typical use levels: Silanes are commonly used at approximately 0.5–2.0% on binder solids (substrate and formulation dependent). Phosphate/phosphonate adhesion promoters are often effective at 0.5–2.0% on polymer solids; dosage should be optimized to minimize water uptake. Crosslinker levels depend on chemistry and supplier recommendations; start at the low end of the range and validate performance under relevant humidity, temperature, and salt-spray conditions.

16.58 Case Study 57—Water Resistance Testing

Background: A masonry paint underwent 7-day water immersion (ISO 2812-2) and condensation chamber testing (ISO 6270-1). While immersion showed no blistering, condensation caused microblisters due to trapped surfactants, leading to reformulation with lower-foaming wetting agents.

Cause: Surfactant migration under prolonged condensation leads to localized osmotic pressure and microblister formation, even if immersion tests show no defects.

Solution: Replaced high-foaming surfactants with low-foaming, polymerizable alternatives and improved pigment wetting balance to reduce surfactant content.

Lesson learned: Condensation testing is a more sensitive indicator of surfactant stability than immersion alone, highlighting the need for surfactant selection based on long-term water exposure.

Technical insight: ISO 6270-1 simulates cyclic condensation, which exacerbates surfactant migration by alternating swelling and drying cycles. Surfactant-resistant formulations often require surfmers and hydrophobic coalescents.

Example trade names (for illustration only): Low-foam wetting agent: TEGO® Wet 280; Surfmers: BASF Hydropalat® WE 3225.

16.59 Case Study 58—Particle Size via Laser Diffraction

Background: A VAE copolymer latex showed a D50 of 0.25 μm and a narrow distribution (span 1.2) by laser diffraction. After 3 months at 35 °C storage, D50 shifted to 0.40 μm with increased span, indicating partial coagulation due to surfactant desorption.

Cause: Elevated storage temperature accelerates surfactant migration from the particle surface, reducing colloidal stability and promoting particle flocculation and growth.

Solution: Incorporated polymerizable surfactants and increased the level of protective colloid to stabilize particle surfaces under thermal stress.

Lesson learned: Particle size monitoring during aging provides an early indicator of emulsion stability. Maintaining surfactant integrity is crucial for shelf life.

Technical insight: Laser diffraction measures particle size distribution by analyzing light scattering, providing span and D50 values as indicators of stability. Coagulation often starts with bridging flocculation, which broadens the size distribution before visible sedimentation occurs.

Example trade names (for illustration only): Surfmers: BASF Hydropalat® WE 3320; Protective colloid: Kuraray Mowiol® PVOH.

16.60 Case Study 59—Gloss Measurement and PVC Influence

Background: Two acrylic paints with pigment volume concentrations (PVC) of 20% and 40% were tested at a 60° gloss angle per ASTM D523. Gloss values were 82 and 18 units, confirming the inverse correlation between PVC and gloss.

Cause: Higher PVC increases surface roughness due to pigment crowding, reducing light reflectance and lowering gloss.

Solution: Selected optimized pigment dispersions and formulated PVC at the critical level for balancing opacity and gloss for the intended application.

Lesson learned: Gloss is highly sensitive to PVC, making pigment volume control a key parameter in the design of high-sheen and matte coatings.

Technical insight: Gloss depends on surface microtexture, binder film continuity, and pigment packing. High-gloss finishes require smooth surfaces with binder-rich top layers and controlled pigment size.

Example trade names (for illustration only): Gloss additive: BYK-378; Acrylic binder: BASF Joncryl® 1972.

16.61 Case Study 60—Freeze–Thaw Stability

Background: A low-MFFT VAE binder passed five freeze–thaw cycles per ASTM D2243 with 1% propylene glycol as stabilizer. Reducing the stabilizer to 0.3% caused coagulation after two cycles, confirming the minimum additive requirement.

Cause: Freeze–thaw cycling forms ice crystals that exclude polymer particles and concentrate electrolytes, destabilizing the dispersion. Adequate cryoprotectant minimizes osmotic shock and prevents irreversible flocculation.

Solution: Established a minimum glycol level for protection and included a secondary stabilizer to improve performance under severe conditions.

Lesson learned: Systematic evaluation of freeze–thaw resistance is essential for products intended for cold storage and distribution environments.

Technical insight: Propylene glycol lowers freezing point and protects colloidal stability by reducing ice crystal formation. Nonionic surfactants and associative thickeners can synergize with glycols to enhance stability.

Example trade names (for illustration only): Cryoprotectant: Dow Propylene Glycol USP; Binder: Wacker VINNAPAS® EF series 2021.

16.62 Case Study 61—Mechanical Stability Testing

Background: Mechanical stability of a PVAc copolymer latex was tested using a high-speed disperser at 4,000 rpm for 30 min. Posttest particle size and viscosity showed no change, confirming good stabilizer efficiency and suitability for high-shear tinting applications.

Cause: High-shear forces can disrupt latex stability by stripping surfactants or collapsing protective colloid layers, leading to flocculation. A stable latex resists such stress due to robust electrosteric stabilization.

Solution: Maintained a balanced protective colloid system and ensured adequate molecular weight for steric stabilization.

Lesson learned: Mechanical stability is essential for tinting and dispersion processes where latex is subjected to intense agitation.

Technical insight: ISO 3219 and ASTM D5621 provide guidelines for shear stability tests. Protective colloids like polyvinyl alcohol or hydroxyethyl cellulose offer superior resistance to flocculation under shear compared to purely surfactant-stabilized systems.

Example trade names (for illustration only): Protective colloid: Kuraray Mowiol® PVOH;

Disperser: IKA Ultra-Turrax® T50.

16.63 Case Study 62—Incorrect Pump Selection for High-viscosity Latexes

Background: A newly commissioned 5-ton plant experienced cavitation and flow instability in its latex transfer line. The engineering team had specified a standard centrifugal pump, assuming a viscosity similar to water. The actual polymer latex had a viscosity exceeding 2,000 cP. The centrifugal pump failed to maintain a steady flow, resulting in air entrainment and microfoaming. The pump was replaced with a lobe pump designed for viscous emulsions, solving the problem.

Cause: Centrifugal pumps are designed for low-viscosity fluids. High-viscosity emulsions increase hydraulic resistance, causing cavitation and flow breakdown.

Solution: Replaced the centrifugal pump with a positive displacement lobe pump capable of handling viscous latexes without inducing foam.

Lesson learned: Correct pump specification based on actual fluid rheology is critical in polymer latex plants.

Technical insight: Positive displacement pumps, such as lobe or diaphragm types, provide smooth, pulsation-free flow at high viscosities. Elastomer compatibility and cleaning considerations are vital in waterborne systems.

Example trade names (for illustration only): Lobe pump: Alfa Laval OptiLobe®; Seal system: John Crane Type 9.

16.64 Case Study 63—Cooling System Undersizing Leading to Thermal Runaway Risk

Background: During summer operation, the jacket cooling system could not maintain reaction temperature below the critical threshold during VAM polymerization. The undersized chiller unit, originally selected for cooler climates, led to rapid temperature rise and near-gelation. An emergency monomer feed cutoff was initiated. Postevent analysis revealed that the heat exchanger surface area was insufficient for peak exotherm loads. The cooling system was retrofitted with a larger chiller and auxiliary cooling coil.

Cause: Heat removal capacity was underestimated during design, failing to account for high ambient conditions and maximum heat release during peak polymerization.

Solution: Upgraded chiller capacity and installed an auxiliary cooling loop with a plate heat exchanger to handle worst-case thermal loads.

Lesson: Cooling system design must account for worst-case heat removal scenarios to prevent runaway reactions.

Technical insight: Heat load calculations should include polymerization heat of reaction, monomer feed strategy, and ambient conditions. Safety margins of 30–50% are standard practice for reactor cooling design.

Example trade names (for illustration only): Chiller: Trane RTWD Series; Plate heat exchanger: Alfa Laval M10.

16.65 Case Study 64—Blocked Filter in Filling Line Causing Production Halt

Background: A clogged inline filter in the filling line caused a complete halt in production mid-batch. Investigation showed that a coagulum formed due to inadequate filtration after postreaction blending. The plant upgraded to a dual-basket filter with automatic switchover to prevent downtime during filter cleaning.

Cause: Single inline filter without redundancy allowed coagulum accumulation to block flow, interrupting filling operations.

Solution: Installed a duplex filter system enabling continuous operation during filter changeover.

Lesson learned: Filtration redundancy is critical in preventing downtime, especially for large-scale production runs.

Technical insight: Dual-basket filters or self-cleaning filters improve reliability. Incorporating prescreening or final filtration during blending reduces coagulum carryover to packaging.

Example trade names (for illustration only): Dual-basket filter: Eaton Model 53BTX; Filter bags: Pall Ultipleat® series.

16.66 Case Study 65—Instrumentation Calibration Neglect

Background: Pressure and temperature sensors in the main reactor were not recalibrated for over 18 months. A faulty temperature probe underreported actual reaction temperature by 4 °C, delaying cooling water activation and causing off-spec batches.

Cause: Calibration drift occurs over time due to thermal cycling, sensor aging, and fouling. Inaccurate readings impair process control, especially in exothermic reactions.

Solution: Implemented a quarterly calibration schedule and installed a redundant temperature sensor for safety-critical readings.

Lesson learned: Regular calibration and redundancy of critical instruments ensure reliable process control and prevent runaway risks.

Technical insight: Temperature measurement errors of even 2–4 °C can significantly alter reaction kinetics in VAM polymerization. ISO 9001 and IEC 61511 standards recommend scheduled calibration and independent verification for critical loops.

Example trade names (for illustration only): Resistance Temperature Detector (RTD) sensors, such as Endress+Hauser iTHERM®, are commonly used for accurate reactor temperature measurement. Calibration of the temperature measurement loop may be performed using a multifunction process calibrator, such as the Fluke 724 Multifunction Calibrator, which verifies the RTD signal and transmitter output.

16.67 Case Study 66—Poor Agitator Design Leading to Hot Spots

Background: In a semibatch polymerization process, localized overreaction occurred due to poor mixing at the reactor base. The single-pitched blade turbine provided insufficient axial flow, leading to temperature gradients.

Cause: Inadequate agitator design results in poor macro-mixing, leaving stagnant zones prone to local monomer accumulation and thermal hotspots.

Solution: Retrofitted the agitator with a dual-impeller system (one axial-flow and one radial-flow impeller) to ensure complete circulation and uniform heat distribution.

Lesson learned: Agitator design must be tailored to reactor geometry and viscosity profile to prevent process nonuniformity.

Technical insight: Dual-impeller systems or pitched blade turbines combined with baffles enhance both axial and radial flow, improving heat and mass transfer in semibatch reactors.

Example trade names (for illustration only): Agitator system: Ekato Isojet®; Mixing simulation tool: Chemineer CFD software.

16.68 Case Study 67—Vacuum Stripping Efficiency Loss Due to Sparger Blockage

Background: Operators reported high residual monomer content despite extended vacuum stripping time. Inspection revealed nitrogen sparger ports were partially blocked with coagulum, reducing stripping efficiency and increasing solvent odor in the final product.

Cause: Coagulum accumulation inside sparger ports restricted gas distribution, impairing stripping efficiency and prolonging residual monomer removal.

Solution: Installed a removable sparger head for easy cleaning and implemented routine inspection and clean-in-place (CIP) procedures.

Lesson learned: Regular maintenance of gas distribution systems ensures consistent stripping performance and compliance with residual monomer specifications.

Technical insight: Vacuum stripping efficiency depends on bubble size distribution and surface area. Sparger fouling reduces interfacial contact, slowing monomer desorption. Removable spargers allow quicker turnaround during maintenance.

Example trade names (for illustration only): Gas sparging systems, such as the Koch-Glitsch MicroMix® sparger, are commonly used to introduce stripping gas and enhance removal of residual monomers during vacuum stripping. Reactor cleaning after stripping operations may be performed using a Clean-in-Place (CIP) system, such as the Alfa Laval CIP Module, which circulates cleaning solutions through the reactor and associated piping without dismantling the equipment.

16.69 Case Study 68—Emergency Relief Valve Set Too High

Background: During an exothermic incident, the reactor's emergency relief valve failed to open at the intended pressure. Investigation revealed the set point was 15% above design due to incorrect spring replacement during maintenance.

Cause: Improper maintenance of safety-critical components caused deviation from set specifications, compromising overpressure protection.

Solution: Revised maintenance SOPs, introduced torque and set point verification procedures, and mandated third-party validation after servicing.

Lesson learned: Safety device calibration and independent verification are essential to comply with process safety standards.

Technical insight: Relief valves must meet ASME Boiler and Pressure Vessel Code and API 520/521 standards. Incorrect spring stiffness alters blow-off pressure, delaying relief during runaway events.

Example trade names (for illustration only): Safety valve: Leser High-Performance Series; Calibration bench: Crosby Valve Test Bench.

16.70 Case Study 69—Steam Condensate Contamination in Process Water

Background: A heat exchanger leak allowed steam condensate containing corrosion inhibitors to enter the process water supply. This contamination destabilized the emulsion and increased particle size variability.

Cause: Leaking tube bundles in heat exchangers caused ingress of treated steam condensate, introducing ionic contaminants that disrupted colloidal stability.

Solution: Installed a steam condensate trap and initiated routine water quality checks as part of the prebatch checklist.

Lesson learned: Process water purity is critical for emulsion polymerization stability; contamination control must be integrated into maintenance programs.

Technical insight: Cationic corrosion inhibitors, such as amines, destabilize anionic emulsions by charge neutralization, leading to flocculation. Water quality should be monitored per ASTM D1193 guidelines.

Example trade names (for illustration only): Condensate trap: Spirax Sarco FT Series; Heat exchanger: Alfa Laval Compabloc®.

16.71 Case Study 70—PVAc Innovation via Functional Monomers and Crosslinking

Background: A manufacturer operating for many years with a basic polyvinyl acetate (PVAc) homopolymer system was producing commodity white glues with limited performance. Without functional groups, such as adhesives, they lack wet strength, alkali resistance, and adhesion to challenging substrates. The strategic upgrade begins by introducing carboxylic functionality through small quantities of acrylic acid (AA) or methacrylic acid (MAA). These groups enable both external and internal crosslinking chemistry, creating high-value products from the same reactor (Table 16.44).

Table 16.44 Typical PVAc homopolymer formulation used before process optimization.

Solution:

1) *Upgrade through functional monomers:* Adding 0.3–1.0% of acrylic acid or methacrylic acid introduces reactive carboxyl groups that significantly improve colloidal stability and wet adhesion. These sites also enable efficient crosslinking, which is essential for applications requiring water resistance and durability.
2) *Upgrade through crosslinking pathways:* Exterior crosslinking systems employ post-added crosslinkers after latex production.
   Carbodiimide dispersions, used at approximately 0.5–2.0% on latex solids, react with carboxyl groups to form N-acylurea linkages, increasing wet strength while maintaining workable pot life.
   Zirconium complexes, at approximately 0.1–0.5%, provide coordination bridges that improve wet adhesion and blocking resistance.
   Internal reactive functionality is created during polymerization with the use of acetoacetoxyethyl methacrylate (AAEM) at approximately 0.5–3.0% introducing beta dicarbonyl groups that can be crosslinked at ambient or low bake using dihydrazides such as adipic dihydrazide. Glycidyl methacrylate (GMA)

Table 16.44 Typical PVAc homopolymer formula used before.

Component	Typical range/notes
Vinyl acetate (VAc)	100 parts as base monomer
Water	Adjust for 50–55% solids
Polyvinyl alcohol (PVOH, 87–89% hydrolysis)	6–12 parts as a 10–15% solution
Initiator	0.2–0.6% on monomer as ammonium persulfate or redox
Optional emulsifier	0.2–0.8% to enhance stability if required
Chain transfer agent	0–0.2% for molecular weight control
pH	Maintain 4.5–5.5 for hydrolysis control

at approximately 0.5–2.0% introduces epoxy groups that react with carboxyl and hydroxyl functionalities within the polymer matrix or with complementary amine functionality, allowing thermally assisted or catalyzed cure to a water-resistant network.

Table 16.45 Key parameters and quality control for upgraded grades.

Parameter	Target/typical	Method/notes
Total solids	50–55%	Gravimetric at 105 °C
Viscosity	3,000–6,000 mPa s at 25 °C	Brookfield RV, spindle 4, 20 rpm
pH	4.5–5.5	Adjust with acetic or ammonia as required
Minimum film formation temperature	≤10 °C	Film formation plate or MFFT bar
Open time	90–180 s	Application dependent on substrate and coat weight
Tack development	Rapid green strength	Thumb twist or loop tack as applicable
Peel strength	Application specific	ASTM D903, 180° peel on selected substrates
Shear or creep	Stable under service	ASTM D3654 or equivalent
Blocking resistance	No block at 40 °C stack	24 h under defined load
Water resistance	Meets D3 class when crosslinked	EN 204 as applicable
Storage stability	No gel at 40 °C for 4 weeks	Closed container test

3) *Upgrade through tackified PVOH:* For lamination or tissue applications that require excellent tack and fiber tear, tackified PVOH such as rosin-modified or polyol-blended PVOH, can be incorporated. This approach increases green strength and extends open time while maintaining latex stability when introduced carefully into the letdown stage (Table 16.45).

Table 16.45 summarizes the key parameters and quality control requirements for upgraded PVAc grades.

Lesson for formulators: Introducing acrylic acid or methacrylic acid at low levels transforms a simple PVAc glue into a versatile platform. The functional monomers enable both external crosslinking and internal reactivity through AAEM or GMA, which allows the same reactor to serve multiple high-value adhesive grades. The combination of tuned crosslink density and optional tackified PVOH yields wet-resistant grades that maintain workable flexibility and processability without additional capital expenditure.

16.72 Case Study 71–High-Performance VAc Copolymers: Acrylic and VeoVa Systems

Background: For many years, a manufacturer relied on a general-purpose vinyl acetate–based copolymer platform, producing mid-tier adhesives and coatings with limited performance. The portfolio showed weak scrub resistance, low gloss retention, and insufficient water resistance in exterior service. The technical limitation arose from insufficient reactive functionality, suboptimal wet adhesion, and surfactant migration.

Solution: The platform was upgraded without additional capital by introducing low levels of acrylic acid or methacrylic acid for reactive carboxyl groups, by adopting formaldehyde-free crosslinking chemistries, by incorporating polymerizable surfactants to reduce migration, and by using reactive grades of polyvinyl alcohol in the protective colloid package. These measures enabled durable crosslinking, improved substrate wetting, and reduced water uptake, which translated to higher scrub resistance, better gloss retention, and strong exterior durability (Tables 16.46 and 16.47).

Table 16.46 summarizes the baseline VAc–acrylic copolymer formulation used for the system discussed in this section.

Table 16.47 summarizes the baseline VAc–VeoVa copolymer formulation used for enhanced alkali resistance and exterior durability.

Table 16.46 Baseline VAc acrylic copolymer formula.

Component	Typical range/notes
Vinyl acetate	60–85 parts as primary monomer
n-Butyl acrylate or 2-ethylhexyl acrylate	5–25 parts to reduce MFFT and increase flexibility
Methyl methacrylate	0–20 parts as hard segment for block and scrub
Acrylic acid or methacrylic acid	0.3–1.0% on monomer to provide carboxyl functionality
Water	Adjust for 50–55% total solids
Polymerizable surfactant	0.3–1.0% total for migration-resistant stabilization
Initiator	0.2–0.6% on monomer as persulfate or redox
pH	Target 4.8–6.2 during polymerization and let down

Table 16.47 Baseline VAc VeoVa copolymer formula.

Component	Typical range/notes
Vinyl acetate	70–90 parts
VeoVa 10 or VeoVa 11	5–20 parts for alkali and exterior durability
n-Butyl acrylate	0–10 parts as optional flexibility modifier
Acrylic acid or methacrylic acid	0.3–0.8% on monomer for carboxyl functionality
Polymerizable surfactant	0.2–0.8% total
Initiator	0.2–0.6% on monomer
pH	Target 4.5–5.8

16.72.1 Functionalization and Crosslinking

Internal reactivity can be built with acetoacetoxyethyl methacrylate (AAEM) at approximately 0.5–3.0% on monomer, which introduces beta dicarbonyl sites that crosslink at ambient or low bake with dihydrazides such as adipic dihydrazide. Glycidyl methacrylate (GMA) at approximately 0.5–2.0% provides epoxy groups that react with carboxyl and hydroxyl functionality in the latex or with suitable amine functionality, enabling thermally assisted cure to a water-resistant network.

16.72.2 Optimization with Acrylic acid or Methacrylic Acid

A carboxyl level in the range of approximately 0.3–0.8% on monomer improves colloidal stability, pigment wetting, and wet adhesion while providing sites for carbodiimide or zirconium-based crosslinking. Higher levels increase hydrophilicity and may raise water uptake, therefore dosage is kept within the lower range for exterior service.

External crosslinkers are selected according to application and process constraints. Water-dispersible polycarbodiimides used at approximately 0.5–2.0% on latex solids convert carboxyl groups to N-acylurea linkages and raise wet strength with workable pot life. Zirconium complexes used at approximately 0.1–0.5% generate coordination bridges and improve blocking resistance.

16.72.3 Optimization with Polymerizable Surfactants

Reactive surfactants that contain a polymerizable double bond are fed with the monomer preemulsion at approximately 0.2–0.6% of the total formulation. They become covalently anchored to the particle surface, which reduces free surfactant migration, lowers water whitening, and improves scrub resistance. A small fraction of conventional surfactant can be retained in the seed to ensure robust nucleation, after which the reactive surfactant dominates the preemulsion feed.

16.72.4 Optimization with Reactive Polyvinyl Alcohol

The protective colloid package can incorporate a reactive grade of polyvinyl alcohol, such as partially acetoacetylated PVOH. Replacing approximately one-third to one-half of the total PVOH with the reactive grade enables keto hydrazide crosslinking when a dihydrazide is added during let down. Typical PVOH solution usage is maintained at 6–12 parts as a 10–15% solution, with dihydrazide dosing verified by gel fraction and wet adhesion testing (Table 16.48).

Table 16.48 summarizes the key parameters and quality control criteria used to evaluate the performance of the optimized copolymer systems.

Summary insight: By combining low-level carboxyl functionality, formaldehyde-free crosslinkers, polymerizable surfactants, and reactive polyvinyl alcohol in the colloid package, a standard vinyl acetate–based platform is upgraded into a high-performance family for coatings and laminating adhesives. The critical controls are comonomer ratio, distribution of functional groups through the particle, and the balance between crosslink density and target flexibility.

Table 16.48 Key parameters and quality control.

Parameter	Target or typical	Method or notes
Total solids	50–55%	Gravimetric at 105 °C
Viscosity at 25 °C	3,000–6,000 mPa s	Brookfield RV, spindle 4, 20 rpm
pH	4.8–6.2	Adjust with suitable acid or base
MFFT	≤10 °C for exterior coatings	MFFT bar or film plate
T_g	Set per application	Differential scanning calorimetry per internal protocol
Water whitening	Minimal after 24 h immersion	Visual rating on drawdown
Scrub resistance	Meets premium class target	ASTM D2486 or ISO 11998
Gloss retention	≥80% after accelerated exposure	QUV or Xenon arc per internal method
Blocking resistance	No block at 40 °C stack	24 h under defined load
Wet adhesion	No failure on alkali-primed panel	Cross-hatch tape after soak
Storage stability	No gel or skin after 4 weeks at 40 °C	Closed container test

17

Regulations, Market Trends, and Sustainability in Emulsion Polymer Technologies

17.1 Introduction: Why Regulations and Market Trends Matter

The global market for emulsion polymers, waterborne adhesives, and coatings is being reshaped by forces that extend beyond pure technical performance. Regulatory compliance, customer expectations, and corporate sustainability targets increasingly determine which technologies reach the market and how long they remain viable. Environmental and occupational health frameworks now act as de facto design constraints, steering formulators away from legacy chemistries and toward safer, lower-emission systems. At the same time, downstream brand owners and converters are raising the bar on compliance documentation, food-contact assurance, low-odor and low-volatile organic compound (VOC) profiles, and end-of-life compatibility with recycling streams. Competitive differentiation therefore hinges as much on proactive regulatory strategy and sustainability engineering as on adhesion, open time, and mechanical strength.

For manufacturers of polyvinyl acetate (PVAc) homopolymers, vinyl acetate monomer (VAM)–based copolymers, and related waterborne binders, this shift is especially consequential. Decisions about monomer selection, protective colloids, surfactants, crosslinkers, coalescents, defoamers, and wetting agents are no longer made solely on the basis of performance and cost. They must also reflect evolving restrictions on substances of very high concern (SVHCs), micro- and nanoscale materials, residual monomers, and nonintentionally added substances (NIAS) (see Annex 17.A for details), biocide use patterns, and labeling under the Globally Harmonized System of Classification and Labeling of Chemicals (GHS). Food-contact and indirect-food-additive rules, migration testing, and documentation packages are now routine requirements in many adhesive applications. In construction and architectural markets, green-building programs and ecolabels increasingly reward verified low-emission and low-hazard binders,

Table 17.1 Scope 1–3 emission categories (GHG protocol).

Scope	Definition	Examples
Scope 1	Direct emissions from company-owned or controlled sources	Boilers, process fuel, company vehicles
Scope 2	Indirect emissions from purchased energy	Grid electricity, purchased steam/heating/cooling
Scope 3	All other indirect emissions in the value chain	Raw materials, supplier energy, transport, product use, end-of-life

Note: GHG, greenhouse gas; values are typically reported as CO_2-equivalent (CO_2e).

while industrial customers prioritize supply security for compliant raw materials and transparent change-control processes (Table 17.1).

Sustainability goals add a further layer of direction. Corporate net-zero pathways and Scope 3 targets—covering value-chain greenhouse gas (GHG) emissions beyond a company's direct operations (Scope 1) and purchased energy (Scope 2)—push suppliers to reduce cradle-to-gate carbon intensity, optimize energy and water footprints, and disclose life-cycle impacts through standardized methods such as life-cycle assessment (LCA). Formulation work is consequently broadening to include bio-based or circular inputs where they do not compromise durability; the elimination of alkylphenol ethoxylates and formaldehyde-releasing components; and the adoption of crosslinking chemistries that deliver water resistance without introducing problematic residuals or reactive hazards. Design-for-recycling in packaging drives adhesives that bond reliably in service yet delaminate or wash out in recycling processes. These trends reward systems thinking: the best solutions align compliance, performance, processing efficiency, and credible environmental attributes.

This chapter equips the formulator and technical manager with a practical map of the terrain. It explains how major regulatory frameworks and product standards translate into actionable formulation constraints and documentation needs; how market trends in packaging, building products, wood assembly, and industrial coating guide binder and additive choices; and how to integrate sustainability metrics into routine decision-making. Readers will find guidance on building compliant raw-material portfolios, selecting crosslinkers and auxiliaries that balance performance with hazard and exposure profiles, designing test plans that satisfy customer and regulatory scrutiny, and preparing dossiers that withstand audits and specification reviews. The emphasis throughout is practical: what to avoid, what to favor, how to justify choices, and how to manage change without disrupting validated performance.

Table 17.2 The dual role of regulations in market success.

Role	Impact on manufacturers	Strategic opportunity
Compliance barrier	Restricts use of certain raw materials, imposes costly testing	Develop compliant formulations early to avoid market disruption
Market differentiator	Products meeting ecolabel or green certification gain customer preference	Use compliance as a selling point to access premium markets
Innovation driver	Forces R&D to explore renewable and low-VOC alternatives	Innovate ahead of regulations to be first-to-market with next-generation products

The remainder of Chapter 17 is organized to support real-world decision flow. A concise survey of the regulatory landscape relevant to emulsion polymers and waterborne adhesives is followed by application-specific considerations for packaging, wood and furniture, construction, and industrial coating. A section on sustainability methods outlines how to use life-cycle thinking, carbon accounting, and safer-by-design principles to guide raw-material selection and process improvements. A design-for-compliance playbook then translates requirements into concrete choices for monomers, stabilizers, surfactants, wetting agents, crosslinkers, coalescents, and preservatives. The chapter concludes with documentation checklists and claim language conventions that improve customer confidence and shorten qualification cycles, together with a forward-looking view of the policies and market signals most likely to influence the next generation of waterborne binders and adhesives.

To frame how regulation shapes commercial outcomes, Table 17.2 summarizes its dual role as a barrier that must be cleared and as an engine for differentiation and innovation. This lens sets up the global drivers that follow (Section 17.2) and the first deep-dive topic on VOC regulation (Section 17.3).

17.2 Global Drivers of Regulatory Change

Climate and environmental policy: Global agreements such as the Paris Agreement are pushing industry toward lower carbon footprints and reduced emissions across manufacturing, logistics, and the use phase.

Health and safety concerns: Evidence linking VOC exposure to adverse health effects is driving tighter indoor air-quality limits and stricter controls on solvent use, residual monomer content, and emissions.

Circular-economy goals: Policies, particularly in the EU and parts of Asia, are accelerating requirements for recyclability, recycled and renewable content, and end-of-life stewardship in packaging and coatings.

Technological transparency: Digital product declarations and regulatory databases, such as the European Union (EU)'s SCIP database (Substances of Concern in articles, as such or in complex objects [Products]), are increasing obligations to disclose complete chemical content and to document supply-chain traceability.

17.3 VOC (Volatile Organic Compounds) Regulations and Market Impact

Volatile organic compounds (VOCs) are a major focus of environmental regulations worldwide due to their contribution to ground-level ozone formation and their adverse health effects. Governments, environmental agencies, and industry associations have imposed increasingly strict VOC limits on paints, coatings, and adhesives. These regulations have pushed manufacturers toward developing low-VOC and zero-VOC formulations, influencing raw-material selection, binder technology, and product performance, labeling, and claim substantiation. In practical terms, formulators are reducing or eliminating traditional coalescents and solvent carriers, favoring waterborne or high-solids systems, selecting lower-volatility plasticizers and wetting aids, and using crosslinking chemistries that achieve film formation and early strength without excess volatile content. Compliance also requires validated test methods for VOC content and emissions over time, clear cutoffs for "low-VOC" and "zero-VOC" claims as defined in the target market, and documentation that aligns safety data sheets (SDS), technical data sheets, and marketing language. For adhesives and coatings used in sensitive applications such as packaging and interior building products, buyers increasingly request both content-based VOC data and chamber-test emissions profiles, reinforcing the shift toward inherently low-emission formulations (Table 17.3).

17.4 Chemical Substance Regulations

Compliance with chemical control laws is a fundamental requirement for the sale and distribution of emulsion polymers, adhesives, and coatings in global markets. These regulations define the conditions under which chemical substances can be manufactured, imported, or used, often including premarket registration, safety assessments, and reporting obligations. Key international frameworks include the following:

1) *REACH (Registration, Evaluation, Authorization and Restriction of Chemicals)—European Union:* Requires manufacturers and importers to register substances

Table 17.3 Typical VOC limits in major markets.

Region/standard	VOC limit (g/L)	Applicable products	Enforcement/ notes
USA—EPA National Rule	250 (flat), 380 (nonflat)	Architectural coatings	Federal limit; states may impose stricter rules
California (CARB/ SCAQMD)	50 (flat), 100 (nonflat)	Architectural coatings	Most stringent in the USA; drives national trends
European Union (Directive 2004/42/EC)	30–130 depending on category	Decorative paints, varnishes	Limits vary by product type
Japan (JIS K 5660)	50 (interior waterborne paints)	Interior decorative coatings	VOC limits tied to indoor air quality; promotes water-based paints
China GB/T 9756	120 (interior), 200 (exterior)	Architectural coatings	Enforcement growing stronger
Middle East (Saudi SASO, UAE ESMA)	50–150	Decorative and industrial coatings	Alignment with EU standards increasing

placed on the EU market in quantities above 1 ton/year. Demands detailed technical dossiers, safety data, and risk assessments. Maintains the SVHC list, which is updated regularly and can result in use restrictions or phase-outs.

2) *TSCA (Toxic Substances Control Act)—United States:* Enforced by the EPA, TSCA requires premanufacture notification (PMN) for new chemicals. Existing chemicals may be subject to risk evaluations and significant new use rules (SNURs). Manufacturers must maintain accurate records and comply with reporting under the Chemical Data Reporting (CDR) rule.
3) *China MEE Order 12—Measures for the Environmental Management Registration of New Chemical Substances:* Applies to new chemicals not listed on the Inventory of Existing Chemical Substances in China (IECSC). Requires hazard classification, risk assessment, and environmental management reporting (Table 17.4).

17.5 Ecolabels and Green Certification Schemes

Beyond legal compliance, many customers and procurement bodies require ecolabel certifications as proof of environmental responsibility. These labels act as market differentiators, signaling reduced environmental impact and safer

Table 17.4 Comparison of major chemical control laws.

Regulation and region	Scope and key requirements	Strategic considerations
REACH (EU)	All substances ≥1 ton/year; registration, safety dossier, SVHC compliance	Complex but provides EU-wide market access
TSCA (USA)	All new and existing chemicals; PMN for new chemicals, risk evaluations, SNURs	Early engagement with EPA speeds approval
China MEE Order 12	New chemicals not on IECSC; hazard classification, environmental risk management	Plan for translation and local agent support

Table 17.5 Criteria for major ecolabels.

Ecolabel	VOC limit	Banned substances	Other requirements
EU Ecolabel	≤30 g/L for interior paints	Heavy metals, formaldehyde, phthalates	Durability, washability testing
Blue Angel	≤50 g/L for coatings	VOCs, carcinogens	Emission testing (indoor air)
Green Seal	≤50 g/L for flat coatings	Carcinogens, reproductive toxins	Performance standards
China Environmental Labelling	≤100 g/L	Hazardous solvents	Resource efficiency proof

formulations. Notable schemes include EU Ecolabel, Blue Angel (Germany), Green Seal (USA), and China Environmental Labeling (Table 17.5).

17.6 Worker Health and Safety Compliance

Worker protection in chemical manufacturing is governed by occupational exposure limits (OELs), hazard communication standards, and workplace safety protocols. Compliance with these requirements not only protects employees but also reduces liability and improves operational reliability (Table 17.6).

Table 17.6 Typical OEL values for common raw materials.

Substance	OEL (8-h TWA)	Source
Vinyl acetate monomer	10 ppm (35 mg/m^3)	ACGIH TLV
Methyl methacrylate	50 ppm (205 mg/m^3)	OSHA PEL
Ammonia	25 ppm (17 mg/m^3)	ACGIH TLV
Butyl acrylate	10 ppm (55 mg/m^3)	ACGIH TLV
Styrene	20 ppm (85 mg/m^3)	OSHA PEL
Formaldehyde	0.75 ppm (0.92 mg/m^3); Ceiling 2 ppm	OSHA PEL
Acrylic acid	2 ppm (5.9 mg/m^3)	ACGIH TLV
Ethylene glycol (vapor)	50 ppm (127 mg/m^3)	OSHA PEL
Isothiazolinone (CMIT/MIT)	0.2 mg/m^3	Internal industry guideline

17.7 Food-contact and Packaging Regulations

For PVAc-based adhesives used in food packaging, regulatory compliance is critical to ensure consumer safety and avoid costly recalls. Food-contact compliance governs both the chemical composition and the migration behavior of adhesives and coatings used in materials that will come into direct or indirect contact with food.

17.7.1 Key Elements

US FDA 21 CFR: Lists permitted substances and establishes migration limits for indirect food-contact materials, including PVAc emulsions and EVA hot-melt adhesives.

EU Framework Regulation (EC) No 1935/2004: Requires that materials do not transfer constituents to food in quantities that could endanger human health or alter the composition, taste, or odor of the food.

EU Regulation (EU) No 10/2011: Applies to plastic food-contact materials and is relevant when PVAc-based adhesives form part of a multilayer structure.

Testing requirements: Include overall migration and specific migration testing in approved food simulants, supported by documentation for declarations of compliance.

Traceability systems: Maintain the ability to trace raw materials and finished products throughout the supply chain to support regulatory investigations (Table 17.7).

Table 17.7 PVAc-based adhesive chemistries and food-contact compliance.

Adhesive type	Typical use	Regulatory status
PVAc homopolymer emulsion	Paper-to-paper laminates in dry food packaging	FDA-compliant grades available under 21 CFR 175.105
VAE emulsion	Laminating adhesive for bakery and confectionery cartons	FDA-compliant formulations possible with low migration profile
Modified PVAc emulsion	Improved water resistance in paperboard packaging	Must meet FDA/EU migration limits
EVA-based hot melt	Carton sealing for dry and semimoist foods	FDA-compliant grades available under 21 CFR 175.105
PVAc-based pressure-sensitive adhesive	Labels for food jars and bottles	Compliance required for adhesives in direct food label contact

Notes

1) This table lists common raw materials relevant to emulsion polymers, adhesives, and coatings. Readers should consult local regulatory databases for complete and up-to-date exposure limits for additional substances of interest.
2) Food-contact and packaging regulations set strict conditions on the substances that may migrate from adhesives, coatings, and polymer dispersions into packaged goods. In addition to intentionally added substances covered by approvals and inventories, regulatory frameworks also recognize the presence of **nonintentionally added substances (NIAS)**. These are by-products, impurities, or degradation products that are not deliberately introduced but may nevertheless be detected in the final adhesive or polymer film.

Note: A detailed discussion of NIAS, including representative sources and screening approaches, is provided in Annex 17A (see Tables A17.1 and A17.2).

17.8 Regional Differences and Export Considerations

Although regulatory harmonization is improving, significant differences remain between regions in how emulsion polymers and PVAc-based adhesives are regulated. Exporters must be aware that a formulation compliant in one market may not automatically meet the requirements of another. Failure to adapt can result in shipment rejections, market delays, or the need for costly reformulations.

Table 17.8 Examples of regional regulatory differences.

Region/market	Example difference	Impact on PVAc-based adhesives
European Union	Ban on certain isothiazolinone levels above 15 ppm in mixtures	Requires reformulation of in-can preservative package
United States	Different VOC calculation method (EPA Method 24)	Allows some coalescents restricted in the EU
China	Mandatory product testing at certified local labs	Additional lead time and cost for product registration
Middle East (GCC)	Alignment with EU VOC limits but variable enforcement	Possible competitive advantage for compliant products
Japan	Stringent indoor air quality standards for formaldehyde	Limits the use of certain crosslinkers in PVAc adhesives

17.8.1 Key Considerations

Different substance restrictions: A preservative permitted in the EU may be prohibited in China, and a plasticizer accepted in the United States may exceed EU migration limits.

VOC and additive standards: Definitions and test protocols vary by jurisdiction, affecting allowable coalescents, defoamers, and other additives in PVAc-based products.

Dual-formulation strategies: Parallel formulations are often required to satisfy multiple regulatory regimes without compromising performance.

Local testing and certification: Some markets require in-country laboratory testing or certification even when the product meets international norms.

Labeling and language requirements: Product labels, SDS, and compliance documentation may need to be prepared in the official language(s) of the importing country (Table 17.8).

17.9 Future Regulatory and Market Trends

The business environment for emulsion polymers and PVAc-based adhesives is shaped by evolving regulatory requirements and shifting market dynamics. Understanding these trends helps manufacturers align product development with compliance obligations while capturing new growth opportunities. This section integrates current and forward-looking perspectives, combining regulatory developments and market forces.

17.9.1 Regulatory Trends

Regulatory changes are increasingly influencing the formulation and commercialization of PVAc-based adhesives. The most prominent trends include tighter VOC limits, expanding SVHC lists, restrictions on microplastics, mandatory carbon-footprint disclosures, and the digitization of compliance documentation.

VOC reduction: Many regions—particularly the EU and the California Air Resources Board (CARB)—are progressively lowering permissible VOC levels in adhesives, including water-based systems. Although PVAc emulsions typically have low VOC content, additives such as coalescents, defoamers, and preservatives can still contribute. Under REACH (Registration, Evaluation, Authorization and Restriction of Chemicals), the SVHC list continues to expand, requiring substitution or detailed customer notifications where applicable.

Microplastics control: Regulators are advancing measures to limit the intentional addition of microplastics to industrial and consumer products. While PVAc emulsions are generally not classified as microplastics in the wet state, dried polymer films may contribute to environmental microplastic load upon degradation, triggering disclosure or environmental-fate considerations.

Carbon-footprint disclosures: Governments and industry bodies are increasingly mandating greenhouse-gas (GHG) reporting and life-cycle assessment (LCA) data for chemical products. Producers employing renewable feedstocks or energy-efficient reactor designs gain a competitive advantage in environmentally sensitive markets.

Digital compliance: Regulators and large end users are adopting digitized systems for safety documentation, including quick-response (QR) codes on packaging that link to up-to-date SDS and regulatory statements. This improves transparency, facilitates audits, and ensures the latest hazard information is readily accessible (Table 17.9).

17.9.2 Market Trends

Market forces are creating both challenges and opportunities for PVAc adhesive manufacturers. Key trends include sustainability as a differentiator, cost competitiveness, performance-driven innovation, growth in packaging demand, and regional localization.

Sustainability: Global demand is increasing for adhesives with lower environmental footprints, including bio-based PVAc emulsions, reduced-VOC formulations, and products with biodegradability certifications. Cost Competitiveness: Volatility in VAM pricing drives the need for cost-optimized

Table 17.9 Regulatory areas and their impact on PVAc-based adhesives.

Regulatory area	Potential change	Impact on PVAc-based adhesives
VOC reduction	Lowering limits to <20 g/L for interior products	Requires reformulation using ultra-low VOC coalescents
SVHC expansion	Addition of new monomers and additives to restriction lists	May necessitate substitution with compliant raw materials
Microplastics control	Ban on intentional use of microsized polymer particles	Could impact specialty thickener and filler choices
Carbon reporting	Mandatory GHG disclosure for manufactured goods	May drive shift to bio-based or low-carbon raw materials
Digital compliance	QR code-linked SDS and supply chain data	Improves transparency and speeds regulatory checks

formulations and process efficiencies. Performance Enhancement: End users seek improved water resistance, heat resistance, and adhesion to challenging substrates, leading to greater use of copolymerization with VeoVa, acrylates, and functional monomers. E-commerce Growth: Rising demand for corrugated board adhesives and fast-setting formulations supports expansion in packaging applications. Regional Expansion: Emerging markets in Asia, Africa, and the Middle East offer opportunities for localized production and compliance with regional regulations (Table 17.10).

Table 17.10 Key market trends and implications for PVAc-based adhesives.

Market trend	Driver	Implication for PVAc Adhesives
Sustainable packaging	Environmental regulations and consumer demand	Development of repulpable and compostable adhesive grades
Flexible packaging growth	Convenience food demand and material innovation	Optimization of VAE emulsions for bond strength and compliance
Food safety focus	Stricter global food-contact regulations	Expansion of certified low-migration PVAc adhesive portfolio
Local sourcing	Raw material cost volatility	Investment in regional production facilities
Emerging markets growth	Urbanization and industrialization	Targeted marketing of cost-effective, compliant adhesive systems

17.10 Sustainability of Emulsion Polymers and PVAc-based Adhesives

Sustainability has become a defining priority for manufacturers, regulators, and end users in the adhesives and coatings sector. For PVAc-based adhesives produced via emulsion polymerization, sustainability encompasses the environmental performance of both the manufacturing process and the final product throughout its life cycle. This includes sourcing of raw materials, energy efficiency in production, minimization of waste and emissions, and consideration of end-of-life impacts such as recyclability or biodegradability.

Life-cycle assessment (LCA): It provides a structured framework to evaluate the total environmental footprint of an adhesive product, from raw material extraction to manufacturing, transportation, use, and disposal. For PVAc emulsions, key LCA stages include raw material sourcing, manufacturing phase, packaging and distribution, use phase, and end-of-life considerations. Raw material sourcing is particularly critical because VAM production is energy-intensive and typically petrochemical-based. Shifting to bio-based acetic acid and ethanol-derived VAM can reduce upstream GHG emissions.

During manufacturing, reactor energy consumption, cooling-water usage, and auxiliary equipment operation significantly contribute to the carbon footprint. Optimizing reaction temperature profiles and implementing heat-recovery systems can improve energy efficiency. Packaging and distribution strategies such as bulk supply in reusable containers and localized manufacturing reduce transportation-related emissions. At the use phase, low-VOC formulations improve indoor air quality and reduce occupational exposure. End-of-life considerations include ensuring that adhesives in paper-based packaging are repulpable and that those used in construction do not impede material recycling.

17.10.1 Environmental Impact Considerations

PVAc-based adhesives already have an advantage over solvent-based systems because they are waterborne; however, several environmental considerations remain. Even low-VOC adhesives may contain residual monomers, plasticizers, or preservatives that impact health and the environment. Dried adhesive films can contribute to microplastic pollution over long degradation periods, a risk that can be mitigated by selecting biodegradable additives and polymers. Manufacturing wastewater containing surfactants and trace monomers requires effective treatment before discharge. Reliance on fossil fuel-derived electricity or steam further increases the carbon footprint, which can be reduced by transitioning to renewable energy sources.

Integration of green chemistry principles: The Twelve Principles of Green Chemistry provide a roadmap for reducing the environmental footprint of chemical processes. For PVAc adhesive manufacturing, this includes the use of renewable feedstocks such as bio-based VAM and co-monomers, designing energy-efficient polymerization reactions at lower temperatures, and minimizing waste through closed-loop cleaning systems. Water as the continuous phase already aligns with green chemistry principles, while any co-solvents used should be biodegradable. Designing adhesives for degradation ensures that polymer structures and additives break down into nontoxic substances after use.

17.10.2 Keys for Manufacturers

Manufacturers should conduct comprehensive LCAs for major adhesive product lines to identify high-impact stages, explore bio-based monomer supply chains to reduce fossil dependency, improve reactor and utility efficiency to cut energy consumption by at least 10–15%, transition to preservatives and additives with lower environmental toxicity, and design adhesives compatible with recycling or composting processes (Table 17.11).

Table 17.11 Sustainability areas, recommended strategies, and potential benefits.

Sustainability area	Recommended strategy	Potential benefit
Raw materials	Use bio-based VAM and renewable co-monomers	Reduce upstream GHG emissions
Energy efficiency	Optimize reactor heating/cooling cycles; implement heat recovery	Lower operating costs and carbon footprint
Packaging	Supply in reusable totes or bulk tankers	Reduce packaging waste
Additives	Select biodegradable plasticizers and low-toxicity preservatives	Minimize environmental hazard
End-of-life	Formulate for repulpability or biodegradability	Support recycling and composting

17.11 Environmental Aspects

Environmental management in emulsion polymerization and adhesive production requires an integrated approach to minimize emissions, reduce waste, and ensure compliance with regional and global standards. This section discusses pollution prevention, disposal practices, recovery of resources, water reuse, and compliance frameworks for PVAc-based adhesive manufacturing.

17.11.1 Pollution Prevention in Emulsion Polymerization

To reduce VOC emissions, PVAc manufacturing facilities implement closed-loop monomer handling systems and nitrogen-blanketed storage for VAM. Secondary condensers and scrubbers capture fugitive vapors, allowing VAM recovery for reuse. Surfactant dosing should be optimized to minimize chemical load in process effluents while maintaining polymer stability.

17.11.2 Disposal of Unreacted Monomers

Residual VAM and acrylate monomers in wastewater must be polymerized before disposal to reduce environmental hazards. Incineration should only occur in facilities equipped with effective acid-gas scrubbers. Compliance with EU Industrial Emissions Directive (IED), US Environmental Protection Agency (EPA) standards, and Chinese GB regulations is essential to maintain permissible limits for monomer content in effluents.

17.11.3 Recovery and Recycling of Surfactants

Advanced recovery technologies such as membrane ultrafiltration enable the reuse of anionic and nonionic surfactants from cleaning solutions or wash water. Nanofiltration can achieve more than 80% recovery for sodium lauryl ether sulfate. Recovered surfactants are suitable for noncritical production batches, reducing discharge loads and operating costs.

17.11.4 Water Reuse and Effluent Treatment

Effluent treatment systems must address chemical oxygen demand (COD) and biological oxygen demand (BOD) through biological treatment methods such as activated sludge with aeration. Electrocoagulation is effective for removing surfactant-rich contaminants prior to discharge. Closed-loop rinse water systems, particularly those used in clean-in-place (CIP) operations, conserve water and reduce effluent volume.

17.11.5 Compliance Frameworks and Best Practices

Environmental compliance in the adhesives industry aligns with multiple international and national frameworks, including the EU IED, US EPA regulations under 40 CFR Part 63, and Chinese GB standards. Adoption of best available techniques (BAT), real-time VOC and effluent monitoring, and implementation of environmental management systems (EMS) such as ISO 14001 certification represent best practices for sustainable operations.

17.11.6 Strategic Keys

Key strategies for PVAc manufacturing include (i) adopting closed-loop handling systems to minimize VOC emissions, (ii) integrating real-time monitoring for emissions and wastewater quality, (iii) recovering surfactants through advanced membrane technologies, and (iv) aligning internal environmental management systems with ISO 14001 to ensure continuous compliance and certification readiness (Table 17.12).

Table 17.12 Environmental control measures, primary benefits, and examples in PVAc plants.

Environmental control measure	Primary benefit	Example in PVAc plant
Closed-loop monomer handling	Reduces VOC emissions	VAM storage under nitrogen blanket
Secondary condensers/scrubbers	Captures fugitive vapors	VAM recovery for reuse
Membrane ultrafiltration	Recovers surfactants	Reuse in CIP rinse cycles
Electrocoagulation	Removes surfactant load	Effluent polishing before discharge
Biological treatment	Reduces COD/BOD	Activated sludge with aeration
ISO 14001 EMS	Improves environmental compliance	Annual certification and audits

17.12 Summary and Strategic Keys

This chapter has outlined the complex interplay between regulatory requirements, market trends, sustainability imperatives, and environmental stewardship affecting PVAc-based adhesives and emulsion polymers. By aligning technical innovation with compliance, integrating green chemistry and life-cycle thinking, implementing robust pollution prevention measures, and anticipating shifts in demand, manufacturers can strengthen their competitive position while reducing the risk of noncompliance, reputational harm, or market exclusion.

17.12.1 Strategic Keys for Industry Stakeholders

Regulatory foresight: Establish a continuous monitoring process for changes in VOC limits, SVHC listings, food-contact regulations, wastewater discharge standards, and emerging environmental reporting requirements across all target markets. This proactive approach ensures timely compliance and minimizes operational and financial risks.

Formulation flexibility: Maintain adaptable formulations that can be modified to meet differing regional requirements without compromising performance, environmental compatibility, or sustainability attributes. Flexibility in recipe design reduces lead times for compliance-driven changes and supports customer-specific adaptations.

Sustainability integration: Prioritize the development of low-carbon, recyclable, biodegradable, and repulpable adhesive solutions in line with green procurement standards, International Organization for Standardization (ISO) 14001 environmental management systems, and circular economy objectives. Embedding sustainability as a design principle enhances long-term competitiveness and customer trust.

Environmental impact reduction: Implement BAT for emission control, residual monomer minimization, and surfactant recovery. Effluent treatment optimization and water reuse systems should meet or exceed discharge standards established by the EU, US EPA, and Chinese GB regulations.

17.12.2 Market Diversification

Balance business exposure between mature and emerging markets to mitigate regional economic, regulatory, and political risks. Identify high-growth sectors aligned with sustainable adhesive demand, such as eco-friendly packaging and green construction materials.

17.12.3 Supply Chain Resilience

Build robust relationships with local raw material suppliers and integrate renewable feedstocks where feasible. Consider partial vertical integration for critical inputs to enhance supply security while reducing transportation-related carbon footprint.

Customer-centric compliance: Provide clients with ready-to-use compliance declarations, environmental impact assessments, sustainability data, and life-cycle summaries to support their regulatory obligations and environmental, social, and governance (ESG) reporting requirements.

17.12.4 Technology Investment

Adopt digital compliance tools, environmental monitoring systems, and advanced testing capabilities to stay ahead of regulatory audits and strengthen competitiveness. Technology-driven transparency enhances customer confidence and ensures alignment with evolving standards.

By applying these strategic keys, companies in the PVAc-based adhesive and emulsion polymer sector can navigate regulatory complexity, reduce environmental impact, respond to evolving market demands, and secure sustainable growth.

Annex 17A Nonintentionally Added Substances (NIAS)–Definitions, Examples, and Screening Approaches

Scope and definition. NIAS are substances present in the finished product that were not deliberately added as functional ingredients. They include impurities in starting materials, reaction intermediates and by-products, degradation products, contaminants from process equipment or recycled inputs, and substances formed during storage or use. This annex provides representative examples relevant to emulsion polymers, adhesives, and coatings used in packaging and industrial applications, together with a concise screening workflow (Tables A17.1 and A17.2).

Table A17.1 Representative NIAS by source and typical control measures.

Source category	Example NIAS (typical)	Where encountered	Typical mitigation/control
Polymerization oligomers	Cyclic PET/PBT oligomers (dimer, trimer); acrylic/methacrylic oligomers	From PET/PBT films, acrylic inks/varnishes, coatings	Optimize cure and postbake; select low-oligomer grades; verify with LC–MS nontarget scans
Antioxidant degradation	Irgafos 168-oxide; 2,4-di-*tert*-butylphenol; BHT derivatives	Polyolefin layers, primers, tie-layers	Use stabilized grades; control residence time/temperature; monitor by GC–MS
Photoinitiator residues/by-products	Benzophenone, 4-MBP, ITX; 1-phenyl-2-butanone; TMBA	UV-cured inks/varnishes; set-off into adhesive	Use low-migration PI packages; optimize UV dose; storage before winding; barrier interlayers
PU adhesive amines	Primary aromatic amines (e.g., 4,4′-MDA; 2,4-/2,6-TDA)	Aromatic-PU laminating adhesives under hydrolysis	Prefer aliphatic systems; ensure full cure; moisture control; migration testing
Residual monomers/by-products	Traces of VAM, AA, MMA; aldehydes; glyoxal residues	Emulsion binders, crosslinkable systems, paper coatings	Tight monomer specs; vacuum stripping; alternative crosslinkers; pH/oxidation control
Solvent residues	Ethyl acetate, IPA, MEK (trace)	Solvent-processed layers; cleaning carryover	Drying validation; solvent balance; headspace GC checks
Recycled content contaminants	MOSH/MOAH; legacy plasticizers; surfactant residues	PCR polyolefins, rPET, recycled paper	Supplier qualification; barrier layers; NIAS screening; specification limits
Metal catalysts/salts	Antimony (from PET), tin, cobalt, aluminum	Polyester films; driers; catalyst residues	Grade selection; chelation/filtration; confirm by ICP-OES/ICP-MS
Biocide transformation products	Isothiazolinone degradation fragments	Preserved aqueous systems; aged stocks	Dose control; storage time/temperature limits; LC-HRMS monitoring
Ink/varnish set-off (non-UV)	Mineral oils (MOSH/MOAH); plasticizer traces	Printed stacks; secondary transfer into adhesive	Low-migration inks; controlled stacking; functional barriers; QC migration panels

Table A17.2 NIAS analytical screening workflow (practical outline).

Step	Purpose	Typical techniques/notes
Sample plan	Representative worst-case selection	Highest contact ratio, longest contact time, smallest food simulant volume, or highest temperature as applicable
Extraction/prep	Liberate volatiles/ semivolatiles and nonvolatiles	Headspace/HS-SPME for volatiles; solvent extraction for semi-/nonvolatiles; QuEChERS variants where relevant
Target analysis	Quantify known markers	GC–MS for photoinitiators/solvents; LC–MS for oligomers/antioxidant derivatives; ICP-OES/ICP-MS for metals
Nontarget/suspect screening	Detect unknowns/NIAS	LC-HRMS and GC–HRMS with library/database and accurate-mass workflows; isotope patterns; in silico fragmentation
Identification confidence	Classify findings	Adopt identification levels (1–5) and flag tentatives for confirmation with standards
Migration and exposure estimates	Compare to limits/thresholds	Use specific migration limits where available; otherwise TTC or internal AOET thresholds with uncertainty factors
Risk management actions	Close the loop	Raw-material change, process adjustment, barrier layer, or specification limit; document in compliance dossier

Note: The examples are illustrative and not exhaustive. Actual NIAS profiles depend on raw-material choices and process conditions; confirm with case-specific screening and migration testing.

18

Toward the Future—Waterborne Alkyd Emulsions

18.1 Introduction—Urgency, Opportunity, and the Alkyd We Still Want

For decades, alkyd resins have remained a mainstay of protective and decorative coatings. Their gloss, leveling, and film hardness continue to set benchmarks for performance. Yet their traditional reliance on organic solvents has become increasingly incompatible with modern regulatory frameworks and consumer expectations. The urgency is evident: solvent-borne alkyds face a shrinking license to operate. At the same time, the opportunity is equally clear: waterborne alkyd emulsions promise to deliver the properties that the market still values—durability, gloss, and processability—without the environmental compromises. The tension between legacy performance and regulatory pressure is precisely what makes the transition to waterborne alkyd emulsions one of the most compelling futures in coating and adhesive technology.

18.1.1 The Signals You Cannot Ignore

The momentum toward waterborne alkyds is unmistakable. Regulators are steadily tightening solvent limits; consumers are demanding low-odor, eco-friendly alternatives; producers are investing in hybrid strategies that blend alkyd chemistry with emulsification, urethane modification, or acrylic synergy; and emerging markets are leapfrogging directly to waterborne solutions rather than retracing the path of solvent-based development. These signals confirm that the waterborne alkyd is no longer experimental—it is rapidly becoming the new baseline.

18.2 What This Book Will Do for You

The present volume has provided a practical understanding of emulsion polymers and PVAc-based adhesives, with close attention to chemistry, process design, troubleshooting, and industrial applications. It also establishes the foundation for appreciating how adjacent technologies, particularly waterborne alkyd emulsions, can be understood through the same principles of stabilization, particle-size control, and formulation balance. Reading forward, the reader is equipped to connect today's emulsion science with the resin technologies of tomorrow. The roadmap to waterborne alkyds begins here.

18.3 The Big Anxieties—And How We Defuse Them

Every transition invites uncertainty, and the move from solvent-borne to waterborne alkyds is no exception. The most frequent concerns focus on whether waterborne alkyds can truly replicate the gloss and leveling associated with solvent systems, whether emulsions can maintain stability under diverse climate conditions, and whether the cost structure of raw materials and production is sustainable. The answers, as demonstrated throughout this book, lie in careful formulation and process design. Surfactant packages, protective colloids, crosslinkers, and reactor strategies all offer practical levers for addressing these concerns. By applying the same rigor that has long underpinned emulsion polymer science, these anxieties can be methodically defused.

18.4 How to Read This Book When You're Behind Schedule

Readers approaching this book under time pressure may appreciate a structured pathway through the material. The case studies in Chapter 16 provide real-world lessons that can be assimilated quickly. The frequently asked questions in Chapter 15 offer immediate troubleshooting support for common production issues. The formulation tables dispersed throughout the chapters serve as ready-to-use templates. The regulatory and sustainability framework in Chapter 17 functions as a compass for aligning development work with compliance and market positioning. Taken together, these elements allow even the most time-constrained reader to extract maximum value from the book. The transition to waterborne alkyd emulsions, in this sense, requires not a reinvention of practice but the adaptation of established logic into a new technological domain.

18.5 Roadmap–What Comes Next

The next stage of this journey will be examined in detail in the forthcoming volume *Waterborne Alkyd Emulsions and Hybrid Coatings*. That work will chart the chemistry of alkyd emulsification and hybrid design, describe the equipment and process modifications required for production, and provide formulations for decorative, protective, and industrial coatings that satisfy global benchmarks for durability and VOC compliance. It will also demonstrate how sustainability can be integrated with performance, showing how bio-based feedstocks and circular design can coexist with uncompromising gloss and hardness. Chapter 18 therefore functions as both a conclusion and a point of departure. It closes the discussion of PVAc-based emulsions and adhesives while opening the pathway toward the next era of waterborne resins.

Bibliography

Part A. Cited References (Chapter 2)

Chapter 2

General-purpose encyclopedias were consulted only for preliminary orientation; all terminology in this chapter is verified against—and cited to—the authoritative sources listed below.

De Bruyn, H. (1999). *The Emulsion Polymerization of Vinyl Acetate*. Boca Raton, FL: CRC Press.

Ebnesajjad, S. (ed.) (2021). *Handbook of Adhesives and Surface Preparation: Technology, Applications and Manufacturing*, 2e. Oxford: Elsevier.

Erbil, Y.H. (2000). *Vinyl Acetate Emulsion Polymerization and Copolymerization with Acrylic Monomers*. Boca Raton, FL: CRC Press.

Gilbert, R.G. (1995). *Emulsion Polymerization: A Mechanistic Approach*. London: Academic Press.

Holmberg, K., Jönsson, B., Kronberg, B., and Lindman, B. (2003). *Surfactants and Polymers in Aqueous Solution*, 2e. Chichester: Wiley.

Kroschwitz, J.I. and Seidel, A. (ed.) (2004). *Kirk-Othmer Encyclopedia of Chemical Technology*, 5e. Hoboken, NJ: Wiley.

Myers, D. (2020). *Surfaces, Interfaces, and Colloids: Principles and Applications*, 3e. Weinheim: Wiley-VCH.

Pizzi, A. and Mittal, K.L. (ed.) (2018). *Handbook of Adhesive Technology*, 3e. Boca Raton, FL: CRC Press.

Rosen, M.J. and Kunjappu, J.T. (2012). *Surfactants and Interfacial Phenomena*, 4e. Hoboken, NJ: Wiley.

Urban, D. and Takamura, K. (ed.) (2002). *Polymer Dispersions and Their Industrial Applications*. Weinheim: Wiley-VCH.

Part B. Background Handbooks and Reference Sources (Chapters 1 and 3–17)

Chapter 1

The references listed here are established handbooks and monographs used solely to provide historical background, core definitions, and foundational principles for emulsion polymerization and adhesives. They frame the field in this Introduction. Current specifications, performance data, and regulatory values are documented with up-to-date primary sources, standards, and supplier data in later chapters.

De Bruyn, H. (1999). *The Emulsion Polymerization of Vinyl Acetate*. Boca Raton, FL: CRC Press.

Ebnesajjad, S. (2021). *Adhesives Technology Handbook*, 3e. Oxford: William Andrew Publishing.

Erbil, Y.H. (2000). *Vinyl Acetate Emulsion Polymerization and Copolymerization with Acrylic Monomers*. Boca Raton, FL: CRC Press.

Flick, E.W. (2002). *Water-Based Paint Formulations*, 3e. Park Ridge, NJ: Noyes Publications.

Lovell, P.A. and El-Aasser, M.S. (1997). *Emulsion Polymerization and Emulsion Polymers*. Chichester: Wiley.

Mittal, K.L. (ed.) (2002). *Adhesion Science and Engineering: Surfaces, Chemistry and Applications*, vol. 1. Amsterdam: Elsevier.

Petrie, E.M. (2007). *Handbook of Adhesives and Sealants*, 2e. New York, NY: McGraw-Hill.

Pizzi, A. and Mittal, K.L. (ed.) (2017). *Handbook of Adhesive Technology*, 3e. Boca Raton, FL: CRC Press.

Chapter 3

Allcock, H.R., Lampe, F.W., and Mark, J.E. (2003). *Contemporary Polymer Chemistry*. Upper Saddle River, NJ: Pearson.

Anonymous (2000). Styrene. In: *Kirk-Othmer Encyclopedia of Chemical Technology*, 5the (ed. J.I. Kroschwitz and A. Seidel). Hoboken, NJ: Wiley.

Bamford, C.H. and Toms, H.A.C. (1974). *Emulsion Polymerization*. London: Academic Press.

BASF SE (2021). *Rheology Modifiers for Water-Based Systems—Technical Literature*. Ludwigshafen: BASF SE.

Brandrup, J., Immergut, E.H., and Grulke, E.A. (ed.) (1999). *Polymer Handbook*. New York, NY: Wiley-Interscience.

Consulting, S.R.I. (2008). *Vinyl Acetate Monomer — Chemical Economics Handbook*. Menlo Park, CA: SRI Consulting.

Erbil, Y.H. (2000). *Vinyl Acetate Emulsion Polymerization and Copolymerization with Acrylic Monomers*. Boca Raton, FL: CRC Press.

European Chemicals Agency (ECHA) (2006). *REACH Regulation (EC) No 1907/2006*. Helsinki: ECHA.

Hexion Inc. (n.d.). *VeoVa Vinyl Esters—Technical Literature*. Columbus, OH: Hexion Inc.

Hofmann, D. and Schaefer, A. (2002). *Process Development in Vinyl Acetate Production*. Weinheim: Wiley-VCH.

James, D.H. and Castor, W.M. (2005). Styrene. In: *Ullmann's Encyclopedia of Industrial Chemistry* (ed. F. Ullmann et al.), 684–688. Weinheim: Wiley-VCH.

Lovell, P.A. and El-Aasser, M.S. (1997). *Emulsion Polymerization and Emulsion Polymers*. Chichester: Wiley.

Marten, F.L. (2000). Vinyl alcohol polymers. In: *Kirk-Othmer Encyclopedia of Chemical Technology* (ed. J.I. Kroschwitz and A. Seidel), 489, 517. Hoboken, NJ: Wiley.

NIOSH (2019). *Pocket Guide to Chemical Hazards*. Cincinnati, OH: Centers for Disease Control and Prevention.

Odian, G. (2004). *Principles of Polymerization*. Hoboken, NJ: Wiley-Interscience.

OSHA (2020). *Occupational Safety and Health Standards*. Washington, DC: U.S. Department of Labor.

OSPAR Commission (2011). *Background Document on Neodecanoic Acid, Ethenyl Ester (Vinyl Neodecanoate; "VeoVa-10")*. London: OSPAR Commission.

Salamone, J.C. (ed.) (1996). *Polymeric Materials Encyclopedia*. Boca Raton, FL: CRC Press.

Seymour, R.B. and Carraher, C.E. (2000). *Polymer Chemistry: An Introduction*. Boca Raton, FL: CRC Press.

Thielking, H. (2006). Cellulose ethers. In: *Ullmann's Encyclopedia of Industrial Chemistry* (ed. F. Ullmann et al.). Weinheim: Wiley-VCH. https://doi.org/10.1002/14356007.a05_461.pub2.

Chapter 4

Abd El-Ghaffar, M.A. (2016). Novel high solid content nano-siliconated poly(VeoVa-acrylate) terpolymer latex. *Progress in Organic Coatings* 101: 230–239.

Alexandridis, P. and Hatton, T.A. (1995). Poly(ethylene oxide)–poly(propylene oxide) block copolymer surfactants in aqueous solutions and at interfaces: thermodynamics, structure, dynamics, and modeling. *Colloids and Surfaces A* 96: 1–46.

BASF (2018). *Surfactant Selection Guide for Emulsion Polymerization*. Ludwigshafen, Germany: BASF SE (Technical Bulletin).

Bhattacharjee, S., Chen, J., Elimelech, M. et al. (2001). Synergistic effects in blending nonionic and anionic surfactants for emulsion polymerization. *Langmuir* 17 (10): 3213–3220.

Daniels, E.S., Sudol, E.D., and El-Aasser, M.S. (ed.) (2001). *Polymer Colloids: Science and Technology of Latex Systems (ACS Symposium Series)*. Washington, DC, USA: American Chemical Society.

EP 3387176 B1 (2022). Vinyl acetate/ethylene emulsions with controlled surfactant ratios. European Patent Office.

Farn, R.J. (2006). *Chemistry and Technology of Surfactants*. Oxford, UK: Wiley-Blackwell.

Goddard, E.D. and Ananthapadmanabhan, K.P. (1993). *Interactions of Surfactants with Polymers and Proteins*. Boca Raton, FL, USA: CRC Press.

Griffin, W.C. (1954). Calculation of HLB values of non-ionic surfactants. *Journal of the Society of Cosmetic Chemists* 5: 249–256.

Guyot, A. (2003). Polymerizable surfactants. In: *Novel Surfactants: Preparation, Applications and Biodegradability*, 2e (ed. K. Holmberg). Boca Raton, FL, USA: CRC Press.

European Commission (2000). *Alternatives to alkylphenol ethoxylates (APEs/NPEs) in surfactant applications*. Environment Series. Brussels: European Commission.

Hexion (2025). VeoVa vinyl esters in emulsion polymers: technical overview. https://www.hexion.com (accessed 20 September 2025).

Hill, R.M. (2006). *Silicone Surfactants*. Boca Raton, FL: CRC Press.

Holmberg, K., Jönsson, B., Kronberg, B., and Lindman, B. (2003). *Surfactants and Polymers in Aqueous Solution*, 2e. Chichester, UK: Wiley.

Karsa, D.R. and Davies, W. (1994). *Industrial Applications of Surfactants*. Cambridge, UK: Royal Society of Chemistry.

Kissa, E. (2001). *Surfactant Science and Technology*. New York, NY, USA: Marcel Dekker.

Landfester, K., Bechthold, N., Förster, S., and Antonietti, M. (1999). Polymerizable surfactants in miniemulsion polymerization. *Macromolecular Rapid Communications* 20 (2): 81–85.

Lovell, P.A. and El-Aasser, M.S. (1997). *Emulsion Polymerization and Emulsion Polymers*. Chichester, UK: Wiley.

Myers, D. (2006). *Surfactant Science and Technology*. Hoboken, NJ, USA: Wiley-Interscience.

Okamoto, Y. et al. (2010). Nonionic EO-based surfactants: comparison of linear, branched, and narrow-range ethoxylates in coatings. *Journal of Surfactants and Detergents* 13 (2): 123–131.

Okubo, M. and Saito, N. (2001). Polymerizable surfactants and their applications in emulsion polymerization. *Colloid & Polymer Science* 279 (10): 882–888.

Rieger, J. (2001). The application of HLB in the selection of surfactants for the preparation of emulsions. *Cosmetics & Toiletries* 116 (1): 59–64.

Rosen, M.J. and Kunjappu, J.T. (2012). *Surfactants and Interfacial Phenomena*, 4e. Hoboken, NJ, USA: Wiley.

Sasol Performance Chemicals (2025). Surfactants for emulsion polymerization (Technical Brochure). https://www.sasol.com (accessed 20 September 2025).

Schott, H. (1995). Narrow-range ethoxylates in emulsion polymerization. *Journal of Coatings Technology* 67 (846): 45–52.

Schramm, L.L. (2000). *Surfactants: Fundamentals and Applications in the Petroleum Industry*. Cambridge, UK: Cambridge University Press.

Tadros, T.F. (2005). *Applied Surfactants: Principles and Applications*. Weinheim, Germany: Wiley-VCH.

Tauer, K. (1997). Reactive surfactants. In: *Polymeric Dispersions: Principles and Applications* (NATO ASI Series E, Vol. 335) (ed. J.M. Asua). Dordrecht, Netherlands: Springer.

Texter, J. (ed.) (2001). *Reactions and Synthesis in Surfactant Systems* (Surfactant Science Series, Vol. 100). Boca Raton, FL, USA: CRC Press.

Witucki, G.L. (1991). A silane primer: chemistry and applications of alkoxy silanes. *Journal of Coatings Technology* 63 (803): 57–60.

Zhang, H. and Zhu, S. (2000). Emulsion polymerization of styrene using polymerizable surfactants. *Journal of Polymer Science Part A: Polymer Chemistry* 38 (7): 1149–1155.

Chapter 5

Myers, D. (2006). *Surfactant Science and Technology*, 3e. Hoboken, NJ: Wiley-Interscience.

Rosen, M.J. and Kunjappu, J.T. (2012). *Surfactants and Interfacial Phenomena*, 4e. Hoboken, NJ: Wiley.

Holmberg, K., Jönsson, B., Kronberg, B., and Lindman, B. (2002). *Surfactants and Polymers in Aqueous Solution*, 2e. Chichester: Wiley.

Goodwin, J. (2009). *Colloids and Interfaces with Surfactants and Polymers: An Introduction*, 2e. Chichester: Wiley.

Lovell, P.A. and El-Aasser, M.S. (1997). *Emulsion Polymerization and Emulsion Polymers*. Chichester: Wiley.

van Herk, A.M. (ed.) (2013). *Chemistry and Technology of Emulsion Polymerisation*, 2e. Oxford: Wiley-Blackwell.

Tadros, T.F. (2005). *Applied Surfactants: Principles and Applications*. Weinheim: Wiley-VCH.

Griffin, W.C. (1949). Classification of surface-active agents by HLB. *Journal of the Society of Cosmetic Chemists*. 1: 311–326.

Rieger, J. (2001). The application of HLB in the selection of surfactants for the preparation of emulsions. *Cosmetics & Toiletries.* 116 (1): 59–64.

Joos, P. (1999). *Dynamic Surface Phenomena*. Utrecht: VSP International Science Publishers.

Schramm, L.L. (2000). *Surfactants: Fundamentals and Applications in the Petroleum Industry*. Cambridge: Cambridge University Press.

Urban, D. and Takamura, K. (2002). *Polymer Dispersions and Their Industrial Applications*. Weinheim: Wiley-VCH.

Winnik, M.A. (1997). Film formation from latex. *Advances in Colloid and Interface Science.* 68: 1–23.

Keddie, J.L. (1997). Film formation of latex. *Materials Science and Engineering: R: Reports.* 21: 101–170.

Steward, P.A., Hearn, J., and Wilkinson, M.C. (2000). An overview of polymer latex film formation and properties. *Advances in Colloid and Interface Science.* 86: 195–267.

Zosel, A. (1991). Surfactants and their role in film formation. *Progress in Organic Coatings* 19: 121–140.

Karsa, D.R. and Davies, W. (ed.) (1994). *Industrial Applications of Surfactants.* Cambridge: Royal Society of Chemistry.

Karsa, D.R. (ed.) (1999). *Industrial Applications of Surfactants IV*. Cambridge: Royal Society of Chemistry.

Fainerman, V.B., Miller, R., and Möhwald, H. (2001). *Surfactants: Chemistry, Interfacial Properties, Applications*. Amsterdam: Elsevier.

Hill, R.M. and Surfactants, S. (1999). *Surfactant Science Series.* New York, NY: Marcel Dekker/CRC Press.

Paint & Coatings Industry (PCI) Magazine (2007). Surfactant selection and its impact on emulsion polymer properties. January 2007.

Sasol Performance Chemicals (2015). Surfactants for emulsion polymerization. Technical brochure.

Dow Chemical Company (2014). DOWFAX anionic surfactants in emulsion polymerization. Technical bulletin.

BASF (2018). Surfactant selection guide for emulsion polymerization. Technical bulletin. Ludwigshafen, Germany.

European Parliament and Council. Regulation (EC) No. 1907/2006 (2006). (REACH), Annex XVII—restrictions on the manufacture, placing on the market and use of certain dangerous substances, mixtures and articles: entries for nonylphenol and nonylphenol ethoxylates. Brussels: Official Journal of the European Union; as amended.

Sandoval, R. (2017). *Early-rain-resistant, surfactant-leaching-resistant binder—field guidance*. Technical communication. American Coatings Conference (ACC), April 2017, Orlando, FL.

American Coatings Association (2018). Early-rain-resistant, surfactant-leaching-resistant binder—field guidance. Technical communication.

Guyot, A. et al. (2003). Surfactants and film defects in coatings. *Progress in Organic Coatings.* 48 (1): 1–27.

Zana, R. and Xia, J. (ed.) (2004). *Gemini Surfactants: Synthesis, Interfacial Properties, and Applications.* New York, NY: CRC Press.

Chapter 6

Allport, D.C. and Gilbert, D.S. (1990). *Aqueous Emulsion Polymers: Chemistry, Properties and Applications.* London, UK: Blackie Academic & Professional. (practical initiator choices; processing).

Carl Roth GmbH + Co. KG. *Potassium persulfate ≥99 %, extra pure.* Safety Data Sheet, Article No. 6795. Revised 19 September 2024. Karlsruhe, Germany.

BASF. (n.d.) Alcohols, glycol ethers, and solvents. Online technical information. https://chemicals.basf.com/global/en/Petrochemicals/alcohols-glycol_ethers-and-solvents (accessed 14 January 2026).

Borisov, A.N., Nazarov, A.P., Kiselev, V.I. et al. (2015). Mechanism and kinetics of ammonium persulfate decomposition in aqueous solutions. *Russian Journal of General Chemistry* 85: 2067–2072. (APS rate vs pH/temperature).

Chern, C.-S. (2008). *Principles and Applications of Emulsion Polymerization.* Hoboken, NJ: Wiley. (industrial selection of initiators and auxiliaries).

Dow. (n.d.) UCAR™ Filmer IBT. Product information page (via SpecialChem). https://www.specialchem.com/coatings/product/dow-ucar-filmer-ibt (accessed 29 October 2024).

PCI Magazine (2014). Redox for main polymerization of emulsion polymers. PCI Magazine, 1 August 2017. https://www.pcimag.com/articles/103744-redox-for-main-polymerization-of-emulsion-polymers.

Eastman Chemical Company. Texanol ester alcohol—coalescent for latex paints, 20 March 2025. https://productcatalog.eastman.com/tds/ProdDatasheet.aspx?product=71000148.

European Parliament and Council (2006). Regulation (EC) No 1907/2006 (REACH). Official Journal of the European Union; 2006 (and amendments). (regulatory).

European Parliament and Council (2008). Regulation (EC) No 1272/2008 (CLP). Official Journal of the European Union; 2008 (and amendments). (classification/labeling).

Evonik Active Oxygens. (2024). Hydrogen peroxide—safe handling and storage. Technical brochure; latest edition (35%/50% H_2O_2 specs, stabilizers, storage). https://products.evonik.com/assets/95/41/Evonik_Hydrogen_Peroxide_Brochure_EN_Asset_2439541.pdf.

Feng, X. and Zhu, S. (2000). Emulsion polymerization with water-soluble azo initiators: kinetics and particle formation. *Journal of Polymer Science Part A: Polymer Chemistry* 38: 978–986. (AIBA/V-50 effects, nucleation).

Furman, O.S., Teel, A.L., and Watts, R.J. (2010). Mechanism of base activation of persulfate. *Environmental Science & Technology* 44: 6423–6428. (pH effects; radical species).

Gächter, R. and Müller, H. (ed.) (1993). *Plastics Additives Handbook*, 4e. Munich, Germany: Hanser. (auxiliaries: antioxidants, UV absorbers, plasticizers).

Gilbert, R.G. (1995). *Emulsion Polymerization: A Mechanistic Approach*. London, UK: Academic Press. (particle nucleation, radical entry, initiator partitioning).

Gilbert, R.G. and Napper, D.H. (1997). Radical entry and phase location (selected chapters). In: *Emulsion Polymerization and Emulsion Polymers* (ed. P.A. Lovell and M.S. El-Aasser). Chichester: Wiley.

van Herk, A.M. (ed.) (2013). *Chemistry and Technology of Emulsion Polymerisation*, 2e. Chichester, UK: Wiley-Blackwell. (comprehensive EP including initiator systems).

Hutchinson, R.A. (2008). Modeling of emulsion polymerization processes. *Progress in Polymer Science* 33: 454–486. (onset time, heat release, dosing strategies).

Keddie, J.L. (1997). Film formation of latex. *Materials Science and Engineering: R: Reports* 21: 101–170. (coalescence/auxiliaries context).

Koleske, J. V., ed. (2012). Paint and Coating Testing Manual: 15th ed. ASTM Manual MNL 17. West Conshohocken, PA: ASTM International (film defects, discoloration, test methods).

Kroschwitz, J.I. (ed.) (2004–2007). *Kirk-Othmer Encyclopedia of Chemical Technology*, 5e. Hoboken, NJ: Wiley. (entries: initiators incl. H_2O_2, TBHP, persulfates, inhibitors).

Lovell, P.A. and El-Aasser, M.S. (ed.) (1997). *Emulsion Polymerization and Emulsion Polymers*. Chichester, UK: Wiley. (initiator selection, Smith–Ewart, intervals, film formation context).

Odian, G. (2004). *Principles of Polymerization*, 4e. Hoboken, NJ: Wiley. (radical generation, initiator kinetics, disproportionation/termination).

Routh, A.F. (2013). Drying of thin colloidal films. *Reports on Progress in Physics* 76: 046603. (coalescent–morphology interplay with initiator-driven Mw).

Schork, F.J., Luo, Y., Smulders, W. et al. (2005). Miniemulsion polymerization. *Advances in Polymer Science* 175: 129–255. (context for initiator partitioning and droplet nucleation).

Wesslau, M. (1994). Polymerization with tert-butyl hydroperoxide. *Macromolecular Chemistry and Physics* 195: 1137–1145. (TBHP mechanisms).

Wicks, Z.W., Jones, F.N., Pappas, S.P., and Wicks, D.A. (2006). *Organic Coatings: Science and Technology*, 4e. Hoboken, NJ: Wiley. (coalescents, plasticizers, defects).

Witt, D. (2007). Recent developments in redox initiation of radical polymerization. *Progress in Polymer Science* 32: 1291–1311. (advanced redox pairs, selection).

Zhang, H. and Zhu, S. (2002). Redox initiation in aqueous free-radical polymerizations: an overview. *Journal of Polymer Science Part A: Polymer Chemistry* 40: 885–901. (temperature control, low-temp starts).

Chapter 7

Allcock, H.R., Lampe, F.W., and Mark, J.E. (2003). *Contemporary Polymer Chemistry*, 3e. Upper Saddle River, NJ: Pearson.

Bamford, C.H. and Toms, H.A.C. (1974). *Emulsion Polymerization*. London: Academic Press.

Borisov, A.N., Nazarov, A.P., Kiselev, V.I. et al. (2015). Mechanism and kinetics of ammonium persulfate decomposition in aqueous solutions. *Russian Journal of General Chemistry* 85: 2067–2072.

de Bruyn, H. (1999). *The Emulsion Polymerization of Vinyl Acetate*. Atlanta, GA: TAPPI Press.

Daniels, E.S., Sudol, E.D., and El-Aasser, M.S. (2001). *Polymer Colloids: Science and Technology of Latex Systems (ACS Symposium Series)*. Washington, DC: American Chemical Society.

El-Aasser, M.S. and Sudol, E.D. (1997). Semi-continuous emulsion polymerization processes: vinyl acetate systems. *Polymer International* 43: 153–162.

Erbil, Y.H. (2000). *Vinyl Acetate Emulsion Polymerization and Copolymerization with Acrylic Monomers*. Boca Raton, FL: CRC Press.

Furman, O.S., Teel, A.L., and Watts, R.J. (2010). Mechanism of base activation of persulfate. *Environmental Science & Technology* 44: 6423–6428.

Gilbert, R.G. (1995). *Emulsion Polymerization: A Mechanistic Approach*. London: Academic Press.

Gilbert, R.G. and Napper, D.H. (1982). Effect of protective colloids in emulsion polymerization of vinyl acetate. *Journal of Polymer Science: Polymer Chemistry Edition* 20: 1989–2002.

Goodwin, J.W. (2009). *Colloids and Interfaces with Surfactants and Polymers: An Introduction*, 2e. Chichester: Wiley.

van Herk, A.M. (ed.) (2013). *Chemistry and Technology of Emulsion Polymerisation*, 2e. Chichester: Wiley-Blackwell.

Hiemenz, P.C. and Rajagopalan, R. (1997). *Principles of Colloid and Surface Chemistry*, 3e. New York, NY: CRC Press.

Hunter, R.J. (1981). *Zeta Potential in Colloid Science: Principles and Applications*. London: Academic Press.

Hutchinson, R.A. (2008). Modeling of emulsion polymerization processes. *Progress in Polymer Science* 33: 454–486.

Israelachvili, J.N. (2011). *Intermolecular and Surface Forces*, 3e. San Diego, CA: Academic Press.

Kostansek, E. (1997). Semi-batch emulsion polymerization optimization for PVAc. *Progress in Organic Coatings* 32: 123–140.

Lovell, P.A. and El-Aasser, M.S. (1997). *Emulsion Polymerization and Emulsion Polymers*. Chichester: Wiley.

Lyklema, J. (2005). *Fundamentals of Interface and Colloid Science, Vol. II: Solid–Liquid Interfaces*. Amsterdam: Elsevier.

Napper, D.H. (1983). *Polymeric Stabilization of Colloidal Dispersions*. London: Academic Press.

Odian, G. (2004). *Principles of Polymerization*, 4e. Hoboken, NJ: Wiley.

Paul, E.L., Atiemo-Obeng, V.A., and Kresta, S.M. (2004). *Handbook of Industrial Mixing: Science and Practice*. Hoboken, NJ: Wiley.

Smith, W.V. and Ewart, R.H. (1948). Kinetics of emulsion polymerization. *The Journal of Chemical Physics* 16 (6): 592–599.

Tadros, T.F. (2007). *Colloid Stability: The Role of Surface Forces*. Weinheim: Wiley-VCH.

Ugelstad, J. and Hansen, F.K. (1976). Particle nucleation in emulsion polymerization of vinyl acetate. *Journal of Polymer Science: Polymer Chemistry Edition* 14: 1953–1977.

Urban, D. and Takamura, K. (ed.) (2002). *Polymer Dispersions and Their Industrial Applications*. Weinheim: Wiley-VCH.

Chapter 8

Asua, J.M. (2012). *Emulsion Polymerization: Industrial Synthesis and Applications*. Chichester: Wiley.

Beshah, K. and Devonport, W. (2013). A study of acetoacetoxyethyl methacrylate hydrolysis in acrylic latex polymers as a function of pH. *Journal of Coating Technology and Research* 10 (6): 821–828.

Blank, W.J., He, Z.A., and Picci, M. (2002). Catalysis of the epoxy–carboxyl reaction. *Journal of Coatings Technology* 74 (926): 33–41.

Brüggemann, K.G. (2017). Redox for main polymerization of emulsion polymers. *PCI Magazine*.

Cheaburu-Yilmaz, C.N., Özkan, C.K., and Yilmaz, O.S. (2022). Synthesis and application of reactive acrylic latexes: effect of particle morphology. *Polymers (Basel)* 14: 2187.

Chen, X., Soria-Carrera, H., Zozulia, O., and Boekhoven, J. (2023). Suppressing catalyst poisoning in the carbodiimide-fueled reaction cycle. *Chemical Science* 14: 11157–11166.

Chern, C.-S. (2008). *Principles and Applications of Emulsion Polymerization*. Hoboken, NJ: Wiley.

Cordeiro, C.F. (2000). Vinyl acetate polymers. In: *Kirk-Othmer Encyclopedia of Chemical Technology*. Hoboken, NJ: Wiley (updated).

Creton, C. (2003). Pressure-sensitive adhesives: an overview. *MRS Bulletin* 28 (6): 434–439.

Ebnesajjad, S. (2011). *Handbook of Adhesives and Surface Preparation*. Oxford: Elsevier.

Ellis, J.W. and Smith, M. (1994). Application of poly(vinyl acetate) copolymer adhesives in packaging. *International Journal of Adhesion and Adhesives* 14 (4): 223–230.

Esser, R.J., Devona, J.E., Setzke, D.E., and Wagemans, L. (1999). Water-based crosslinkable surface coatings. *Progress in Organic Coating* 36: 45–52.

European Committee for Standardization (CEN) (2001). EN 204:2001—Classification of thermoplastic wood adhesives for non-structural applications. Brussels: CEN.

Gadhave, R.V. (2024). Glycidyl methacrylate copolymer emulsion-based heat- and water-resistant wood adhesive. *Journal of Adhesion Science and Technology* 1–22.

González, I., Asua, J.M., and Leiza, J.R. (2006). Crosslinking in acetoacetoxy functional waterborne crosslinkable latexes. *Macromolecular Symposia* 243: 53–62.

Jacob, L.I., Rusu, L., and Rusu, G. (2022). Influence of redox initiator component ratios on the properties of polymers obtained by emulsion polymerisation. *Polymers (Basel)* 14 (10): 2062.

Lovell, P.A. and El-Aasser, M.S. (ed.) (1997). *Emulsion Polymerization and Emulsion Polymers*. Chichester: Wiley.

Luxfer MEL Technologies (2024). Bacote® Series—Stabilized ammonium zirconium carbonate (AZC), Technical Datasheet DS-101.

Machotová, J., Kalendová, A., Zlámaná, B. et al. (2020). Waterborne coating binders based on self-crosslinking acrylic latex with embedded nanoparticles: comparison of nanostructured ZnO and MgO as crosslink density-enhancing agents. *Coatings* 10 (4): 339.

Derksen, A.J. Polycarbodiimides as classification-free and easy-to-use crosslinkers for water-based coatings. Stahl International bv (Stahl Polymers), Technical White Paper. Stahl-Polymers-White-paper.

Mittal, K.L. (2000). *Polymer Surface Modification: Relevance to Adhesion*. Utrecht: VSP.

NIOSH. (2019). Acrylic acid: pocket guide to chemical hazards. Centers for Disease Control and Prevention. https://www.cdc.gov/niosh/npg/npgd0013.html (accessed 31 August 2025).

NIOSH (2019). Methacrylic acid: pocket guide to chemical hazards. Centers for Disease Control and Prevention. https://www.cdc.gov/niosh/npg/npgd0386.html.

NIOSH. (2019). Vinyl acetate: pocket guide to chemical hazards. Centers for Disease Control and Prevention. https://www.cdc.gov/niosh/npg/npgd0656.html (accessed 31 August 2025).

Park, Y.-J., Monteiro, M.J., van Es, S., and German, A.L. (2001). Effect of ambient crosslinking on the mechanical properties and film morphology of

PSTY-P(BA-co-AAEMA) reactive composite latexes. *European Polymer Journal* 37: 965–973.

Pham, H.H. and Winnik, M.A. (1999). Polymer interdiffusion vs cross-linking in carboxylic acid–carbodiimide latex films. *Macromolecules* 32: 7692–7695.

Pham, H.H. and Winnik, M.A. (2000). Synthesis, characterization, and stability of carbodiimide groups in carbodiimide-functionalized latex dispersions and films. *Journal of Polymer Science Part A: Polymer Chemistry* 38: 855–869.

Pham, H.H. and Winnik, M.A. (2006). Polymer interdiffusion vs cross-linking in carboxylic acid–carbodiimide latex films—effect of annealing temperature. *Macromolecules* 39: 1425–1434.

Pizzi, A. (1983). *Wood Adhesives: Chemistry and Technology*. Boca Raton, FL: CRC Press.

Pizzi, A. and Mittal, K.L. (2018). *Handbook of Adhesive Technology*, 3e. Boca Raton, FL: CRC Press/Taylor & Francis.

Reis, A.V., Fajardo, A.R., Schuquel, I.T.A. et al. (2009). Reaction of glycidyl methacrylate at hydroxyl and carboxylic groups of poly(vinyl alcohol) and poly(acrylic acid). *The Journal of Organic Chemistry* 74: 3750–3757.

Satas, D. and Tracton, A. (ed.) (2001). *Handbook of Pressure-Sensitive Adhesive Technology*, 3e. Boca Raton, FL: CRC Press.

Skeist, I. (ed.) (1990). *Handbook of Adhesives*, 3e. New York, NY: Van Nostrand Reinhold.

Taylor, J.W., Collins, M.-J., and Bassett, D.R. (1999). A study on the chemistry of alkylcarbodiimide ethyl methacrylates as reactive monomers for acrylic- and vinyl ester-based latexes. *Progress in Organic Coating* 35: 215–221.

Wake, W.C. (1982). *Adhesion and the Formulation of Adhesives*. London: Applied Science Publishers.

Wang, S. and Choi, P. (2016). Redox-initiated emulsion polymerization using H_2O_2/ascorbic acid/Fe^{2+}: kinetics and particle characteristics. *Journal of Applied Polymer Science* 133 (40): 43037.

Chapter 9

VAM-Based Copolymers with Acrylates and VeoVa 10 in Paints and Coatings.

Erbil, Y.H. (2000). *Vinyl Acetate Emulsion Polymerization and Copolymerization with Acrylic Monomers*. Boca Raton, FL: CRC Press.

de Bruyn, H. (1999). *The Emulsion Polymerization of Vinyl Acetate*. Atlanta, GA: TAPPI Press.

Odian, G. (2004). *Principles of Polymerization*, 4e. Hoboken, NJ: Wiley.

Gilbert, R.G. (1995). *Emulsion Polymerization: A Mechanistic Approach*. London: Academic Press.

Lovell, P.A. and El-Aasser, M.S. (ed.) (1997). *Emulsion Polymerization and Emulsion Polymers*. Chichester: Wiley.

Smith, W.V. and Ewart, R.H. (1948). Kinetics of emulsion polymerization. *The Journal of Chemical Physics* 16 (6): 592–599.

Allcock, H.R., Lampe, F.W., and Mark, J.E. (2003). *Contemporary Polymer Chemistry*, 3e. Upper Saddle River, NJ: Pearson.

Choi, J. and Kim, J. (2012). Effect of comonomer type on polymerization rate and molecular weight in vinyl acetate copolymerization. *Journal of Applied Polymer Science* 124 (3): 2287–2293.

Kim, H.S. and Lee, S.B. (2015). Role of surfactants and chain transfer agents in vinyl acetate–acrylic emulsion polymerization. *Progress in Organic Coating* 88: 1–10.

Lin, X. and Sui, X. (2018). Copolymer structure–property relationships of vinyl acetate-based emulsions. *Journal of Polymer Research* 25: 102.

Rao, M. and Zha, W. (2015). Reactivity ratios of vinyl acetate with various acrylic monomers in emulsion copolymerization. *Polymer-Plastics Technology and Engineering* 54 (4): 430–439.

Fineman, M. and Ross, S.D. (1950). Linear method for determining monomer reactivity ratios in copolymerization. *Journal of Polymer Science* 5 (2): 259–262.

Kelen, T. and Tüdös, F. (1975). Analysis of linear methods for determining copolymerization reactivity ratios. *Journal of Macromolecular Science* A9 (1): 1–27.

Fox, T.G. (1956). Influence of diluent and copolymer composition on the glass transition temperature. *Bulletin of the American Physical Society* 1: 123.

Winnik, M.A. and Gilbert, R.G. (1996). Polymer film formation from latex dispersions. *Advances in Colloid and Interface Science* 64: 167–243.

Gulmine, J.V. and Akcelrud, L. (2005). Weathering and hydrolysis of vinyl polymers. *Polymer Degradation and Stability* 87 (2): 211–219.

Ahn, K.H. and Lee, J.H. (2019). Advances in emulsion polymers for low-VOC architectural coatings. *Journal of Coating Technology and Research* 16 (5): 1247–1255.

Farwaha, R. and de Sousa Gomes, L. (2010). Vinyl acetate–versatic acid vinyl ester copolymer for masonry coatings. CoatingsTech.

Hexion (2020). VeoVa 10 product literature. Technical Datasheet.

Celanese Corporation (2022). Acrylic–VeoVa–VAM binder design: benefits and limitations. Technical Datasheet.

Arkema (2020). Acrylic vs VeoVa–vinyl copolymers: strengths, limitations, and use guidance. Technical Datasheet.

Arkema Coating Resins (2021). Roof coatings for harsh climates – formulation guide.

Eastman Chemical Company (2022). Coalescents for low-VOC waterborne coatings. Technical Bulletin.

Holmberg, K., Jönsson, B., Kronberg, B., and Lindman, B. (2003). *Surfactants and Polymers in Aqueous Solution*, 2e. Chichester: Wiley.

Napper, D.H. (1983). *Polymeric Stabilization of Colloidal Dispersions*. London: Academic Press.

Hiemenz, P.C. and Rajagopalan, R. (1997). *Principles of Colloid and Surface Chemistry*, 3e. New York: CRC Press.

Gradzielski, M. (2022). Polymer–surfactant interaction for controlling the rheological properties of aqueous surfactant solutions. *Current Opinion in Colloid & Interface Science* 60: 101662.

Dow (2020). ACRYSOL™ ASE-60ER rheology modifier for HEC-free formulations. Technical Data Sheet.

BASF (2019). Rheovis™ ASE and HEUR rheology modifiers. Technical Information.

European Chemicals Agency (ECHA) (2021). REACH Regulation (EC) No. 1907/2006.

U.S. Environmental Protection Agency (EPA) (2023). VOC compliance guidelines for architectural coatings.

Chapter 10

Bartoline (2017). PVA adhesive & sealer—technical data sheet (bonds polystyrene and other porous substrates).

East, A.J. and Robson, D.J. (2005). Glycidyl methacrylate and epoxide crosslinkers in high-performance coatings. *Progress in Organic Coating* 52 (2): 115–123.

Eastman Chemical Company (2019). Spectrum® tackifier resins – formulating options. Brochure.

EN 204:2016 (2016). Classification of thermoplastic wood adhesives for non-structural applications. Brussels: CEN.

Finch, C.A. (1992). *Polyvinyl Alcohol: Properties and Applications*. New York, NY: John Wiley & Sons.

Hassan, C.M. and Peppas, N.A. (2000). Structure and applications of poly(vinyl alcohol) hydrogels produced by conventional crosslinking or by freezing/thawing methods. *Advances in Polymer Science* 153: 37–65.

Henkel (Polybit) (2018). Polybond PVA—polyvinyl acetate based adhesive/bonding agent. Technical Data Sheet. (adhesive for fixing polystyrene tiles).

Hesselmans, L.C.J., Derksen, A.J., and van den Goorbergh, J.A.M. (2006). Polycarbodiimide crosslinkers. *Progress in Organic Coating* 55: 183–190.

Höfer, R. and Bigorra, J. (2007). Green chemistry: a sustainable approach to adhesives. *Macromolecular Symposia* 262: 167–172.

Insulfoam (2016). TB-26: EPS-compatible adhesives—guidance & product list. Technical Bulletin.

Kleinschmidt, F. and Müller, M. (2005). Crosslinking agents for waterborne coatings. *Macromolecular Materials and Engineering* 290 (4): 360–366.

Kraton (2025). SYLVATAC™ rosin ester tackifiers – application overview. Brochure.

Kraton (2025). Tackifiers for adhesives – polarity and application guide. Brochure.

Kuraray (2025). KURARAY POVAL™ R – PVOH with silanol groups. Product Brief.
Kuraray (2025). KURARAY POVAL™ / EXCEVAL™ overview. Technical Datasheet.
Luxfer MEL Technologies (2025). Bacote® AZC series (ammonium zirconium carbonate). Datasheet DS-101.
Park, H.-H., Kim, C.H., Lee, T.G. et al. (2024). Effects of zirconium-based crosslinkers on coating film performance. *Applied Sciences* 14: 9187.
Petersen, H.O. (1992). PVAc–PVOH blends in wood adhesives. *International Journal of Adhesion and Adhesives* 12 (1): 12–19.
Pizzi, A. and Mittal, K.L. (ed.) (2018). *Handbook of Adhesive Technology*, 3e. Boca Raton, FL: CRC Press.
Plueddemann, E.P. (1991). *Silane Coupling Agents*, 2e. Berlin: Springer.
Posthumus, W., Derksen, A.J., van den Goorbergh, J.A.M., and Hesselmans, L.C.J. (2007). Crosslinking by polycarbodiimides. *Progress in Organic Coating* 58: 61–65.
Satas, D. and Tracton, A.A. (ed.) (2001). *Handbook of Pressure-Sensitive Adhesive Technology*, 3e. Boca Raton, FL: CRC Press.
Sekisui Specialty Chemicals (2013). Selvol™ polyvinyl alcohol for adhesive and emulsion applications. Brochure.
Skeist, I. (ed.) (1990). *Handbook of Adhesives*, 3e. New York, NY: Springer.
Soarus LLC (Mitsubishi Chemical) (2025). GOHSENX™ T – carboxyl-modified PVOH. Technical Data Sheet.
Soarus LLC (Mitsubishi Chemical) (2025). GOHSENX™ Z – acetoacetylated reactive PVOH. Technical Data Sheet.
SpecialChem (2025). Selecting tackifiers for adhesives and sealants – guide. Online Resource.
Taylor, J.W. (1997). The application of carbodiimide chemistry to coatings. In: *ACS Symposium Series*. Washington, DC: American Chemical Society.
U.S. Patent 4,560,604 (1985). Tackified polyvinyl alcohol adhesive composition for clear label applications.
U.S. Patent 5,948,849 (1999). Water-soluble polymer composition containing PVOH and plasticizers for remoistenable adhesives.
U.S. Patent 6,534,598 (2003). PVAc/PVOH-based adhesive with modified polyol for paper-to-foil lamination.
Weiss Chemie (2024). COSMO® DS-440.120—special dispersion adhesive (PVAc/acrylate family) for surface bonding of PS-foam to absorbent substrates. Technical Data Sheet.
United States Patent US20160215179A1 (2016). Glue-bonded multi-ply absorbent sheet and polyvinyl alcohol ply bonding adhesive.
United States Patent US20040138739A1 (2004). Tail seal adhesive for rolled paper products.
United States Patent US20100236639A1 (2010). Adhesive compositions for paper roll end sealing.

United States Patent US8263689B2 (2012). Water-based adhesives for industrial applications.

Chapter 11

Scott, P.J., Penlidis, A., and Rempel, G.L. (1993). Ethylene–vinyl acetate semi-batch emulsion copolymerization: Experimental design and preliminary screening experiments. *Journal of Polymer Science Part A: Polymer Chemistry*. https://doi.org/10.1002/pola.1993.080310213.

American Coatings Association (2012). New developments in VAE copolymers for architectural coatings. *CoatingsTech/PCI* 9 (1): 30–33.

Blackley, D.C. (1997). Polymer latices: science and technology. In: *Types of Latices*, 2e, vol. 2. Dordrecht: Springer.

Blackley, D.C. (1997). Polymer latices: science and technology. In: *Applications of Latices*, 2e, vol. 3. Dordrecht: Springer.

Blank, W.J., He, Z., and Picci, M. (2002). Catalysis of the epoxy–carboxyl reaction. *Journal of Coatings Technology and Research* 74 (926): 33–41.

Celanese. (2026). EcoVAE™—vinyl acetate ethylene emulsions (family overview). Irving, TX: Celanese Corp.; current online resources. https://www.celanese.com/products/ecovae-vinyl-acetate-ethylene.

Celanese. (2026). Vinamul®—emulsion polymers (VAE lines). Irving, TX: Celanese Corp.; current online resources. https://www.celanese.com/products/vinamul

Dairen Chemical Corporation (DCC) (2025). DA-series—VAE emulsions (catalog/TDS). Taipei: DCC; current online resources.

East, A.J. and Robson, D.J. (2005). Glycidyl methacrylate and epoxide crosslinkers in high-performance coatings. *Progress in Organic Coatings* 52 (2): 115–123.

Gruber, B.A., Vratsanos, M.S., and Smith, C.D. (2000). Effects of colloidal stabilizer on vinyl acetate–ethylene copolymer emulsions and films. *Macromol Symposia* 155: 163–170.

Hesselmans, L.C.J., Derksen, A.J., and van den Goorbergh, J.A.M. (2006). Polycarbodiimide crosslinkers. *Progress in Organic Coatings* 55: 183–190.

Liang, R., Zhang, P., Wei, C. et al. (2017). Spontaneous formation of bimodal particle size distributions in VAE via one-step semi-batch route. *Colloids and Surfaces A* 110: 86–96.

Park, H.-H., Lee, S.-H., Choi, J.-H. et al. (2024). Effects of zirconium-based crosslinkers on coating film performance. *Applied Sciences* 14: 9187.

Pizzi, A. and Mittal, K.L. (ed.) (2018). *Handbook of Adhesive Technology*, 3e. Boca Raton, FL: CRC Press.

Poljanšek, I., Fabjan, E., Burja, K., and Kukanja, D. (2013). Emulsion copolymerization of VAE in a high-pressure reactor—inline FTIR characterization. *Progress in Organic Coatings* 76: 1798–1804.

Posthumus, W., Derksen, A.J., van den Goorbergh, J.A.M., and Hesselmans, L.C.J. (2007). Crosslinking by polycarbodiimides. *Progress in Organic Coatings* 58: 231–236.

Salehi-Mobarakeh, H. and Rostami Daronkola, M.R. (2005). Effect of parameters in emulsion copolymerization of vinyl acetate–ethylene in a 7.5-L high-pressure reactor. *Iranian Polymer Journal* 14 (6): 579–587.

Scott, P.J. and Penlidis, A. (1993). Semi-batch emulsion copolymerization of vinyl acetate–ethylene: factorial design and process optimization. *Journal of Polymer Science Part A: Polymer Chemistry* 31.

Scott, P.J., Penlidis, A., Rempel, G.L. et al. (1994). Reactor design for gas–liquid emulsion copolymerization of vinyl acetate–ethylene. *Chemical Engineering Science* 49: 1573–1583.

Haitung Chemical Co., Ltd.Sinopec. (n.d.) Sinopec VAE emulsion CW40 series. Online distributor product page. https://www.haitungchemical.com/sinopec-vae-cw40-916/.

SpecialChem (2018). Developments in VAE emulsions for coatings—review article.

Tan, C.B. (2005). Emulsion polymerization of ethylene–vinyl acetate–branched vinyl ester copolymers: reactor design and synthesis. PhD thesis, Nanyang Technological University.

Vinyl esters. In: *Ullmann's Encyclopedia of Industrial Chemistry*. Weinheim: Wiley-VCH latest online ed. Urban, D. and Takamura, K. (ed.) (2002). *Polymer Dispersions and Their Industrial Applications*. Weinheim, Germany: Wiley-VCH.

US 4,510,274 (1985) Vinyl acetate–ethylene copolymer emulsion adhesive composition improved with epoxy resin.

Wacker Chemie AG. (2025). VINNAPAS® VAE dispersions—product family overview. Munich: Wacker Chemie AG; current online resources.

Wacker Chemie AG. ETONIS®—polymer modifiers for bitumen/construction (VAE-based). Munich: Wacker Chemie AG; current online resources.

Warson, H. and Finch, C.A. (2001). *Applications of Synthetic Resin Latices. Vol. 1: Fundamental Chemistry & Applications in Adhesives; Vol. 2: Latices in Surface Coatings; Vol. 3: Latices in Diverse Applications*. Chichester: Wiley.

WO 2016/061110 A9 (2016). Vinyl acetate–ethylene/acrylic polymer emulsions and methods of making same.

Chapter 12

AkzoNobel Paints (2019). "Smart exterior coating technologies: easy-to-clean surfaces for harsh weather." Technical Memo.

American Coatings Association (2022, 2022). Paint weatherability: predicting durability with accelerated and natural testing. *CoatingsTech*, August/September.

ASTM D4214. (2014). Standard Test Methods for Evaluating the Degree of Chalking of Exterior Paint Films.

Doherty, K.T. (2004). Environmental effects on waterborne coatings Paint & Coatings Industry Magazine.

Erbil, H.Y. (2000). *Surface Chemistry of Solid and Liquid Interfaces.* Wiley-Interscience.

Zhao, C.-L., Winnik, M.A., Feng, X., and Hinderberger, D. (2006). Direct observation of freeze–thaw instability of latex coatings by fluorescence microscopy. *Journal of Coatings Technology and Research* 3 (4): 321–329.

Fei, X. and Fang, J. (2013). Optimization of cold weather waterborne paints. *Journal of Coatings Technology* 85 (10): 42–50.

Ghosh, S. (2018). Cost-effective binder systems for wall putty applications. *International Journal of Building Materials Research* 34 (2): 117–123.

Gupta, P. and Banerjee, S. (2020). Cement-modified wall putties for diverse climates. *Journal of Construction Chemistry* 14 (1): 33–40.

Jang, H. and Choi, J. (2016). Cold weather-compatible hydrophobic additives for exterior emulsion paints. *Progress in Organic Coatings* 94: 72–78.

Johnston, J.H. (2013). Cold climate construction and paint performance. *Architectural Coatings Journal* 45 (3): 22–28.

Koleske, J.V. (ed.) (1995). *Paint and Coating Testing Manual: 15th Edition of the Gardner-Sward Handbook*. ASTM International.

Koleske JV. (2002). Mechanisms of film degradation and chalking in exterior coatings. ASTM STP Publication.

Kovacs, A. (2017). Fast-drying binders for cold-region applications. Eastern European Coatings Conference.

Kumar, R. and Jain, A. (2015). Formulation and development of wall putty for extreme conditions. *PaintIndia* 65 (6): 58–63.

Larsson, R. and Persson, T. (2018). Overcoming mfft barriers in VeoVa-based exterior emulsions. *Progress in Organic Coatings* 123: 45–53.

Machotová, J., Kalendová, A., Zlámaná, B. et al. (2020). Waterborne coating binders based on self-crosslinking acrylic latex with embedded nanoparticles: comparison of nanostructured ZnO and MgO as crosslink density-enhancing agents. *Coatings* 10 (4): 339.

Meyer, R.N. and Blankenship, M.L. (2006). Coatings for cold weather applications. *Journal of Coatings Technology* 78 (978): 41–48.

Mezger, T.G. (2014). *The Rheology Handbook*. Vincentz Network.

Palmans, R. (2011). Coalescence and internal plasticization in cold-cure latex paints. *Journal of Applied Polymer Science* 120 (6): 3345–3353.

Bagley, E.B. and Nelson, G.L. (1971). Water sensitivity of latex films during early stages of film formation. *Journal of Paint Technology* 43 (561): 50–58.

Pappas, S.P. (1993). Zinc compounds in paint formulations: roles in UV resistance and water blocking. *Coatings Technology* 65 (826): 33–38.

Roper, R. (2008). Coalescing agents in waterborne paints. Paint & Coatings Industry Magazine.

Roper, R. (2016). Best practices for painting in cold weather conditions. *Professional Painter's Journal* 48 (1): 15–20.

Down, J.L. (2016). The effect of modifiers on the stability of a vinyl acetate–ethylene copolymer dispersion. *Studies in Conservation* 61 (4): 190–200.

Tsavalas, J.G. and Mitsoulis, E. (2010). Fundamentals of latex film formation. *Progress in Organic Coatings* 67 (3): 186–193.

U.S. Army Corps of Engineers (USACE) (2002). *Engineer Manual (EM) 1110-2-1612: Ice Engineering*. Washington, DC: Headquarters, USACE.

Wacker Chemie AG (2020). SILRES® BS 50 – Silicone resin for highly durable paints and plasters [Technical Data Sheet].

Kellogg, Shelby. 2018. Developing and Evaluating Early Rain Resistance in Exterior Architectural Coatings. CoatingsTech.

Daisey, George. Early Rain Resistance Marks Next Step in Evolution of Elastomeric Roof Coatings. Paint & Coatings Industry. December 3, 2012. https://www.pcimag.com/articles/97140-early-rain-resistance-marks-next-step-in-evolution-of-elastomeric-roof-coatings.

Chapter 13

ASTM D1308-13 (2020). Standard Test Method for Effect of Household Chemicals on Clear and Pigmented Organic Finishes. ASTM International, West Conshohocken, PA.

ASTM D1475-13 (2020). Standard Test Method for Density of Paint, Varnish, Latex, and Related Products. ASTM International, West Conshohocken, PA.

ASTM D1644-01 (2017). Standard Test Methods for Nonvolatile Content of Varnishes. ASTM International, West Conshohocken, PA.

ASTM D1849-16 (2021). Standard Test Method for Package Stability of Paint. ASTM International, West Conshohocken, PA.

ASTM D2196-20 (2020). Standard Test Methods for Rheological Properties of Non-Newtonian Materials by Rotational (Brookfield) Viscometer. ASTM International, West Conshohocken, PA.

ASTM D2240-15 (2021). Standard Test Method for Rubber Property—Durometer Hardness. ASTM International, West Conshohocken, PA.

ASTM D2243-19 (2020). Standard Test Method for Freeze-Thaw Resistance of Water-Borne Coatings. ASTM International, West Conshohocken, PA.

ASTM D2354-10 (2018). Standard Test Method for Minimum Film Forming Temperature (MFFT) of Emulsion Paints. ASTM International, West Conshohocken, PA.

ASTM D2486-17 (2025). Standard Test Methods for Scrub Resistance of Wall Paints. ASTM International, West Conshohocken, PA.

ASTM D3330/D3330M-04 (2015). Standard Test Method for Peel Adhesion of Pressure-Sensitive Tape. ASTM International, West Conshohocken, PA.

ASTM D3359-17 (2023). Standard Test Methods for Rating Adhesion by Tape Test. ASTM International, West Conshohocken, PA.

ASTM D3363-22. (2022). Standard Test Method for Film Hardness by Pencil Test. ASTM International, West Conshohocken, PA.

ASTM D3418-21. (2021). Standard Test Method for Transition Temperatures and Enthalpies of Fusion and Crystallization of Polymers by DSC. ASTM International, West Conshohocken, PA.

ASTM D3985-05 (2017). Standard Test Method for Oxygen Gas Transmission Rate Through Plastic Film and Sheeting Using a Coulometric Sensor. ASTM International, West Conshohocken, PA.

ASTM D412-16 (2021). Standard Test Methods for Vulcanized Rubber and Thermoplastic Elastomers—Tension. ASTM International, West Conshohocken, PA.

ASTM D4747-87 (2017). Standard Test Method for Determining Unreacted Monomer Content of Latexes Using Gas-Liquid Chromatography. ASTM International, West Conshohocken, PA.

ASTM D4827-88 (2022). Standard Test Method for Determining the Unreacted Monomer Content of Latexes Using Capillary Column Gas Chromatography. ASTM International, West Conshohocken, PA.

ASTM D4946-89(2017). Standard Test Method for Blocking Resistance of Architectural Paints. ASTM International, West Conshohocken, PA.

ASTM D522/D522M-17 (2001). Standard Test Methods for Mandrel Bend Test of Attached Organic Coatings. ASTM International, West Conshohocken, PA.

ASTM D523-14 (2020). Standard Test Method for Specular Gloss. ASTM International, West Conshohocken, PA.

ASTM D870-15 (2020). Standard Practice for Testing Water Resistance of Coatings Using Water Immersion. ASTM International, West Conshohocken, PA.

ASTM D882-18. (2018). Standard Test Method for Tensile Properties of Thin Plastic Sheeting. ASTM International, West Conshohocken, PA.

ASTM D903-98 (2017). Standard Test Method for Peel or Stripping Strength of Adhesive Bonds. ASTM International, West Conshohocken, PA.

ASTM E70-19 (2024). Standard Test Method for pH of Aqueous Solutions With the Glass Electrode. ASTM International, West Conshohocken, PA.

ASTM E96/E96M-22. (2022). Standard Test Methods for Water Vapor Transmission of Materials. ASTM International, West Conshohocken, PA.

Celina, M. (2013). Review of polymer oxidation and its relationship with materials performance and lifetime prediction. *Chemical Reviews* 113 (8): 7030–7078. https://doi.org/10.1021/cr300295x.

Chelazzi, D., Giorgi, R., and Baglioni, P. (2014). Poly(vinyl acetate)-based adhesives: characterization of degradation pathways by FTIR and DSC during artificial ageing. *Polymer Degradation and Stability* 109: 151–157. https://doi.org/10.1016/j.polymdegradstab.2014.07.013.

EPA Method 312A (2017). Determination of Styrene in Aqueous Samples by Gas Chromatography/Flame Ionization Detection. U.S. Environmental Protection Agency, Washington, DC.

ISO 13320:2020 (2020). Particle Size Analysis—Laser Diffraction Methods. International Organization for Standardization, Geneva.

ISO 2115:1996 (1996). Plastics—Polymer dispersions—Determination of white point temperature and minimum film-forming temperature. International Organization for Standardization, Geneva.

ISO 22412:2017 (2017). Particle Size Analysis—Dynamic Light Scattering (DLS). International Organization for Standardization, Geneva.

ISO 2409:2020 (2020). Paints and Varnishes—Cross-cut Test. International Organization for Standardization, Geneva.

ISO 2555:2018 (2018). Plastics—Resins in the Liquid State or as Emulsions or Dispersions—Determination of Apparent Viscosity by Brookfield Viscometer. International Organization for Standardization, Geneva.

ISO 3219-1:2021 (2021). Rheology—Part 1: Vocabulary and Symbols. International Organization for Standardization, Geneva.

ISO 3219-2:2021 (2021). Rheology—Part 2: General Principles of Rotational and Oscillatory Rheometry. International Organization for Standardization, Geneva.

ISO 3251:2019 (2019). Paints and Varnishes—Determination of Non-volatile-matter Content. International Organization for Standardization, Geneva.

Lazzari, M. and López-Morán, T. (2024). Ageing of a polyvinyl acetate-based white glue and its durability in contemporary artworks. *Polymers* 16 (12): 1712. https://doi.org/10.3390/polym16121712.

Martin, J. and Gonzalez, A. (2019). Accelerated weathering and ageing tests for emulsion polymers and paints: methods and limitations. *Progress in Organic Coatings* 131: 121–135. https://doi.org/10.1016/j.porgcoat.2019.02.012.

Viana, C., Lourenço, A., and Schilling, M. (2024). Durability of polyvinyl acetate adhesives: impact of formulation and ageing conditions. *Polymers* 16 (19): 2841. https://doi.org/10.3390/polym16192841.

Chapter 14

Calgavin (2021). Heat exchangers for batch reactors: external recirculation loop design. Technical Article.

CCPS (2007). *Guidelines for Risk Based Process Safety*. Hoboken: Wiley-AIChE.

CCPS (2010). *Guidelines for Safe Handling of Reactive Chemicals*. Hoboken: Wiley-AIChE.

CCPS (2012). *Guidelines for Engineering Design for Process Safety*, 2e. Hoboken: Wiley-AIChE.

CCPS (2022). *Guidelines for Process Hazard Analysis (PHA), Revalidations, and Recommendations*. Hoboken: Wiley-AIChE.

CCPS (Center for Chemical Process Safety) (2021). *Guidelines for Process Safety in Polymer Reaction Systems*. Hoboken: Wiley-AIChE.

Chardon Laboratories. How do cooling towers work with chillers? Chardon Laboratories (accessed 25 September 2025).

Chemical Engineering Magazine (2021). Case studies in polymer pumping systems.

Dhodapkar, S., Jain, P., and Villa, C. (2016). Designing polymerization reaction systems. *Chemical Engineering Progress (CEP)* 112 (11): 28–35.

De Dietrich Process Systems (2018–2024). Telescopic dip pipes and combination baffle/dip pipe for glass-lined reactors. Technical Notes.

Emerson Process Management (2019). Flow meter selection for batch dosing systems. Emerson Automation Solutions.

Endress+Hauser (2021). Instrumentation solutions for polymer production reactors.

Endress+Hauser (2022). Load cell integration in polymer blenders.

Energy-Models.com. Chiller plant design. Energy-models (accessed 25 September 2025).

EP 4230289 A1 (2022). Emulsion Polymerization Management—External Plate-and-Frame Heat Exchanger with Closed-Loop Control. European Patent Office.

Evans P. (2019). Chiller condenser—explained. The engineering mindset (accessed 25 September 2025).

Harnby, N., Edwards, M.F., and Nienow, A.W. (1997). *Mixing in the Process Industries*, 2e. Oxford: Butterworth-Heinemann.

Hensen, R.M. (2005). *Engineering Design for Polymer Plants*. Boca Raton: CRC Press.

Hermann, F. (2018). *Steam Stripping in Polymer Dispersions: Process Design and Optimization*. Weinheim: Wiley-VCH.

Honeywell Process Solutions (2017). Batch control using experion PKS.

HRS Heat Exchangers (2022). Correct design for batch heat exchanger systems. Case Study.

ISA (2017). *Practical Guide to Industrial Control Systems*. Research Triangle Park, NC: ISA Press.

ISA-88 (2010). Batch control: models and terminology.

JMP. Cooling tower and condenser water design—Part 1: The refrigeration cycle. Engineered Systems—JMP (accessed 25 September 2025).

Kister, H.Z. (1990). *Distillation Operation*. New York: McGraw-Hill.

Kister, H.Z. (1992). *Distillation Design*. New York: McGraw-Hill.

Kostansek, E.C., Lovell, P.A., and el-Aasser, M.S. (1997). Pre-emulsification equipment for emulsion polymerization. In: *Emulsion Polymerization and Emulsion Polymers* (ed. P.A. Lovell and M.S. El-Aasser), 189–208. Chichester: Wiley.

Kresta, S.M., Etchells, A.W., and Dickey, D.S. (2015). *Advances in Industrial Mixing*. Hoboken: Wiley.

Lees, F.P. and Mannan, S. (2012). *Lees' Loss Prevention in the Process Industries: Hazard Identification, Assessment and Control*, 4e. Oxford: Elsevier.

Rodionov, Y.V., Selivanov, Y.T., Nikitin, D.V. et al. (2021). Design of liquid-ring vacuum pump with adjustable degree of internal compression. *Chemical and Petroleum Engineering* 57 (5–6): https://doi.org/10.1007/s10556-021-00962-5 (English translation published September 2021; Russian original Nos. 5–6, May–June 2021.).

Lipták, B.G. (2013). *Instrument Engineers' Handbook, Vol 1: Process Measurement and Analysis*. Boca Raton: CRC Press.

Myers, R.H., Montgomery, D.C., and Anderson-Cook, C.M. (2016). *Response Surface Methodology: Process and Product Optimization Using Designed Experiments*, 4e. Hoboken: Wiley.

Nagata, S. (1975). *Mixing: Principles and Applications*. New York: Wiley.

National Instruments (NI) (2018). Best practices for designing distributed control systems (DCS).

Paul, E.L., Atiemo-Obeng, V.A., and Kresta, S.M. (2004). *Handbook of Industrial Mixing: Science and Practice*. Hoboken: Wiley.

Perry, R.H. and Green, D.W. (2007a). *Perry's Chemical Engineers' Handbook*, 8e. New York: McGraw-Hill.

Perry, R.H. and Green, D.W. (2007b). Pump selection guidelines. In: *Perry's Chemical Engineers' Handbook*, 8e. New York: McGraw-Hill.

Perry, R.H. and Green, D.W. (2007c). Agitation and mixing for liquid systems. In: *Perry's Chemical Engineers' Handbook*, 8e. New York: McGraw-Hill.

Perry, R.H. and Green, D.W. (2007d). Heat exchangers and cooling load calculations. In: *Perry's Chemical Engineers' Handbook*, 8e. New York: McGraw-Hill.

Perry, R.H. and Green, D.W. (2007e). Vacuum and stripping techniques in polymer reactors. In: *Perry's Chemical Engineers' Handbook*, 8e. New York: McGraw-Hill.

Perry, R.H. and Green, D.W. (2007f). Instrumentation and controls. In: *Perry's Chemical Engineers' Handbook*, 8e. New York: McGraw-Hill.

Perry, R.H. and Green, D.W. (2007g). Handling and storage of hazardous liquids. In: *Perry's Chemical Engineers' Handbook*, 8e. New York: McGraw-Hill.

Perry, R.H. and Green, D.W. (2007h). Filtration equipment and design. In: *Perry's Chemical Engineers' Handbook*, 8e. New York: McGraw-Hill.

Perry, R.H. and Green, D.W. (2007i). Packaging and dispensing systems. In: *Perry's Chemical Engineers' Handbook*, 8e. New York: McGraw-Hill.

Perry, R.H. and Green, D.W. (2007j). Nozzle design for foam-sensitive materials. In: *Perry's Chemical Engineers' Handbook*, 8e. New York: McGraw-Hill.

Rockwell Automation (2020). *FactoryTalk View SE Troubleshooting Manual.* Milwaukee, WI: Rockwell Automation, Inc.

Rockwell Automation (Allen-Bradley) (2020). FactoryTalk View SE Station and ControlLogix Solutions for batch process automation.

Schmidt, J. (ed.) (2012). *Process and Plant Safety: Applying Computational Fluid Dynamics*. Weinheim: Wiley-VCH.

Seavey, K.C. (2000). Design and operation of chemical plants. In: *Encyclopedia of Chemical Processing and Design*. New York: Marcel Dekker.

Siemens AG (2020). SIMATIC PCS 7 process control system manual.

Spitzer, D.W. (2001). *Flow Measurement: Practical Guides for Measurement and Control*. Research Triangle Park, NC: ISA.

Stoessel, F. (2008). *Thermal Safety of Chemical Processes: Risk Assessment and Process Design*, 2e. Weinheim: Wiley-VCH.

Tarleton, E.S. (2011). Equipment selection and process design for solid/liquid separation. In: *Filtech Proceedings*, 61–77. Wiesbaden: Filtech Exhibitions.

Tarleton, S. and Wakeman, R. (2006). *Solid/Liquid Separation: Equipment Selection and Process Design*, 2e. Oxford: Elsevier.

Toshniwal Instruments (2019). Instrumentation and agitation control in reactor systems.

Towler, G. and Sinnott, R. (2012). *Chemical Engineering Design: Principles, Practice and Economics of Plant and Process Design*. Oxford: Elsevier.

Trane (2021). Comprehensive chilled-water system design. Engineers Newsletter, APP-PRC006A-EN. Trane Technologies.

Tubular Exchanger Manufacturers Association (TEMA) (2019). *Standards of the Tubular Exchanger Manufacturers Association*, 10e. Tarrytown, NY: TEMA.

U.S. Department of Energy. (2023). EnergyPlus engineering reference: plant and condenser loops. Washington, DC: U.S. DOE (accessed 25 September 2025).

U.S. Department of Energy. (2023). EnergyPlus plant application guide: condenser loop (cooling tower). Washington, DC: : U.S. DOE (accessed 25 September 2025).

Uhl, V.W. and Gray, J.B. (1966). *Mixing: Theory and Practice*. New York: Academic Press.

Walas, S.M. (1996). *Chemical Process Equipment: Selection and Design*. Boston: Butterworth-Heinemann.

Wang, L. (2009). *Modeling, Control, and Optimization of Polymer Processes*. Hoboken: Wiley.

WO2006048408A2 (2006). Loop Reactor for Emulsion Polymerisation. World Intellectual Property Organization.

Tobita H.; Hamielec A. E., *Polymerization Processes, 2. Modeling of Processes and Reactors.*

Dimitratos, J., Eliçabe, G., and Georgakis, C. (1994). Control of emulsion polymerization reactors. *AIChE Journal.*

SpecialChem Editorial Team. *Storage Stability in Paints and Coatings*. 2025.

Green, D.W. and Southard, M.Z. (ed.) (2019). *Perry's Chemical Engineers' Handbook*, 9ee. New York, NY, McGraw-Hill Education.

Karassik, I.J., Messina, J.P., Cooper, P., and Heald, C.C. (ed.) (2008). *Pump Handbook*, 4ee. New York, NY: McGraw-Hill Education.

Biegler, L.T., Grossmann, I.E., and Westerberg, A.W. (1997). *Systematic Methods of Chemical Process Design*. Upper Saddle River, NJ: Prentice Hall PTR.

Ulrich, G.D. and Vasudevan, P.T. (2004). *Chemical Engineering Process Design and Economics: A Practical Guide*, 2ee. Durham, NH: Process Publishing.

Crowl, D.A. and Louvar, J.F. (2011). *Chemical Process Safety: Fundamentals with Applications*, 3ee. Upper Saddle River, NJ: Prentice Hall.

Roth, A. (1990). *Vacuum Technology*, 3ee. Amsterdam: Elsevier Science.

Towler, G. and Sinnott, R. (2013). *Chemical Engineering Design: Principles, Practice and Economics of Plant and Process Design*, 2ee. Oxford: Butterworth-Heinemann.

Brown, R.N. (1998). *Practical Vacuum Systems Design*. Hoboken, NJ: Wiley.

Wakeman, R.J. and Tarleton, E.S. (2005). *Solid/Liquid Separation: Equipment Selection and Process Design*. Oxford: Butterworth-Heinemann.

Purchas, D.B. and Sutherland, K. (2002). *Handbook of Filter Media*, 2ee. Oxford: Elsevier.

Green, D.W. and Southard, M.Z. (ed.) (2019). *Perry's Chemical Engineers' Handbook*, 9ee. New York, NY: McGraw-Hill Education.

Jacob, H.-J. Advanced dispersing systems for water-based coatings and polymer dispersions. ystral GmbH, Ballrechten-Dottingen, Germany. PCI (Paint & Coatings Industry), November 1, 2016.

Paul, E.L., Atiemo-Obeng, V.A., and Kresta, S.M. (ed.) (2004). *Handbook of Industrial Mixing: Science and Practice*. Hoboken, NJ: Wiley.

Yokogawa Electric (2021). Precision dosing in polymer reactors using Coriolis Technology. Yokogawa Corporation of America.

Chapter 15

ASTM International (latest version). ASTM D3359—Standard Test Methods for Adhesion by Tape Test. ASTM International, West Conshohocken, PA.

ASTM International (latest version). ASTM D4946—Standard Test Method for Blocking Resistance of Organic Coatings. ASTM International, West Conshohocken, PA.

ASTM International (latest version). ASTM D2196—Standard Test Methods for Rheological Properties of Non-Newtonian Materials by R Viscometer. ASTM International, West Conshohocken, PA.

ASTM International (latest version). ASTM D1210—Standard Test Method for Fineness of Dispersion of Pigment-Vehicle Systems by Hegman gauge. ASTM International, West Conshohocken, PA.

ASTM International (latest version). ASTM D2578—Standard Test Method for Wetting Tension of Polyethylene and Polypropylene Films. ASTM International, West Conshohocken, PA.

BASF (2015). Thickener compatibility matrix for acrylic systems (internal guide).

Bovey, F. A., Tiers, G. V. D., and Hampton, E. M. (1993). Field problems in latex manufacturing. Adhesives Age Magazine.

Clariant (2017). Foam control in emulsion polymerization (Technical Bulletin).

Dow Coating Materials (2016). Troubleshooting guide: emulsion paints.

Erbil, Y.H. (2000). *Vinyl Acetate Emulsion Polymerization and Copolymerization with Acrylic Monomers*. CRC Press.

Fleer, G.J., Cohen Stuart, M.A., Scheutjens, J.M.H.M. et al. (1993). *Polymers at Interfaces*. Chapman & Hall.

Ghosh, P. (2001). *Polymer Science and Technology: Plastics, Rubbers, Blends and Composites*. Tata McGraw-Hill.

Gilbert, R.G. (1995). *Emulsion Polymerization: A Mechanistic Approach*. Academic Press.

Gilbert, R.G. and Winnik, M.A. (2000). Latex film formation and cracking phenomena. *Advances in Colloid and Interface Science* 88 (1–2): 1–36.

Kuo, S.W. (2004). Effects of shear and monomer feed irregularities in semi-batch emulsion polymerization. *Journal of Applied Polymer Science* 91 (5): 3215–3222.

Lovell, P.A. and El-Aasser, M.S. (1997). *Emulsion Polymerization and Emulsion Polymers*. Wiley.

Mayer, J.W. and Hansen, C.M. (2010). *Paint and Coating Testing Manual*, 15e. ASTM International.

Myers, D. (2005). *Surfactant Science and Technology*, 3e. Wiley-Interscience.

Paint & Coatings Industry (PCI) Magazine. (2017). pH drift in waterborne latex systems: Causes and remedies.

Piirma, I. (1982). *Emulsion Polymerization*. Academic Press.

Routh, A.F. (2013). Drying of thin colloidal films. *Reports on Progress in Physics* 76 (4): 046603.

TAPPI Journal (2015). *Optical Brighteners in Paper Coatings: Performance and Aggregation Control*. TAPPI Press.

Tracton, A.A. (2005). *Coatings Technology Handbook*. CRC Press.

Urban, D. and Takamura, K. (2002). *Polymer Dispersions and Their Industrial Applications*. Wiley-VCH.

Winnik, M.A. and Gilbert, R.G. (1996). Polymerization-induced phase separation in aqueous dispersions. *Macromolecules* 29 (6): 1718–1729.

Blackley, D.C. (1975). Latex-based adhesives. In: *Polymer Latices*, 474, 543. London: Applied Science Publishers.

Chapter 16

Alfa Laval (2021). *OptiLobe® Rotary Lobe Pumps for Hygienic Applications*. Lund: Alfa Laval.

Alfa Laval (2021). *M-Series Plate Heat Exchangers Technical Data*. Lund: Alfa Laval.

Alfa Laval (2021). *CIP Modules for Hygienic Industries—Technical Guide*. Lund: Alfa Laval.

Alfa Laval (2021). *Compabloc® Welded Heat Exchanger — Technical Data*. Lund: Alfa Laval.

American Coatings Association (n.d.). *Vinyl Acetate–Versatic Acid Vinyl Ester Copolymer for Masonry Coatings*. CoatingsTech Archive.

Angus Chemical Company (2021). *AMP-95 (2-Amino-2-Methyl-1-Propanol) — Technical Data Sheet*. Buffalo Grove, IL: Angus.

API (latest edition). *API Standard 520/521: Sizing, Selection, and Installation of Pressure-Relieving Devices*. Washington, DC: American Petroleum Institute.

Arkema (2021). VeoVa 10 and its role in exterior coatings (Technical Bulletin).

Arkema (2021). VeoVa 10 applications in exterior coatings (Technical Bulletin).

ASTM International (latest version). *ASTM D2196—Standard Test Methods for Rheological Properties of Non-Newtonian Materials by Rotational Viscometer*. West Conshohocken, PA: ASTM International.

ASTM International (latest version). *ASTM D3330—Standard Test Method for Peel Adhesion of Pressure-Sensitive Tape*. West Conshohocken, PA: ASTM International.

ASTM International (latest). ASTM D1208—Standard Test Method for pH of Aqueous Solutions of Emulsion Polymers.

ASTM International (latest). ASTM D1183—cyclic humidity conditions for adhesives.

ASTM International (latest). ASTM D903—peel or stripping strength of adhesive bonds.

ASTM International (latest). ASTM D2765—Standard Test Methods for Determination of Gel Content and Swell Ratio of Crosslinked Ethylene Plastics.

ASTM International (latest). ASTM D4214—Standard Test Methods for Evaluating the Degree of Chalking of Exterior Paint Films.

ASTM International (latest). ASTM G154/G155—UV exposure of non-metallic materials (fluorescent/arc).

ASTM International (latest). ASTM D3330—peel adhesion of pressure-sensitive tape.

ASTM International (latest). ASTM D2979—pressure-sensitive tack of adhesives using an inverted probe machine.

ASTM International (latest). ASTM D3654—shear adhesion of pressure-sensitive tapes.

ASTM International (latest). ASTM E2041—Standard Guide for Fire Hazard Assessment of Reaction Systems.

ASTM International (latest). ASTM D2243—Standard Test Method for Freeze–Thaw Resistance of Waterborne Coatings.

ASTM International (latest). *ASTM D2369—Standard Test Method for Volatile Content of Coatings*. West Conshohocken, PA: ASTM.

ASTM International (latest edition). *ASTM D562—Standard Test Method for Stormer Viscosity of Paints*. West Conshohocken, PA: ASTM.

ASTM International (latest edition). *ASTM E1356—Standard Test Method for Glass Transition Temperatures by Differential Scanning Calorimetry (DSC)*. West Conshohocken, PA: ASTM.

ASTM International (latest edition). *ASTM D2486—Standard Test Method for Scrub Resistance of Wall Paints*. West Conshohocken, PA: ASTM.

ASTM International (latest edition). *ASTM D4541—Standard Test Method for Pull-Off Strength of Coatings Using Portable Adhesion Testers*. West Conshohocken, PA: ASTM.

ASTM International (latest edition). *ASTM D523—Standard Test Method for Specular Gloss*. West Conshohocken, PA: ASTM.

ASTM International (latest edition). *ASTM D2243—Standard Test Method for Freeze–Thaw Resistance of Emulsion Paints*. West Conshohocken, PA: ASTM.

ASTM International (latest edition). *ASTM D5621—Standard Practice for Sonic Shear Stability of Hydraulic Fluids*. West Conshohocken, PA: ASTM.

ASTM International (latest edition). *ASTM D1193—Standard Specification for Reagent Water*. West Conshohocken, PA: ASTM.

BASF (2019). Glycidyl methacrylate in functional acrylics and PVAc systems (Technical Bulletin).

BASF (2019). *Acronal® Technical Bulletin: Exterior Masonry Systems*. Ludwigshafen: BASF SE.

BASF (2020). Redox initiator systems: best practices for stability and onset control (Technical Bulletin).

BASF (2020). Functional monomers for adhesion promotion.

BASF (2021). *Adhesive Solutions Guide*. Tackifiers for Label Adhesives.

BASF (2021). *Joncryl® Acrylic Resins Technical Data Sheets*. Ludwigshafen: BASF SE.

BASF (2021). Luwipal® crosslinking resins and carbodiimide additives for waterborne adhesives (Technical Bulletin).

BASF Adhesive Solutions. (2020). NMA and functional monomer optimization for D3/D4 adhesives.

BASF Dispersions & Resins (2020). Foam control and surfactant selection in pressure-sensitive adhesives (Technical Report).

Borchers (2021). Silane coupling agents in waterborne systems (Technical Datasheet).

Borchers (2022). Zirconium chelate crosslinkers for waterborne adhesives (Technical Datasheet).

Bovey, F.A., Tiers, G.V.D., and Hampton, E.M. (1993). Field problems in latex manufacturing. *Adhesives Age Magazine*.

Brookfield Engineering (2021). *Viscosity Measurement in Coatings: Best Practices*. Middleboro, MA: Brookfield.

de Bruyn, H. (1999). *The Emulsion Polymerization of Vinyl Acetate*. Science and Technology.

BYK Additives (2021). *Gloss and Surface Control Additives: Technical Data*. Wesel: BYK.

BYK Additives & Instruments (2020). *Adhesion Promoters for Waterborne Coatings*. Wesel: BYK.

CEPE (2016). Guidelines on efflorescence in decorative and protective coatings.

Chemineer (2021). *Mixing CFD Simulation for Chemical Reactors*. Dayton, OH: Chemineer.

Clariant (2017). Foam control in emulsion polymerization (Technical Bulletin).

Plueddemann, E.P. (1991). *Silane Coupling Agents*, 2ee. New York, NY: Springer.

Gelest, Inc. *Silane Coupling Agents. Technical bulletin*. Morrisville, PA: Gelest, Inc. https://www.gelest.com/wp-content/uploads/Silane_Coupling_Agents.pdf (accessed 2024).

DIN EN 204:2016 (2016). Classification of thermoplastic wood adhesives for non-structural applications.

Dow (2019). Foam management in high-shear polymerization systems (Technical Bulletin).

Dow (2021). Borated PVOH adhesives for tissue core applications—storage stability guidelines.

Dow Chemical (2020). *Optibase™ Buffer Systems — Technical Data Sheet*. Midland, MI: Dow.

Dow Chemical (2021). Polyisocyanate crosslinkers for wood adhesives (Technical Data Sheet).

Dow Chemical (2021). *Propylene Glycol USP for Coating Formulations*. Midland, MI: Dow.

Dow Chemical. (2025). Texanol™ and glycol-based freeze–thaw stabilizers (Technical Data).

Dow Coating Materials (2021). *Rhoplex™ and Loxanol® Additives for Scrub Resistance*. Midland, MI: Dow.

Dow Coating Materials (2021). Formulating with acrylic and VAc copolymers for exterior durability and gloss retention.

Dow Coating Materials (2022). *Cold-Weather Paint Formulation Guide*. Midland, MI: Dow.

Dufresne, A. (2012). *Surface Treatments for Polymers: Adhesion, Wettability, and Bonding*. CRC Press.

Eastman Chemical Company (2021). *Texanol™ Ester Alcohol for Low-Temperature Coalescence (Technical Bulletin)*. Kingsport, TN: Eastman.

Eastman Chemical Company (2021). *Texanol™ Ester Alcohol — Coalescent Technical Bulletin*. Kingsport, TN: Eastman.

Eaton Filtration (2021). *Model 53BTX Duplex Basket Strainer Technical Bulletin*. Cleveland, OH: Eaton.

Ekato Group (2021). *Agitator Design and Scale-Up for Polymerization Reactors*. Schopfheim: Ekato.

Enderlin, G. and Plevachuk, A. (2017). Carbodiimide crosslinkers for waterborne adhesives and coatings. *Journal of Coatings Technology and Research* 14 (4): 927–936.

Endress+Hauser (2021). *iTHERM® Temperature Sensors — Technical Data*. Reinach: Endress+Hauser.

Erbil, Y.H. (2000). *Vinyl Acetate Emulsion Polymerization and Copolymerization with Acrylic Monomers*. CRC Press.

European Chemicals Agency (ECHA) (2018). Restriction of nonylphenol ethoxylates (NPEs) under REACH.

Evonik Industries. TEGO® additives guide: wetting and dispersing agents for adhesives and coatings. Essen, Germany: Evonik Industries AG.

Gilbert, R.G. (1995). *Emulsion Polymerization: A Mechanistic Approach*. Academic Press.

van Herk, A.T.M. (2005). *Chemistry and Technology of Emulsion Polymerisation*. Blackwell Publishing.

Hexion Inc (2023). VeoVa™ vinyl ester monomers (Technical Brochure).

IEC (2016). *IEC 61511: Functional Safety—Safety Instrumented Systems for the Process Industry Sector*. Geneva: IEC.

IKA-Werke GmbH & Co (2021). *Ultra-Turrax® High-Shear Mixers: Technical Data*. Staufen: IKA.

ISO 11998:2006 (2006). *Paints and Varnishes — Determination of Wet Scrub Resistance and Cleanability of Coatings*. Geneva: ISO.

ISO 15528 (latest). *Paints and Varnishes—Sampling*. International Organization for Standardization.

ISO 2409:2020 (2020). *Paints and Varnishes—Cross-Cut Test*. Geneva: ISO.

ISO 2812-2:2018 (2018). *Paints and Varnishes—Determination of Resistance to Liquids—Part 2: Water Immersion*. Geneva: ISO.

ISO 3219:1993 (1993). *Plastics—Polymers/Resins in the Liquid State or as Emulsions or Dispersions—Determination of Viscosity Using a Rotational Viscometer with Defined Shear Rate*. Geneva: ISO.

ISO 6270-1:2017 (2017). *Paints and Varnishes — Determination of Resistance to Humidity — Part 1: Condensation (Single-Sided Exposure).* Geneva: ISO.

John Crane (2020). *Mechanical Seals for Industrial Pumps.* Morton Grove, IL: John Crane.

Koch-Glitsch (2020). *Gas Spargers for Stripping and Mass Transfer Applications.* Wichita, KS: Koch-Glitsch.

Kostansek, E.C. (2003). Polymerizable surfactants in emulsion polymerization. *Progress in Polymer Science* 28 (8): 1091–1120.

Kuraray (2021). Mowiol® polyvinyl alcohol grades and modified PVOH for adhesives (Technical Guide).

Kuraray Co., Ltd (2021). *Mowiol® Polyvinyl Alcohol Technical Data Sheets.* Tokyo: Kuraray.

Leser GmbH & Co (2020). *High-Performance Safety Valves—Technical Brochure.* Hamburg: Leser.

Lovell, P.A. and El-Aasser, M.S. (1997). *Emulsion Polymerization and Emulsion Polymers.* Wiley.

Magazine, P.C.I. (2020). *Managing Thermal Initiator Dosage and Exotherm Control in Latex Polymerization.* Paint & Coatings Industry.

Malvern Panalytical (2021). *Laser Diffraction Particle Size Analysis: Technical Overview.* Malvern, UK: Malvern Panalytical.

Mittal, K.L. (ed.) (2008). *Polymer Surface Modification: Relevance to Adhesion*, vol. 4. Brill.

Myers, D. (2005). *Surfactant Science and Technology*, 3e. Wiley-Interscience.

Nouryon (2021). Polymerizable surfactants (surfmers) for emulsion polymers (Application Guide).

Odian, G. (2004). *Principles of Polymerization*, 4e. Wiley.

Pall Corporation (2021). *Ultipleat® Filter Elements Technical Specifications.* Port Washington, NY: Pall.

PCI Magazine (2013). *Branched Vinyl Ester Monomers for High-Performance Emulsion Polymers.* Paint & Coatings Industry.

PCI Magazine (2017). Practical approaches to foam control in waterborne coatings.

PCI Magazine (2017). *pH drift in waterborne latex systems: Causes and remedies.* Paint & Coatings Industry.

PCI Magazine (2018). *Mixed Surfactant Strategies in Emulsion Polymerization.* Paint & Coatings Industry.

PCI Magazine (2019). *Surfactant Migration and Film Defects in Coatings.* Paint & Coatings Industry.

PCI Magazine (2019). *Surface Treatment and Primer Strategies in PVC-to-Wood Bonding.* Paint & Coatings Industry.

PCI Magazine (2020). *Optimizing Polymerization Temperature for Latex Stability.* Paint & Coatings Industry.

PCI Magazine (2020). *Storage Stability of Crosslinkable PVAc Adhesives: Challenges and Solutions*. Paint & Coatings Industry.

PCI Magazine (2020). *Foam Control in Emulsion Polymers: Best Practices for Defoamer Selection*. Paint & Coatings Industry.

PCI Magazine (2020). *High-Speed Adhesive Application: Wet Tack Optimization Strategies*. Paint & Coatings Industry.

PCI Magazine (2021). *Managing Foam in Emulsion Polymerization Processes*. Paint & Coatings Industry.

PCI Magazine (2021). *Initiator Residues and Their Impact on Color Stability in Coatings*. Paint & Coatings Industry.

PCI Magazine (2021). *Pressure-sensitive adhesives: Tack retention strategies in VAM-based systems*. Paint & Coatings Industry.

PCI Magazine (2021). *Surfactant Leaching and Efflorescence in Architectural Coatings: Causes and Solutions*. Paint & Coatings Industry.

Perry, R.H. and Green, D.W. (2008). *Perry's Chemical Engineers' Handbook*, 8e. New York: McGraw-Hill.

Pizzi, A. and Mittal, K.L. (2003). *Handbook of Adhesive Technology*, 2e. CRC Press.

Pizzi, A. and Mittal, K.L. (ed.) (2011). *Handbook of Adhesive Technology*, 2e. CRC Press.

Sasol Performance Chemicals (2016). Technical bulletin: alcohol ethoxylates in emulsion polymerization.

Sasol Performance Chemicals (2019). Surfactant selection and particle size control in emulsion polymerization (Technical Bulletin).

Sasol Performance Chemicals (2021). Low-foam nonionic surfactants for adhesive and coating applications (Technical Bulletin).

Sasol Performance Chemicals (2021). Low-foam nonionic surfactants for adhesive applications (Technical Bulletin).

Sasol Performance Chemicals (2021). Surfactants for emulsion polymerization: low-foam strategies (Technical Guide).

Spirax Sarco (2021). *FT Series Float & Thermostatic Steam Traps*. Cheltenham, UK: Spirax Sarco.

Tadros, T.F. (2013). *Emulsion Formation and Stability*. Wiley-VCH.

TAPPI T 821. (latest). Moisture resistance of paperboard adhesive bonds.

Trane Inc (2020). *RTWD Series Water-Cooled Chillers*. La Crosse, WI: Trane.

Urban, D. and Takamura, K. (2002). *Polymer Dispersions and Their Industrial Applications*. Wiley-VCH.

Wacker Chemie AG (2020). Silane–siloxane emulsions for water repellency (Technical Data Sheets).

Wacker Chemie AG (2021). *VINNAPAS® Dispersions for Masonry and Exterior Paints (Technical Bulletin)*. Munich: Wacker.

Wacker Chemie AG (2021). *VINNAPAS® EP Series: Polymerizable Surfactants for Improved Leaching Resistance*. Munich: Wacker.

Wacker Chemie AG (2021). *VINNAPAS® dispersions for low-temperature application*. Munich: Wacker.

Wicks, Z.W., Jones, F.N., Pappas, S.P., and Wicks, D.A. (2007). *Organic Coatings: Science and Technology*, 3e. Wiley-Interscience.

Wypych, G. (2019). *Handbook of Adhesion Promoters*. ChemTec Publishing.

Chapter 17

Anastas, P.T. and Warner, J.C. (1998). *Green Chemistry: Theory and Practice*. Oxford University Press.

BASF (2021). *Sustainability in Emulsion Polymers – White Paper*. Ludwigshafen: BASF SE.

Dodiuk, H. and Goodman, S.H. (2013). *Handbook of Thermoset Plastics*, 3e. Elsevier. [Contains sections on bio-based and low-VOC adhesives].

ECHA (2008). Candidate list of SVHC (REACH). European Chemicals Agency. https://www.echa.europa.eu/candidate-list-table (accessed 2024).

Erbil, Y.H. (2000). *Vinyl Acetate Emulsion Polymerization and Copolymerization with Acrylic Monomers*. CRC Press. [Discusses market applications and performance requirements relevant to regulatory context].

European Chemicals Agency (ECHA) (2021). Guidance on information requirements and chemical safety assessment. ECHA.

European Commission (2018). *Best Available Techniques (BAT) Reference Document for the Manufacture of Organic Fine Chemicals (including polymer emulsions)*. Seville: Joint Research Centre. ISBN 978-92-79-89492-3.

FDA. (2024). Indirect Food Additives: Adhesives and Components of Coatings; Paper and Paperboard Components; Polymers; Adjuvants, Production Aids, and Sanitizers. https://www.ecfr.gov/current/title-21/chapter-I/subchapter-B/part-175/subpart-B/section-175.105 (accessed 2024).

FDA. (2024). Inventory of Food Contact Substances (21 CFR Parts 175–178). U.S. Food and Drug Administration. https://www.fda.gov/food/packaging-food-contact-substances-fcs/inventory-food-contact-substances-listed-21-cfr.

FDA. (2007). Food Contact Notification (FCN) and Threshold of Regulation (TOR) Guidance. U.S. Food and Drug Administration. https://www.fda.gov/regulatory-information/search-fda-guidance-documents/guidance-industry-preparation-premarket-submissions-food-contact-substances-chemistry (accessed 2025).

U.S. Food and Drug Administration. 21 CFR § 175.105—Adhesives. Code of Federal Regulations, Title 21, Part 175—Indirect Food Additives: Adhesives and Components of Coatings, Subpart B—Substances for Use Only as Components of Adhesives (Subpart A [Reserved]). Source: 42 FR 14534, March 15, 1977. Authority:

21 U.S.C. 321, 342, 348, 379e. https://www.ecfr.gov/current/title-21/section-175.105 (accessed 2025).
GB 8978-1996 (1996). Integrated Wastewater Discharge Standard. Ministry of Ecology and Environment of the People's Republic of China.
International Organization for Standardization (ISO) (2015). ISO 14001:2015 – Environmental Management Systems – Requirements with Guidance for Use. Geneva: ISO.
ISO 14040:2006 (2006). Environmental Management – Life Cycle Assessment – Principles and Framework. International Organization for Standardization.
Mohanty, A.K., Misra, M., and Drzal, L.T. (2005). *Natural Fibers, Biopolymers, and Biocomposites*. CRC Press.
OECD (2011). *Sustainable Chemistry: Indicators and Measurement*. OECD Publishing.
OECD (2019). Guidance on sustainable chemistry – waste reduction in polymer production. Organisation for Economic Co-operation and Development.
Patel, M., von Thienen, N., Jochem, E., and Worrell, E. (2000). Sustainable development and the chemical industry – an analysis of the emergent eco-efficiency concept. *Journal of Industrial Ecology* 4 (2): 105–124.
Pizzi, A. and Mittal, K.L. (2017). *Handbook of Adhesive Technology*, 3e. Boca Raton, FL: CRC Press.
REACH SVHC Expansion News—Candidate List Updates (2024). *Compliance & risks*. https://www.complianceandrisks.com/blog/eu-reach-five-new-svhc-additions/ (accessed 2025).
REACH (2024). Total SVHC count (240 substances as of Jan 2024). Enviropass. https://getenviropass.com/240-svhc/ (accessed 2025).
Shogren, R.L. and Glasser, W.G. (2003). Sustainable polymers from renewable resources. *Polymer Reviews* 43 (4): 299–356.
U.S. EPA (2020). *Safer Choice Standard*. U.S. Environmental Protection Agency.
United States Environmental Protection Agency (EPA) (2022). 40 CFR Part 63 – National emission standards for hazardous air pollutants: miscellaneous organic chemical manufacturing (MON).
Wypych, G. (2017). *Handbook of Solvents*, Volume 1: Properties. ChemTec Publishing. [*Contains* VOC emission control strategies for polymer plants].

Annex 17.A

Alanazi, S., Tabassum, H., Abudawood, M. et al. (2023). Quantification of methylisothiazolinone and methylchloroisothiazolinone by HPLC: method and survey data. *Molecules* 28: 969. https://doi.org/10.3390/molecules28030969.
BfR (2017). Frequently Asked Questions about printing inks and primary aromatic amines (PAAs) in food contact materials. Berlin: German Federal Institute for Risk Assessment (BfR). https://www.bfr.bund.de/cm/349/frequently-asked-questions-

about-printing-inks-and-primary-aromatic-amines-in-food-contact-materials.pdf (accessed 2025).

Colombo, G., Corredig, M., Unalan, I.U. et al. (2024). Untargeted screening of NIAS and cyclic oligomers in virgin and recycled PET by UHPLC-qTOF-MS. *Food Packaging and Shelf Life* 39: 101282.

Commission Regulation (EC) No 2023/2006 on good manufacturing practice for materials and articles intended to come into contact with food. OJ L 384, 29 Dec 2006, pp. 75–78.

Commission Regulation (EU) No 10/2011 on plastic materials and articles intended to come into contact with food (consolidated text). OJ L 12, 15 Jan 2011.

Commission Regulation (EU) 2020/1245 amending and correcting Regulation (EU) No 10/2011 (incl. PAA detection limits). OJ L 288, 3 Sept 2020.

Diamantidou, D., Tsochatzis, E., Kalogiannis, S. et al. (2023). Analysis of migrant cyclic PET oligomers in olive oil and food simulants using UHPLC-qTOF-MS. *Food* 12 (14): 2739. https://doi.org/10.3390/foods12142739.

EFSA (2009). Statement on 4-methylbenzophenone found in breakfast cereals. *EFSA Journal* 7 (6): 243r.

EFSA AFC Panel (2005). Opinion related to 2-isopropyl thioxanthone (ITX) used in printing inks. *EFSA Journal* 293: 1–26.

EFSA CONTAM Panel (2023). Update of the risk assessment of mineral oil hydrocarbons (MOH) in food. *EFSA Journal* 21 (9): 8215. https://doi.org/10.2903/j.efsa.2023.8215.

EuPIA (2021). Guidance for Risk Assessment of Non-Intentionally Added Substances (NIAS) and Non-Listed Substances (NLS) in Printing Inks for Food Contact Materials. EuPIA, 2017; rev. 2021.

EuPIA (2023). Guideline on Printing Inks applied to Food Contact Materials. EuPIA, May 2023.

European Parliament and Council (2004). Regulation (EC) No 1935/2004 on materials and articles intended to come into contact with food. OJ L 338, 13 Nov 2004, pp. 4–17.

Goodier, M.C., Siegel, P.D., Zang, L.-Y., and Warshaw, E.M. (2018). Isothiazolinone in residential interior wall paint: a high-performance liquid chromatographic–mass spectrometry analysis. *Dermatitis* 29 (6): 332–338. https://doi.org/10.1097/DER.0000000000000410.

Haldimann, M., Alt, A., Blanc, A. et al. (2013). Migration of antimony from PET trays into food simulant and foods. *Food Additives and Contaminants: Part A* 30 (3): 587–595.

ILSI Europe (2016). Guidance on Best Practices on the Risk Assessment of NIAS in Food Contact Materials and Articles. Brussels, 72 pp.

Joint Research Centre (JRC) (2011). Technical Guidelines on Testing the Migration of Primary Aromatic Amines (PAAs). EUR 24815 EN.

Markley, L.C., Gonzalez-Bonet, A., Ogungbesan, A., and Patton, G. (2023). Safety assessment for Irgafos 168 used as an antioxidant in food-contact applications. *Food and Chemical Toxicology* 178: 113877.

Schreier, V.N., Çörek, E., Appenzeller-Herzog, C. et al. (2023). Evaluating the food safety and risk assessment evidence for oligomers in food contact materials. *Environment International* 173: 107878.

U.S. FDA (2021). Updated safety assessment of tris(2,4-di-tert-butylphenyl) phosphite (Irgafos 168) used in food-contact applications. FDA Science Forum.

Welle, F. and Franz, R. (2011). Migration of antimony from PET bottles into beverages: activation energy of diffusion and migration modelling. *Food Additives and Contaminants: Part A* 28: 115–126.

Index

f

j

k

l

m

Printed and bound by CPI Group (UK) Ltd, Croydon, CR0 4YY

17/09/2026

14964235-0008